FOOD PROTEOMICS

FOOD PROTEOMICS

Technological Advances, Current Applications and Future Perspectives

Edited by

MARÍA LÓPEZ PEDROUSO

Department of Zoology, Genetics and Physical Anthropology, University of Santiago de Compostela, Santiago de Compostela, Spain

DANIEL FRANCO RUIZ

Head of Research at the Centro Tecnológico de la Carne de Galicia, Ourense, Spain

JOSE M. LORENZO

Head of Research at the Centro Tecnológico de la Carne de Galicia, Ourense, Spain; Associate Professor at the University of Vigo, Spain

ELSEVIER

ACADEMIC PRESS

An imprint of Elsevier

elsevier.com/books-and-journals

Academic Press is an imprint of Elsevier
125 London Wall, London EC2Y 5AS, United Kingdom
525 B Street, Suite 1650, San Diego, CA 92101, United States
50 Hampshire Street, 5th Floor, Cambridge, MA 02139, United States
The Boulevard, Langford Lane, Kidlington, Oxford OX5 1GB, United Kingdom

ISBN: 978-0-323-90889-4

For Information on all Academic Press publications visit our website at
https://www.elsevier.com/books-and-journals

Publisher: Nikki Levy
Acquisitions Editor: Nina Bandeira
Editorial Project Manager: Kathrine Esten
Production Project Manager: Paul Prasad Chandramohan
Cover Designer: Vicky Pearson Esser

Typeset by Aptara, New Delhi, India

Contents

II

Applications of proteomic in food sciences

4. Proteomic advances in crop improvement

Rubén Agregán, Noemí Echegaray, María López Pedrouso, Mirian Pateiro, Daniel Franco Ruiz, Jose M. Lorenzo

5. Proteomic advances in seafood and aquaculture

Robert Stryiński, Elżbieta Łopieńska-Biernat, Mónica Carrera

6. Proteomics advances in beef production

Mohammed Gagaoua, Yao Zhu

7. Proteomic advances in poultry science

Xue Zhang, Surendranath Suman, M. Wes Schilling

Contributors

Ana G. Abril Department of Microbiology and Parasitology, Faculty of Pharmacy, University of Santiago de Compostela, Santiago de Compostela, Spain; Department of Food Technology, Spanish National Research Council, Marine Research Institute, Vigo, Spain

Rubén Agregán Centro Tecnológico de la Carne de Galicia, Ourense, Spain

Rituparna Banerjee ICAR-National Research Centre on Meat, Chengicherla, Hyderabad, Telangana, India

Jorge Barros-Velázquez Department of Analytical Chemistry, Nutrition and Food Science, School of Veterinary Sciences, University of Santiago de Compostela, Lugo, Spain

Subhasish Biswas Department of Livestock Products Technology, West Bengal University of Animal and Fishery Sciences, Kolkata, India

S.B. Bravo Proteomic Unit, Instituto de Investigaciones Sanitarias-IDIS, Complejo Hospitalario Universitario de Santiago de Compostela (CHUS), Santiago de Compostela, Spain

Pilar Calo-Mata Department of Analytical Chemistry, Nutrition and Food Science, School of Veterinary Sciences, University of Santiago de Compostela, Lugo, Spain

Mónica Carrera Department of Food Technology, Spanish National Research Council, Marine Research Institute, Vigo, Spain

M.D.P. Chantada-Vázquez Research Unit, Hospital Universitario, Lucus Augusti (HULA), Servizo Galego de Saúde (SERGAS), Lugo, Spain; Proteomic Unit, Instituto de Investigaciones Sanitarias-IDIS, Complejo Hospitalario Universitario de Santiago de Compostela (CHUS), Santiago de Compostela, Spain

Anand Raj Dhanapal Department of Biotechnology, Karpagam Academy of Higher Education (Deemed to be University), Coimbatore, Tamil Nadu, India

Ruben Dominguez Centro Tecnológico de la Carne de Galicia, Ourense, Spain

Noemí Echegaray Centro Tecnológico de la Carne de Galicia, Ourense, Spain

Natalya Fedoseeva Russian State Agrarian Correspondence University, Department of Zootechnii, Production and Processing of Livestock Products, 50 Shosse Entuziastov, Balashikha, Russian Federation

Daniel Franco Ruiz Centro Tecnológico de la Carne de Galicia, Ourense, Spain

Mohammed Gagaoua Food Quality and Sensory Science Department, Teagasc Food Research Centre, Ashtown, Dublin, Ireland

Francisco Javier González-Barcala Department of Medicine, Universidad de Santiago de Compostela. Department of Respiratory Medicine, Complejo Hospitalario Universitario de Santiago de Compostela (CHUS). Head of TRIAD Research Group, Instituto de Investigación Sanitaria de Santiago de Compostela (IDIS). A Coruña, Spain

María López Pedrouso Department of Zoology, Genetics and Physical Anthropology, University of Santiago de Compostela, Santiago de Compostela, Spain

Jose M. Lorenzo Centro Tecnológico de la Carne de Galicia, Ourense, Spain; University of Vigo, Spain

Naveena Basappa Maheswarappa ICAR-National Research Centre on Meat, Chengicherla, Hyderabad, Telangana, India

Kiran Mohan Department of Livestock Products Technology, Veterinary College, KVAFSU, Bidar, Karnataka, India

Paulo E.S. Munekata Centro Tecnológico de la Carne de Galicia, Ourense, Spain

Juan José Nieto-Fontarigo Department of Biochemistry and Molecular Biology, Centro de Investigaciones Biológicas de la Universidad de Santiago (CIBUS), Universidad de Santiago de Compostela. TRIAD Research Group, Instituto de Investigación Sanitaria de Santiago de Compostela (IDIS). A Coruña, Spain; Respiratory Immunopharmacology unit, Department of Experimental Medical Science, Lund University, Sweden

C. Núñez Research Unit, Hospital Universitario, Lucus Augusti (HULA), Servizo Galego de Saúde (SERGAS), Lugo, Spain

Mirian Pateiro Centro Tecnológico de la Carne de Galicia, Ourense, Spain

Maksim Rebezov V. M. Gorbatov Federal Research Center for Food Systems of Russian Academy of Sciences, 26 Talalikhina St., Moscow, Russian Federation; K.G. Razumovsky Moscow State University of Technologies and Management (The First Cossack University), 73, Zemlyanoy Val St., Moscow, Russian Federation; Faculty of Biotechnology and Food Engineering, Ural State Agricultural University, 42 Karl Liebknecht str., Yekaterinburg, Russian Federation

Francisco Javier Salgado Castro Department of Biochemistry and Molecular Biology, Centro de Investigaciones Biológicas de la Universidad de Santiago (CIBUS), Universidad de Santiago de Compostela. TRIAD Research Group, Instituto de Investigación Sanitaria de Santiago de Compostela (IDIS). A Coruña, Spain

M. Wes Schilling Department of Food Science, Nutrition and Health, Promotion, Mississippi State University, MS, United States

Enrique Sentandreu Instituto de Agroquímica y Tecnología de Alimentos (CSIC), Calle Agustín Escardino 7, Paterna, Valencia, Spain

Miguel Ángel Sentandreu Instituto de Agroquímica y Tecnología de Alimentos (CSIC), Calle Agustín Escardino 7, Paterna, Valencia, Spain

Mohammad Ali Shariati K.G. Razumovsky Moscow State University of Technologies and Management (The First Cossack University), 73, Zemlyanoy Val St., Moscow, Russian Federation

Robert Stryiński Department of Biochemistry, Faculty of Biology and Biotechnology, University of Warmia and Mazury in Olsztyn (UWM), Michała Oczapowskiego 1a, Olsztyn, Poland

Surendranath Suman Department of Animal and Food Sciences, University of Kentucky, Lexington, KY, United States

Muthu Thiruvengadam Department of Applied Bioscience, College of Life and Environmental Science, Konkuk University, Seoul, Republic of Korea

Baskar Venkidasamy Department of Biotechnology, Sri Shakthi Institute of Engineering and Technology, Coimbatore, Tamil Nadu, India

Tomás G. Villa Department of Microbiology and Parasitology, Faculty of Pharmacy, University of Santiago de Compostela, Santiago de Compostela, Spain

Sol Zamuz Centro Tecnológico de la Carne de Galicia, Ourense, Spain

Xue Zhang Department of Food Science, Nutrition and Health, Promotion, Mississippi State University, MS, United States

Yao Zhu Food Quality and Sensory Science Department, Teagasc Food Research Centre, Ashtown, Dublin, Ireland

Elżbieta Łopieńska-Biernat Department of Biochemistry, Faculty of Biology and Biotechnology, University of Warmia and Mazury in Olsztyn (UWM), Michała Oczapowskiego 1a, Olsztyn, Poland

About the editors

María López Pedrouso is a researcher in the field of food proteomics. She holds an MSc degree in advanced chemistry and the PhD degree in biochemistry and molecular biology, both from the University of Santiago de Compostela (Spain), where she is currently a technical assistant in the field of biology and medicine. Her career began improving her skills in gel-based proteomics to study the proteins and their post-translational modifications. Particularly, the behavior of seed storage proteins and their phosphorylation during germination were underpinned by these proteomic tools. In her postdoctoral stage, she broadened her technical knowledge on other gel-free proteomic technologies, and her research is currently focused on the search of protein biomarkers and extending the knowledge at the molecular level in relation to food quality. She also has wide experience in vegetable, fish, and meat products throughout her career. In meat quality, she led a significant number of proteomic studies to understand the quality in relation to color, tenderness, and water holding capacity. Moreover, she has been a great achievement in the search for bioactive peptides from animal waste products promoting greater sustainability of the food industry. In this sense, she is an author of 42 scientific papers in recognized international journals, 6 book chapters, and 9 posters at international congresses. She is an associate editor in *Frontiers in Animal Science Journal*, and she has also edited in several special issues such as "Food Proteins and Peptides Focused on Functional and Bioactive Properties" in Molecules and "Advances in Natural Antioxidants for Food Improvement" in Antioxidants.

Daniel Franco Ruiz is a head of Research at the Centro Tecnológico de la Carne de Galicia, Ourense, Spain, and received the PhD degree in chemical sciences from the University of Santiago (Spain). His research lines are focused on understanding physicochemical, biochemical, and microbial changes during the technological processes applied to meat products and in identifying proteomic, peptidomic, and genetic biomarkers associated with food quality, using molecular techniques for protein separation and subsequent identification and quantification applying spectrometry techniques. Within meat industry applied lines, he is involved in research works regarding extension of food shelf life using natural extracts with antioxidant and antimicrobial capacities and also in the development of novel-healthier meat products based on fat and salt reduction, replacement of animal fat, or incorporating functional compounds.

He has had authored 170 peer-reviewed publications in well-recognized peer-reviewed international journals. Furthermore, he has participated in more than 200 communications to congresses, mostly international and in more than 60 research projects. He had two Spanish patents. Moreover, he has written more than 20 book chapters in international and national books. He has co-directed four doctoral theses. Currently, he belongs to the editorial board of several prestigious journals as *Food Chemistry Advances, Biology, Antioxidants, Foods* and *Frontiers in Animal Science,* in which he collaborates in the regular review process and also editing special issues.

José Manuel Lorenzo Rodriguez is a head of Research at the Centro Tecnológico de la Carne de Galicia, Ourense, Spain, and associate professor at the University of Vigo, Spain. His short but intense career gives him a great experience in the meat sector, recognized in 2021 by the Expertscape ranking (https://expertscape.com/) as the world's leading expert on meat, meat products, and food technology. In addition, it gives him the second place in "Food" and the seventh "Food preservation." Moreover, José M. Lorenzo was elected as the top 2% of scientists—Ranking University of Standford (https://elsevier.digitalcommonsdata.com/datasets/btchxktzyw/3), who ranks 64th in the field of "Agriculture, Fisheries & Forestry" and 29th in the subfield "Food Science." He also holds the third place as the most cited researcher among all Spanish researchers in this field. He is the author of more than 620 scientific articles, and more than 300 communications to congresses, mostly international. He has edited 12 international books and one national. Moreover, he has written 76 chapters in international and national books. He is chief editor in *Frontiers in Animal Science Journal,* and associate editor of five prestigious journals: *Food Analytical Methods, Journal of the Science Food and Agriculture, Animal Science Journal, Canadian Journal of Animal Science,* and *Food Research International.* He has also edited several special issues for these high-impact journals to create more discussion around key aspects of functional food development: innovative technologies, functional ingredients, and food safety and quality.

Preface

In recent years, food science is undergoing a great change encouraged by technological developments. Specifically, food biotechnology will be a great impact on new production methods and novel ingredients and additives. All these tools must be used to achieve better health, consumer safety, and the protection of the environment. It also clear that globalization creates new challenges in the field of preservation and control quality, as well as the sustainability of the food industry and animal welfare have become relevant challenges. To achieve all these goals, proteins as one of the most important components of food are decisive for improving the nutritional and functional properties of foodstuffs.

Proteins consist of 21 amino acids in different sequences and post-translational modifications resulting in a wide variety of conformations. This complex heterogeneity of proteins further complicates its study. Considering the living organism of food origin, the proteins through enzymes and hormones lead the vital functions of the animal or vegetal. Changes in feeding, environmental stress, and/or diseases of animals unleash a cascade of events altering the proteins. In the case of vegetables, changes of proteins associated with climate change such as drought and salinity are the most studied. On the other hand, proteins are very species-specific which can provide good clues of breeding animal and plant variety through the food chain. A further important point to consider is the nutritional aspect of food whereby the proteins also play a key role. The nutritional quality of proteins together with the digestibility and bioavailability of proteins should be considered by the food industry. The processing of food is often designed to increase food safety, but this fact could affect dramatically protein digestibility. For all these reasons, protein research is an important field within food science.

Current advances in food proteomics and its main applications may contribute to the development and innovation of the food industry. The qualitative and quantitative analysis of food proteomes derived from complex matrix provides accurate measurements of proteins, protein–protein interactions, and interactions with other food components. All these factors play a key role in the quality traits of raw and processed foods. However, the rapid developments of mass spectrometry instruments as well as new bioinformatics tools require us to update knowledge for researchers and professionals. Accordingly, this book aims to organize the necessary information about new protein biomarkers and proteomic approaches of classical issues as allergenicity, authentication, or food safety.

This book is divided into an introduction and three sections. In the introduction, the role of proteomics in the field of food science as well as the conceptual background is described. Following, the first section methodological tackle aspects and bioinformatic tools employed in the field of food proteomics practically. The second section includes proteomic studies collected from the most relevant animal and vegetable species in food production. Finally, important food challenges from a proteomic point of view will be discussed and analytical tools will be described to introduce technical innovations in the food industry.

Thanks to all the authors for their great effort and dedication to this book. The editors were truly impressed by the result, and we hope to collaborate again with all of them in the future.

María, Daniel, and Lorenzo

CHAPTER

1

Introduction

Food proteomics: technological advances, current applications and future perpectives

María López Pedrouso[a], *Jose M. Lorenzo*[b,c], *Daniel Franco Ruiz*[b]

[a]Department of Zoology, Genetics and Physical Anthropology, University of Santiago de Compostela, Santiago de Compostela, Spain
[b]Centro Tecnológico de la Carne de Galicia, Ourense, Spain
[c]University of Vigo, Spain

1.1 Importance of the food industry and emerging trends in food science

The food industry plays a major role in the global economy. According to USDA, the food system is described as the whole food industry—from farming and food production, packaging, and distribution, to retail and catering (USDA, 2021). This includes farmers, food processors, wholesalers, retailers, and food service establishments among others. In 2021, the revenue in the food market amounted to US$8,049,240 m and it is expected a compound annual growth rate of 3.14% during 2021–2025.

Current times also bring the main challenge of ensuring food security and food safety. Both terms are closely linked and concern society. Food security relates to have physical and economic access to sufficient, safe, and nutritious food for the population, meeting their dietary needs, and food preferences for a healthy life. Nevertheless, food safety refers to handling, preparation, and storage of food with proper conditions or practices which reduce the risk of individuals becoming sick from foodborne illnesses (FAO, 2008). The economic growth of the food industry

Food Proteomics
DOI: https://doi.org/10.1016/B978-0-323-90889-4.00004-X

will be further reinforced by the world population increase. Indeed, it should be noted that the world population is expected to reach nine billion people by 2050 (United Nations, 2019). Hence, the security of the food supply will be a challenging task maintaining the current level of food consumption of about 338.9 kg/person in 2021 (Statista, 2021). To meet this demand, agriculture in 2050 will need to double its production to generate almost 50% more food, feed, and biofuel than it did in 2012 (FAO, 2017). Apart from the growth of the population, food security will be also seriously threatened by climate change. Particularly, the primary sector (agriculture, livestock, and aquaculture) is expected to rise production in a more sustainable way (Cole, Augustin, Robertson, and Manners, 2018). Overall, a huge technological development will be required to enhance processing, distribution, and retailing.

The globalization of the food industry faces important nutritional and economic challenges. Emerging food processing technologies are advancing in preservation and control quality boosted by growing consumer demands and price war at the global level. Food processing and new technologies play a key role to achieve the long shelf life of food commodities traveling long distances. Certainly, improvement in the food shelf-life is necessary, because in the world approximately one-third of all food produced is lost or wasted along the food chain (HLPE, 2012), indicating an inefficiency of current food systems. In this sense, to provide the best answer to the rising demands of society, the food industry has to face increasingly complex challenges that require the best available science and technology (e.g., smart and active packaging).

Furthermore, environmental issues should be considered from farms to processors. The environmental impact and sustainability should be assessed and enhanced at every stage. Animal-based foods have a higher environmental impact than vegetal products and this fact is influencing consumer behavior. In developed countries, a huge amount of food wastes has been produced and recycling of by-products should be effectively carried out. In this sense, the sustainability of the food industry is an essential goal (Cucurachi, Scherer, Guinée, and Tukker, 2019). Subsequently, the main ecological aspects like biodiversity loss, nitrogen cycle acceleration, and carbon cycle acceleration should be taken into consideration. These facts could be behind currently dietary changes, toward diets based on plant protein products (Aiking and de Boer, 2020).

There is a close relationship between what we eat (food) and our health. In other words, food quality and its relationship with human health in terms of its nutritional value (i.e., protein and fatty acid profile) play a key role. Therefore, food is currently considered not only a source of energy, macro-, and micro-nutrients, but also one of the best strategies to prevent future diseases. Indeed, the increasing evidence of food-related disorders and chronic diseases have prompted consumers to bring about vital changes in their diet and lifestyle, making them more health-conscious than ever. However, paradoxically, the triple burden (undernutrition, micronutrient deficiencies, and overweight) of malnutrition remains a global health emergency in developing states in contrast with nutrition problems present in developed countries. In the latter ones, processed foods are linked to an increased risk of death from heart disease, diabetes, or other illnesses.

In the field of the agri-food industry, there is a need to ensure food designing new crops within sustainable agriculture. Simultaneously, strategies for maximizing production with minimal effect on the environment improving yield, nutritional quality, the efficiency of resources, and tolerance of biotic and abiotic stress are priority objectives (Tian, Wang, Li, and Han, 2021). Moreover, food safety should give an upgraded level for producers and retailers of fruits and vegetables. A change in consumer behavior is evidenced and a new market for fresh and minimally processed fruits and vegetables is emerging. Packaging and preservation are being enhanced using novel chemicals and the latest developments without compromising the product quality (De Corato, 2020). The mechanical operations of cutting and peeling are the most sensitive steps, disinfection and washing procedures avoiding the growth of pathogenic and spoilage microorganisms should be implemented (Ali, Yeoh, Forney, and Siddiqui, 2018). The market of ready-to-eat leafy green salads and other fresh products is demanding higher microbiological quality in industrialized countries (Arienzo et al., 2020). But increasing demand for vegan or vegetarian products requires further research to tackling new solutions for new situations.

Despite the fact that animal products such as meat, milk, and eggs are essential components of our diets. There is a great concern about health problems associated with red and processed meat intake and a strategy to reduce saturated fat and cholesterol consumption is being carried out. On the contrary, poultry and pork production has been growing exponentially during the last few years, mainly in less developed countries. It is important, however, to bear in mind also the link between an extensive intake of sugars and the occurrence of chronic diseases. In any case, efforts to improve meat quality from a nutritional point of view are constantly in the meat industry via reducing/replacing fat, salt, and additives and incorporating new sources of protein and fiber in the meat products. Furthermore, healthier meat products are being developed via the incorporation of bioactive compounds into them to elaborate functional meat products (Pogorzelska-Nowicka, Atanasov, Horbańczuk, and Wierzbicka, 2018; Ruiz-Capillas and Herrero, 2021). On other hand, environmental concerns of meat consumption and other ethical issues related to animal welfare are changing the consumer attitude (López-Pedrouso et al., 2020a; Sanchez-Sabate and Sabaté, 2019). Additionally, the potential impact of animal diseases on human health is magnified further by increasing levels of resistance in bacteria, parasites, viruses, and fungi to antimicrobial drugs, such as antibiotics, antifungals, antivirals, antimalarials, and anthelmintics. Today, some 700,000 people die of drug-resistant infections every year (FAO, 2017). Hence, the livestock industry should seriously consider all these issues in the future.

Food from aquatic environments is important in the human diet, providing significant health benefits. In this sense, aquaculture production is a growing industry versus traditional fisheries. Its production is continuously increasing due to issues related to the lack of sustainability of capture fisheries and climate change. As in other farm production, the main goal is to achieve a maximum growth rate and at the same time a minimum production cost. To achieve this purpose, strategies of maximizing food conversion with formulated diets, ensuring fish welfare are developed, as a quality indicator of the product. Moreover, the utilization of fish wastes (heads, skin, trimmings, fins viscera, frames, and others) to generate bioactive protein hydrolysates is a current challenge, that can economically help the global process. Indeed, these fish protein hydrolysates could be used for food, nutraceuticals, and dietary supplement industries (Gao et al., 2021; López-Pedrouso et al., 2020a). Simultaneously to these production aspects, aquatic food production must be guaranteed high standards of food safety. In this sense, one of the main constraints in fish and seafood consumption is the allergenicity induced by parvalbumin, tropomyosin, and arginine kinase which must be reduced and, in any case, labeled (López-Pedrouso, Lorenzo, Gagaoua and Franco, 2020b). The second limitation relates to aquatic pollution, which poses a risk for seafood safety in the consumption of contaminated fish species (Okpala, Sardo, Vitale, Bono, and Arukwe, 2018).

To sum up, emerging concerns as globalization, food security, and sustainability pose the main challenges to the food industry and a knowledge gap between food technologies and new market demands should be fill in recent years. At the same time, new opportunities including healthy products, long-life products, and ready to eat should be exploited by food product developers.

1.2 An overview of technological applications based on food proteins

Fig. 1.1 presents the main challenges in the food industry directly related to the protein field. The following aspects will be briefly discussed in the next lines.

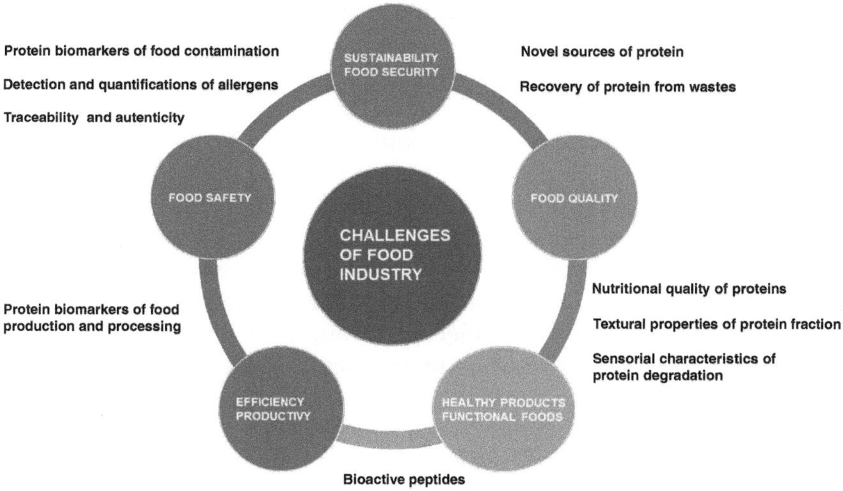

FIG. 1.1 Main currently food challenges in the food industry related to the protein field.

1.2.1 The impact of proteins on food quality and safety

Growing consumer demands in a globalized market lead the food industry to seek the highest standards of food quality. Adding nutritional value to food products without diminishing their quality or compromising on taste and flavor, maintaining the cost-effectiveness and the shelf-life adds to the long list of current challenges in the food industry. Proteins are one of the most important components in our diet playing a key role in human nutrition and the functional properties of foodstuffs. The nutritional quality of proteins, formed by 21 different amino acids, is very heterogeneous, and our dietary requirements of essential amino acids depend on body weight, physiological state, and physical activity level among others. Moreover, protein digestibility is an important factor for the evaluation of their nutritive value, as well as, their bioaccessibility and bioavailability, bearing in mind that their effects on human health will be determined by the food processing (López-Pedrouso, Lorenzo, Zapata, and Franco, 2019). It should be noted also that functional properties of proteins (e.g., gel formation, emulsifying effect, viscosity, solubility, foam formation), not only influence textural aspects because protein colloidal structures can alter the nutritional properties and biological activity of foodstuffs (Foegeding and Davis, 2011).

In the context of globalization, food safety has still unresolved challenges. Indeed, transboundary pests and plant and animal diseases are producing highly contagious epidemic outbreaks that spread rapidly across national borders, causing high rates of death and illness. The risk of serious outbreaks is increasing as more people, animals, plants, and agricultural products move across international borders, and as animal production systems become more intensive (FAO, 2017). This requires routine food safety analysis from the primary sector to the final consumer, rapid, in-field, and low-cost techniques are needed. The contaminants which cause great concern to the population are microbes, mycotoxins, pesticides, packaging components, seafood toxins, veterinary drugs and preservatives. These pollutants alter the biological process of the organic matrix, inducing changes in the proteome as a response to these environmental factors. The detection of protein biomarkers to control food safety could be performed by immunoassay (ELISA or western blotting) measuring the concentration of biomolecule through an antigen or antibody interaction. The search for protein biomarkers constitutes a previous step to implement other technologies. A great level of sensitivity and selectivity of antibodies could have a practical application of surface-enhanced Raman spectroscopy (SERS). Aptamers as an emergent biomaterial have also been described to detect alterations in the food matrix (Hermann, Duerkop, and Baeumner, 2019; Yaseen, Pu, and Sun, 2018). In recent years, it is important to highlight that the nanotechnology field, including magnetic nanoparticles, quantum dots, carbon nanotubes, and nanosensors is being developed to detect and biomonitoring foodborne diseases many using protein biomarkers (Krishna et al., 2018). These biomarkers are mainly based on protein-related molecules (e.g. nucleic acids, proteins, antigens, and metabolites) present in the food matrix.

1.2.2 Bioactive peptides from food proteins

Technological advances are focused on the extraction and identification of new bioactive compounds. These compounds present in small quantities from food have a beneficial effect on human health including the prevention of cardiovascular disease and cancer (Kris-Etherton et al., 2002). In the case of peptides, the fragmentation of proteins increases their bioactivity in numerous physiological functions. Indeed, amino acids and small peptides (4-6 amino acids) were demonstrated to have high digestibility and fast absorption with relevant biological activities (López-Pedrouso et al., 2020a). Therefore, protein hydrolysates obtained from chemical and enzymatic digestion are being investigated and, isolation and purification of the biopeptides is a priority of the food industry. It is certainly possible that protein hydrolysates with biological activity (e.g. antihypertensive, antioxidant, antithrombotic and antimicrobial) give it an added value to new foodstuffs formulations (Borrajo, López-Pedrouso, Franco, Pateiro, and Lorenzo, 2020).

The food industry is an important source of by-products and wastes, that need to be recycling and/or recovery. For instance, discards from the fish industry (fish frame, skin, and viscera) are used to extract polyunsaturated fatty acids, collagen, gelatin, and bioactive peptides. Specifically for bioactive peptides, antioxidant and angiotensin-converting enzyme (ACE) inhibitory effects have been described (Atef a Mahdi Ojagh, 2017). Regarding processed meats, it has been described that endogenous enzymes and microbial peptidases act during the ripening process of dry-cured hams and other dry-fermented meat products affecting the taste and flavor and generating bioactive peptides (Mora, Gallego, Reig, and Toldrá, 2017). Another relevant source of biopeptides is milk and dairy products from mammalian animals. Fragments from casein (αs1-, αs2-, b-, and κ-casein) and whey proteins (α-lactalbumin, β-lactoglobulin, and lactoferrin) have been demonstrated to be active, exhibiting antimicrobial, antihypertensive, and antioxidant functions among others. In this sense, fermentation is the elected process to generate the biopeptides (Egger and Ménard, 2017; Nielsen, Beverly, Qu, and Dallas, 2017). The future commercialization of these biopeptides will be introduced in nutraceuticals and functional foods although the low concentration of the peptide in the food matrix decreases the bioactivity. Other disadvantages of biopeptides are their hydrophobic nature and the sensory issues which hamper the formulating products (Li-Chan, 2015), hence further research will be needed to implement their production and functionality.

1.2.3 Allergenicity of food proteins

New formulations and industrial process, as well as, the novel sources of proteins, requires more technological advances in the field of food allergens. They represent a major concern and a global food challenge because food allergies are increasingly important in western countries, particularly in children. In the period from 2014 to 2016, 6.9% and 7.6% of children suffer from food allergies in Europe and United States, respectively (Jiang, Warren, and Gupta, 2020). The most abundant allergens are proteins that trigger a mediated or non-mediated IgE response. The severity of symptoms including alterations in skin and oral mucosa, respiratory, gastrointestinal, cardiovascular, and anaphylactic shock as well as its prevalence results in high economic losses for the health system. In this regard, the detection and quantification of the allergens for correct food labeling seem of utmost relevance. The food industry needs to considerably improve in this field. A large quantity of these allergens are proteins with different biological functions. In the case of plant-based food, peanuts, soy, tree nuts, and wheat trigger allergies in connection with gluten proteins, globulin, albumin, and other storage proteins. On the other hand, milk, eggs, and fish also have a significant number of protein allergens such as lactalbumin, lactoglobulin, and parvalbumin (López-Pedrouso et al., 2020a). The food industry must identify the proper and routine method to avoid the allergic component or label it and this issue must be considered an important priority. After a thorough knowledge of food allergens, numerous industrial applications of the proteins could be introduced like biosensors of allergens (Zhou et al., 2019).

1.2.4 Food authenticity and traceability based on proteomic profiles

In the global food chain, traceability and authenticity of origin and food processing must be monitored to meet the consumer requirements. The definition of food traceability is the ability to follow the movement of food and ingredients across all steps in the supply chain according to the FDA (https://www.fda.gov/food/new-era-smarter-food-safety/tracking-and-tracing-food). Thus, tracking the food information through the food supply is critical to improve consumer trust and purchase willingness. The knowledge of the product life cycle to ensure the safety, sustainability, and quality of the product must be strongly enhanced. One of the food categories with the highest risk of adulteration or substitution is meat and several cases have been reported (Li et al., 2020). Additionally, in the case of meat, ethical values (organic food, fair trade), cultural issues (vegan, halal, kosher) are highly important for numerous populations groups. Thus, technological advances including isotope analysis, chemometrics, and NIR and DNA coding are being developed to increase traceability (Badia-Melis, Mishra, and Ruiz-García, 2015). From a molecular point of view, for example, DNA-based methods and rapid evaporative ionization mass spectrometry could be used to identify several species of animals in minced meat matrices (Cavin, Cottenet, Cooper, and Zbinden, 2018). Equally important and with the problem of a higher number of commercial species the fishery products are also susceptible to food frauds, hence analytical methods including DNA-based, chromatographic and spectroscopic technologic are being developed (Fiorino et al., 2018). In this sense, proteins could be used as the major target for mass spectrometry, high-performance liquid chromatography, and immunoassays. For this purpose, proteomic fingerprint detecting differences in protein/DNA structure is a valuable tool for quantitative and qualitative analysis (Esteki, Regueiro, and Simal-Gándara, 2019).

1.3 Why proteomics?

Food proteomics is becoming more frequent to address issues in the field of food quality and food safety. The need to develop new analytical strategies for proteins makes food proteomics an essential tool. Food protein composition has a key influence on the nutritional quality and technological properties of foods. Therefore, high-throughput research in

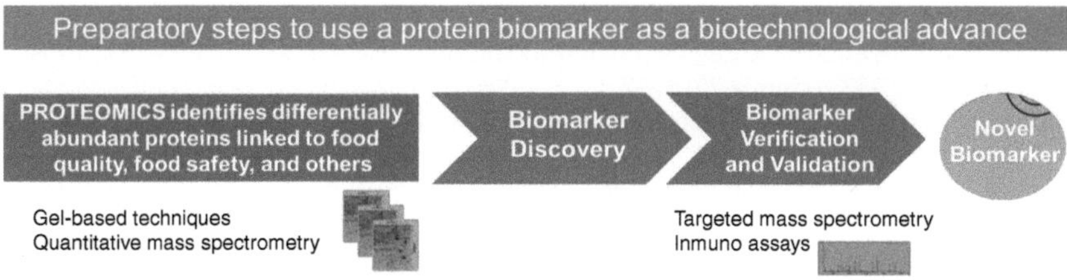

FIG. 1.2 Main stages to discover protein biomarkers.

proteins for food traceability as well as food quality and safety detecting allergens or microbial contaminants is the way forward (Cunsolo, Muccilli, Saletti, and Foti, 2014). To fulfill this goal, the proteomics has been developed to find biomarkers for predicting sensory and technological quality traits of foodstuffs along with a better understanding of molecular mechanisms (Gagaoua et al., 2021; Huang et al., 2020). A biomarker is related to a functional, physiological or biochemical process at the cellular and molecular level influencing a specific feature (Strimbu and Tavel, 2010). A protein biomarker is an indispensable tool to monitor and predict these quality traits but a large amount of information and preparatory steps including, verification and validation, should be achieved (Fig. 1.2).

Further, proteomic technologies also give information about protein isoforms, posttranslational modifications, molecular interactions which define the quality features. The understanding of molecular mechanisms and their relationship with food quality drive current proteomic research.

References

Aiking, H., & de Boer, J. (2020). The next protein transition. *Trends in Food Science and Technology, 105*(July 2018), 515–522. https://doi.org/10.1016/j.tifs.2018.07.008.

Ali, A., Yeoh, W. K., Forney, C., & Siddiqui, M. W. (2018). Advances in postharvest technologies to extend the storage life of minimally processed fruits and vegetables. *Critical Reviews in Food Science and Nutrition, 58*(15), 2632–2649. https://doi.org/10.1080/10408398.2017.1339180.

Arienzo, A., Murgia, L., Fraudentali, I., Gallo, V., Angelini, R., & Antonini, G. (2020). Microbiological quality of ready-to-eat leafy green salads during shelf-life and home-refrigeration. *Foods, 9*(10), 1421.

Atef, M., & Mahdi Ojagh, S. (2017). Health benefits and food applications of bioactive compounds from fish byproducts: A review. *Journal of Functional Foods, 35*, 673–681. https://doi.org/10.1016/j.jff.2017.06.034.

Badia-Melis, R., Mishra, P., & Ruiz-García, L. (2015). Food traceability: New trends and recent advances. A review. *Food Control, 57*, 393–401. https://doi.org/10.1016/j.foodcont.2015.05.005.

Borrajo, P., López-Pedrouso, M., Franco, D., Pateiro, M., & Lorenzo, J. M. (2020). Antioxidant and antimicrobial activity of porcine liver hydrolysates using flavourzyme. *Applied Sciences (Switzerland), 10*(11), 1–14. https://doi.org/10.3390/app10113950.

Cavin, C., Cottenet, G., Cooper, K. M., & Zbinden, P. (2018). Meat vulnerabilities to economic food adulteration require new analytical solutions. *Chimia, 72*(10), 697–703. https://doi.org/10.2533/chimia.2018.697.

Cole, M. B., Augustin, M. A., Robertson, M. J., & Manners, J. M. (2018). The science of food security. *Npj Science of Food, 2*(1), 1–8. doi:10.1038/s41538-018-0021-9.

Cucurachi, S., Scherer, L., Guinée, J., & Tukker, A. (2019). Life cycle assessment of food systems. *One Earth, 1*(3), 292–297. https://doi.org/10.1016/j.oneear.2019.10.014.

Cunsolo, V., Muccilli, V., Saletti, R., & Foti, S. (2014). Mass spectrometry in food proteomics: A tutorial. *Journal of Mass Spectrometry, 49*(9), 768–784. https://doi.org/10.1002/jms.3374.

De Corato, U. (2020). Improving the shelf-life and quality of fresh and minimally-processed fruits and vegetables for a modern food industry: A comprehensive critical review from the traditional technologies into the most promising advancements. *Critical Reviews in Food Science and Nutrition, 60*(6), 940–975. https://doi.org/10.1080/10408398.2018.1553025.

Egger, L., & Ménard, O. (2017). Update on bioactive peptides after milk and cheese digestion. *Current Opinion in Food Science, 14*, 116–121. https://doi.org/10.1016/j.cofs.2017.03.003.

Esteki, M., Regueiro, J., & Simal-Gándara, J. (2019). Tackling fraudsters with global strategies to expose fraud in the food chain. *Comprehensive Reviews in Food Science and Food Safety, 18*(2), 425–440. https://doi.org/10.1111/1541-4337.12419.

FAO (2008). *An Introduction to the basic concepts of food security*, Social indicators research 95. http://www.ncbi.nlm.nih.gov/pubmed/21812205%5Cnhttp://jn.nutrition.org/content/140/1/153S.abstract%5Cnhttp://link.springer.com/10.1007/s11205-009-9455-4%5Cnhttp://www.fao.org/docrep/013/al936e/al936e00.pdf%5Cnwww.andes.info.ec/es/noticias/fao-ecuador-bolivia-son. (Accessed 3 September 2020).

FAO (2017). *The future of food and agriculture: trends and challenges*, The future of food and agriculture: Trends and challenges 4. www.fao.org/publications%0Ahttp://www.fao.org/3/a-i6583e.pdf%0Ahttp://siteresources.worldbank.org/INTARD/825826-1111044795683/20424536/Ag_ed_Africa.pdf%0Awww.fao.org/cfs%0Ahttp://www.jstor.org/stable/4356839%0Ahttps://ediss.uni-goettingen.de/bitstream/han. (Accessed 3 September 2020).

Fiorino, G. M., Garino, C., Arlorio, M., Logrieco, A. F., Losito, I., & Monaci, L. (2018). Overview on untargeted methods to combat food frauds: A focus on fishery products. *Journal of Food Quality, 2018*. doi:10.1155/2018/1581746.

Foegeding, E. A., & Davis, J. P. (2011). Food protein functionality: A comprehensive approach. *Food Hydrocolloids, 25*(8), 1853–1864. https://doi.org/10.1016/j.foodhyd.2011.05.008.

Gagaoua, M., Terlouw, E. M. C., Maria, A., Franco, D., Warner, R. D., Lorenzo, J. M., et al. (2021). Molecular signatures of beef tenderness : underlying mechanisms based on integromics of protein biomarkers from multi-platform proteomics studies. *Meat Science, 172*(September 2020), Article 108311. https://doi.org/10.1016/j.meatsci.2020.108311.

Gao, R., Yu, Q., Shen, Y., Chu, Q., Chen, G., Fen, S., et al. (2021). Production, bioactive properties, and potential applications of fish protein hydrolysates: Developments and challenges. *Trends in Food Science and Technology, 110*(October 2020), 687–699. doi:10.1016/j.tifs.2021.02.031.

Hermann, C. A., Duerkop, A., & Baeumner, A. J. (2019). Food safety analysis enabled through biological and synthetic materials: A critical review of current trends. *Analytical Chemistry, 91*(1), 569–587. https://doi.org/10.1021/acs.analchem.8b04598.

HLPE. (2012). Food security and climate change. https://www.ifpri.org/publication/food-security-and-climate-change-report-high-level-panel-experts-hlpe-food-security-and. (Accessed 3 September 2020

Huang, C., Hou, C., Ijaz, M., Yan, T., Li, X., Li, Y., et al. (2020). Proteomics discovery of protein biomarkers linked to meat quality traits in post- mortem muscles: Current trends and future prospects: A review. *Trends in Food Science & Technology, 40*(3), 82–85. https://doi.org/10.4213/faa748.

Jiang, J., Warren, C. M., & Gupta, R. S. (2020). Correction to: epidemiology and racial/ethnic differences in food allergy. *Pediatric Food Allergy*, C1. doi:10.1007/978-3-030-33292-1_20. -C1.

Kris-Etherton, P. M., Hecker, K. D., Bonanome, A., Coval, S. M., Binkoski, A. E., Hilpert, K. F., et al. (2002). Bioactive compounds in foods: Their role in the prevention of cardiovascular disease and cancer. *American Journal of Medicine, 113*(9 SUPPL. 2), 71–88. https://doi.org/10.1016/s0002-9343(01)00995-0.

Krishna, V. D., Wu, K., Su, D., Cheeran, M. C. J., Wang, J. P., & Perez, A. (2018). Nanotechnology: Review of concepts and potential application of sensing platforms in food safety. *Food Microbiology, 75*, 47–54. https://doi.org/10.1016/j.fm.2018.01.025.

Li-Chan, E. C. Y. (2015). Bioactive peptides and protein hydrolysates: Research trends and challenges for application as nutraceuticals and functional food ingredients. *Current Opinion in Food Science, 1*(1), 28–37. https://doi.org/10.1016/j.cofs.2014.09.005.

Li, Y. C., Liu, S. Y., Meng, F. B., Liu, D. Y., Zhang, Y., Wang, W., et al. (2020). Comparative review and the recent progress in detection technologies of meat product adulteration. *Comprehensive Reviews in Food Science and Food Safety, 19*(4), 2256–2296. https://doi.org/10.1111/1541-4337.12579.

López-Pedrouso, M., Lorenzo, J. M., Cantalapiedra, J., Zapata, C., Franco, J. M., & Franco, D. (2020a). Aquaculture and by-products: Challenges and opportunities in the use of alternative protein sources and bioactive compounds. *Advances in Food and Nutrition Research, 92*, 127–185. https://doi.org/10.1016/bs.afnr.2019.11.001.

López-Pedrouso, M., Lorenzo, J. M., Gagaoua, M., & Franco, D. (2020b). Current trends in proteomic advances for food allergen analysis. *Biology, 9*(9), 1–13. https://doi.org/10.3390/biology9090247.

López-Pedrouso, M., Lorenzo, J. M., Zapata, C., & Franco, D (2019). Proteins and amino acids. In *Innovative thermal and non-thermal processing, bioaccessibility and bioavailability of nutrients and bioactive compounds* (pp. 139–169). Woodhead Publishing.

López-Pedrouso, M., Rodríguez-Vázquez, R., Purriños, L., Oliván, M., García-Torres, S., Sentandreu, M.Á., et al. (2020c). Sensory and physicochemical analysis of meat from bovine breeds in different livestock production systems, pre-slaughter handling conditions, and ageing time. *Foods, 9*(2), 1–17. https://doi.org/10.3390/foods9020176.

Mora, L., Gallego, M., Reig, M., & Toldrá, F. (2017). Challenges in the quantitation of naturally generated bioactive peptides in processed meats. *Trends in Food Science and Technology, 69*, 306–314. https://doi.org/10.1016/j.tifs.2017.04.011.

Nielsen, S. D., Beverly, R. L., Qu, Y., & Dallas, D. C. (2017). Milk bioactive peptide database: A comprehensive database of milk protein-derived bioactive peptides and novel visualization. *Food Chemistry, 232*, 673–682. https://doi.org/10.1016/j.foodchem.2017.04.056.

Okpala, C. O. R., Sardo, G., Vitale, S., Bono, G., & Arukwe, A. (2018). Hazardous properties and toxicological update of mercury: From fish food to human health safety perspective. *Critical Reviews in Food Science and Nutrition, 58*(12), 1986–2001. https://doi.org/10.1080/10408398.2017.1291491.

Pogorzelska-Nowicka, E., Atanasov, A. G., Horbańczuk, J., & Wierzbicka, A. (2018). Bioactive compounds in functional meat products. *Molecules (Basel, Switzerland), 23*(2), 1–19. https://doi.org/10.3390/molecules23020307.

Ruiz-Capillas, C., & Herrero, A. M. (2021). Sensory analysis and consumer research in new product development. *Foods, 10*(3). doi:10.3390/foods10030582.

Sanchez-Sabate, R., & Sabaté, J. (2019). Consumer attitudes towards environmental concerns of meat consumption: A systematic review. *International Journal of Environmental Research and Public Health, 16*(7). doi:10.3390/ijerph16071220.

Statista. (2021). https://www.statista.com/outlook/cmo/food/worldwide?currency=chf (Accessed 17 July 2021)

Strimbu, K., & Tavel, J. A. (2010). What are biomarkers? *Current Opinion in HIV and AIDS, 5*(6), 463–466. https://doi.org/10.1097/COH.0b013e32833ed177.

Tian, Z., Wang, J. W., Li, J., & Han, B. (2021). Designing future crops: challenges and strategies for sustainable agriculture. *Plant Journal, 105*(5), 1165–1178. https://doi.org/10.1111/tpj.15107.

United Nations. (2019). Department of economic and social affairs, population division. *World Population Prospects, 2019*. https://population.un.org/wpp/Publications/Files/WPP2019_Highlights.pdf (Accessed 4 August 2021).

USDA. (2021). World agricultural production. https://www.fas.usda.gov/data/world-agricultural-production (Accessed 17 July 2021).

Yaseen, T., Pu, H., & Sun, D. W. (2018). Functionalization techniques for improving SERS substrates and their applications in food safety evaluation: A review of recent research trends. *Trends in Food Science and Technology, 72*(January), 162–174. https://doi.org/10.1016/j.tifs.2017.12.012.

Zhou, J., Qi, Q., Wang, C., Qian, Y., Liu, G., Wang, Y., & Fu, L. (2019). Surface plasmon resonance (SPR) biosensors for food allergen detection in food matrices. *Biosensors and Bioelectronics, 142*(April), Article 111449. https://doi.org/10.1016/j.bios.2019.111449.

SECTION I

Technological advances in food proteomics

CHAPTER

2

Quantitative proteomics by mass spectrometry in food science

M.D.P. Chantada-Vázquez[a,b], C. Núñez[a], S.B. Bravo[b]

[a]Research Unit, Hospital Universitario, Lucus Augusti (HULA), Servizo Galego de Saúde (SERGAS), Lugo, Spain
[b]Proteomic Unit, Instituto de Investigaciones Sanitarias-IDIS, Complejo Hospitalario Universitario de Santiago de Compostela (CHUS), Santiago de Compostela, Spain

2.1 Introduction

The proteome is defined as the total protein content of one biological system. The proteome is highly dynamic and is constantly changing according to different stimuli. Proteomics is the science that performs a large-scale analysis of proteins in a particular biological system at a particular time (Pandey & Mann, 2000). Moreover, proteomics includes the structural and functional knowledge of proteins, their post-translational modifications (PTMs), and interaction between proteins, protein localization, and their abundance quantification. Although for a long time the genome become a more important area in disease knowledge, today is known that proteins changes are involved in the disease process. Therefore, in the actual postgenomic era, high number of mass spectrometry (MS)-based proteome analysis methods have been improved, to provide valuable data into proteins and their expression levels directly not only from complex tissue or body fluid samples but also in food science (Aebersold & Mann, 2016; Walther & Mann, 2010). This technology implying the identification and quantization of proteins is essential for understanding biological processes involved in both health and disease.

Food Proteomics
DOI: https://doi.org/10.1016/B978-0-323-90889-4.00006-3

In the process of food production, proteins can act as indicators for any changes in properties associated with food quality, composition, or origin. Therefore, proteome analysis can be applied for the systematic search for new marker proteins/peptides, thus accelerating the development of assays to detect adulteration and deceptive practices. Subsequently, accurate and reliable analytical methodologies can be developed and validated to detect the previously identified marker proteins/peptides.

On the other hand, a precise *quantification* of proteins in tissues or body fluids provides valuable data and allow the validation of potential biomarkers for, e.g., for disease stages (Bendixen, Danielsen, Hollung, Gianazza, & Miller, 2011). For this purpose, there are several applications as metabolic or isobaric protein labeling or label-free techniques as sequential window acquisition of all theoretical fragment ion spectra mass spectrometry (SWATH-MS) that allow the quantification of most of the proteins expressed by one cell, tissue, body fluid in a specific condition vs control. Other techniques (targeted techniques) as selected reaction monitoring (SRM) allows the unique protein quantification by runor multiple-reaction monitoring (MRM) that allows multiple protein quantification by a run. More information about these techniques can be found in the next points.

In food sciences, protein quantification is especially important too. To have good control and characterization of foods is necessary to understand diverse organism systems from a biological point of view. Therefore, in the last decade, proteomics methodologies have suffered a great advance necessary to improve the development of new food safety and quality control in food production processes in meat or milk or other protein-based foods (Almeida et al., 2014).

Thus, better knowledge about the structure, function, and levels of proteins present in both raw materials and final food products is essential to perform an optimization of processing methodologies. This is due to the involvement of proteins in quality parameters such as texture, color, crops yield, or important changes induced in response to stress in meat food.

This chapter is a comprehensive and updated overview of the application, that the quantitative proteomics approach, can offer as a tool to analyze food. Additionally, the proteomic workflows, technologies, and tools that are being assessed in proteomics-related studies, followed by a review of the specific applications regarding food quality and safety, are also collected. The revision is centered in papers where SWATH-MS or Isobaric tags for relative and absolute quantitation (iTRAQ) labeling were used in the two recent years.

2.2 Proteomics

Proteomic analysis can be divided into different areas.

2.2.1 Protein identification and characterization

The first type of proteomic analysis is the *qualitative analysis* that allows the identification and characterization of the complete set of proteins present in one sample. This analysis can include the characterization of the protein PTMs. This analysis can be performed in complete proteomes (complex samples) (e.g., all the proteins comprised in a sample that can be a food product) or in a specific type of proteins (e.g., cytoplasmatic proteins; glycosylated proteins; phosphorylated proteins).

The most common methods for protein identification are peptide mass fingerprinting (Pappin, Hojrup, & Bleasby, 1993) and peptide fragmentation fingerprinting (Eng, McCormack, & Yates, 1994). In both approaches, it is necessary the previously enzymatic digestion of the proteins (bottom–up) (see next section), generally using trypsin as an enzyme of digestion. One alternatively, is the use of intact protein (without a digest) to obtain the fragmentation (MS/MS)data, (top–down) proteomics (McLafferty et al., 2007). In both cases, the search is performed against the databases where the protein sequence and other important information are available.

In the case of the food analysis, there are some problems. In some cases, many peptides/proteins of the animal or plant species used by the food industry are not present in the current databases because the complete proteome is not available. This issue combined with the different ingredients used in food makes food proteomics in a continuous process of change to improve protein identification.

On the other hand, it is known that protein PTMs play a crucial role both in protein activity and stability. Although only a few PTMs are well characterized, there are more than 300 different types of PTMs, such as phosphorylation, acetylation, glycosylation, or oxidation (Virág et al., 2020). In food proteins, PTMs may be caused by different biological signals or due to some of the food processing methods as cooking or preservation treatments. This makes it interesting too, in food analysis the study of this protein PTMs.

2.2.2 Differential/quantitative proteomics

Quantitative proteomics gives us information about the protein levels (Feng, Cappelletti, & Picotti, 2017). This information can be the relative abundance of a specific protein between different group of samples, or the absolute amount of the one protein. These data are immensely helpful to look for differences between different conditions, for example, in the search of disease biomarkers. In the case of the food, the different protein levels provide information about different technological treatments of food products. Relative quantification can be performed using different proteomic, gel-based, label-based, or label-free approaches (Matthiesen & Carvalho, 2013; Pappireddi et al., 2019).

In the *gel-based methods*, proteins were separated by two-dimensional electrophoresis (by their isoelectric point and their molecular weight in the first and second dimension, respectively). After the separation, gel is staining with sypro, or other staining types, and the protein abundance were obtained by comparison of the spot volume between different samples using specific software (PDquest, etc.). In this approach, each sample is run on a different gel. This makes the gels can run differently and sometimes this protein comparison is very tediously.

One more recent alternatively is the differential sample labeled, using fluorophores as Cy3, Cy5, and Cy2. In this case, the two samples and the internal control (equal mix between both samples) can be run on the same gel making quite easy the protein comparison and increasing confidence in the detection and quantification of differences in protein abundance. This technology is named 2D-DIGE(difference gel electrophoresis) (Meleady, 2018). For this analysis, it is also necessary to have an specific software to perform the comparison as DeCyder, SameSpots, or Dimension 3 (Hmmier & Dowling, 2018). In food proteomics, this in gel-based method has been the most used since now. It is because this technology offers more information not only about protein abundances but also about PTMs as glycosylations or phosphorylations.

In *label-based methods*, proteins or peptides are previously labeled using different labeling types (Fig. 2.1) and relative quantification is obtained

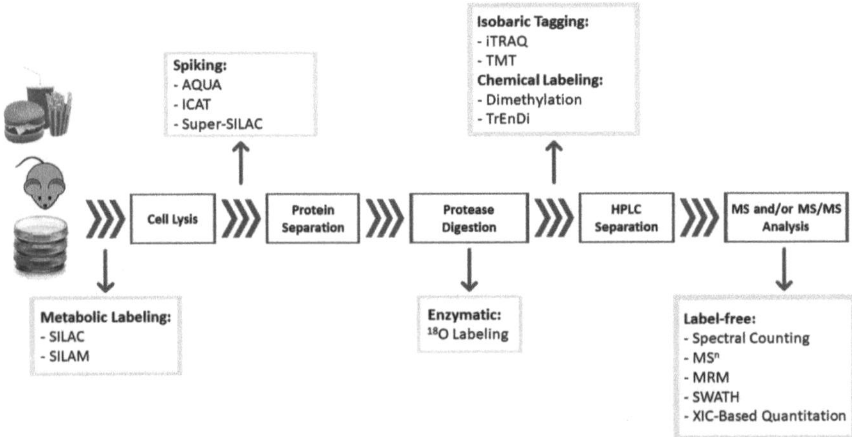

FIG. 2.1 Schematic representation of the different labeled-based and label-free strategies for the protein quantification in the proteomics field.

from the MS or MS/MS data depending on the labeling used, Fig. 2.1. One of the first label-based methods used in proteomics is named SILAC (Stable-isotope labeling with amino acids in cell culture) because urea growth medium containing different isotopically labeled amino acids (Gingras, Aebersold, & Raught, 2005; Wasinger, Zeng, & Yau, 2013). These amino acids are incorporated into proteins as cells divide. The complete incorporation is performed in a few generations of cell division. Considering that typical proteomic workflows involve tryptic digestion, isotopically labeled arginine, and lysine are the most common amino acid for incorporate the label. After digestion, they can yield peptides that terminate with one or both of these two label amino acids (Ong & Mann, 2006).

Initially, the use of SILAC labeling was limited to studies involving cell culture; however, recent developments have allowed their use on in vivo studies. This new technique, the Super-SILAC method, consist in generate isotopically labeled cell lysates that can be mixed with tissue lysates without labeling to make a relative quantization. One of the first paper using this technique were made by Monetti, Nagaraj, Sharma, and Mann (2011) to quantify 10,000 phosphorylation sites associated with the insulin response in mouse hepatocytes. Other SILAC-based strategies are also being developed to apply in different research field (spike-in SILAC, etc.).

Other methods (see Fig. 2.1) for relative quantization involve the introduction of staple isotope tagging at the protein level, including isotope-coded affinity tags (ICAT). ICATs reagent has a biotin tag with a light or heavy (deuterium or ^{13}C) linker chain and a reactive group. This group binds covalently to cysteine residues. The process of quantification is performed starting to be labeling the two samples to be compared with the two forms of ICAT and mixing an equal amount of the samples before protein digestion and LC-MS/MS. The relative quantification of peptides is performed by comparing the relative abundance of light and heavy ICAT-tagged peptides (Bantscheff, Schirle, Sweetman, Rick, & Kuster, 2007; Wasinger et al., 2013). This technique shows the advantage that the tag enables the separation of biotin-labeled peptides before LC-MS/MS analysis to reduce sample complexity. On the other hand, the disadvantage of this method is that only allows for 2 samples comparisons and only cysteine-containing peptides.

Another method that allows compare up to four different samples is the labeling with iTRAQ. iTRAQ is a stable isobaric reagent with reaction specifically with amine and the labeling is performed after the digestion step. The iTRAQ reagents have a peptide-reactive group that binds covalently to lysine and an N terminal moieties of peptides, a reporter group with a variably sized and a balanced group (Bantscheff et al., 2007; Wasinger et al., 2013). Therefore, iTRAQ has the advantage that labels all peptides in all samples. The balance groups allow that all peptide has an equal mass in MS analysis (Bantscheff et al., 2007; Wasinger et al., 2013). The advantage of this labeling method over the ICAT is that in MS analysis peptides labeled with the different iTRAQ reagents will be obtained as a single peak but in MS/MS analysis the balance group is lost and the reporter groups are detected in the low m/z region between 113 and 119. Therefore, the comparison of the intensity of the reporter group allows relative accurately quantification.

Other quantitative methods as tandem mass tag (TMT) or stable isotope incorporation via enzyme reaction (i.e., ^{18}O) are available too for protein quantification (Smith, Martins-De-Souza, & Fioramonte, 2019) (Fig. 2.1).

Quantitative proteomics based on stable isotope-coding strategies requires labeling reagents, high number of samples, multiple sample preparation steps to purify the sample for the correct labeling, etc. Subsequently, in this process, there are a considerable sample loss and a reduction of the detection sensitivity. Label-free LC-MS/MS methods represent good alternatives, because they are available for all types of biological samples, simple, reproducible, cost-effective, and without side reactions related to the labeling process.

In some *label-free methods*, protein amount is normally calculated based on the MS extracted ion current signal of the peptides/proteins during a liquid chromatography (LC) run. Quantitative proteomic methodologies recently have been improved with the introduction of selected reaction monitoring (SRM) experiments or multiple reaction monitoring (MRM) experiments, which is considered a highly sensitive LC-MS/MS acquisition mode used in biomedicine research to validate possible biomarker proteins (Mermelekas, Vlahou, & Zoidakis, 2015). This technology is known as targeted proteomics.

Therefore, SRM canquantify from attomolar to millimolar amounts of a specific protein in a body fluid or a range of 50 to >1 million copies of a protein in a cell (Pappireddi et al., 2019; Vidova & Spacil, 2017). In SRM methods are necessary the use of a triple-quadrupole mass spectrometer (Fig. 2.2) as a mass filter, to perform the analysis only of one desiderate targeted proteolytic peptides from the protein of interest. This means that in a triple quadrupole machine in Q1 is performed a precursor ions selection and in Q3 is performed a selection of one specific fragment ions as a

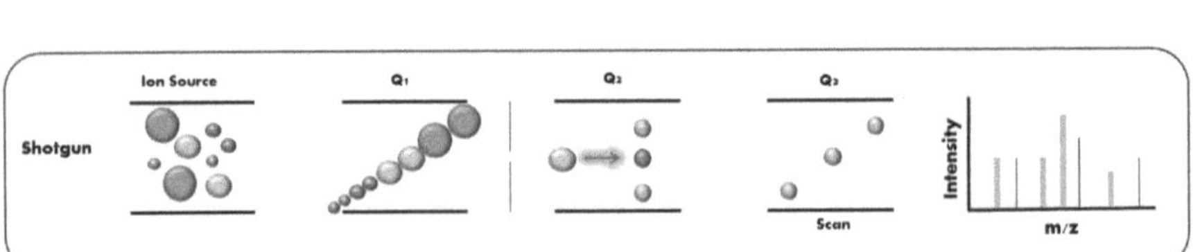

FIG. 2.2 Schematic representation of the operation of a triple quadrupole mass spectrometer.

mass to charge (m/z) values (Pappireddi et al., 2019; Vidova & Spacil, 2017) (in Q2 an ion fragmentation is made). Therefore, the intensities of the transitions signal of this peptide can be viewed over the time and this can be converted in a measure of the quantity of a protein of interest. This technique shows a high sensitivity, reproducibility, and broad dynamic range, making SRM a powerful tool for both absolute and relative quantification in biological samples, which is normally applied in the biomarker search. However, this technique has a low throughput because is necessary to perform the search of one per one protein, and it difficulties the analysis of a high number of proteins (only between 10 and 50 proteins can be quantified simultaneously in a complex sample).

Recently, this inconvenience was solved by the improvement of the sequential window acquisition of all theoretical spectra (SWATH-MS) method based on data-independent acquisition (DIA). This technique is considered a novel SRM-like analysis (Ludwig et al., 2018). SWATH-MS was considered as a new and faster alternative to SRM-MS, which involves the analysis of peptides and their fragments over time and allows both, the identification and quantification of specific proteins in complex tissue body fluid, cells, or food products. SRM and SWATH-MS are technologies that present important differences. SWATH-MS can select all MS/MS fragment ion spectra for all precursor ions found in a complex sample using m/z windows. This important issue makes this technology useful in the biomarker discovery (Narasimhan, Kannan, Chawade, Bhattacharjee, & Govekar, 2019). On the other hand, both SRM and SWATH-MS methods are used for the quantification of proteins (Meyer & Schilling, 2017). However, as MS-based methods, this technology is species and tissue-specific. Therefore, the linearity method and the limit of quantification of a protein and peptide depend on the complexity of the biological sample.

In label-free approaches, there are other strategies. One of these strategies was the quantization based on the frequency of the identification; that is named *spectral counting*; this approach is based on counting the total number of spectra identified in an MS/MS experiment for protein. Other approaches perform quantization based on *peak intensity*, in this case, are considered that the peak areas of peptides correlate to the concentration of a protein (Tate, Larsen, Bonner, & Gingras, 2013; Zhu, Smith, & Huang, 2010). More recently, there are other more accurately techniques named SWATH-MS where the quantification is based on the extracted area from 10 peptides per protein and seven fragments per peptide (Huang et al., 2015).

The development of MS-based protein quantification for other species different to human, including those of farm animals, is now less available. Thus, such MS methods must be developed specifically for each farm animal species, making them available for food studies and farm animal health (Wu, Ruitong, & Bendixen, 2019a). Food analysis, as a more complex system, needs some specific method to detect each food component.

Moreover, in food proteomics, it is necessary to mention that the relative number of proteins can change due to food composition, the technology used in food processing, and biological variability in the food components.

2.2.3 Functional proteomics

Functional proteomics studies the interaction between proteins (interactomes), or between a protein and other molecules, and the important consequences of these interactions (Doerr, 2012). This part of the proteomics science is very usefully for biomedical research because the understanding of protein–protein interactions, through network analysis, will be crucial to understand the physiopathology of many diseases. Moreover, in food quality the knowledge of protein–protein interaction is essential to know the food component relation.

There is another proteomics area related to functional proteomics that is *activity-based proteomics*. This area studies the specific activities of the proteins in a sample, such as function or its inhibition (Doerr, 2012). In this sense, proteomics techniques as MS imaging, which is based on imaging modes and allows mapping proteins in a specific tissue or sample section (Buchberger, DeLaney, Johnson, & Li, 2018), have proved to be an important tool for functional proteomics. This is due to the different protein isoform's location can help to understand their possible roles.

As a small summary indicate that in food, the main applications of the quantitative (as well qualitative) proteomic technique are: (1) study changes in food quality (Gašo-Sokač et al., 2010), (2) identifying bioactive compounds and different ingredients in functional foods (Valdés, Cifuentes, & León, 2017), (3) identification of allergic proteins (Bräcker & Brockmeyer, 2018).

2.3 Proteomic workflows

To perform a complete proteome analysis, it is necessary a correct sample preparation, using one or more separation steps. Therefore, the common proteomic workflows include separation at protein and/or peptide level before the mass spectrometer analysis (MS). Commonly, this separation step is based on gel electrophoresis (gel-based approach) and/or LC (gel-free approach).

In the case of the proteins in food-related samples, the great diversity and high dynamic range make the MS analysis a great challenge. For this reason, sample preparation usually includes enrichment of proteins of interest. Moreover, when using complex samples, the most common protocols include depletion of the most abundant interfering proteins, selective enrichment of the low abundance proteins, or partial purification of the target proteins (Bodzon-Kulakowska et al., 2007).

Depending on if the MS analysis is carried out on the intact proteins or the peptide fragments, two different proteomic workflows can be selected namely *top–down* and *bottom–up*, respectively.

In the bottom–up approach (Fig. 2.3), proteins are initially enzymatically digested, commonly using trypsin which promotes the formation of peptides with an ideal size to be analyzed by MS (Pandey & Mann, 2000). This approach is known as peptide-based proteomics. On the other hand, bottom-up approaches can be divided into two groups depending on the fractionation step selected to reduce the sample complexity: (1) the separation is performed at the protein level, before the digestion step or (2) the separation is performed at the peptide level, after the digestion step (Fig. 2.3).

One gel-based approach, the two-dimensional electrophoresis (2-DE), is the method most used to perform the separation at the protein level. In this case, proteins are separated based on PI and MR and can be individually excised from the gel and digested into peptides. After that, generated peptides are analyzed by MS for the protein identification.

In the second approach, the protein mixture is first enzymatically digested, and the resulting peptides mixture are separated by high-throughput LC, normally Reverse Phase HPLC (RP-HPLC) or ultra-high performance liquid chromatography. Eluted peptides are analyzed using MS, which is usually coupled on-line with the LC system. Depending on the peptide mixtures complexity, multidimensional LC is sometimes needed to increase the separation power of the chromatographic separation. In this case, SCX columns coupled with RP LC are typically used. This approach commonly used in the proteomic analysis is named "*shotgun*" proteomics (Zhang, Fonslow, Shan, Baek, & Yates, 2013).

The *top–down* proteomics approach (Donnelly et al., 2019) (Fig. 2.3) characterizes the fragments produced by the dissociation of the *intact proteins* directly inside the MS, without protein digestion. This intact protein analysis is currently available due to the recent advances in new dissociation mechanisms and the high mass accuracy obtained in the high-resolution MS machines. Moreover, this approach needs new improvements so that its use is more widespread.

2.4 Sample preparation

Protein extraction (Fig. 2.4) is one of the most important steps in proteomic analysis (Feist & Hummon, 2015; Ye et al., 2019). In the last years

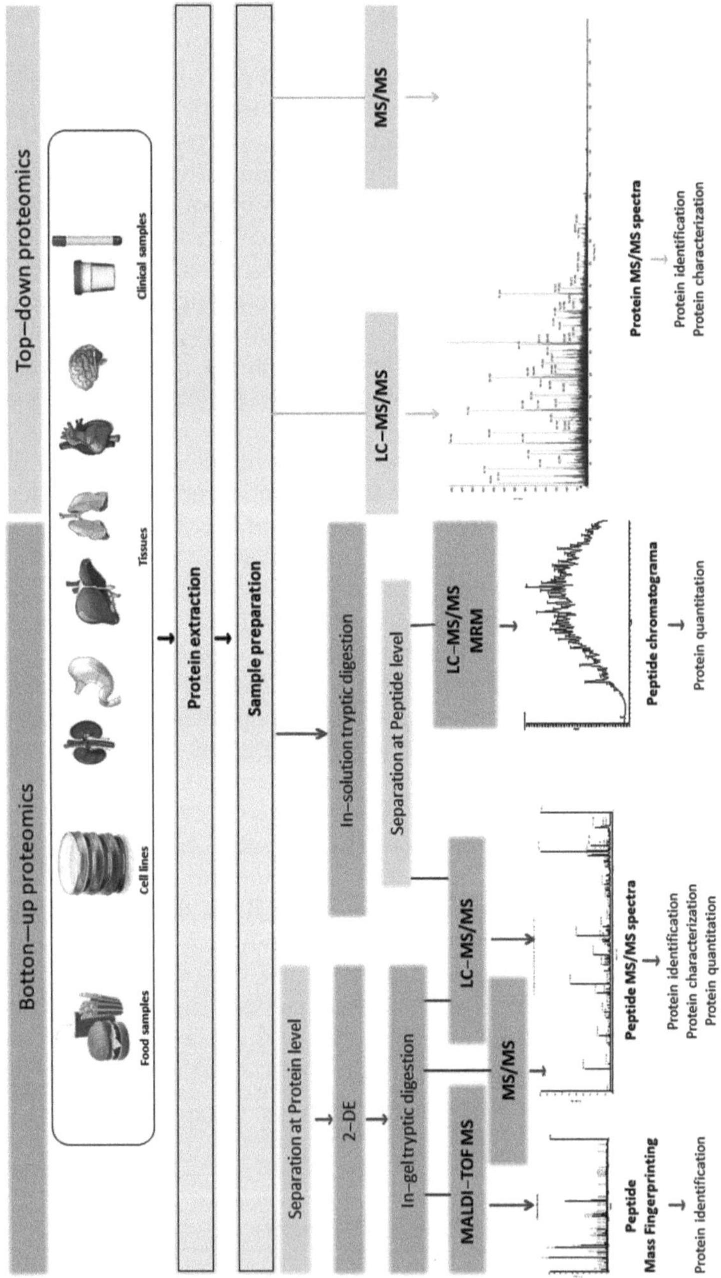

FIG. 2.3 Proteomic workflow commonly used in food proteomics.

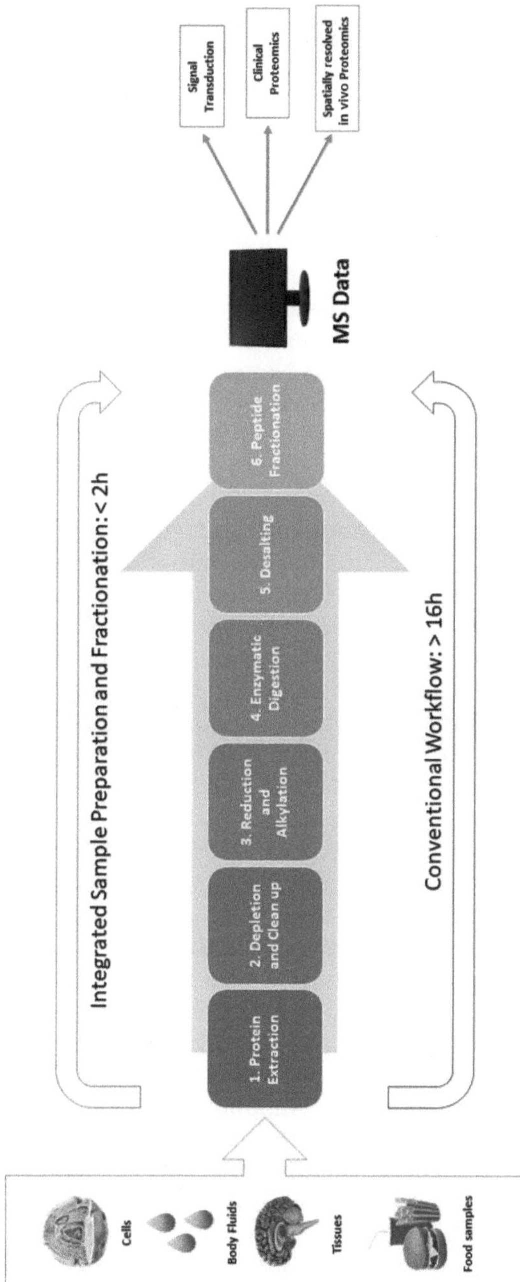

FIG. 2.4 Common proteomic steps are used for the sample preparation/treatment before the proteomic analysis.

(Fig. 2.7), this technology has been used on fruit juices (orange and apple fruit juices) for their authentication (Xu, Xu, Kelly, & Liao, 2020; Xu, Xu, Strashnov, & Liao, 2020; Xu, Xu, Wang, Wang, & Liao, 2020)

Samples can be obtained by different methods: biopsies, blood draws, etc., using the corresponding ethics guidelines. Once the sample is obtained, the proteins must be obtained via lysis and extraction from the cells in the sample. There are numerous methods for lysis and extraction; lysis buffers join to mechanical disruption strategies have been reviewed in several papers (Lee, 2007; Reddy et al., 2014). These methods must be different in function of the sample type. The most common methods use chemical lysis with an extraction agent; join to some mechanical method that makes disruption of the cell. This allows the chemical agent to solubilize the present protein (Islam, Aryasomayajula, & Selvaganapathy, 2017). The most common detergents, including Nonionic, Zwitterionic, and Anionic detergents are Triton X-100 (nonionic), NP-40(nonionic), Tween 20(nonionic), CHAPS (Zwitterionic) or Deoxycholate (Anionic). Therefore, highly efficient lysis and extraction have been obtained using different composition lysis buffer, with the mechanical disruption that includes gentle rocking or cell scraping to sonication or French pressing. The choice of one lysis buffer and one mechanical disruption method depends on the sample precedence, sample size, protein target of extraction, and in some cases from the expertise of the technician in preparation methods.

Food proteomics is necessary to optimize a specific protocol for each food sample, to maximize protein recovery and minimize the proteolysis process and the apparition of protein modifications (Martinović, Šrajer Gajdošik, & Josić, 2018). Therefore, it is difficult to obtain a universal extraction buffer composition that can be used in all cases. In all extraction buffers are added different compounds for stabilizing and solubilizing the proteins, such as pH regulators (e.g., Tris, Hepes, MOPS), reducing agents (e.g., dithiothreitol, 2-mercaptoethanol), and denaturing compounds (e.g., 8 M urea, SDS, CHAPS). In most cases is necessary the elimination of contaminants such as nucleic acids by adding in the buffer reagents as DNases. The use of protease inhibitors, such as PMSF or EDTA, is necessary to avoid protein proteolysis. Some of the compounds found in the extraction buffer are not fully compatible with MS and thus it is necessary prior MS analysis a cleanup step. However, these compounds found in the protein extraction buffers are compatible with gel-based proteomics approaches as gel electrophoresis both 1D-PAGE and 2D-PAGE.

Regarding the methods to sample purification, this can commonly include one or more processes as (Finoulst, Pinkse, Van Dongen, & Verhaert, 2011; Martínez-Maqueda, Hernández-Ledesma, Amigo, Miralles, & Gómez-Ruiz, 2013): (1) methods for proteins clarify and concentrate, such as protein precipitation (e.g., ammonium sulfate, TCA/Acetone, chloroform), centrifugation, filtration (e.g., dialysis, ultrafiltration), lyophilization; (2) procedures for protein enrichment, depletion, or fractionation to reduce sample complexity and therefore increase the proportion of the desirer proteins as chromatography (e.g., RP; SCX; SEC) and SDS-PAGE (1-DE or 2-DE). Although chromatography or SDS-PAGE are the most common techniques in some cases is possible to use off-gel isoelectric focusing. For high-abundance proteins depletion or protein enrichment, there are a lot of commercial immunoaffinity depletion kits. These kits generally are specifically recommended for albumin and other human plasma proteins removal, reflecting the difficulty found in clinical proteomics like food proteomics. In the case of food samples, the use of bead-based combinatorial peptide ligand libraries such as ProteoMiner allows the compression of the dynamic range of the protein concentration (Li, 2015; Piras, Roncada, Rodrigues, Bonizzi, & Soggiu, 2016).

Depletion and fractionation procedures allow obtain protein concentration and therefore improving the sensitivity of MS analysis. However, these strategies using a column in sample fractionation have low sample throughput limitations, since the fractions obtained must be analyzed individually and this produces a poor protein recovery date. Moreover, antibody columns used for depletion of high abundance proteins (e.g., albumin), have the problem due to non-specific binding of the target proteins. This can induce loss of the protein of interest and poor protein identification (Basel New York and Owusu-Apenten, 2002; de Jesus, da Silva Fernandes, de Souza Pessôa, Raimundo, & Arruda, 2017).

It is necessary to mention that, in bottom–up proteomics, a critical step is the protein digestion, which is performed after electrophoresis ("in-gel" digestion) for gel-based approaches, or on the protein mixture ("in-solution" digestion) (Hildonen, Halvorsen, & Reubsaet, 2014). Different proteolytic enzymes can be used, although generally trypsin is used as a common digestion enzyme because it offers some advantages: (1) give us peptides with a good amino acid composition for MS (2) the peptides generate using trypsin gives correct charge state and homogeneous fragmentation by collision-induced dissociation in MS analysis (Kolsrud et al., 2012). This process of digestion is the most time-consuming step in proteomic analysis because the typical digestion times range from 6 to 24 h. In some cases to accelerate protein digestion have been proposed, other more rapid protocols such as the use of microwaves, infrared radiation, high-intensity focused ultrasounds, modified trypsin, and immobilized trypsin (e.g., in columns or nanoparticles) (Zheng & DeMarco, 2017).

2.5 Gel-based or mass spectrometry based approaches

2.5.1 Gel-based proteomics approaches

Although in recent years, there is a tendency to make gel-free workflows in proteomics studies, electrophoresis has been more used and continues to be used at different levels in proteomics food studies. This is due that data provided by electrophoresis is often complementary to those obtained from LC-MS/MS approaches. Therefore techniques as SDS-PAGE, IEF and 2DE-PAGE protein electrophoretic patterns have been extensively used for the development of methods to obtain a good spots resolution to study not only the post-transcriptional modifications (Bernal et al., 2019; Mato et al., 2019) but also the proteomics distance to see the quantitative proteomic distance between species (Rodríguez-Vázquez et al., 2020). But it is necessary to point that in food, although these electrophoretic methodologies are widely used, they show some limitations: (1) difficulty to analyze hydrophobic and poorly soluble proteins, (2) limited dynamic range of the detection methods, (3) high complexity of the proteins in the food sample, (4) degradation of proteins induced by the food processing methods, (5) closely related species used in food preparation, (6) poorly identification when using samples with a mixture of species, and (7) the necessary use of reference samples. This makes that in food some aspects of food authentication are studied commonly by MS-based methods.

2.5.2 MS-based proteomics approaches

MS is recently the most useful method used in the high-throughput identification, characterization, and quantification of proteins. This approach allows the analysis of thousands of proteins in one unique experiment. Moreover, this analysis, using adequate bioinformatics tools, gives not only the protein identification but also a lot of information about the peptide sequence, abundance, possible protein–protein interaction, possible pathways, etc.

Therefore, in MS-based proteomics approaches there is the possibility of change the analysis mode to perform two different approaches, data-dependent analysis (DDA) and/or data-independent analysis (DIA or IDA) (Fig. 2.5). The most common method for the acquisition of the MS/MS data is the approach known as DDA. This approach is used when the objective is to analyze/identify all the proteins present in the sample. However, this approach has inconvenient because detected peptides can be biased toward those coming from the highest abundance of proteins. To avoid this inconvenience, there is another acquisition mode named data-DIA. In this mode, the MS doesn´t select one specific ion for fragmentation; however, all peptide ions present in the mass analyzer are fragmented.

This last approach is currently being used in a lot of investigation due to recent advances introduced in MS instruments, such as Sequential Window Acquisition of all Theoretic Mass Spectra (SWATH-MS) (Ludwig et al., 2018) acquisition modes, which allows good protein quantification without a protein/peptide labeling. On the other hand, it is necessary to

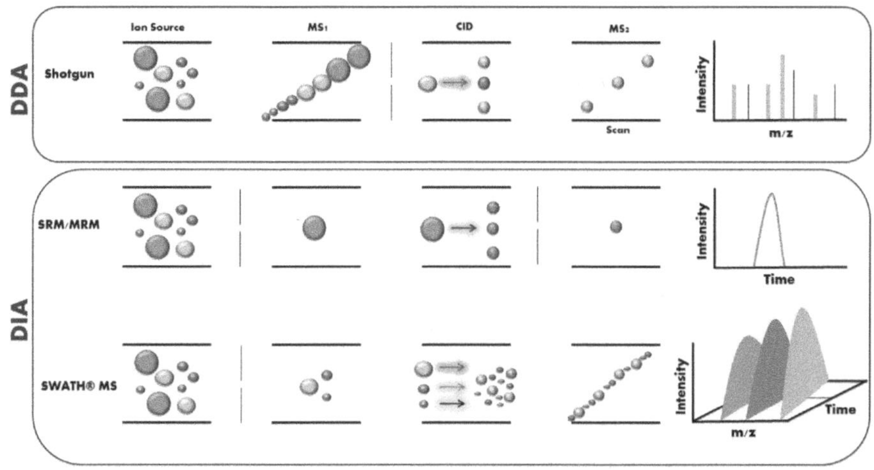

FIG. 2.5 A schematic representation of different proteomic approaches: DDA approach: Shotgun, DIA approach: SRM/MRM and SWATH-MS. DDA is used for the massive protein identification, and DIA is used as a quantitative targeted method in protein quantification. *DDA*, data-dependent analysis; *DIA*, data-independent acquisition.

say that if the protein or peptide of interest is known, a targeted proteomics analysis can be done using an SRM or MRM analysis (Mermelekas et al., 2015). SRM is considered one of the best methods, due to its high sensitivity and specificity, for the protein's quantification.

2.6 Quantitative proteomics in food

Proteomes is quite different among individuals, cell types, and within the same cell, depending on the state or cell activity. The great advance in proteomic techniques increasing their sensitivity and accuracy join to potent bioinformatic tools have allowed the development of more accurately methodologies not only for the identification but also for the quantification of biomarkers and study the proteome expression data. Therefore, an important challenge in proteomics is the quantitative difference in concentration in all expressed proteins (from the most- to the least-abundant proteins). The increase in proteomics studies applied to different samples and problems has generated a need for apply quantitative analysis methods. This made many different proteomic techniques became to transform into quantitative techniques both gel-based or gel-free techniques to quantify levels of proteins (Feng et al., 2017; Ong, Foster, & Mann, 2003; Pappireddi et al., 2019; Yan & Chen, 2005).

Between the proteomics quantitative techniques, comparative proteomics is the method of choice used in the assessment of both food quality and safety (Mora, Gallego, & Toldrá, 2018).

It is necessary to mention, again, that in food proteomics, many different factors significantly can complicate proteomics and have to be taken into account when a proteomic assay is being planned (Korte & Brockmeyer, 2017): (1) no complete proteome data are available for the totality of foods types; (2) food processing can introduce overly complex modifications in food proteins, e.g., via the Maillard reaction that normally are not considered in proteomics approaches; (3) storage conditions and processing of food often induce a (and often unknown) dynamic change in food proteome by degradation or modification; (4) food matrices can complicate proteome analysis due to interferences or interactions of food proteins with other compounds as storage lipids or carbohydrates, connective tissues, fiber, and polyphenols.

But on the other hand, it is true, that a large variety of different aspects in food authenticity, food safety, and food quality has gained more knowledge using the high amount of information available in databases from the proteome of food or the protein interaction in food, consumer, and the present microorganisms. Authenticity and traceability of food products are especially important at all levels of the production process, from raw materials to finished products. And all of this is possible to make thanks to the great advance in proteomics techniques.

2.6.1 Food quality

During the last decade, proteomics has been the method selected to develop methodologies that allow quality control in food production and food safety (Jagadeesh et al., 2017). Therefore, it is particularly important to know the structure and function of proteins present in raw food materials and final food products. This is because changes in protein structure occurring at specific amino acid residues during processing events can help to predict the quality and stability of food product. This knowledge can not only help to optimize the processing methodologies but also to ensure good quality in parameters such as texture, color, crops yield, or in meet, changes derived from the stress response.

In this chapter, we expose several applications of quantitative proteomics to food quality (Fig. 2.6).

2.6.1.1 Taste, flavor consistence

Meat represents the food of animal origin rich in protein content, therefore, the evaluation of taste, flavor consistency from meat is one of the most representative examples in the food quality evaluation. The meat quality is influenced by several factors such as genetics, environment, and

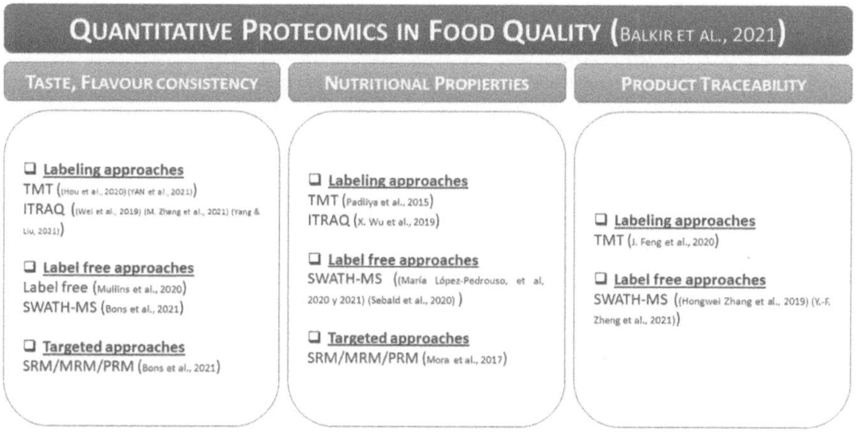

FIG. 2.6 Schematic representation of the quantitative proteomic techniques used in food quality, including references.

post-mortem processing. Therefore, there are a lot of proteomics studies evaluating meat quality in many types of meat such as beef, pork, lamb, and chicken (Nair & Zhai, 2019). Most of these studies are performed using in gel-based approaches that allow obtain the information about differences in protein isoforms (López-Pedrouso et al., 2019; López-Pedrouso et al., 2020c).

In the last years, there are studies using label techniques as TMT labeling to explore the molecular mechanisms of meat quality (Hou et al., 2020) or to investigate the differences in the proteomes of yak in different muscles (Yan, Li, Hu, Ma, & Lu, 2021). Studies in muscle from M longissimus dorsi by label-free proteomic approached, identified key proteins involved in the expression of compensatory growth (Mullins et al., 2020). In last year's studies performed by other quantitative label proteomics technique as iTRAQ in meat quality were published too. The group of Wei et al. study the differences in the proteomes of high- and low-quality goat and bovine muscles (Wei, Li, Zhang, & Liu, 2019). Therefore, it is known that lipid oxidation is the main cause of quality deterioration in meat and meat-derived products. Some studies performed by quantitative iTRAQ approaches identified candidate molecular biomarkers linked to lipid oxidation in duck muscle (Zhang, Wang, Xu, Xu, & Zhou, 2021). In 2021, a study by iTRAQ gives a proteomic profile to explain the beef poor meat quality. In this work, the author identified protein markers of beef quality during post-mortem storage at 4°C (Yang & Liu, 2021).

The group of Bons et al. (2021)) in the same year used a combination of quantitative proteomic techniques as SWATH-MS, parallel reaction monitoring, and SRM to quantify ten potential biomarkers of beef meat tenderness or marbling in a cohort of 64 muscle samples.

2.6.1.2 Nutritional properties

In the nutritional properties of food there is an important part in the identification and characterization of natural bioactive peptides. In some cases, the use of in silico approaches combined with proteomics is important to obtain good results. Therefore, targeted quantitative methods as the MRM MS methodology are used in the quantitation of active peptides and biomarkers and allow absolute quantitation or semi quantitation of this bioactive peptide (Mora, Gallego, Reig, & Toldrá, 2017). The bioactive properties of the peptides are determined by the inherent amino acid composition and sequence of the peptide.

Chicken egg yolk is a rich source of nutrients that have high-quality proteins, vitamins, minerals, carotenoids, and antioxidants. In 2015, a proteomic study was made by TMT using to known protein markers from chicken egg yolk fertilized or unfertilized. In total, 225 protein were identified and they found that some proteins upregulated in fertilized egg yolk play an important role in angiogenesis (pleiotrophin, histidine-rich glycoprotein) and defense against pathogens (mannose-binding lectin, β-defensin 11, serum amyloid P-component, ovostatin) (Padliya et al., 2015).

As antioxidant peptides are increasingly used as food additives due to their effects on the body human, regulating oxidative stress against oxidation of lipids and proteins in the last years there are some papers using SWATH-MS to characterize/quantify and identify these bioactive peptides. In 2021, one paper has been published using SWATH-MS to identification and quantified bioactive peptides, produced by enzymatic hydrolysis, as health-promoting agents (López-Pedrouso, Borrajo, Amarowicz, Lorenzo, & Franco, 2021). In 2020, the same group use specifically porcine liver to prepare hydrolysates with antioxidant activity employing proteolytic enzymes such as alcalase, bromelain, papain, and flavourzyme. In this hidrolysates, they are able to determinate by SWATH-MS bioactive peptides (López-Pedrouso, Borrajo, Pateiro, Lorenzo, & Franco, 2020a).

The peptidome of fermented, protein-rich food is extraordinarily complex, and high accurately proteomic technique must be used to map the bitter-tasting peptidome of fermented foods. In this paper, author used SWATH-MS on two fresh cheese samples (Sebald, Dunkel, & Hofmann, 2020).

Another bioactive peptide are Ovalbumin peptides. In 2019, the group of Wu et al. make a proteomics characterization using isobaric tags for the relative and absolute quantification labeling technique (iTRAQ) in Tibetan chicken egg ovalbumin peptides, compared to ordinary eggs (Wu, Liu, Li, and Chi, 2019b).

2.6.1.3 *Product traceability*

The proteomic profiling of food can play an important role in food traceability. This food quality is considered of interest because of the increasing demands of cost and time-effective procedures that will be able to determine the food origin.

Proteomic techniques can detect differences related not only to quality, species, or treatment, but also for geographical variation.

In 2019, a paper reports a method to discriminate of dried sea cucumber Apostichopus japonicus (A. japonicus) products from different geographical regions in China (Xiapu, Jiaonan, Weihai, Yantai, Dalian) using SWATH-MS (Wei et al., 2019). In 2020, another paper report a method using TMT labeling to characterize the proteome of sea cucumbers from five main Chinese origins, including Xiamen (XM), Dalian (DL), Weihai (WH), Yantai (YT), and Qingdao (QD) (Feng, Zhang, Xia, Hu, & Zhou, 2020). In 2021, a paper was published where SWATH-MS was used to discriminate honey products from different geographical and botanical sources (Zheng et al., 2021).

2.6.2 Food safety

Food safety is a challenging field in which proteomic techniques as quantitative proteomic techniques must provide accurate, precise, and robust methods to determine any harmful compounds (allergens, etc) or organisms that may contaminate different types of food and that may be present at very low concentrations (Herrero, Simó, García-Cañas, Ibáñez, & Cifuentes, 2012) (Fig. 2.6).

The use of exogenous contaminants in food is common practice in agriculture and farming to avoid possible issues that might stop the correct growth of crops and animals. However, there is an exhaustive control with a limit in the quantity of use of these compounds, to protect the health of the consumers. Moreover, the different countries' legislature establish the maximum residue levels (MRLs)—defined as the maximum amount of a particular compound that might reach the final food product (*Legislation* | Food Safety, n.d.).

In recent years SWATH-MS has been used to obtain a method to prevent and control meat safety. In one paper, used marker peptides for discriminating non-edible meat of fur-producing animals (including fox, silver fox, blue fox, raccoon dog, ussuri raccoon dog, mink, and American mink) from the meat of food-providing animals (including pig, cattle, sheep, and donkey) (Qiu et al., 2020). In the same way, two more papers published in 2021 using iTRAQ study the effect of citral for the preservation of grains and other agricultural products (Wang et al., 2021) or a lactobionic acid

(LBA) that is a natural polyhydroxy acid widely used in the food industry (Cao & Zheng, 2021).

2.6.2.1 Microbial contaminants (pathogens and toxins)

The identification, confirmation, and the quantification of bacteria, bacterial toxins, or other contaminants in food are especially important problems in food safety. The most common bacteria that can cause food intoxication are *Staphylococcus aureus*, *Campylobacter jejuni*, some *Salmonella* and *Staphylococcus* species, some *Bacillus strains* and *Escherichia coli* O157:H7 strain (Gaso-Sokač, Kovač, & Josić, 2010a, b).

There are sensitive and well-established methods for the detection of bacteria and their toxins, by immunochemical methods. However, proteomics technologies offer more sensitive and specific methods not only for the identification of microbial food contaminants, but also for monitoring the elimination of contamination in food (Gaso-Sokač et al., 2010a, b).

Other important food contaminant is prions. All prion diseases or transmissible spongiform encephalopathies are characterized by the apparition of a different conformation (PrPSc) of protein normal protein PrPC in neural tissues both in humans and animals.

These different protein conformations were associated with different physicochemical properties (Sitammagari & Masood, 2021). Therefore, PrPC is a relatively soluble and protease-sensitive protein, while, on the other hand, PrPSc is relatively insoluble and protease-resistant. The human form of this infectious disease is known as Creutzfeldt–Jakob disease and is caused by the consumption of meat or meat products of prion-infected animals (Sitammagari & Masood, 2021).

A great review on proteomics techniques to determinants microbial contamination on food you can see Martinović, Andjelković, Gajdošik, Rešetar, and Josić (2016))

Some investigations by quantitative label-free LC-MS/MS and bioinformatic analysis were performed to determine the microorganism contaminant (Lasch, Schneider, Blumenscheit, & Doellinger, 2020). Using this proteomic approach was detected several differently expressed proteins in different bacteria. These proteins were the majority enzymes and cofactors involved in protein synthesis and energy metabolism as well as chaperones including heat shock. Another quantitative proteomics approach was used to study differences in bacterial expression after treatment with an antimicrobial agent totarol (Reddy et al., 2015). In this case, the quantitative method used was an iTRAQ labeling analyses (Reddy et al., 2015). The study showed downregulation in the expression of several central metabolic primary dehydrogenases (e.g., glyceraldehyde 3 phosphate dehydrogenase and succinatedehydrogenase) and among up-regulated related to in anaerobic respiration(e.g., nitrate reductase and lactate dehydrogenase), heme biosynthesis, and cell homeostasis (Reddy et al., 2015).

Other food pathogens are gram-positive bacteria Listeria monocytogenes and the gram-negative Yersinia enterocolitica. Therefore, the first bacteria have de ability to survive a broad range of temperatures; therefore, the contamination with this microorganism is of principal problem in processed food products. Studies by proteomics can evaluate the type of growth and virulence (Santos, Viala, Chambon, Esbelin, & Hébraud, 2019; Shamloo et al., 2019). Other studies by LC-MS/MS or quantitative label-free LC-MS/MS (Reddy et al., 2015) can evaluate the changes in the bacteria proteome under the influence of biologically active peptides and synthetic inhibitors of bacterial growth. On the other hand, the proteome of Yersinia enterocolitica was analyzed not only by 2D electrophoresis coupled to MALDI-TOF identification but also recently by quantitative eLC-MS/MS of different bacterial lysates (Li, Murugaiyan, Thomas, Alter, & Riedel, 2020).

In recent years, a new quantitative proteomics technique as SWATH-MS has been used for microbial or other food pathogens quantification to ensure food safety. In this sense, Campylobacter jejuni is a major cause of food-borne gastroenteritis. A paper published in 2020 (Man et al., 2020) using LC-MS/MS identified proteins associated with the growth of these bacteria in 0.1% sodium deoxycholate (DOC, a component of gut bile salts). Moreover, this group perform a validation by data-independent acquisition (DIA-SWATH-MS) (Man et al., 2020). In 2019, some papers study the fungal disease of barley that is responsible for substantial yield loss and lowers seed quality (Kerr et al., 2019) or the antibacterial mechanism of LBA against methicillin-resistant staphylococcus aureus (Kang et al., 2019) using this label-free proteomics approaches.

iTRAQ label was used recently to study the antibacterial mechanism of PLA (Phenyllactic acid) and LA (lactic acid) was investigated against Bacillus cereus (Ning et al., 2021). Other papers (Yao et al., 2020) were performed to study citrus greening, which is a destructive disease continuing to devastate citrus production worldwide.

2.6.2.2 Food allergens

Food allergy prevalence is rising (Croote & Quake, 2016); therefore, the economic costs of food allergies have also grown. Proteins from peanut, milk, egg, soy, wheat, fish, shellfish, and tree nuts are the major food allergens that must be indicated in the food if they were added(*Allergen Labelling for Food Manufacturers* | Food Standards Agency, n.d.). The most common allergic reaction is anaphylaxis. However, it is important to mention that, although not intentionally, a cross-contamination of allergens into foods may still occur in shared production facilities, on shared equipment, or along the supply chain(*Allergen Guidance for Food Businesses* | Food Standards Agency, n.d.). Therefore, it was necessary to establish robust analytical workflows capable of performing an accurate allergen quantization. In the last decade, a lot of proteomic studies were made to identify and quantify allergens on food of plant origin or animal origin (Hoffmann-Sommergruber, 2016; López-Pedrouso et al., 2020b). Therefore, in the last years, a new discipline named foodomics, which studies food and nutrition areas using advanced –omics techniques to improve consumer confidence and health, has been reported (Balkir et al., 2021). In this, new discipline are involved, high-throughput technological –omics approaches as genomics, transcriptomics, proteomics, lipidomics, and metabolomics combined with bioinformatics and systems biology.

There are two perfect established methods for quantifying allergenic protein in food: an Elisa and MS-based method. This last method has an advantage with respect to Elisa making a more sensitive multiplexed quantitation of allergenic proteins (Croote & Quake, 2016).

However other methods as peak integration and spectral counting in comparison with a protein standard were used by Stevenson, Chu, Ozias-Akins, and Thelen (2009) for the identification of peanut allergens.

The MS method most extensively used was a targeted technique as SRM (Zhang et al., 2011), or other as multiple reactions monitoring (MRM). SRM has been widely used to quantitate a lot of proteins in complex matrices such as tissue (Fukuda et al., 2017), plasma (Wang et al., 2017), and cell lysate (Manes, Mann, & Nita-Lazar, 2015), and is also the principal technique for allergen detection and quantitation (Carrera, Cañas, & Gallardo, n.d.; Croote et al., 2017).

Thus, some studies are using an LC-SRM/MRM assay for specific detection and quantitation of the presence of: gluten (Alves, D'Almeida, & Ferreira, 2017, 2019; Martínez-Esteso et al., 2016), eggs (Monaci et al., 2020; Nitride et al., 2019; Sharma, Khuda, Parker, Eischeid, & Pereira, 2016), milk (Le, Poulsen, Kristiansen, & Larsen, 2020; Martinez-Esteso et al., 2020; Monaci et al., 2020; Nitride et al., 2019; Sharma et al., 2016), soy (Jia, Zhou, Zhu, Shen, & He, 2019; Krager, 2020), hazelnut (Mills et al., 2019; Pilolli, De Angelis, & Monaci, 2017), walnut (Korte & Brockmeyer, 2016; Sealey-Voyksner, Zweigenbaum, & Voyksner, 2016), almond (Korte & Brockmeyer, 2016; Planque, Arnould, & Gillard, 2017; Sealey-Voyksner et al., 2016) or peanuts (Careri et al., 2007; Sealey-Voyksner et al., 2016) in complex food products. Moreover, there are too, several strategies for the multi-allergen detection by LC-SRM/MRM (Boo, Parker, & Jackson, 2018; Gu et al., 2018; Hoffmann, Münch, Schwägele, Neusüß, & Jira, 2017; Huschek, Bönick, Löwenstein, Sievers, & Rawel, 2016; Mills et al., 2019; Pilolli et al., 2017; Planque et al., 2019).

Four of these methods (Gu et al., 2018; Hoffmann et al., 2017; Huschek et al., 2016; Planque et al., 2019) had been validated with high precision

(% CV) and recovery, different from other reported methods, having a linear regression R2 > 99% in the calibration curves.

Using more recent proteomics approaches as DIA mode such as sequential window acquisition of all theoretical mass spectra (SWATH-MS) the group of Monaci, Losito, De Angelis, Pilolli, and Visconti (2013)) found limits values of detection (0.4 to 1.1 μg/mL) for traces of egg-white powders and caseinate in white wines. Other work using SWATH-MS approach was made to analyze tropomyosin (TM) which is considered the major shrimp allergen that could trigger anaphylactic reactions (Xu, Zhang, Zhang, Lin, & Guo, 2020).

iTRAQ label approach was employed in a paper published in 2021 (Liu et al., 2021), to analyze the seeds of two genetically modified (GM) rice lines, T2A-1 and T1C-19, and their nontransgenic isogenic variety, MH63, to investigate the effects of genetic modification.

Another quantitative method as absolute quantitation by AQUA-LC-MRM were too employed to the allergens quantification (Ahsan, Rao, Gruppuso, Ramratnam, & Salomon, 2016; Houston, Lee, Stevenson, Ladies, & Bannon, 2011)

2.6.2.3 *Food authenticity and detection of adulterations*

Adulteration of food products is a great problem not only for economic fraud, but also for safety reasons. The non-declared introduction of ingredients in food, such as toxic or allergenic products, may be very harmful to consumer's health. This represents an important public health risk (Ortea et al., 2016; Ruhland & Klinger, 2019). The authentication of food etiquette labeling must be guaranteed, and therefore is necessarily accurate and reliable analytical methods to verify the components used in a food product to ensure that they are of nature or correct quality demanded by the purchaser.

Proteomic approaches for the authentication of food have significant importance in the last years. The great progress in MS allows the highly specific and sensitive identification of biomarker peptides or intact protein fingerprint spectra. Consequently, targeted and non-targeted proteomic approaches have been developed.

The direct spectral comparison of proteomics data is a technique commonly used for the authentication of food samples. Proteome profiling of authentic food samples is important for the peptide marker identification and, therefore, the generation of a good, targeted MRM-approach as still not or highly incomplete proteomics data are available for a great number of different foods. Some studies were made using this quantitative proteomic targeted method as studies in milk to ensure the qualitative detection and quantitative determination of cow, sheep, and goat milk in the raw materials used in dairy products (Chen et al., 2016).

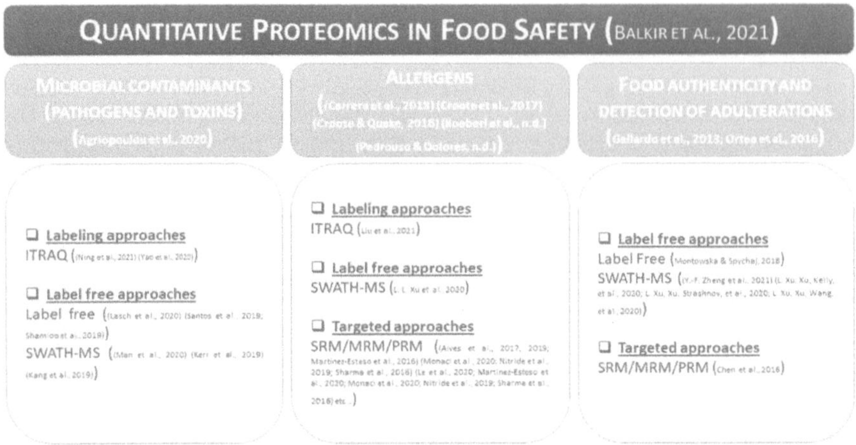

FIG. 2.7 Schematic representation of the quantitative proteomic techniques used in food safety, including references.

Label-free quantification combined with high-resolution infusion-based MS was used to evaluate the authenticity of 'horse sausages' made from horse meat and pork (Montowska & Spychaj, 2018).

In this year, one paper has been published using SWATH-MS to discriminate honey products from different geographical and botanical sources. This work considers the importance of the use of proteomics in the search of honey products containing distinct forms of syrup employed in honey adulteration (Zheng et al., 2021). In the last years, this technology has been used on fruit juices (orange and apple fruit juices) for their authentication (Xu, Xu, Kelly, & Liao, 2020; Xu, Xu, Strashnov, & Liao, 2020; Xu, Xu, Wang, Wang, & Liao, 2020) (Fig. 2.7).

2.7 Conclusions and future trends

Quantitative proteomic techniques are very useful not only for the assessment of raw materials and final products but also for control, optimization, and development of new processes in food technology and biotechnology.

However, most proteomic analyses are performed using comparative 2D electrophoresis, although recently were develop, faster, and more effective methods such as quantitative isotope labeling or label-free quantitative proteomics that give more accurate results. Therefore, the use of these methods combined with other analytical strategies will let better in-process control and characterization of batch-to-batch variations. Moreover, these proteomics techniques can help to answer some questions in food science as detection of food contaminants and allergens, and evaluation of the safety of foods.

In the last years, a great advance has been made in label-free techniques as SWATH-MS analysis combined with a great advance in bioinformatics tools. Therefore, the proteomic techniques used in the study, characterization, and food control will take advantage of these new versatile and high accuracy approaches to improve food knowledge.

References

Aebersold, R., & Mann, M. (2016). *Mass-spectrometric exploration of proteome structure and function* Nature *537* (pp. 347–355). London: Nature Publishing Group. https://doi.org/10.1038/nature19949.

Ahsan, N., Rao, R. S. P., Gruppuso, P. A., Ramratnam, B., & Salomon, A. R. (2016). Targeted proteomics: Current status and future perspectives for quantification of food allergens. *Journal of Proteomics, 143*, 15–23. https://doi.org/10.1016/j.jprot.2016.04.018.

Allergen guidance for food businesses | Food Standards Agency (n.d.). https://www.food.gov.uk/business-guidance/allergen-guidance-for-food-businesses (accessed March 31, 2021)

Allergen labelling for food manufacturers | Food Standards Agency. (n.d.). https://www.food.gov.uk/business-guidance/allergen-labelling-for-food-manufacturers (accessed March 31, 2021)

Almeida, A. M., Bassols, A., Bendixen, E., Bhide, M., Ceciliani, F., Cristobal, S., et al. (2014). Animal board invited review: Advances in proteomics for animal and food sciences. *Animal, 9*(1), 1–17. https://doi.org/10.1017/S1751731114002602.

Alves, T. O., D'Almeida, C. T. S., & Ferreira, M. S. L. (2017). Determination of gluten peptides associated with celiac disease by mass spectrometry. In *Celiac Disease and Non-Celiac Gluten Sensitivity*. London, UK: InTech. https://doi.org/10.5772/67547.

Alves, T. O., D'Almeida, C. T. S., Scherf, K. A., & Ferreira, M. S. L. (2019). Modern Approaches in the Identification and Quantification of Immunogenic Peptides in Cereals by *LC-MS/MS. Frontiers in Plant Science 10* (p. 1470). London, UK: Frontiers Media S.A. https://doi.org/10.3389/fpls.2019.01470.

Balkir, P., Kemahlioglu, K., & Yucel, U. (2021). Foodomics: A new approach in food quality and safety. *Trends in Food Science and Technology 108* (pp. 49–57). Amsterdam: Elsevier Ltd. https://doi.org/10.1016/j.tifs.2020.11.028.

Bantscheff, M., Schirle, M., Sweetman, G., Rick, J., & Kuster, B. (2007). Quantitative mass spectrometry in proteomics: A critical review. *Analytical and Bioanalytical Chemistry, 389*(4), 1017–1031. https://doi.org/10.1007/s00216-007-1486-6.

Bendixen, E., Danielsen, M., Hollung, K., Gianazza, E., & Miller, I. (2011). Farm animal proteomics—A review. *Journal of Proteomics, 74*(3), 282–293. https://doi.org/10.1016/j.jprot.2010.11.005.

Bernal, J., Mouzo, D., López-Pedrouso, M., Franco, D., García, L., & Zapata, C. (2019). The major storage protein in potato tuber is mobilized by a mechanism dependent on its phosphorylation status. *International Journal of Molecular Sciences, 20*(8):1889–1905. doi:10.3390/ijms20081889.

Bodzon-Kulakowska, A., Bierczynska-Krzysik, A., Dylag, T., Drabik, A., Suder, P., Noga, M., et al. (2007). Methods for samples preparation in proteomic research. *Journal of Chromatography B: Analytical Technologies in the Biomedical and Life Sciences, 849*(1–2), 1–31. https://doi.org/10.1016/j.jchromb.2006.10.040.

Bons, J., Husson, G., Chion, M., Bonnet, M., Maumy-Bertrand, M., Delalande, F., et al. (2021). Combining label-free and label-based accurate quantifications with SWATH-MS:

Comparison with SRM and PRM for the evaluation of bovine muscle type effects. *Proteomics, 21*(10), e2000214. doi:10.1002/pmic.202000214.

Boo, C. C., Parker, C. H., & Jackson, L. S. (2018). A targeted LC-MS/MS method for the simultaneous detection and quantitation of egg, milk, and peanut allergens in sugar cookies. *Journal of AOAC International, 101*(1), 108–117. https://doi.org/10.5740/jaoacint.17-0400.

Bräcker, J., & Brockmeyer, J. (2018). Characterization and detection of food allergens using high-resolution mass spectrometry: Current status and future perspective. *Journal of Agricultural and Food Chemistry 66* (pp. 8935–8940). Washington, D.C: American Chemical Society. https://doi.org/10.1021/acs.jafc.8b02265.

Buchberger, A. R., DeLaney, K., Johnson, J., & Li, L. (2018). Mass spectrometry imaging: A review of emerging advancements and future insights. *Analytical Chemistry 90* (pp. 240–265). Washington, D.C.: American Chemical Society. https://doi.org/10.1021/acs.analchem.7b04733.

Cao, J., & Zheng, Y. (2021). ITRAQ-based quantitative proteomic analysis of the antimicrobial mechanism of lactobionic acid against: Staphylococcus aureus. *Food and Function, 12*(3), 1349–1360. https://doi.org/10.1039/d0fo02491k.

Careri, M., Costa, A., Elviri, L., Lagos, J. B., Mangia, A., Terenghi, M., et al. (2007). Use of specific peptide biomarkers for quantitative confirmation of hidden allergenic peanut proteins Ara h 2 and Ara h 3/4 for food control by liquid chromatography-tandem mass spectrometry. *Analytical and Bioanalytical Chemistry, 389*(6), 1901–1907. https://doi.org/10.1007/s00216-007-1595-2.

Carrera, M., Cañas, B., & Gallardo, J. M. (n.d.). *Advanced proteomics and systems biology applied to study food allergy*.

Chen, Q., Ke, X., Zhang, J. S., Lai, S. Y., Fang, F., Mo, W. M., et al. (2016). Proteomics method to quantify the percentage of cow, goat, and sheep milks in raw materials for dairy products. *Journal of Dairy Science, 99*(12), 9483–9492. https://doi.org/10.3168/jds.2015-10739.

Croote, D., Braslavsky, I., & Quake, S. R. (2017). Matrix-corrected mass spectrometry enables sensitive detection of food allergens and reveals widespread soy contamination of processed foods. (p. 231266). bioRxiv. doi:10.1101/231266

Croote, D., & Quake, S. R. (2016). Food allergen detection by mass spectrometry: The role of systems biology. *npj Systems Biology and Applications 2* (pp. 1–10). London: Nature Publishing Group. https://doi.org/10.1038/npjsba.2016.22.

de Jesus, J. R., da Silva Fernandes, R., de Souza Pessôa, G., Raimundo, I. M., & Arruda, M. A. Z. (2017). Depleting high-abundant and enriching low-abundant proteins in human serum: An evaluation of sample preparation methods using magnetic nanoparticle, chemical depletion and immunoaffinity techniques. *Talanta, 170*, 199–209. https://doi.org/10.1016/j.talanta.2017.03.091.

Doerr, A. (2012). Proteomics: Interactomes by mass spectrometry. *Nature Methods, 9*(11), 1043. https://doi.org/10.1038/nmeth.2235.

Donnelly, D. P., Rawlins, C. M., DeHart, C. J., Fornelli, L., Schachner, L. F., Lin, Z., et al. (2019). Best practices and benchmarks for intact protein analysis for top-down mass spectrometry. *Nature Methods, 16*(7), 587–594. https://doi.org/10.1038/s41592-019-0457-0.

Eng, J. K., McCormack, A. L., & Yates, J. R. (1994). An approach to correlate tandem mass spectral data of peptides with amino acid sequences in a protein database. *Journal of the American Society for Mass Spectrometry, 5*(11), 976–989. https://doi.org/10.1016/1044-0305(94)80016-2.

Feist, P., & Hummon, A. B. (2015). Proteomic challenges: Sample preparation techniques for Microgram-Quantity protein analysis from biological samples. *International Journal of Molecular Sciences, 16*(2), 3537–3563. https://doi.org/10.3390/ijms16023537.

Feng, J., Zhang, L., Xia, X., Hu, W., & Zhou, P. (2020). Effect of geographic variation on the proteome of sea cucumber (Stichopus japonicus). *Food Research International, 136,* 109498. doi:10.1016/j.foodres.2020.109498.

Feng, Y., Cappelletti, V., & Picotti, P. (2017). Quantitative proteomics of model organisms. *Current Opinion in Systems Biology 6* (pp. 58–66). Amsterdam: Elsevier Ltd. https://doi.org/10.1016/j.coisb.2017.09.004.

Finoulst, I., Pinkse, M., Van Dongen, W., & Verhaert, P. (2011). Sample preparation techniques for the untargeted LC-MS-based discovery of peptides in complex biological matrices. *Journal of Biomedicine and Biotechnology, 2011,* 1–14. doi:10.1155/2011/245291.

Fukuda, T., Nomura, M., Kato, Y., Tojo, H., Fujii, K., Nagao, T., et al. (2017). A selected reaction monitoring mass spectrometric assessment of biomarker candidates diagnosing large-cell neuroendocrine lung carcinoma by the scaling method using endogenous references. *Plos One, 12*(4), e0176219. https://doi.org/10.1371/journal.pone.0176219.

Gaso-Sokač, D., Kovač, S., & Josić, D. (2010a). Application of proteomics in food technology and food biotechnology. *9862*(3), 284–295. http://bib.irb.hr/datoteka/479763.2448_corrected.doc. (Accessed 5 July 2021

Gašo-Sokač, D., Kovač, S., & Josić, D. (2010b). Application of proteomics in food technology and food biotechnology: Process development, quality control and product safety. *Food Technology and Biotechnology, 48*(3), 284–295.

Gingras, A., Aebersold, R., & Raught, B. (2005). Advances in protein complex analysis using mass spectrometry. *The Journal of Physiology, 563*(Pt 1), 11–21. https://doi.org/10.1113/jphysiol.2004.080440.

Gu, S., Chen, N., Zhou, Y., Zhao, C., Zhan, L., Qu, L., et al. (2018). A rapid solid-phase extraction combined with liquid chromatography-tandem mass spectrometry for simultaneous screening of multiple allergens in chocolates. *Food Control, 84,* 89–96. https://doi.org/10.1016/j.foodcont.2017.07.033.

Herrero, M., Simó, C., García-Cañas, V., Ibáñez, E., & Cifuentes, A. (2012). Foodomics: MS-based strategies in modern food science and nutrition title (short version): MS methodologies in Foodomics. *Mass Spectrometry Reviews, 31*(1), 49–69. http://digital.csic.es/bitstream/10261/49035/1/Foodomics_review_CIFUENTES.pdf.

Hildonen, S., Halvorsen, T. G., & Reubsaet, L. (2014). Why less is more when generating tryptic peptides in bottom-up proteomics. *Proteomics, 14*(17–18), 2031–2041. https://doi.org/10.1002/pmic.201300479.

Hmmier, A., & Dowling, P. (2018). DIGE analysis software and protein identification approaches. *Methods in Molecular Biology, 1664* (pp. 41–50). New Jersey: Humana Press Inc. https://doi.org/10.1007/978-1-4939-7268-5_4.

Hoffmann, B., Münch, S., Schwägele, F., Neusüß, C., & Jira, W. (2017). A sensitive HPLC-MS/MS screening method for the simultaneous detection of lupine, pea, and soy proteins in meat products. *Food Control, 71,* 200–209. https://doi.org/10.1016/j.foodcont.2016.06.021.

Hoffmann-Sommergruber, K. (2016). Proteomics and its impact on food allergy diagnosis. *EuPA Open Proteomics, 12,* 10–12. https://doi.org/10.1016/j.euprot.2016.03.016.

Hou, X., Liu, Q., Meng, Q., Wang, L., Yan, H., Zhang, L., et al. (2020). TMT-based quantitative proteomic analysis of porcine muscle associated with postmortem meat quality. *Food Chemistry, 328,* 127133. https://doi.org/10.1016/j.foodchem.2020.127133.

Houston, N. L., Lee, D. G., Stevenson, S. E., Ladies, G. S., Bannon, G. A., et al. (2011). Quantitation of soybean allergens using tandem mass spectrometry. *Journal of Proteome Research, 10*(2), 763–773. https://doi.org/10.1021/pr100913w.

Huang, Q., Yang, L. U., Luo, J. I., Guo, L., Wang, Z., Yang, X., et al. (2015). SWATH enables precise label-free quantification on proteome scale. *Proteomics, 15*(7), 1215–1223. https://doi.org/10.1002/pmic.201400270.

Huschek, G., Bönick, J., Löwenstein, Y., Sievers, S., & Rawel, H. (2016). Quantification of allergenic plant traces in baked products by targeted proteomics using isotope marked peptides. *LWT - Food Science and Technology, 74*, 286–293. https://doi.org/10.1016/j.lwt.2016.07.057.

Islam, M. S., Aryasomayajula, A., & Selvaganapathy, P. R. (2017). A review on macroscale and microscale cell lysis methods. *Micromachines. 8*. Switzerland: MDPI AG. https://doi.org/10.3390/mi8030083.

Jagadeesh, D. S., Kannegundla, U., & Reddy, R. K. (2017). Application of proteomic tools in food quality and safety. *Advances in Animal and Veterinary Sciences, 5*(5), 213–225. https://doi.org/10.17582/journal.aavs/2017/5.5.213.225.

Jia, H., Zhou, T., Zhu, H., Shen, L., & He, P. (2019). Quantification of gly m 5.0101 in soybean and soy products by liquid chromatography-tandem mass spectrometry. *Molecules, 24*(1), 68. doi:10.3390/molecules24010068.

Kang, S., Kong, F., Liang, X., Li, M., Yang, N., Cao, X., et al. (2019). Label-free quantitative proteomics reveals the multitargeted antibacterial mechanisms of lactobionic acid against methicillin-resistant staphylococcus aureus (MRSA) using SWATH-MS technology. *Journal of Agricultural and Food Chemistry, 67*(44), 12322–12332. https://doi.org/10.1021/acs.jafc.9b06364.

Kerr, E. D., Phung, T. K., Caboche, C. H., Fox, G. P., Platz, G. J., & Schulz, B. L. (2019). The intrinsic and regulated proteomes of barley seeds in response to fungal infection. *Analytical Biochemistry, 580*, 30–35. https://doi.org/10.1016/j.ab.2019.06.004.

Kolsrud, H., Malerod, H., Ray, S., Reubsaet, L., Lundanes, E., & Greibrokk, T. (2012). A critical review of trypsin digestion for LC-MS based proteomics. In *Integrative Proteomics*. London: InTech. https://doi.org/10.5772/29326.

Korte, R., & Brockmeyer, J. (2016). MRM3-based LC-MS multi-method for the detection and quantification of nut allergens. *Analytical and Bioanalytical Chemistry, 408*(27), 7845–7855. https://doi.org/10.1007/s00216-016-9888-y.

Korte, R., & Brockmeyer, J. (2017). Novel mass spectrometry approaches in food proteomics. *TrAC - Trends in Analytical Chemistry 96* (pp. 99–106). Amsterdam: Elsevier B.V. https://doi.org/10.1016/j.trac.2017.07.010.

Krager, J. (2020). *DigitalCommons @ University of Nebraska - Lincoln Assessing the quantification of soy protein in incurred matrices using targeted LC-MS /MS*.

Lasch, P., Schneider, A., Blumenscheit, C., & Doellinger, J. (2020). Identification of microorganisms by liquid chromatography-mass spectrometry (LC-MS1) and in silico peptide mass libraries. *Molecular and Cellular Proteomics, 19*(12), 2125–2138. https://doi.org/10.1074/mcp.TIR120.002061.

Le, T. T., Poulsen, N. A., Kristiansen, G. H., & Larsen, L. B. (2020). Quantitative LC-MS/MS analysis of high-value milk proteins in Danish Holstein cows. *Heliyon, 6*(9), e04620. https://doi.org/10.1016/j.heliyon.2020.e04620.

Lee, C. (2007). Protein extraction from mammalian tissues. *Methods in Molecular Biology, 362*(1), 385–389. https://doi.org/10.1385/1-59745-257-2:385.

Legislation | Food Safety. *Legislation | Food Safety* (n.d.). https://ec.europa.eu/food/safety/chemical_safety/contaminants/legislation_en (Accessed 23 March 2021)

Li, C., Murugaiyan, J., Thomas, C., Alter, T., & Riedel, C. (2020). Isolate specific cold response of yersinia enterocolitica in transcriptional, proteomic, and membrane physiological changes. *Frontiers in Microbiology, 10*, 3037. https://doi.org/10.3389/fmicb.2019.03037.

Li, L. (2015). Dynamic range compression with proteominer^{TM}: Principles and examples. *Methods in Molecular Biology, 1295*, 99–107. https://doi.org/10.1007/978-1-4939-2550-6_9.

Liu, W., Chen, H., Li, L., Dong, M., Zhang, Z., Wan, Y., et al. (2021). Proteomic analysis of the seeds of transgenic rice lines and the corresponding nongenetically modified isogenic variety. *Journal of the Science of Food and Agriculture, 101*(5), 1869–1878. https://doi.org/10.1002/jsfa.10802.

López-Pedrouso, M., Borrajo, P., Amarowicz, R., Lorenzo, J. M., & Franco, D. (2021). Peptidomic analysis of antioxidant peptides from porcine liver hydrolysates using SWATH-MS. *Journal of Proteomics, 232*, 104037. doi:10.1016/j.jprot.2020.104037.

López-Pedrouso, M., Borrajo, P., Pateiro, M., Lorenzo, J. M., & Franco, D. (2020a). Antioxidant activity and peptidomic analysis of porcine liver hydrolysates using alcalase, bromelain, flavourzyme and papain enzymes. *Food Research International, 137*, 109389. doi:10.1016/j.foodres.2020.109389.

López-Pedrouso, M., Lorenzo, J. M., Gagaoua, M., & Franco, D. (2020). *Current trends in proteomic advances for food allergen analysis* Biology *9* (pp. 1–13). Switzerland: MDPI AG. https://doi.org/10.3390/biology9090247.

López-Pedrouso, M., Pérez-Santaescolástica, C., Franco, D., Carballo, J., Zapata, C., & Lorenzo, J. M. (2019). Molecular insight into taste and aroma of sliced dry-cured ham induced by protein degradation undergone high-pressure conditions. *Food Research International, 122*, 635–642. https://doi.org/10.1016/j.foodres.2019.01.037.

López-Pedrouso, M., Rodríguez-Vázquez, R., Purriños, L., Oliván, M., García-Torres, S., Sentandreu, M.Á., et al. (2020c). Sensory and physicochemical analysis of meat from bovine breeds in different livestock production systems, pre-slaughter handling conditions, and ageing time. *Foods*, 9(2), 176. doi:10.3390/foods9020176.

Ludwig, C., Gillet, L., Rosenberger, G., Amon, S., Collins, B. C., & Aebersold, R. (2018). A data-independent acquisition-based global phosphoproteomics system enables deep profiling. *Molecular Systems Biology, 14*(8), e8126–e8140. https://doi.org/10.15252/msb.20178126.

Man, L., Dale, A. L., Klare, W. P., Cain, J. A., Sumer-Bayraktar, Z., Niewold, P., et al. (2020). Proteomics of Campylobacter jejuni growth in deoxycholate reveals Cj0025c as a cystine transport protein required for wild-type human infection phenotypes. *Molecular and Cellular Proteomics, 19*(8), 1263–1280. doi:10.1074/mcp.RA120.002029.

Manes, N. P., Mann, J. M., & Nita-Lazar, A. (2015). Selected reaction monitoring mass spectrometry for absolute protein quantification. *Journal of Visualized Experiments, 2015*(102), 1–15. https://doi.org/10.3791/52959.

Martínez-Esteso, M. J., Nørgaard, J., Brohée, M., Haraszi, R., Maquet, A., & O'Connor, G. (2016). Defining the wheat gluten peptide fingerprint via a discovery and targeted proteomics approach. *Journal of Proteomics, 147*, 156–168. https://doi.org/10.1016/j.jprot.2016.03.015.

Martinez-Esteso, M. J., O'Connor, G., Nørgaard, J., Breidbach, A., Brohée, M., Cubero-Leon, E., et al. (2020). A reference method for determining the total allergenic protein content in a processed food: The case of milk in cookies as proof of concept. *Analytical and Bioanalytical Chemistry, 412*(30), 8249–8267. https://doi.org/10.1007/s00216-020-02959-0.

Martínez-Maqueda, D., Hernández-Ledesma, B., Amigo, L., Miralles, B., & Gómez-Ruiz, J.Á. (2013). Extraction/fractionation techniques for proteins and peptides and protein digestion. In *Proteomics in Foods: Principles and Applications* (pp. 21–50). New York: Springer US. https://doi.org/10.1007/978-1-4614-5626-1_2.

Martinović, T., Andjelković, U., Gajdošik, M.Š., Rešetar, D., & Josić, D. (2016). Foodborne pathogens and their toxins. *Journal of Proteomics, 147*, 226–235. https://doi.org/10.1016/j.jprot.2016.04.029.

Martinović, T., Šrajer Gajdošik, M., & Josić, D. (2018). Sample preparation in foodomic analyses. *Electrophoresis 39* (pp. 1527–1542). New Jersey: Wiley-VCH Verlag. https://doi.org/10.1002/elps.201800029.

Mato, A., Rodríguez-Vázquez, R., López-Pedrouso, M., Bravo, S., Franco, D., & Zapata, C. (2019). The first evidence of global meat phosphoproteome changes in response to pre-slaughter stress. *BMC Genomics, 20*(1), 590. https://doi.org/10.1186/s12864-019-5943-3.

Matthiesen, R., & Carvalho, A. S. (2013). Methods and algorithms for quantitative proteomics by mass spectrometry. *Methods in Molecular Biology, 1007*, 183–217. https://doi.org/10.1007/978-1-62703-392-3_8.

McLafferty, F. W., Breuker, K., Jin, M., Han, X., Infusini, G., Jiang, H., et al. (2007). Top-down MS, a powerful complement to the high capabilities of proteolysis proteomics. *FEBS Journal, 274*(24), 6256–6268. https://doi.org/10.1111/j.1742-4658.2007.06147.x.

Meleady, P. (2018). Two-dimensional gel electrophoresis and 2D-DIGE. *Methods in Molecular Biology 1664* (pp. 3–14). New Jersey: Humana Press Inc. https://doi.org/10.1007/978-1-4939-7268-5_1.

Mermelekas, G., Vlahou, A., & Zoidakis, J. (2015). SRM/MRM targeted proteomics as a tool for biomarker validation and absolute quantification in human urine. *Expert Review of Molecular Diagnostics 15* (pp. 1441–1454). England: Taylor and Francis Ltd. https://doi.org/10.1586/14737159.2015.1093937.

Meyer, J. G., & Schilling, B. (2017). Clinical applications of quantitative proteomics using targeted and untargeted data-independent acquisition techniques. *Expert Review of Proteomics 14* (pp. 419–429). England: Taylor and Francis Ltd. https://doi.org/10.1080/14789450.2017.1322904.

Mills, E. N. C., Adel-Patient, K., Bernard, H., De Loose, M., Gillard, N., Huet, A. C., et al. (2019). Detection and quantification of allergens in foods and minimum eliciting doses in food-Allergic individuals (ThRAll). *Journal of AOAC International, 102*(5), 1346–1353. https://doi.org/10.5740/jaoacint.19-0063.

Monaci, L., Angelis, E. De, Guagnano, R., Ganci, A. P., Garaguso, I., Fiocchi, A., et al. (2020). Validation of a MS based proteomics method for milk and egg quantification in cookies at the lowest VITAL levels: An alternative to the use of precautionary labeling. *Foods, 9*(10), 1–13. https://doi.org/10.3390/foods9101489.

Monaci, L., Losito, I., De Angelis, E., Pilolli, R., & Visconti, A. (2013). Multi-Allergen quantification of fining-related egg and milk proteins in white wines by high-resolution mass spectrometry. *Rapid Communications in Mass Spectrometry, 27*(17), 2009–2018. https://doi.org/10.1002/rcm.6662.

Monetti, M., Nagaraj, N., Sharma, K., & Mann, M. (2011). Large-scale phosphosite quantification in tissues by a spike-in SILAC method. *Nature Methods, 8*(8), 655–658. https://doi.org/10.1038/nmeth.1647.

Montowska, M., & Spychaj, A. (2018). Quantification of species-specific meat proteins in cooked and smoked sausages using infusion mass spectrometry. *Journal of Food Science and Technology, 55*(12), 4984–4993. https://doi.org/10.1007/s13197-018-3437-y.

Mora, L., Gallego, M., Reig, M., & Toldrá, F. (2017). Challenges in the quantitation of naturally generated bioactive peptides in processed meats. *Trends in Food Science and Technology, 69*, 306–314. https://doi.org/10.1016/j.tifs.2017.04.011.

Mora, L., Gallego, M., & Toldrá, F. (2018). New approaches based on comparative proteomics for the assessment of food quality. *Current Opinion in Food Science 22* (pp. 22–27). Amsterdam: Elsevier Ltd. https://doi.org/10.1016/j.cofs.2018.01.005.

Mullins, Y., Keogh, K., Kenny, D. A., Kelly, A., O' Boyle, P., & Waters, S. M. (2020). Label-free quantitative proteomic analysis of M. longissimus dorsi from cattle during dietary restriction and subsequent compensatory growth. *Scientific Reports, 10*(1), 1–13. https://doi.org/10.1038/s41598-020-59412-6.

Nair, M. N., & Zhai, C. (2019). Application of proteomic tools in meat quality evaluation. In *Meat Quality Analysis: Advanced Evaluation Methods, Techniques, and Technologies* (pp. 353–368). Amsterdam: Elsevier. https://doi.org/10.1016/B978-0-12-819233-7.00019-7.

Narasimhan, M., Kannan, S., Chawade, A., Bhattacharjee, A., & Govekar, R. (2019). Clinical biomarker discovery by SWATH-MS based label-free quantitative proteomics: Impact of criteria for identification of differentiators and data normalization method. *Journal of Translational Medicine, 17*(1), 184. https://doi.org/10.1186/s12967-019-1937-9.

Ning, Y., Fu, Y., Hou, L., Ma, M., Wang, Z., Li, X., et al. (2021). iTRAQ-based quantitative proteomic analysis of synergistic antibacterial mechanism of phenyllactic acid and lactic acid against Bacillus cereus. *Food Research International, 139*, 109562. doi:10.1016/j.foodres.2020.109562.

Nitride, C., Nørgaard, J., Omar, J., Emons, H., Esteso, M. J. M., & O'Connor, G. (2019). An assessment of the impact of extraction and digestion protocols on multiplexed targeted protein quantification by mass spectrometry for egg and milk allergens.

Analytical and Bioanalytical Chemistry, 411(16), 3463–3475. https://doi.org/10.1007/s00216-019-01816-z.

Ong, S.-E., Foster, L. J., & Mann, M. (2003). Mass spectrometric-based approaches in quantitative proteomics. *Methods (San Diego, Calif.), 29*(2), 124–130. http://www.ncbi.nlm.nih.gov/pubmed/12606218.

Ong, S.-E., & Mann, M. (2006). A practical recipe for stable isotope labeling by amino acids in cell culture (SILAC). *Nature Protocols, 1*(6), 2650–2660. https://doi.org/10.1038/nprot.2006.427.

Ortea, I., O'Connor, G., & Maquet, A. (2016). Review on proteomics for food authentication. *Journal of Proteomics, 147*, 212–225. https://doi.org/10.1016/j.jprot.2016.06.033.

Owusu-Apenten, R. K., Basel New York (2002). *Food protein analysis quantitative effects on processing*. Florida: CRC Press, 488e. doi:10.1201/9780203910580. http://www.dekker.com.

Padliya, N. D., Qian, M., Mimi Roy, S., Chu, P., Zheng, H., Tess, A., et al. (2015). The impact of fertilization on the chicken egg yolk plasma and granule proteome 24 hours post-lay at room temperature: Capitalizing on high-pH/low-pH reverse phase chromatography in conjunction with tandem mass tag (TMT) technology. *Food and Function, 6*(7), 2303–2314. https://doi.org/10.1039/c5fo00304k.

Pandey, A., & Mann, M. (2000). Proteomics to study genes and genomes. *Nature, 405* (pp. 837–846). United Kingdom: Nature Publishing Group. https://doi.org/10.1038/35015709.

Pappin, D. J. C., Hojrup, P., & Bleasby, A. J. (1993). Rapid identification of proteins by peptide-mass fingerprinting. *Current Biology, 3*(6), 327–332. https://doi.org/10.1016/0960-9822(93)90195-T.

Pappireddi, N., Martin, L., & Wühr, M. (2019). A review on quantitative multiplexed proteomics. *Chembiochem, 20*(10), 1210–1224. https://doi.org/10.1002/cbic.201800650.

Pilolli, R., De Angelis, E., & Monaci, L. (2017). Streamlining the analytical workflow for multiplex MS/MS allergen detection in processed foods. *Food Chemistry, 221*, 1747–1753. https://doi.org/10.1016/j.foodchem.2016.10.110.

Piras, C., Roncada, P., Rodrigues, P. M., Bonizzi, L., & Soggiu, A. (2016). Proteomics in food: Quality, safety, microbes, and allergens. *Proteomics, 16*(5), 799–815. https://doi.org/10.1002/pmic.201500369.

Planque, M., Arnould, T., Delahaut, P., Renard, P., Dieu, M., & Gillard, N. (2019). Development of a strategy for the quantification of food allergens in several food products by mass spectrometry in a routine laboratory. *Food Chemistry, 274*, 35–45. https://doi.org/10.1016/j.foodchem.2018.08.095.

Planque, M., Arnould, T., & Gillard, N. (2017). Food allergen analysis: Detection, quantification and validation by mass spectrometry. In *Allergen*. Zanjan: InTech. https://doi.org/10.5772/intechopen.69361.

Qiu, W., Zhang, X., Zhang, H., Liang, C., Xu, J., Gao, H., et al. (2020). Discrimination of meat from fur-producing and food-providing animals using mass spectrometry-based proteomics. *Food Research International, 137*, 109446. doi:10.1016/j.foodres.2020.109446.

Reddy, P. J., Ray, S., Sathe, G. J., Gajbhiye, A., Prasad, T. K., Rapole, S., et al. (2015). A comprehensive proteomic analysis of totarol induced alterations in Bacillus subtilis by multipronged quantitative proteomics. *Journal of Proteomics, 114*, 247–262. https://doi.org/10.1016/j.jprot.2014.10.025.

Reddy, P., Rao, A., Malhotra, D., Sharma, S., Kumar, R., Jain, R., et al. (2014). A simple protein extraction method for proteomic analysis of diverse biological specimens. *Current Proteomics, 10*(4), 298–311. https://doi.org/10.2174/15701646113106660004.

Rodríguez-Vázquez, R., Mato, A., López-Pedrouso, M., Franco, D., Sentandreu, M. A., & Zapata, C. (2020). Measuring quantitative proteomic distance between Spanish beef breeds. *Food Chemistry, 315*, 126293. doi:10.1016/j.foodchem.2020.126293.

Ruhland, M., & Klinger, R. (2019). Food fraud: A simple and efficient LC–MS/MS approach for peptide-based food authentication. *Journal of AOAC INTERNATIONAL, 102*(5), 1303–1308. https://doi.org/10.1093/jaoac/102.5.1303.

Santos, T., Viala, D., Chambon, C., Esbelin, J., & Hébraud, M. (2019). Listeria monocytogenes biofilm adaptation to different temperatures seen through shotgun proteomics. *Frontiers in Nutrition, 6*(89), 1–16. https://doi.org/10.3389/fnut.2019.00089, Listeria monocytogenes Biofil.

Sealey-Voyksner, J., Zweigenbaum, J., & Voyksner, R. (2016). Discovery of highly conserved unique peanut and tree nut peptides by LC-MS/MS for multi-allergen detection. *Food Chemistry, 194*, 201–211. https://doi.org/10.1016/j.foodchem.2015.07.043.

Sebald, K., Dunkel, A., & Hofmann, T. (2020). Mapping taste-relevant food peptidomes by means of sequential window acquisition of all theoretical fragment ion-mass spectrometry. *Journal of Agricultural and Food Chemistry, 68*(38), 10287–10298. https://doi.org/10.1021/acs.jafc.9b04581.

Shamloo, E., Hosseini, H., Moghadam, A. Z., Larsen, H. M., Haslberger, A., & Alebouyeh, M. (2019). Importance of Listeria monocytogenes in food safety: A review of its prevalence, detection, and antibiotic resistance. *Iranian Journal of Veterinary Research 20* (pp. 241–254). Shiraz: Shiraz University. /pmc/articles/PMC6983307/.

Sharma, G. M., Khuda, S. E., Parker, C. H., Eischeid, A. C., & Pereira, M. (2016). Detection of allergen markers in food: Analytical methods. In *Food Safety: Innovative Analytical Tools for Safety Assessment* (pp. 65–121). New York: Wiley. https://doi.org/10.1002/9781119160588.ch4.2018.

Sitammagari, K. K., & Masood, W. (2021). Creutzfeldt Jakob Disease. In *StatPearls*. Treasure Island (FL): StatPearls Publishing. http://www.ncbi.nlm.nih.gov/pubmed/29939637. (Accessed 28 June 2021).

Smith, B. J., Martins-De-Souza, D., & Fioramonte, M. (2019). A guide to mass spectrometry-based quantitative proteomics. *Methods in Molecular Biology 1916* (pp. 3–39). New Jersey: Humana Press Inc. https://doi.org/10.1007/978-1-4939-8994-2_1.

Stevenson, S. E., Chu, Y., Ozias-Akins, P., & Thelen, J. J. (2009). Validation of gel-free, label-free quantitative proteomics approaches: Applications for seed allergen profiling. *Journal of Proteomics, 72*(3), 555–566. https://doi.org/10.1016/j.jprot.2008.11.005.

Tate, S., Larsen, B., Bonner, R., & Gingras, A. C. (2013). Label-free quantitative proteomics trends for protein-protein interactions. *Journal of Proteomics, 81*, 91–101. https://doi.org/10.1016/j.jprot.2012.10.027.

Valdés, A., Cifuentes, A., & León, C. (2017). Foodomics evaluation of bioactive compounds in foods. *TrAC - Trends in Analytical Chemistry 96* (pp. 2–13). Amsterdam: Elsevier B.V. https://doi.org/10.1016/j.trac.2017.06.004.

Vidova, V., & Spacil, Z. (2017). A review on mass spectrometry-based quantitative proteomics: Targeted and data independent acquisition. *Analytica Chimica Acta 964* (pp. 7–23). Switzerland: Elsevier B.V. https://doi.org/10.1016/j.aca.2017.01.059.

Virág, D., Dalmadi-Kiss, B., Vékey, K., Drahos, L., Klebovich, I., Antal, I., et al. (2020). Current trends in the analysis of post-translational modifications. *Chromatographia 83* (pp. 1–10). Switzerland: Springer. https://doi.org/10.1007/s10337-019-03796-9.

Walther, T. C., & Mann, M. (2010). Mass spectrometry-based proteomics in cell biology. *Journal of Cell Biology, 190*(4), 491–500. https://doi.org/10.1083/jcb.201004052.

Wang, Q., Zhang, M., Tomita, T., Vogelstein, J. T., Zhou, S., Papadopoulos, N., et al. (2017). Selected reaction monitoring approach for validating peptide biomarkers. *Proceedings of the National Academy of Sciences of the United States of America, 114*(51), 13519–13524. https://doi.org/10.1073/pnas.1712731114.

Wang, Y., Lin, W., Yan, H., Neng, J., Zheng, Y., Yang, K., et al. (2021). iTRAQ proteome analysis of the antifungal mechanism of citral on mycelial growth and OTA production in Aspergillus ochraceus. *Journal of the Science of Food and Agriculture*, 101(12), 4969–4979. doi:10.1002/jsfa.11140.

Wasinger, V. C., Zeng, M., & Yau, Y. (2013). Current status and advances in quantitative proteomic mass spectrometry. *International Journal of Proteomics, 2013*, 180605. https://doi.org/10.1155/2013/180605.

Wei, Y., Li, X., Zhang, D., & Liu, Y. (2019). Comparison of protein differences between high- and low-quality goat and bovine parts based on iTRAQ technology. *Food Chemistry, 289*, 240–249. https://doi.org/10.1016/j.foodchem.2019.03.052.

Wu, W., Ruitong, D., & Bendixen, E. (2019a). Comparing SRM and SWATH methods for quantitation of bovine muscle proteomes. *Journal of Agricultural and Food Chemistry* 67(5) 1608–1618. doi:10.1021/acs.jafc.8b05459.

Wu, X., Liu, L., Li, J., & Chi, F. (2019b). Proteome analysis using iTRAQ reveals the differentiation between Tibetan and ordinary ovalbumin peptides. *International Journal of Biological Macromolecules, 132*, 722–728. https://doi.org/10.1016/j.ijbiomac.2019.03.075.

Xu, L. L., Zhang, H. W., Zhang, X. M., Lin, H., Guo, Y. M., et al. (2020). Natural Shrimp (Litopenaeus vannamei) tropomyosin shows higher allergic properties than recombinant ones as compared through SWATH-MS-based proteomics and immunological response. *Journal of Agricultural and Food Chemistry, 68*(41), 11553–11567. https://doi.org/10.1021/acs.jafc.0c03840.

Xu, L., Xu, Z., Kelly, S., & Liao, X. (2020). Integrating untargeted metabolomics and targeted analysis for not from concentrate and from concentrate orange juices discrimination and authentication. *Food Chemistry, 329*, 127130. doi:10.1016/j.foodchem.2020.127130.

Xu, L., Xu, Z., Strashnov, I., & Liao, X. (2020). Use of information dependent acquisition mass spectra and sequential window acquisition of all theoretical fragment-ion mass spectra for fruit juices metabolomics and authentication. *Metabolomics, 16*(7), 81. https://doi.org/10.1007/s11306-020-01701-2.

Xu, L., Xu, Z., Wang, X., Wang, B., & Liao, X. (2020). The application of pseudotargeted metabolomics method for fruit juices discrimination. *Food Chemistry, 316*, 126278. doi:10.1016/j.foodchem.2020.126278.

Yan, W., & Chen, S. S. (2005). Mass spectrometry-based quantitative proteomic profiling. *Briefings in Functional Genomics & Proteomics, 4*(1), 27–38. http://www.ncbi.nlm.nih.gov/pubmed/23528454.

Yan, Z., Li, W., Hu, R., Ma, Q., & Lu, Z. (2021). Quantitative proteomic comparison of protein differences in different parts of yak meat. *Food Science and Technology, AHEAD*. doi:10.1590/fst.62020.

Yang, B., & Liu, X. (2021). Application of proteomics to understand the molecular mechanisms determining meat quality of beef muscles during postmortem aging. *Plos One, 16*(3), e0246955. doi:10.1371/journal.pone.0246955.

Yao, L., Yu, Q., Huang, M., Song, Z., Grosser, J., Chen, S., et al. (2020). Comparative iTRAQ proteomic profiling of sweet orange fruit on sensitive and tolerant rootstocks infected by 'Candidatus Liberibacter asiaticus.' *Plos One, 15*(2), e0228876. doi:10.1371/journal.pone.0228876.

Ye, X., Tang, J., Mao, Y., Lu, X., Yang, Y., Chen, W., et al. (2019). Integrated proteomics sample preparation and fractionation: Method development and applications. *TrAC - Trends in Analytical Chemistry 120*. Amsterdam: Elsevier B.V.

Zhang, H., Liu, Q., Zimmerman, L. J., Ham, A.-J. L., Slebos, R. J. C., Rahman, J., et al. (2011). Methods for peptide and protein quantitation by liquid chromatography-multiple reaction monitoring mass spectrometry. *Molecular & Cellular Proteomics : MCP, 10*(6), M110.006593. https://doi.org/10.1074/mcp.M110.006593.

Zhang, H., Zhang, X., Zhao, X., Xu, J., Lin, C., Jing, P., et al. (2019). Discrimination of dried sea cucumber (Apostichopus japonicus) products from different geographical origins by sequential windowed acquisition of all theoretical fragment ion mass spectra (SWATH-MS)-based proteomic analysis and chemometrics. *Food Chemistry, 274*, 592–602. https://doi.org/10.1016/j.foodchem.2018.08.082.

Zhang, M., Wang, D., Xu, X., Xu, W., & Zhou, G. (2021). iTRAQ-based proteomic analysis of duck muscle related to lipid oxidation. *Poultry Science, 100*(4), 101029. https://doi.org/10.1016/j.psj.2021.101029.

Zhang, Y., Fonslow, B. R., Shan, B., Baek, M. C., & Yates, J. R. (2013). *Protein analysis by shotgun/bottom-up proteomics* Chemical Reviews *113* (pp. 2343–2394). Maryland: NIH Public Access. https://doi.org/10.1021/cr3003533.

Zheng, Y. Z., & DeMarco, M. L. (2017). Manipulating trypsin digestion conditions to accelerate proteolysis and simplify digestion workflows in development of protein mass spectrometric assays for the clinical laboratory. *Clinical Mass Spectrometry, 6*, 1–12. https://doi.org/10.1016/j.clinms.2017.10.001.

Zheng, Y.-F., Wu, M.-C., Chien, H.-J., Wang, W.-C., Kuo, C.-Y., & Lai, C.-C. (2021). Honey proteomic signatures for the identification of honey adulterated with syrup, producing country, and nectar source using SWATH-MS approach. *Food Chemistry, 354*, 129590. https://doi.org/10.1016/j.foodchem.2021.129590.

Zhu, W., Smith, J. W., & Huang, C.-M. (2010). Mass spectrometry-based label-free quantitative proteomics. *Journal of Biomedicine & Biotechnology, 2010*, 840518–840524. https://doi.org/10.1155/2010/840518.

CHAPTER

3

Technological developments of food peptidomics

Enrique Sentandreu, Miguel Ángel Sentandreu

Instituto de Agroquímica y Tecnología de Alimentos (CSIC), Calle Agustín Escardino 7, Paterna, Valencia, Spain

3.1 Introduction

3.1.1 Differences between peptidomics and proteomics

To establish differences between proteomics and peptidomics is not an easy task in many cases since both terms are closely interrelated from a biological perspective because peptides are indeed multiple-sized fragments of proteins. Proteomics can be defined as the study of the genome expression products in a biological sample with the objective to obtain an integrated and global vision of the cellular processes at a given time period. Complementary to this, peptidomics is the area of science that studies the set of peptides of an organisms, tissue or cell. As we will see later, in most proteomics studies, peptides are studied as a part of the protein and thus information obtained from each peptide belonging to the same protein is accumulative. In such a case, lack of identification of a particular peptide sequence can be compensated by information obtained for peptides belonging to other regions of the protein. On the contrary, in peptidomics the peptide itself is the object of study and this means that the lack of information about a target peptide would not necessarily be replaced by information obtained from another peptide of the same protein. In proteomics, the peptide can be the way to reach the protein; in peptidomics, the peptide is the final stop of the journey, but this does not mean that it will be an easier trip.

The term peptidomics can already be found in some scientific literature already published in 2001 (Minamino, 2001; Schulz-Knappe et al., 2001). We can say that peptides are short molecules integrating a chain of various amino acids in the range from 2 to 50. This convention is, however, rather

Food Proteomics
DOI: https://doi.org/10.1016/B978-0-323-90889-4.00009-9

arbitrary and flexible, not being possible to establish a clear difference in many cases, for example, between big polypeptides and small proteins. For that reason, it has also been accepted that a peptide has a less complex structure, which can contain helices, sheets, other functional subunits, or modifications of some particular residues but without the more complex tertiary and quaternary structures reached in the case of proteins.

Another important aspect that proteomics and peptidomics have in common is that, nowadays, both rely in analytical strategies based on mass spectrometry (MS) as part of what we understand by "modern Omics technologies," characterized by high throughput, resolution, and sensitivity. However, this has not been always the case. In the 1970s, protein and peptide analysis was associated with the separation proteome/peptidome from a particular cell or tissue by two-dimensional electrophoresis (2-DE) or high performance liquid chromatography (HPLC). This was the beginning of building protein databases, being protein/peptide identification the main limiting step, which mainly trusted on the detection featured by the use of specific antibodies or identification by N-terminal sequencing based on Edman degradation. These analytical approaches had considerable limitations such as their low sensitivity, time-consuming and complicated workflows that made them not suitable for large-scale peptide/protein analysis. In the 1990s, a set of events allowed a completely different way to identify peptides and proteins, which gave rise to a real revolution in proteomics known as the "Post-genomic Revolution" or modern proteomics. Main factors that contributed to this revolution were (1) development of the so-called "soft" ionization techniques, i.e., electrospray ionization (ESI) and matrix-assisted laser desorption/ionization (MALDI), which allowed the ionization of big molecules of big size such as peptides and proteins; (2) the massive genome sequencing and translation of all this information into protein sequences eased the development of protein databases into its current format and easy access via world wide web; and (3) the development of high-throughput mass spectrometers together with powerful bioinformatic tools.

3.2 What type of peptides can we found in food matrices?

There can be different criteria at the time to classify peptides in biological samples. If we focus on the case of food matrices, we can establish two main groups.

3.2.1 Endogenous peptides

They would be those that are naturally present in foods, mainly generated during food storage and/or food processing. Here we could try to establish a difference between those peptides naturally present in the food matrix and those that have been generated during processing. However, to make such a difference is complicated and ambiguous in many cases since it is difficult to define what is the extent of the food itself and what of processing. This could be the case of peptides generated by food storage, for example. From an analytical point of view, peptides naturally present in foods or generated during food storage/processing would share similar biochemical characteristics, and thus they can be all grouped under the concept of "endogenous peptides."

Endogenous peptides would be those peptides naturally present in meat due to the action of several endogenous proteolytic systems, which will remain active after slaughter and during the meat aging process. The consequence will be the degradation of intact myofibrillar proteins into big polypeptides, a process that will decrease the mechanical resistance of meat and hence will contribute to the development of meat tenderness (Sentandreu, Coulis, and Ouali, 2002). In the case of dry-cured meat products, since processing times are considerably longer that meat aging (12 months or even longer) the degradation of muscle proteins will be much more intense and, as consequence, large polypeptides will be further degraded to medium and small peptides, together with free amino acids. All these protein degradation products will be essential at the time to reach the final and highly appreciated texture and flavor characteristics of dry-cured products (Gallego, Mora, and Toldra, 2018a; Zhou et al., 2021). In the case of meat products in which bacterial fermentation is used during processing, degradation of meat proteins will be an even more complex process because the proteolytic machinery of the microorganisms, growing spontaneously or added as starter cultures, will be added to the action of endogenous muscle enzymes, yielding in a particularly complex peptidome, which is still plenty of uncertainties. In that respect, López et al. (2015) firstly reported small peptides (<3 kDa) as quality biomarkers (taste and flavor) of fermented sausages that helped in the characterization of the starter cultures used. Outside taste and flavor contribution, it was demonstrated how lactic acid bacteria induces the generation of bioactive peptides in fermented sausage by combining starter cultures and modifying growing conditions (Gallego et al., 2018b). A very recent application of biopeptide research in fermented sausage is their use to ameliorate organoleptic defects induced by NaCl substituents (i.e., Mg, Ca, or KCl). Combination of starters yielded high release rates of low molecular weight peptides with positive sensorial properties in low-sodium fermented sausages (de Almeida et al., 2018).

Similarly to meat, raw milk also contains a variety of endogenous peptides that will be the product of peptidase activities during milk storage. This proteolytic action of milk peptidases will be notably modulated by the different processing conditions (heat treatment) normally employed by industry to guarantee milk stability and safety over long storage periods (Wölk, Milkovska-Stamenova, & Hoffmann, 2020). In this respect, the main difference between meat and milk peptidome would be the nature and variety of the parent proteins of these peptides. While meat peptides originate from a wide variety of both sarcoplasmic and myofibrillar proteins (Clerens et al., 2016), milk peptidome would be mainly generated from the degradation of alpha- and beta-caseins and, to a lower extent, from k-casein and whey proteins (Dalabasmaz et al., 2019; Dallas, Murray, and Gan, 2015). Similarly to fermented meat products, milk serves as the raw material for the production of a wide variety of fermented products such as cheese (Afshari et al., 2020; Tilocca et al., 2020), yogurt (Kunda et al., 2012), and kefir (Izquierdo-Gonzalez et al., 2019). In such foodstuffs, the enzymes responsible for the hydrolysis of milk proteins will be those present in raw milk, those added during the manufacturing process, as coagulant for cheeses or as starter cultures, or those derived from microorganisms that can growth spontaneously in milk during product processing. This will act as a powerful proteolytic machinery enabling an extensive degradation of milk caseins, generating a wide spectrum of peptides of different size and quantity, ranging from 5 to around 40 amino acids and up to 12 orders of magnitude (Fontenele et al., 2017; Panchaud, Affolter, and Kussmann, 2012). In dairy fermented food, activity of lactic acid bacteria has an essential role in the release of multiple peptides, mainly from caseins, conferring characteristic properties of final products. Complex peptidomic profiling achieved by liquid chromatography-high resolution mass spectrometry (LC-HRMS) analysis led to understand biochemical changes that occurred by fermentation and to identify sensory and authentication biomarkers. Moreover, current demands of consumers for the acquisition of affordable and healthier products induced researchers and industry to create new insights into the optimization of lactic acid fermentation through innovative starters and genetically modified lactic acid bacteria strains (Martini et al., 2021). As a result, different dairy products enriched with biopeptides are obtained with their health properties (i.e., antihypertensive and antioxidant) improved (Gu et al., 2020; Raveschot et al., 2020).

Despite the protein/peptide content of beer is lower than other protein-rich food matrices such as those like meat and milk commented above, they constitute an important fraction of beer components for their role on the technological and sensory characteristics of beer including color, foam properties, haze formation, and stability. They will also influence the texture, carbonation, mouthfeel, and palatability (Picariello et al., 2017). Most of beer peptides originate from the proteolytic breakdown of malt proteins, such as albumins, globulins, and hordelins, either from barley or from wheat since they represent the vast majority of the beer proteome. This peptide fraction has remained uncharacterized until recently because of its complexity in terms of having a wide range of peptide sizes with a high dynamic range and the occurrence of many interfering small compounds, such as amino acids or sugars. Detailed peptide composition was possible with the development high-resolution separation methods such as reversed-phase HPLC (Dale and Young, 1989). Another important limitation was the lack of sequence information about those proteins present in beer. Fortunately, this gap has been progressively filled with proteomic works addressing the barley seed compartments (Finnie and Svensson, 2009; Maeda et al., 2003) but especially with the completion of the barley genome (Mayer et al., 2012). More recently, impressive advances in the analytical capabilities of MS instrumentation have at least partially contributed to understand the intricate beer peptidome. Such complexity arises from the heterogeneity of the proteins from which the peptides originate, the difficulty to know the enzyme specificity responsible for the peptide generation, and the occurrence of peptide modifications induced by the malting and brewing process including glycation, deamidation, reduction, oxidation, and aggregation. Main advances in MS techniques include the incorporation of high resolution/high throughput benchtop LC-MS instruments such as those based on the Orbitrap technology (Picariello et al., 2017). Adoption of these cutting-edge MS technologies allowed the identification of hundreds of peptides belonging to a wide range of proteins including hordeins, gliadins, and high- and low-MW glutenin subunits (Picariello et al., 2015). Interestingly, some researchers pointed out that the complexity of the beer peptidome is strongly dependent on beer brands (Picariello et al., 2011), thus allowing the possibility to use each beer's peptide profile for authentication purposes.

3.2.2 Peptides generated during the digestion of food proteins

These peptides would be produced only after the digestion of food proteins following oral ingestion. For that reason, some authors refer to them as "encrypted peptides" (Tagliazucchi et al., 2016). Digestion of food comprises its physical breakdown during mastication, followed by chemical/enzymatic degradation at both stomach and intestine levels. Gastrointestinal enzymes and brush border membrane peptidases will be in charge to degrade food proteins as much as possible into short peptides and free amino acids with the aim to facilitate their absorption by the small intestine. In addition to the essential nutritive function of peptides and amino acids, research on dietary peptides has attracted much research efforts during the last twenty years because they can exert numerous biological activities in the organisms through the regulation of different physiological functions that can be either beneficial (Mada, Ugwu, and Abarshi, 2020) or detrimental (Alves et al., 2019) for the organism. For a deeper lecture on this topic, we encourage readers to look at chapter 16, which specifically addresses the analysis of bioactive food peptides.

The number of studies considering in vivo digestomic approaches is rather limited, finding those related to negative effects (mainly inflammation) of released peptides from gluten proteins as the most popular (Fasano 2011). In the case of bioactive peptides, those from milk proteins, mainly caseins, were studied by their health-promoting (i.e., antimicrobial activity in infants) properties after ingestion (Beverly et al., 2021; Boutrou et al., 2013). On the other hand, recent studies demonstrated in vivo activities (i.e., antioxidant, antihypertensive, and cardio protective) of peptides from plant-based (i.e., quinoa, soy, and potato) proteins (Daroit and Brandelli, 2021).

Even if the ideal situation to study the generation of peptides during food digestion would be to directly analyze the peptides present in the stomach, intestine, and/or bloodstream, the drawback is that human digestive models are often not affordable due to ethical concerns and/or economical costs. For that reason, researchers have used alternative and more affordable digestion models such as in vitro digestion models that use the main proteolytic enzymes naturally present in the gastrointestinal tract, i.e., pepsin and trypsin (Chen et al., 2021). Conscious of the conceptual error of assuming that a vitro model is comparable to a real human digestion, Sanchon et al. (2018) undertook an investigation for directly compare in vitro digestion results of casein and whey proteins with in vivo digestion finding of same proteins. Interestingly, the authors concluded that, although some differences were found between the two models assayed, the in vitro digestion model would constitute a good approximation to the human gastrointestinal digestion of milk proteins. In vitro gastro-intestinal digestion was used to study bioactive (i.e., antioxidant, antihypertensive, anticancer) peptides of cow, oat, sheep, camel, and buffalo milk as well as, from milk-derived dairy products such as cheeses, yogurt, and ice cream, demonstrating the influence of processing on the release of bioactive agents (Martini et al., 2021). Similarly, in vitro peptide profiling of raw meat and meat-derived products from different animal sources (i.e., pork, chicken, beef, turkey, etc.) were also achieved, highlighting the effect of food processing and cooking on protein digestibility and peptide profile of samples (Li et al., 2017; Martini et al., 2019; Zhao et al., 2020).

Regarding alternative protein sources, in vitro assays helped in the search of marine-derived (i.e., algae, tuna, anchovy, sponges) biopeptides that can be considered as nutraceuticals to potentiate well-being of consumers (Pangestuti and Kim, 2017). In this line, different policies worldwide promoted the study of extensive vegetables (i.e., legumes, soy, and wheat) as an affordable and sustainable source of natural bioactive peptides (Montesano et al., 2020). Recent advances in the design of in vitro protocols (i.e., INFOGEST assay) led to closely mimic naturally occurring digestion. Despite their utility, it is also true that in vitro models are unable to fully reproduce the proteolytic machinery of human digestion and the subsequent resulting peptide profile. An interesting alternative to the human digestion model can be found in the use of in vivo animal digestion models. In this way, Barbe et al. (2014) studied the effect of the food matrix on the digestion kinetics of milk proteins more concretely the generation of peptides into the duodenum of mini-pigs. They concluded that the structure of the dairy matrices had only a minor influence on the nature of the protein cleavage sites, which seem to be highly conserved. However, the food structure showed to have an important influence on the amount of the identified peptides.

During the digestive process, some of the generated peptides can be resistant to further proteolytic action and absorbed intact through the intestinal barrier, with the possibility to exert biological functions either locally in the gastrointestinal tract or at systemic level by reaching the bloodstream and arriving to other tissues (Fernandez-Tome et al., 2018). In general, resistance of peptides to the digestive process is due to their content in proline and glutamine residues within their sequence. Depending on the nature of the resistant peptide, these can exert a positive, i.e., nutraceutical peptides (Cruz-Huerta et al., 2015) or a negative action, i.e., allergen peptides, on the organism.

An intermediate and interesting approach between affordable but (somewhat) incomplete in vitro models and realistic but more ethically and economically challenging in vivo digestion models would be the digestion models that combine in vitro gastrointestinal digestion with intestinal absorption models based on culture of human Caco-2 cells. This cell type is able to spontaneously differentiate into enterocyte-like monolayers, mimicking the characteristics of human absorption through intestinal epithelial cells. This makes adequate as a model to study the bioavailability of digested peptides and their mode of transport through intestinal cells before reaching the bloodstream. This approach was used by Fernandez-Tome et al. (2018) to investigate if lunasin, a peptide naturally present in soybean, was able to resist the action of brush-border peptidases, its susceptibility to intestinal epithelial transport and its potential bioactive function as antiproliferative agent against human adenocarcinoma and colorectal cancer. In previous work, same authors evaluated the in vitro resistance of this peptide to the action of gastrointestinal enzymes due to the presence of naturally present peptidase inhibitors in foods (Cruz-Huerta et al., 2015).

3.3 Identification of food peptides

As commented, proteomics and peptidomics currently rely on analytical strategies based on MS. This is behind the development of common analytical workflows from food matrices but with some particular characteristics according if the target analytes are peptides or proteins. In proteomics, it is usual that proteins are digested into peptides with the objective to facilitate the analysis of the peptide sequence by tandem mass spectrometry (MS/MS). This strategy is known as *bottom–up* proteomics, since the obtained peptide information will be an addition of parts that will reach up to the identification of the parent protein. Contrary to this, in Peptidomics peptides are not the way to find the objective but the objective itself. Then, it will point at characterizing as much as possible the structure and history of the target peptides, trying to get information about the full sequence, the presence of possible post-translational modifications, and the proteolytic processes responsible for their generation. This will be the case of *a top–down* approach, in which the uniqueness of the peptide will make necessary to take special care in checking the quality of the MS/MS spectra and the accurate interpretation of the peptide sequence (Schrader, Schulz-Knappe, and Fricker, 2014).

It is true that cutting proteins into peptides will increase sample complexity; however, the ease to analyze relatively short peptide sequences compared to intact protein sequences largely compensates for this. Another drawback of protein digestion is that information concerning tertiary and quaternary structure will be lost. There are additional problems of applying the *bottom–up* strategy to the characterization of large polypeptides such as, for example, those generated during meat storage commented previously. In many cases, those polypeptides are too large to be directly sequenced by tandem MS and thus, they are normally analyzed by introducing a prior trypsin digestion step before MS analysis as in the case of proteins. As a consequence of the trypsin action, the polypeptide is split into smaller peptides, but most often the fragments corresponding to either N- or C- terminal ends are not identified and consequently it will be quite difficult to elucidate information about cleavage sites responsible for the generation of these polypeptides. This is probably one of the features we lost in the evolution of peptide/protein identification, from N-terminal sequencing using Edman degradation to modern peptidomics/proteomics based on MS sequencing. The work carried out by Muroya et al. (2004) illustrates this. They were able to reveal the identity of five polypeptides ranging from 25 to 32 kDa generated during postmortem storage of bovine muscle. Western blot analysis revealed that these fragments originated from the hydrolysis of troponin T. N-terminal sequencing was carried out for the first eight amino acid residues of these fragments, allowing both to confirm troponin T as the origin of these fragments and to reveal de exact N-terminal cleavage sites responsible for their generation.

3.4 Current workflows in peptide identification

Main strategies that can be employed in food peptidomics are summarized in Fig. 3.1.

1) *Gel-based approaches* use one- or two-dimensional polyacrylamide gel electrophoresis (1- or 2-DE) for fractionate the peptide/protein content of the sample. Main advantages of gel-based fractionation is the possibility

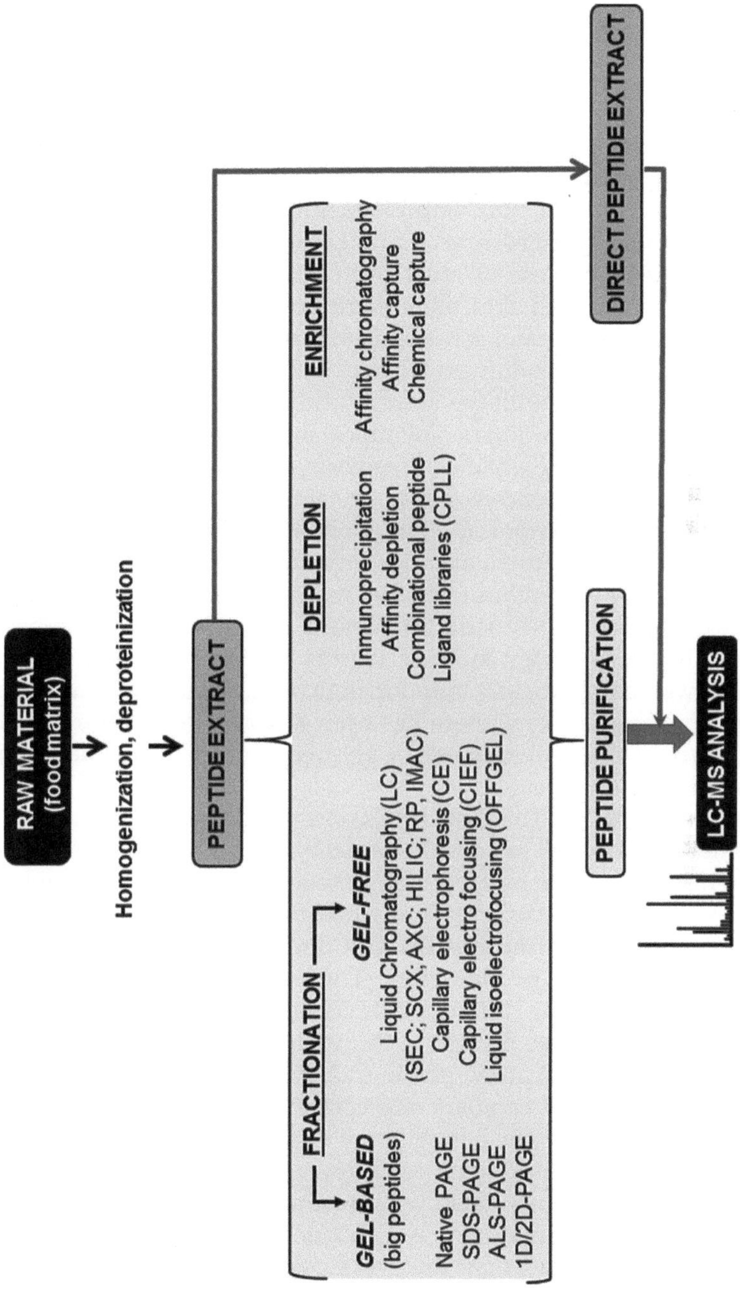

FIG. 3.1 General overview of most common sample preparation processes considered in peptidomic research. SEC: size exclusion; SCX: strong cation exchange; AXC: anion exchange; HILIC: hydrophilic interaction; RP: reverse phase; IMAC: immovilized metal affinity.

to perform simultaneous qualitative and quantitative analyses, allowing visual assessment and compatibility, but without direct hyphenation, with MS and Western blotting techniques. This approach has been particularly useful in the study of large polypeptides, such those that are generated during meat aging (Lametsch et al., 2003; Laville et al., 2009; Morzel et al., 2008). Monitoring the generation of some specific polypeptides can give valuable information about the aging process and hence about the final quality of meat, because they will be directly related to the hydrolysis rate of myofibrillar proteins and improvement of meat tenderness. This approach was used by Polati et al., (2012) to study proteome changes during the ageing process of meat from Charolaise heifers and bulls. These authors concluded that there were no differences between proteomic patterns of male and females *Longissimus dorsi* muscle, with no improvements in meat tenderness when extending the aging process from 12 to 26 days. Together with the identification of spots corresponding to intact proteins, they identified an important number of polypeptides derived from the breakdown of actin, myosin titin and troponinT. Despite their utility in the characterization of large polypeptides, gel-based approaches have some important limitations such as the use of hazardous chemical, the need for manual work, impairing automation and then giving rise to low throughput. It also requires a considerable sample volume to visualize those peptides present in low amount. Gel-based strategies are generally not suitable neither for the characterization of peptides lower than 3–5 kDa nor for high acid, basic or hydrophobic peptides. For these reasons, there has been a considerable effort during the last years in the development of gel-free strategies suitable for the analysis of food peptides.

B) *Gel-free approaches*: The first step is normally the deproteinization of the food matrix. This will remove the proteins and big polypeptides of the sample. This can be done by means of protein precipitation using organic solvents or commercial kits specifically designed for this purpose. Alternatively, sample can be fractionated with filters of a particular pore size according to the target peptide sizes. The set of peptides can be then further fractionated using different techniques such as liquid chromatography (LC, including size exclusion, ion exchange, hydrophobic/hydrophilic interaction, reverse phase, and immobilized metal affinity), capillary electrophoresis (CE) and capillary isoelectrofocusing. Main advantage of LC and CE separations is the possibility of direct hyhenation to an autosampler and an MS/MS detector after column/capillary separation enabling peptide identification, thus notably increasing automation and throughput. Despite these advantages, it is also true that chromatographic and electrophoretic techniques may involve a considerable running time, in addition, to be highly sensitive to the presence of salts and other buffer components. Additional constrains are high operational costs and the need of skilled operators (Lippolis and Nally, 2018). Thus, alternative gel-free MS strategies not including the use of previous LC or CE separation have been developed. This is the case of methods in which sample is directly spotted in a ProteinChip and further analyzed by surface-enhanced laser desorption/ionization mass spectrometry (SELDI-MS) as well as MALDI-MS. In this approach, sample is applied onto a ProteinChip, which will bind certain peptides/proteins according to the chemical characteristics of the chip surface. The chip is then washed with the objective to remove non-specifically bound analytes. Bound compounds will be then mixed with the appropriate matrix (normally α-cyano-4-hydroxycinnamic or sinapinic acid) allowing ionization of peptides prior to MS analysis. This approach was successfully applied by Park et al., (2013) in the differentiation of three different rice cultivars by means of their specific peptide profiles. SELDI-MS has been also applied in the characterization of peptides and proteins in human saliva as potential biomarkers of satiety and overweight (Harthoorn et al., 2007). On the other hand, Piraino et al. (2007) carried out the comparison of the peptide profile of different cheese varieties by the preparation of deproteinized cheese extracts followed by their direct analysis by MALDI-MS. This was a rapid and easy way to study the proteolytic processes of the different cheeses. Both SELDI-MS and MALDI-MS approaches are techniques with great potential to achieve a rapid peptide profiling of food samples. Peptide profiles are generally integrated by spectra containing a considerable number mass values with differing intensities. Appropriate statistical data treatment will allow discriminating among samples. The main drawback of these approaches concerns the available information that is restricted to the intact mass of peptides, thus compromising the elucidation of their amino acid sequence by the absence of MS/MS fragmentation data. Limitation of direct MS analysis of food peptides can be overcome by means of direct infusion (DI)-HRMS analysis, in where peptides are directly and accurately analyzed (MS^1 and MS/MS) skipping peptide separation featured by the LC column (Meyer et al., 2020), boosting the simultaneous qualitative/quantitative analysis of large sample sets.

3.5 Quantification strategies in food peptidomics

3.5.1 Labeled vs label-free quantitation

Strategies addressing shotgun (bottom–up) MS-based quantitative proteomics/peptidomics could be classified according to the use (label-based) or not (label-free) of stable isotope labeling (Fig. 3.2). Looking back in time, the first labeling procedure was reported by Gygi et. al (1999) through a stable isotope introduction in proteins by a chemical labeling procedure named ICATs (isotope-coded affinity tags) coupled to liquid

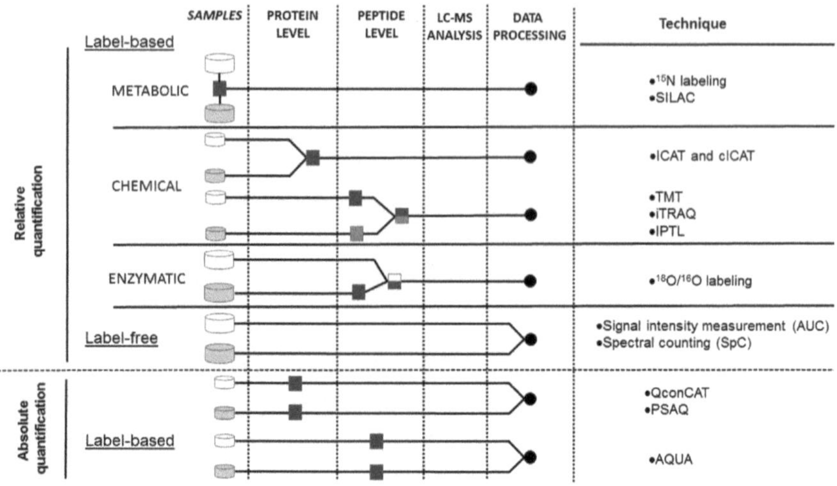

FIG. 3.2 Most common quantitative strategies used in protein research. Steps introducing labeling are illustrated by colored squares (*red* and *green* colors indicate different labeling meanwhile white indicates no derivatization). Figure adapted with permission from Lindemann et al. (2017).

chromatography-tandem MS analysis to perform an untargeted relative quantification study of the Saccharomyces cerevisiae proteome. Such technique evolved to other untargeted relative quantification approaches based on metabolic labeling of proteins (^{15}N; Stable Isotope Labeling of Amino acids in Cell culture, SILAC), chemical labeling of peptides (Tandem Mass Tag, TMT; isobaric Tag for Relative and Absolute Quantitation, iTRAQ; Isobaric peptide termini labeling, IPTL) and enzymatic labeling of peptides (proteolytic catalysis by class-2 proteases to incorporate two ^{18}O atoms in the place of ^{16}O atoms) (Anand et al., 2017). It must be pointed out that metabolic, chemical, and enzymatic labeling imply sample mixing that greatly increases complexity of the resultant matrix, hindering efficiency of the LC-MS analysis (mainly from isolation of co-eluted peptides and impurities) and data processing. Such drawbacks can be overcome by modern quad-based HRMS instruments and efficient algorithms of analysis (Anand et al., 2017). Furthermore, targeted absolute quantification research can be exclusively featured by stable isotope labeling through the incorporation (in a known concentration) of synthetic labeled protein or peptide analogs into sample before (Quantification conCATamers, QconCAT; Protein Standard Absolute Quantification, PSAQ) or after (Absolute QUAntification, AQUA) enzymatic protein digestion previously to LC-MS/MS peptidomic analysis (Andjelković and Josić, 2018). In this case, no sample mixing takes place with the subsequent relief of the instrumental analysis and data post-processing burden.

Compared to label-based strategies, label-free quantification is a straightforward, flexible (suitable for simple and complex samples), reproducible, cost-efficient, and easier technique that considers the measurement of the MS signal (peak area) intensity or the MS/MS spectral counting of each peptide (Anand et al., 2017). Popularity of this relative quantitation strategy was boosted during the last years by the continuous technological evolution of benchtop LC-HRMS analyzers combined with efficient bioinformatic analysis. Main advantage of label-free analysis can be summarized in its ability to measure almost an unlimited number of samples enabling the study of multiple untreated proteins or peptides (Lindemann et al., 2017). However, since analyses are individually prepared in label-free quantitation, variations reached can be significantly higher than those from labeling approaches multiplexing samples.

3.5.2 Label-based peptide quantification strategies

3.5.2.1 Relative quantification

3.5.2.1.1 Metabolic labeling

This technique minimizes experimental variability since labeling occurs at the very early stages of the sampling procedure (Bantscheff et al., 2007). Here, different labels are introduced into the proteome of two or more biologically different samples using an isotopically defined medium during protein metabolism, pooling samples in the same proportion (Iliuk et al., 2009). As a result, two versions of considered peptides are created with different isotopic compositions but having the same sampling and LC-MS response. This strategy was originally proposed to substitute ^{14}N atoms by ^{15}N but exhibited serious problems such as the incorporation of partial labeling and excessive costs (Iliuk et al., 2009).

SILAC was a more efficient alternative firstly introduced by Mann (2006) in where two populations of cells are growing in different mediums containing light (natural) and heavy (isotopically labeled) amino acids, replacing natural H, ^{14}N and ^{12}C to ^{2}H, ^{15}N, and ^{13}C, respectively. Samples are mixed, treated, and analyzed by LC-MS evidencing the presence of different proteomes. Most commonly used amino acids conferring a very effective labeling are leucine, lysine, methionine, and arginine (Iliuk et al., 2009). Main limitation of SILAC concerns the number of cellular states that can be compared in a single study (up to five) but in any case, its great labeling efficiency makes this option the election of choice in most labeling applications addressing the study of the diet-health relationship (Andjelković and Josić, 2018).

3.5.2.1.2 Chemical labeling

Flexibility in the selective introduction of isotopic labels through chemical reactions into desired protein or peptide locations characterizes this strategy. Gygi et al (1999) pioneered ICAT technology consisting in the use of a reagent containing an iodoacetyl reactive group to target cysteinyl residues, a linker region with heavy ($^2H_8$) or light ($^1H_8$) deuterium atoms, and a biotin group for affinity purification. Specific labeling only occurs at cysteine residues of proteins, meanwhile the biotin group facilitates the capture of cysteine-containing peptides, simplifying LC-MS analysis. Some authors stated that the retention time shift caused by the deuterium isotope effect can negatively affect robustness of ICAT assay, finding a recent evolution named cleavable ICAT (cICAT) where deuterium atoms were replaced by acid-cleavable isotopic tags with ^{13}C, $^{13}C_9$, or $^{12}C_9$ (Hsu et al., 2003).

TMT and iTRAQ are isobaric labeling options that introduce isobaric tags (identical mass but with different MS/MS fragmentation patterns) into derivatized peptides from samples, turning them chromatographically indistinguishable and finding a single peak in MS^1 representation of peptides. Differences in the relative abundance of the isobarically tagged peptides appear in the MS/MS spectrum reporting characteristic fragments with different masses (Domon and Aebersold, 2006). Tagging occurs in all primary amine functional groups of the peptides with a peptide reactive group (balancer) and a reporter group, labeling almost all peptides in the sample facilitating their quantification. The reporter group has 4–8 tagged sites, creating a mass difference that is corrected by the balance group, having peptides with similar masses (Iliuk et al., 2009). Fragmentation of TMT and iTRAQ reagents generates reporter groups in different mass ranges of 126–131 and 113–121 Da, respectively. The abundance of the reporter ion released is proportional to the amount of labeled peptides from samples assayed. Main differences between both alternatives are the number of samples than can be simultaneously analyzed (10 for TMT *vs* 8 for iTRAQ), coverage of the study and quantification precision (higher for iTRAQ) (Iliuk et al., 2009). Isobaric labeling is constrained by costs, lack of specificity of reporter ions for some peptides, presence of impurities from tag moieties and the aforementioned high technological requirements (quad-based HRMS detection and efficient software solutions). In any case, TMT and iTRAQ are very popular labeling options to study the proteome of animal and vegetal foodstuffs attending to their high efficiency (in contrast to ICAT/cICAT) in the generation of tagged peptides (He et al., 2017)

Alternatively to most popular TMT and iTRAQ labeling approaches, IPTL is a recent strategy for the identification and quantification of two differentially labeled states in MS/MS analysis (Koehler et al., 2009). C-terminal lysine residues of peptides generated by endoproteinase Lys-C are labeled with either the light version of 2-methoxy-4,5-dihydro-1H-imidazole (MDHI) or with deuterated MDHI-D4 (state 1). Furthermore, their N-termini are also tagged either with succinic anhydride (SA) or deuterated SA-D4 (state 2), finally yielding isobaric peptide pairs in which the deuterium mass increase in the MDHI tag is compensated by the deuterium mass increase of SA tags. Main difference of IPTL compared to TMT/iTRAQ is the presence of quantification locations in all -y and –b ions that are present in the MS/MS spectra, offering more than one quantification data point (pairs of –y and –b ions shifted by 4 mass units) per peptide that significantly increase robustness of the assay.

3.5.2.1.3 Enzymatic labeling

This method is based on the proteolytic catalysis by class-2 proteases (i.e., trypsin) to exchange two ^{16}O for two ^{18}O atoms, resulting in a mass shift (4 Da in singly charged species monitored by MS^1 analysis) that enables differentiation of peptides from two biologically distinct samples (Reynolds et al., 2002). This two-step mechanism simultaneously performs digestion and labeling (^{18}O incorporation into the carboxyl terminus of released peptides). Although it is a simple technique, digestion deserves a special care since the use of immobilized enzymes is highly recommended to avoid back exchange that can occur during sample mixing (labeled and unlabeled), altering $^{16}O/^{18}O$ ratio. Moreover, partial labeling can easily occur and can compromise robustness of determination. These constraints greatly limited the use of enzymatic labeling in protein/peptide quantitative analysis although some applications were developed to study proteome of meat products such as bovine and porcine gelatin (Sha et al., 2014)

3.5.2.2 *Absolute quantification*

AQUA is the most common label-based technique for absolute peptide quantitation through synthetic labeled peptides as internal standards (Gerber et al., 2003). Such tagged species are created under the same digestion conditions than their native counterparts but incorporating stable isotopes, mimicking their LC-MS features (retention time and ionization efficiency). It is also feasible the creation of synthetic peptides with desired modifications (i.e., oxidations). Depending on the heavy amino acids considered, different mass shifts will be achieved between native/synthetic peptide pairs; +4 Da (l-alanine-$^{13}C_3$,^{15}N), +6 Da (l-proline $^{13}C_5$,^{15}N, l-valine $^{13}C_5$,^{15}N), +7 Da (l-isoleucine $^{13}C_6$,^{15}N, l-leucine $^{13}C_6$,^{15}N), +8 Da (L-lysine $^{13}C_6$,$^{15}N_2$), or +10 Da (l-arginine $^{13}C_6$,$^{15}N_4$, l-phenylanaline $^{13}C_9$,^{15}N).

Absolute quantitation is achieved by adding a known concentration of synthetic peptides to the protein sample (before digestion). Quantity of natural peptides is directly proportional to peak area ratios established between native/labeled peptide pairs since MS response of both is the same (relative response factor is 1 for considered pairs). AQUA is feasible for both untargeted (MS^1) and targeted (selected/multiple reaction monitoring, SRM/MRM) analyses, receiving this strategy the alternative denomination of isotope dilution-multiple reaction monitoring-mass spectrometry (SID-MRM-MS) (Chahrour et al., 2015).

Robustness and flexibility characterize AQUA analysis, allowing the absolute quantification of unmodified and post-translational modified species. Moreover, since the spiking of heavy peptides occurs before the enzymatic digestion, absolute loss by sampling procedure does not affect the native/AQUA peptide pair ratio. However, AQUA analysis is mainly constrained by high costs and the limited availability of synthetic peptides, thus focusing this alternative toward targeted approaches such as those addressing food quality and safety (Andjelković and Josić, 2018).

Compared to AQUA, QconCAT and PSAQ are affordable alternatives aiming at minimizing accuracy deviations caused by the protease digestion efficiency. The former uses polypeptide constructs (concatamers) made of many different isotope-labeled peptides that are biologically synthetized, which are spiked in a known concentration to the sample before protease digestion releasing tagged peptides. Then, labeled peptides simultaneously control protease efficiency and play the role of internal standards. Nevertheless, QconCAT is also prone to deviations caused by low protease efficiency, finding PSAQ a more efficient solution that considers the accurate addition of isotope-labeled intact proteins before protein digestion, providing multiple peptide standards for target proteins after protease activity (Andjelković and Josić, 2018).

3.5.3 Label-free relative peptide quantification strategies

Originally, this technique used to combine gel electrophoresis with LC-MS analysis to relatively quantify proteins considering spot intensities from gel bands followed by further identification of digested bands through MS/MS analysis. Fortunately, technical advances in LC-MS analysis (hardware and software) greatly helped in the promotion of label-free qualitative/quantitative bottom–up proteomics through straightforward peptidomic research turning this alternative a so flexible analytical tool addressing the proteomic study of very complex matrices (biological fluids, cell lines and extracts from animal/vegetal tissues).

Label-free quantitation strategies can be divided into two main groups: (1) measurement of peptide signal intensity, also known as area under the curve (AUC), based on the measurement of peptide signal intensities through the extracted ion chromatogram (XIC) (Lindemann et al., 2017) and (2), spectral counting (SpC) that is based on the number of peptide-to-spectrum matches assigned to each identified protein (Neilson et al., 2011).

3.5.3.1 Signal intensity measurement (AUC)

This option is based on full-scan analysis of samples that enables the peptide quantification by integration of ion intensities (represented as chromatographic peaks) of any given peptide by time. The simplest approach is the exploratory (non-targeted) MS^1 analysis that permits the integration of all recorded signals into a considered *m/z* range. Evidently, the efficiency of MS^1 research was extraordinarily empowered by HRMS detection decreasing mass uncertainties, having as a result the improvement of critical quantitative aspects such as specificity and selectivity. A less flexible but very robust alternative, SRM/MRM assay performs a data-dependent MS/MS analysis with a predefined parent mass list and monitoring only their selected transition (fragments) signals. This targeted approach was traditionally used for routine analysis in food quality control through the determination of peptide biomarkers, resulting a very useful tool in the fight against fraud (Sentandreu and Sentandreu, 2011) or to detect undesirable alterations in the food production chain (Sentandreu et al. 2021a).

Advances in LC-HRMS analysis recently proposed an innovative label-free approach consisting on a full-scan MS/MS analysis of fragments but without ion precursor selection. This option was firstly introduced by Thermo Scientific (All Ion Fragmentation, AIF, analysis) for metabolite research, receiving other denominations according to LC-HRMS manufacturers (i.e., All Ions MS/MS, MSE, and MS^{ALL}). In this data independent analysis (DIA), collision energy is applied to all ionized molecules (parent ions) without mass filtering (step 1), monitoring all their high mass accuracy fragmentation data (HRMS MS/MS fragments) by time (step 2). One drawback of this untargeted DIA strategy is the over-fragmentation that can occur when considering too wide mass range windows in step 1 (Sentandreu et al., 2018), mainly affecting reliability of the quantification of fragments from low abundant peptides. This problem can be overcome by sequential windowed acquisition of all theoretical fragment ion mass spectra (SWATH-MS) that considers narrow windows (normally 25 to 50 amu) of intact (MS^1) masses stepped across the complete mass range of interest (Gillet et al., 2012). Non-ion precursor consideration confers an extreme flexibility to MS^1 and AIF HRMS approaches by merging targeted and untargeted analyses, with the additional advantage of enabling recursive retrospective inspection of raw data files at any time, in contrast to rigid SRM/MRM targeted approaches. Applications of SWATH-MS high-throughput peptide analysis are rapidly growing during the last years to study multiple aspects of food proteome (Andjelković and Josić, 2018), finding this alternative as one of the most promising analytical tool for foodomics.

3.5.3.2 *Spectral counting (SpC)*

The basis of SpC is the frequency and abundance of the protein, thus protein size and abundance in samples is proportional to the number of identified MS/MS peptide spectra. The simplest SpC approach is the Protein Abundance Index (PAI) that is the number of detected (observed) peptides respecting the number of theoretically observable tryptic peptides for each protein within the MS^1 mass range considered (Neilson et al., 2011):

$$PAI = \frac{\text{number of observed peptides}}{\text{theoretical number of observable tryptic peptides}}$$

PAI lately varied to other expressions to calculate protein abundance such as emPAI (Shinoda et al., 2010), normalized emPAI (Arike and Peil, 2014), and normalized Spectral Abundance Factor (NSAF, McIlwain et al., 2012). Although SpC mainly aims at protein quantification, its practical translation into peptide research is through peptide scores achieved by the interrogation of MS/MS data from protein digests using protein search engines (score values according to peptide abundance). As an example, Sentandreu et al (2021b) demonstrated utility of the peptide score as a coarse quantitative indicator of peptide abundance in the search of biomarkers in meat quality assessment approached by LC-MS analysis.

3.6 Applications and future trends in food peptidomics

High analytical capacity provided by recent advances in LC-MS hardware (benchtop qTOF and Orbitrap HRMS technologies) and bioinformatics boosted relevance of peptidomic analysis for the creation of new insights into the study and understanding of biological systems. In the particular case of food research, peptidomic efforts mainly aimed at the investigation of food digestion, food processing, biomarker discovery, and food waste revalorization (Martini et al., 2021).

3.6.1 Peptidomics of digestion (digestomics)

Classic digestomic analysis (before apparition of MS-based solutions) was supported by liquid chromatography separation coupled to Edman sequencing providing, for instance, successful results about the perseverance of bioactive milk peptides in the plasma of human infants after milk or milk-formulation ingestion (Chabance et al., 1995). Compared to the efficiency shown by this tedious technique, LC-HRMS analysis allows the simultaneous study of multiple peptides generated by digestion but still having some limitations in the research of small peptides (normally below 5 amino acids), large-size polypeptides and disulfide cross-linked hetero-oligomers (De Cicco et al., 2019). Continuous efforts dealing with bioinformatics and hardware advances (i.e., mass resolution, scan rate, and mass range) are continuously minimizing existing uncertainties and will help in the (almost) entire characterization of complex digestomes.

3.6.2 Peptidomics in food processing

Plant food proteins are getting popular to public opinion as substituents of animal-based foods. Multiple studies were recently released evidencing the lack of sustainability of regular production of meat-derived foodstuffs, emphasizing the negative perception of livestock practices and activity of industry (Tziva et al., 2020). Although consumption of fermented vegetable food was a traditional practice in Eastern and Western cultures, their relevance as a valuable source of biopeptides was revealed in the last years (Rizzello et al., 2016). However, little is known about generation of the characteristic biopeptide patterns in food by lactic bacteria and/or inoculated starters but in any case, results achieved to date are very promising. MS analysis mainly supported the study of peptidomic profiles generated by co-culturing practices in fermented plant foods with special relevance of those from whole grain flours. Thus, there were described biopeptides with antihypertensive activity in fermented wholegrain oats (Wu et al., 2018) and with antioxidant/anti-inflammatory properties in sourdoughs for bread preparation (Luti et al., 2020). In addition to promote generation of biopeptides, sourdough fermentation reduces gliadin allergenicity (Nionelli and Rizzello, 2016), currently deserving a particular attention by food industry to produce safe bread with improved health properties.

3.6.3 Peptidomics in biomarker hunting

Quality is an extensive term that includes sensory/nutritional, safety, and origin properties of food. Such features can be assessed by the food fingerprint profile achieved by peptidomic research. Peptide analysis traditionally addressed the study of bioactive species improving organoleptic and healthy features of foodstuffs but in recent years, safety and authentication became hot topics considering social, health and economic implications of food as exemplified by the horsegate scandal (Brooks et al., 2017). Such factors motivated the development of state-of-the-art analytical methodologies to ensure safety of commercial food products and to provide reliable and sensitive tools in the fight against fraud. It must be highlighted chemical complexity of foodstuffs and the multiple compositional changes that final products can suffer by processing. New insights toward the simplification of the study of such intricate matrices have been mainly encouraged by continuous advances in LC-MS (mostly referred to HRMS) technology and bioinformatics. Nowadays, biomarker hunting supported by high-output qualitative/quantitative HRMS peptidomic analysis is an essential tool for the assessment of food safety and authentication (Martini et al., 2021).

Alimentary fraud arises from intentional or accidentally mislabeling practices in which non-declared ingredients are detected in foodstuffs. Most classical food fraudulent behavior consists in the addition of low quality and less expensive (i.e., chicken meat in pork sausages) and/or unappreciated (i.e., horse meat in beef burgers) ingredients to an officially declared high quality processed product. Then, analytical strategies for food authentication are challenged by the sensitivity (adulterants added in low proportion) and specificity (cross-identifications) of the analysis. Furthermore and as mentioned, food processing can alter integrity of target analytes (i.e., DNA denaturation by cooking) turning impossible viability of the assay. Such drawbacks are clearly surpassed by MS-based approaches addressing the search of peptide biomarkers, having a special relevance in authentication of processed foodstuffs. Peptide biomarker hunting powered by LC-MS analysis has a key role in the fight against food fraud as demonstrated in origin assessment of meat and meat products (Sentandreu and Sentandreu, 2011; Sentandreu and Sentandreu, 2014), dairy products (Sassi et al., 2015), fish and fish products (Lasch et al., 2019), leguminous-based products (Huschek et al., 2018), and in a persistent number of increasing applications.

3.6.4 Peptidomics in food allergens

Similarly to food authentication, search of allergenic peptide biomarkers hyphenating LC-MS analysis and bioinformatics significantly improved safety of raw and processed foodstuffs. A wide variety of allergenic proteins were unveiled through the determination of target peptide biomarkers in milk, soy, cereals, peanut, tree nuts, and egg whose detection in derived products is critical to ensure the health of compromised population segments (Gavage et al., 2019; Planque et al., 2017; Van Vlierberghe et al., 2020).

3.6.5 Peptidomics in food waste valorization

Climate change and unstoppable growing of the world population make food production a critical issue that continuously struggles industry, agriculture and farming systems with the subsequent increase of energy demands. Huge amounts of food waste is daily generated worldwide causing multiple pollution problems with very negative social impact from environmental risks and costs derived from waste management (Imbert, 2017). To ameliorate pollution, costs and energy burden created by food production activity, current trends toward valorization of food waste and by-products aim at the incorporation of "residual" biological active compounds into commercial food formulations (Martini et al., 2021). During the last decade, production of protein hydrolysates from food waste and by-products was massively wronged as an affordable and sustainable solution to obtain bioactive peptides. Many studies demonstrated the social, environmental, and economic relevance of biopeptides (i.e., antihypertensive, antioxidant, antimicrobial, cardio protective) recovered from industrial leftovers of dairy (Pires et al., 2021), meat (Holloway and Wu, 2019), vegetable (Görgüç et al., 2020), fish (Le Gouic et al., 2018), and insect (i.e., from silk industry, Altomare et al., 2020) sources. These findings highlighted relevance of collaborative efforts made by scientific research together with the food industry to promote circular economy policies.

3.6.6 Peptidomic analysis with minimal sample preparation

Inherent technical limitations of analytical platforms traditionally used (i.e., LC-MS ion traps and former low scan-rate HRMS detectors) for peptidomic research forced, in general, the inclusion of purification steps into sampling protocols to enhance sensitivity and selectivity of the detection, but significantly increasing workflow time and costs. Unstoppable refinement of the analytical performance of benchtop LC-HRMS devices has propitiated the creation of a new paradigm in the study of biological systems. Detailed qualitative and quantitative analyses of complex organic matrices can be simultaneously carried out by all-in-one HRMS approaches yielding massive amounts of information (Rochat, 2018). Rapidity, sensitivity, and specificity exhibited by modern qTOF and Orbitrap hardware, combined with powerful bioinformatic solutions, favored the creation of new insights into food peptidomic research through straightforward workflows with minimal sample handling (i.e., protein precipitation with alcohol and subsequent aqueous solubilization). Thus, the peptidomic pipeline could be shortened to only two steps (Fig. 3.1) that are peptide extraction from food matrix and LC-HRMS analysis without further sample purification (Sentandreu et al., 2021b). As a result, these innovative straightforward protocols greatly favor the reduction of time, energy, reagents, and chemical waste in agreement with current guidelines in green analytical chemistry (de Marco et al., 2019). As mentioned at the end of section 3.3, this simple workflow analysis can be even more simplified and optimized through (DI-) HRMS analysis.

It must be highlighted that appropriate optimization of classic targeted SRM/MRM approaches addressing quantification of peptide biomarkers allowed the study of digests from direct protein extracts (Fuente-García et al., 2021). However, the majority of conventional peptidomic pipelines featuring separate exploratory (untargeted) qualitative and targeted quantitative analyses used to incorporate a purification procedure into sampling protocols (Giacometti and Buretić-Tomljanović, 2017).

We can conclude that performance of modern LC-HRMS analysis (hardware and bioinformatics) together with current trends in green analytical chemistry promote simplification of sampling procedures in food peptidomic research. That explains current trends in the creation of innovative, sustainable, effective, and straightforward peptidomic pipelines aiming at the study of peptides directly from food extracts.

Acknowledgments

This work was funded by Spanish Agencia Estatal de Investigación (RTI2018-096162-R-C22) and FEDER funds.

References

Afshari, R., Pillidge, C. J., Dias, D. A., Osborn, A. M., & Gill, H. (2020). Cheesomics: the future pathway to understanding cheese flavour and quality. *Critical Reviews in Food Science and Nutrition, 60*(1), 33–47.

Alves, T. O., D'Almeida, C. T. S., Scherf, K. A., & Ferreira, M. S. L. (2019). Modern approaches in the identification and quantification of immunogenic peptides in cereals by LC-MS/MS. *Frontiers in Plant Science, 10*, 1470.

Anand, S., Samuel, M., Ang, C.-S., Keerthikumar, S., & Mathivanan, S (2017). Label-based and label-free strategies for protein quantitation. In S. Keerthikumar, & S. Mathivanan (Eds.), *Proteome bioinformatics, Methods in Molecular Biology*, (pp. 31–43). New York: Springer.

Altomare, A. A., Baron, G., Aldini, G., Carini, M., & D'Amato, A (2020). Silkworm pupae as source of high-value edible proteins and of bioactive peptides. *Food Science & Nutrition, 8*(6), 2652–2661.

Andjelković, U., & Josić, D. (2018). Mass spectrometry based proteomics as foodomics tool in research and assurance of food quality and safety. *Trends in Food Science & Technology, 77*, 100–119.

Arike, L., & Peil, L. (2014). Spectral counting label-free proteomics. In D. Martins-de-Souza (Ed.), *Shotgun proteomics: Methods and protocols* (pp. 213–222). New York: Springer.

Bantscheff, M., Schirle, M., Sweetman, G., Rick, J., & Kuster, B. (2007). Quantitative mass spectrometry in proteomics: A critical review. *Analytical and Bioanalytical Chemistry, 389*(4), 1017–1031.

Barbe, F., Le Feunteun, S., Remond, D., Menard, O., Jardin, J., Henry, G., et al. (2014). Tracking the in vivo release of bioactive peptides in the gut during digestion: Mass spectrometry peptidomic characterization of effluents collected in the gut of dairy matrix fed mini-pigs. *Food Research International, 63*, 147–156.

Beverly, R. L., Huston, R. K., Markell, A. M., McCulley, E. A., Martin, R. L., & Dallas, D. C. (2021). Differences in human milk peptide release along the gastrointestinal tract between preterm and term infants. *Clinical Nutrition, 40*(3), 1214–1223.

Boutrou, R., Gaudichon, C., Dupont, D., Jardin, J., Airinei, G., Marsset-Baglieri, A., et al. (2013). Sequential release of milk protein–derived bioactive peptides in the jejunum in healthy humans. *The American Journal of Clinical Nutrition, 97*(6), 1314–1323.

Brooks, S., Elliott, C. T., Spence, M., Walsh, C., & Dean, M. (2017). Four years post-horsegate: An update of measures and actions put in place following the horsemeat incident of 2013. *Npj Science of Food, 1*(1), 5.

Chabance, B., Jollès, P., Izquierdo, C., Mazoyer, E., Francoual, C., Drouet, L., et al. (1995). Characterization of an antithrombotic peptide from α-casein in newborn plasma after milk ingestion. *British Journal of Nutrition, 73*(4), 583–590.

Chahrour, O., Cobice, D., & Malone, J. (2015). Stable isotope labelling methods in mass spectrometry-based quantitative proteomics. *Journal of Pharmaceutical and Biomedical Analysis, 113*, 2–20.

Chen, W., Liao, A., Hou, Y., Pan, L., Yu, G., Du, J., et al. (2021). Digestive characteristics and peptide release from wheat embryo proteins in vitro. *Food & Function, 12*(5), 2257–2269.

Clerens, S., Thomas, A., Gathercole, J., Plowman, J. E., Yu, T.-Y., Grosvenor, A. J., et al. (2016). Proteomic and peptidomic differences and similarities between four muscle types from New Zealand raised Angus steers. *Meat Science, 121*, 53–63.

Cruz-Huerta, E., Fernandez-Tome, S., Carmen Arques, M., Amigo, L., Recio, I., Clemente, A., et al. (2015). The protective role of the Bowman-Birk protease inhibitor in soybean lunasin digestion: the effect of released peptides on colon cancer growth. *Food & Function, 6*(8), 2626–2635.

Dalabasmaz, S., Dittrich, D., Kellner, I., Drewello, T., & Pischetsrieder, M. (2019). Identification of peptides reflecting the storage of UHT milk by MALDI-TOF-MS peptide profiling. *Journal of Proteomics, 207*, 103444.

Dale, C. J., & Young, T. W. (1989). Fractionation of small peptides from beer. *Journal of the Institute of Brewing, 95*(1), 35–41.

Dallas, D. C., Murray, N. M., & Gan, J. (2015). Proteolytic systems in milk: Perspectives on the evolutionary function within the mammary gland and the infant. *Journal of Mammary Gland Biology and Neoplasia, 20*(3-4), 133–147.

Daroit, D. J., & Brandelli, A. (2021). In vivo bioactivities of food protein-derived peptides—a current review. *Current Opinion in Food Science, 39*, 120–129.

de Almeida, M. A., Saldaña, E., da Silva Pinto, J. S., Palacios, J., Contreras-Castillo, C. J., Sentandreu, M. A., et al. (2018). A peptidomic approach of meat protein degradation in a low-sodium fermented sausage model using autochthonous starter cultures. *Food Research International, 109*, 368–379.

De Cicco, M., Mamone, G., Di Stasio, L., Ferranti, P., Addeo, F., & Picariello, G. (2019). Hidden "Digestome": Current analytical approaches provide incomplete peptide inventories of food digests. *Journal of Agricultural and Food Chemistry, 67*(27), 7775–7782.

de Marco, B. A., Rechelo, B. S., Tótoli, E. G., Kogawa, A. C., & Salgado, H. R. N. (2019). Evolution of green chemistry and its multidimensional impacts: A review. *Saudi Pharmaceutical Journal, 27*(1), 1–8.

Domon, B., & Aebersold, R. (2006). Mass spectrometry and protein analysis. *Science, 312*(5771), 212–217.

Fasano, A. (2011). Zonulin and its regulation of intestinal barrier function: The biological door to inflammation, autoimmunity, and cancer. *Physiological Reviews, 91*(1), 151–175.

Fernandez-Tome, S., Sanchon, J., Recio, I., & Hernandez-Ledesma, B. (2018). Transepithelial transport of lunasin and derived peptides: Inhibitory effects on the gastrointestinal cancer cells viability. *Journal of Food Composition and Analysis, 68*, 101–110.

Finnie, C., & Svensson, B. (2009). Barley seed proteomics from spots to structures. *Journal of Proteomics, 72*(3), 315–324.

Fontenele, M. A., Bastos, M.d. S. R., dos Santos, K. M. O., Bemquerer, M. P., & do Egito, A. S. (2017). Peptide profile of Coalho cheese: A contribution for protected designation of origin (PDO). *Food Chemistry, 219*, 382–390.

Fuente-García, C., Sentandreu, M. A., Aldai, N., Oliván, M., & Sentandreu, E. (2021). Proteomic pipeline for biomarker hunting of defective bovine meat assisted by liquid chromatography-mass spectrometry analysis and chemometrics. *Journal of Proteomics, 238*, Article 104153.

Gallego, M., Mora, L., & Toldra, F. (2018a). Perspectives in the Use of Peptidomics in Ham. *Proteomics, 18*(18), 1700422.

Gallego, M., Mora, L., Escudero, E., & Toldrá, F. (2018b). Bioactive peptides and free amino acids profiles in different types of European dry-fermented sausages. *International Journal of Food Microbiology, 276*, 71–78.

Gavage, M., Van Vlierberghe, K., Van Poucke, C., De Loose, M., Gevaert, K., Dieu, M., et al. (2019). Selection of egg peptide biomarkers in processed food products by high resolution mass spectrometry. *Journal of Chromatography A, 1584*, 115–125.

Gerber, S. A., Rush, J., Stemman, O., Kirschner, M. W., & Gygi, S. P. (2003). Absolute quantification of proteins and phosphoproteins from cell lysates by tandem MS. *Proceedings of the National Academy of Sciences, 100*(12), 6940–6945.

Giacometti, J., & Buretić-Tomljanović, A. (2017). Peptidomics as a tool for characterizing bioactive milk peptides. *Food Chemistry, 230*, 91–98.

Gillet, L. C., Navarro, P., Tate, S., Röst, H., Selevsek, N., Reiter, L., et al. (2012). Targeted data extraction of the MS/MS spectra generated by data-independent acquisition: A new concept for consistent and accurate proteome analysis*. *Molecular & Cellular Proteomics, 11*(6), Article O111.016717.

Görgüç, A., Gençdağ, E., & Yılmaz, F. M. (2020). Bioactive peptides derived from plant origin by-products: Biological activities and techno-functional utilizations in food developments—a review. *Food Research International, 136*, Article 109504.

Gu, Y., Li, X., Liu, H., Li, Q., Xiao, R., Dudu, O. E., et al. (2020). The impact of multiple-species starters on the peptide profiles of yoghurts. *International Dairy Journal, 106*, Article 104684.

Gygi, S. P., Rist, B., Gerber, S. A., Turecek, F., Gelb, M. H., & Aebersold, R. (1999). Quantitative analysis of complex protein mixtures using isotope-coded affinity tags. *Nature Biotechnology, 17*(10), 994–999.

Harthoorn, L. F., Schipper, R. G., Loof, A., Vereijken, P. F. G., Van Heerde, W. L., & Dransfield, E. (2007). Salivary biomarkers associated with perceived satiety and body mass in humans. *Proteomics Clinical Applications, 1*(12), 1637–1650.

He, H., Chen, S., Liang, W., & Liu, X. (2017). Genome-wide proteomics analysis on longissimus muscles in Qinchuan beef cattle. *Animal Genetics, 48*(2), 131–140.

Holloway, J. W., & Wu, J. (2019). Meat by-products. In J. W. Holloway, & J. Wu (Eds.), *Red meat science and production: Volume 1. The consumer and extrinsic meat character* (pp. 161–166). Singapore: Springer.

Hsu, J.-L., Huang, S.-Y., Chow, N.-H., & Chen, S.-H. (2003). Stable-isotope dimethyl labeling for quantitative proteomics. *Analytical Chemistry, 75*(24), 6843–6852.

Huschek, G., Bönick, J., Merkel, D., Huschek, D., & Rawel, H. (2018). Authentication of leguminous-based products by targeted biomarkers using high resolution time of flight mass spectrometry. *LWT, 90*, 164–171.

Iliuk, A., Galan, J., & Tao, W. A. (2009). Playing tag with quantitative proteomics. *Analytical and Bioanalytical Chemistry, 393*(2), 503–513.

Imbert, E. (2017). Food waste valorization options: Opportunities from the bioeconomy. *Open Agriculture, 2*(1), 195–204.

Izquierdo-Gonzalez, J. J., Amil-Ruiz, F., Zazzu, S., Sanchez-Lucas, R., Fuentes-Almagro, C. A., & Rodriguez-Ortega, M. J. (2019). Proteomic analysis of goat milk kefir: Profiling the fermentation-time dependent protein digestion and identification of potential peptides with biological activity. *Food Chemistry, 295*, 456–465.

Koehler, C. J., Strozynski, M., Kozielski, F., Treumann, A., & Thiede, B. (2009). Isobaric peptide termini labeling for MS/MS-based quantitative proteomics. *Journal of Proteome Research, 8*(9), 4333–4341.

Kunda, P. B., Benavente, F., Catala-Cariana, S., Gimenez, E., Barbosa, J., & Sanz-Nebot, V. (2012). Identification of bioactive peptides in a functional yogurt by micro liquid chromatography time-of-flight mass spectrometry assisted by retention time prediction. *Journal of Chromatography A, 1229*, 121–128.

Lametsch, R., Karlsson, A., Rosenvold, K., Andersen, H. J., Roepstorff, P., & Bendixen, E. (2003). Postmortem proteome changes of porcine muscle related to tenderness. *Journal of Agricultural and Food Chemistry, 51*(24), 6992–6997.

Lasch, P., Uhlig, S., Uhlig, C., Wilhelm, C., Bergmann, N., & Wittke, S. (2019). Development and in-house validation of an LC–MS and LC–MS/MS assay for the determination of food fraud for different fish species. *Journal of AOAC International, 102*(5), 1330–1338.

Laville, E., Sayd, T., Morzel, M., Blinet, S., Chambon, C., Lepetit, J., et al. (2009). Proteome changes during meat aging in tough and tender beef suggest the importance of apoptosis and protein solubility for beef aging and tenderization. *Journal of Agricultural and Food Chemistry, 57*(22), 10755–10764.

Le Gouic, A. V., Harnedy, P. A., & FitzGerald, R. J (2018). Bioactive peptides from fish protein by-products. In J.-M. Mérillon, & K. G. Ramawat (Eds.), *Bioactive Molecules in Food* (pp. 1–35). Cham: Springer.

Li, L., Liu, Y., Zou, X., He, J., Xu, X., Zhou, G., et al. (2017). In vitro protein digestibility of pork products is affected by the method of processing. *Food Research International, 92*, 88–94.

Lindemann, C., Thomanek, N., Hundt, F., Lerari, T., Meyer, H. E., Wolters, D., et al. (2017). Strategies in relative and absolute quantitative mass spectrometry based proteomics. *Biological Chemistry, 398*(5–6), 687–699.

Lippolis, J. D., & Nally, J. E. (2018). Considerations for farm animal proteomic experiments: An introductory view gel-based versus non-gel-based approaches. In A. M. de Almeida, D. Eckersall, & I. Miller (Eds.), *Proteomics in Domestic Animals: From Farm to Systems Biology* (pp. 7–16). Cham: Springer.

López, C. M., Sentandreu, M. A., Vignolo, G. M., & Fadda, S. G. (2015). Proteomic and peptidomic insights on myofibrillar protein hydrolysis in a sausage model during fermentation with autochthonous starter cultures. *Food Research International, 78*, 41–49.

Luti, S., Mazzoli, L., Ramazzotti, M., Galli, V., Venturi, M., Marino, G., et al. (2020). Antioxidant and anti-inflammatory properties of sourdoughs containing selected Lactobacilli strains are retained in breads. *Food Chemistry, 322*, Article 126710.

Mada, S. B., Ugwu, C. P., & Abarshi, M. M. (2020). Health promoting effects of food-derived bioactive peptides: A review. *International Journal of Peptide Research and Therapeutics, 26*(2), 831–848.

Maeda, K., Finnie, C., Ostergaard, O., & Svensson, B. (2003). Identification, cloning and characterization of two thioredoxin H isoforms, HvTrxh1 and HvTrxh2, from the barley seed proteome. *European Journal of Biochemistry, 270*(12), 2633–2643.

Mann, M. (2006). Functional and quantitative proteomics using SILAC. *Nature Reviews Molecular Cell Biology, 7*(12), 952–958.

Martini, S., Conte, A., & Tagliazucchi, D. (2019). Comparative peptidomic profile and bioactivities of cooked beef, pork, chicken and turkey meat after in vitro gastro-intestinal digestion. *Journal of Proteomics, 208*, Article 103500.

Martini, S., Solieri, L., & Tagliazucchi, D. (2021). Peptidomics: New trends in food science. *Current Opinion in Food Science, 39*, 51–59.

Mayer, K. F. X., Waugh, R., Langridge, P., Close, T. J., Wise, R. P., & Graner, A. (2012). A physical, genetic and functional sequence assembly of the barley genome. *Nature, 491*(7426), 711-+.

Meyer, J. G., Niemi, N. M., Pagliarini, D. J., & Coon, J. J. (2020). Quantitative shotgun proteome analysis by direct infusion. *Nature Methods, 17*(12), 1222–1228.

McIlwain, S., Mathews, M., Bereman, M. S., Rubel, E. W., MacCoss, M. J., & Noble, W. S. (2012). Estimating relative abundances of proteins from shotgun proteomics data. *BMC Bioinformatics, 13*(1), 308.

Minamino, N. (2001). Peptidome: the fact-database for endogenous peptides. *Tanpakushitsu kakusan koso, 46*(11 Suppl), 1510–1517.

Montesano, D., Gallo, M., Blasi, F., & Cossignani, L. (2020). Biopeptides from vegetable proteins: New scientific evidences. *Current Opinion in Food Science, 31*, 31–37.

Morzel, M., Terlouw, C., Chambon, C., Micol, D., & Picard, B. (2008). Muscle proteome and meat eating qualities of Longissimus thoracis of "Blonde d'Aquitaine" young bulls: A central role of HSP27 isoforms. *Meat Science, 78*(3), 297–304.

Muroya, S., Kitamura, S., Tanabe, S., Nishimura, T., Nakajima, I., & Chikuni, K. (2004). N-terminal amino acid sequences of troponin T fragments, including 30 kDa one, produced during postmortem aging of bovine longissimus muscle. *Meat Science, 67*(1), 19–24.

Neilson, K. A., Ali, N. A., Muralidharan, S., Mirzaei, M., Mariani, M., Assadourian, G., et al. (2011). Less label, more free: Approaches in label-free quantitative mass spectrometry. *Proteomics, 11*(4), 535–553.

Nionelli, L., & Rizzello, C. G. (2016). Sourdough-based biotechnologies for the production of gluten-free foods. *Foods, 5*(3), 65.

Panchaud, A., Affolter, M., & Kussmann, M. (2012). Mass spectrometry for nutritional peptidomics: How to analyze food bioactives and their health effects. *Journal of Proteomics, 75*(12), 3546–3559.

Pangestuti, R., & Kim, S.-K. (2017). Bioactive peptide of marine origin for the prevention and treatment of non-communicable diseases. *Marine Drugs, 15*(3), 67.

Park, S. J., Lee, Y. H., Nam, Y. J., Baik, B.-K., & Kim, T. W. (2013). Peptide-based biomarker discovery for identification of rice cultivars using surface enhanced laser desorption/ionization time-of-flight mass spectrometry. *Crop Science, 53*(3), 987–995.

Picariello, G., Bonomi, F., Iametti, S., Rasmussen, P., Pepe, C., Lilla, S., et al. (2011). Proteomic and peptidomic characterisation of beer: Immunological and technological implications. *Food Chemistry, 124*(4), 1718–1726.

Picariello, G., Mamone, G., Cutignano, A., Fontana, A., Zurlo, L., Addeo, F., et al. (2015). Proteomics, peptidomics, and immunogenic potential of wheat beer (Weissbier). *Journal of Agricultural and Food Chemistry, 63*(13), 3579–3586.

Picariello, G., Mamone, G., Nitride, C., & Ferranti, P. (2017). Chapter 26: Proteomic analysis of beer. In: Colgrave, M. (Ed.). *Proteomics in food science* (pp. 383–403). Cambridge.

Piraino, P., Upadhyay, V. K., Rossano, R., Riccio, P., Parente, E., Kelly, A. L., et al. (2007). Use of mass spectrometry to characterize proteolysis in cheese. *Food Chemistry, 101*(3), 964–972.

Pires, A. F., Marnotes, N. G., Rubio, O. D., Garcia, A. C., & Pereira, C. D. (2021). Dairy by-products: A review on the valorization of whey and second cheese Whey. *Foods, 10*(5), 1067.

Planque, M., Arnould, T., Dieu, M., Delahaut, P., Renard, P., & Gillard, N. (2017). Liquid chromatography coupled to tandem mass spectrometry for detecting ten allergens in complex and incurred foodstuffs. *Journal of Chromatography A, 1530*, 138–151.

Polati, R., Menini, M., Robotti, E., Millioni, R., Marengo, E., & Novelli, E. (2012). Proteomic changes involved in tenderization of bovine Longissimus dorsi muscle during prolonged ageing. *Food Chemistry, 135*(3), 2052–2069.

Raveschot, C., Cudennec, B., Deracinois, B., Frémont, M., Vaeremans, M., Dugersuren, J., et al. (2020). Proteolytic activity of Lactobacillus strains isolated from Mongolian traditional dairy products: A multiparametric analysis. *Food Chemistry, 304*, Article 125415.

Reynolds, K. J., Yao, X., & Fenselau, C. (2002). Proteolytic 18O labeling for comparative proteomics: Evaluation of endoprotease Glu-C as the catalytic agent. *Journal of Proteome Research, 1*(1), 27–33.

Rizzello, C. G., Tagliazucchi, D., Babini, E., Sefora Rutella, G., Taneyo Saa, D. L., et al. (2016). Bioactive peptides from vegetable food matrices: Research trends and novel biotechnologies for synthesis and recovery. *Journal of Functional Foods, 27*, 549–569.

Rochat, B. (2018). *Quantitative and qualitative LC-high-resolution MS: The technological and biological reasons for a shift of paradigm*. In *Recent advances in analytical chemistry*. London: IntechOpen 2019-04-10.

Sanchon, J., Fernandez-Tome, S., Miralles, B., Hernandez-Ledesma, B., Tome, D., Gaudichon, C., et al. (2018). Protein degradation and peptide release from milk proteins in human jejunum. Comparison with in vitro gastrointestinal simulation. *Food Chemistry, 239*, 486–494.

Sassi, M., Arena, S., & Scaloni, A. (2015). MALDI-TOF-MS platform for integrated proteomic and peptidomic profiling of milk samples allows rapid detection of food adulterations. *Journal of Agricultural and Food Chemistry, 63*(27), 6157–6171.

Schrader, M., Schulz-Knappe, P., & Fricker, L. D. (2014). Historical perspective of peptidomics. *EuPA Open Proteomics, 3*, 171–182.

Schulz-Knappe, P., Zucht, H. D., Heine, C., Jurgens, M., Hess, R., & Schrader, M. (2001). Peptidomics: The comprehensive analysis of peptides in complex biological mixtures. *Combinatorial Chemistry & High Throughput Screening, 4*(2), 207–217.

Sentandreu, M. A., Coulis, G., & Ouali, A. (2002). Role of muscle endopeptidases and their inhibitors in meat tenderness. *Trends in Food Science & Technology, 13*(12), 400–421.

Sentandreu, M. A., & Sentandreu, E. (2011). Peptide biomarkers as a way to determine meat authenticity. *Meat Science, 89*(3), 280–285.

Sentandreu, M. A., & Sentandreu, E. (2014). Authenticity of meat products: Tools against fraud. *Food Research International, 60*, 19–29.

Sentandreu, E., Peris-Díaz, M. D., Sweeney, S. R., Chiou, J., Muñoz, N., & Tiziani, S. (2018). A survey of orbitrap all ion fragmentation analysis assessed by an R metabolist package to study small-molecule metabolites. *Chromatographia, 81*(7), 981–994.

Sentandreu, E., Fuente-García, C., Navarro, J. L., & Sentandreu, M. A. (2021a). A straightforward gel-free proteomics pipeline assisted by liquid isoelectric focusing (OFFGEL) and mass spectrometry analysis to study bovine meat proteome. *Food Science and Technology International, 27*(2), 112–122.

Sentandreu, E., Fuente-García, C., Pardo, O., Oliván, M., León, N., Aldai, N., et al. (2021b). Protein biomarkers of bovine defective meats at a glance: gel-free hybrid quadrupole-orbitrap analysis for rapid screening. *Journal of Agricultural and Food Chemistry, 69*(26), 7478–7487.

Sha, X.-M., Tu, Z.-C., Wang, H., Huang, T., Duan, D.-L., He, N., et al. (2014). Gelatin quantification by oxygen-18 labeling and liquid chromatography–high-resolution mass spectrometry. *Journal of Agricultural and Food Chemistry, 62*(49), 11840–11853.

Shinoda, K., Tomita, M., & Ishihama, Y. (2010). emPAI Calc—For the estimation of protein abundance from large-scale identification data by liquid chromatography-tandem mass spectrometry. *Bioinformatics, 26*(4), 576–577.

Tagliazucchi, D., Helal, A., Verzelloni, E., & Conte, A. (2016). Bovine milk antioxidant properties: effect of in vitro digestion and identification of antioxidant compounds. *Dairy Science & Technology, 96*(5), 657–676.

Tilocca, B., Costanzo, N., Morittu, V. M., Spina, A. A., Soggiu, A., Britti, D., et al. (2020). Milk microbiota: Characterization methods and role in cheese production. *Journal of Proteomics, 210*.

Tziva, M., Negro, S. O., Kalfagianni, A., & Hekkert, M. P. (2020). Understanding the protein transition: The rise of plant-based meat substitutes. *Environmental Innovation and Societal Transitions, 35*, 217–231.

Van Vlierberghe, K., Gavage, M., Dieu, M., Renard, P., Arnould, T., Gillard, N., et al. (2020). Selection of universal peptide biomarkers for the detection of the allergen hazelnut in food trough a comprehensive, high resolution mass spectrometric (HRMS) based approach. *Food Chemistry, 309*, Article 125679.

Wölk, M., Milkovska-Stamenova, S., & Hoffmann, R. (2020). Comprehensive profiling of the native and modified peptidomes of raw bovine milk and processed milk products. *Foods, 9*(12), 1841.

Wu, H., Rui, X., Li, W., Xiao, Y., Zhou, J., & Dong, M. (2018). Whole-grain oats (Avena sativa L.) as a carrier of lactic acid bacteria and a supplement rich in angiotensin I-converting enzyme inhibitory peptides through solid-state fermentation. *Food & Function, 9*(4), 2270–2281.

Zhao, D., Xu, Y., Gu, T., Wang, H., Yin, Y., Sheng, B., et al. (2020). Peptidomic investigation of the interplay between enzymatic tenderization and the digestibility of beef semimembranosus proteins. *Journal of Agricultural and Food Chemistry, 68*(4), 1136–1146.

Zhou, C.-Y., Pan, D.-D., Cao, J.-X., & Zhou, G.-H. (2021). A comprehensive review on molecular mechanism of defective dry-cured ham with excessive pastiness, adhesiveness, and bitterness by proteomics insights. *Comprehensive Reviews in Food Science and Food Safety, 20*, 3838–3857.

SECTION II

Applications of proteomic in food sciences

CHAPTER

4

Proteomic advances in crop improvement

Rubén Agregán[a], Noemí Echegaray[a], María López Pedrouso[b], Mirian Pateiro[a], Daniel Franco Ruiz[a], Jose M. Lorenzo[a,c]

[a]Centro Tecnológico de la Carne de Galicia, Ourense, Spain
[b]Department of Zoology, Genetics and Physical Anthropology, University of Santiago de Compostela, Santiago de Compostela, Spain
[c]University of Vigo, Spain

4.1 Introduction

Since ancient times, humans have introduced plant crops as part of their diet, contributing with an important amount of nutrients, vitamins, secondary metabolites, and other compounds that improve health (Radovich, 2011). Vegetables, such as lettuce, zucchini, tomato, eggplant, cucumber, onion, peas, or potatoes are widely consumed as an important source of starch and cellulose that also contain a wide variety of vitamins and minerals (Butnariu & Butu, 2015). In addition to vegetable crops, cereals, such as corn, sorghum, barley, or oats are the basis of livestock feeding (Kaur, Jha, Sabikhi, & Singh, 2014), and together with those used in human nutrition represent a substantial source of income for many farmers and drives a trade that is the cornerstone of many countries.

Growing plants is a thriving business that supplies food to many countries around the globe. Proteomics has emerged as a tool born from biology that could play a prominent role in the short and medium-term in controlling some of the factors that contribute to the success of this business, such as quality and food safety. Technologies derived from this branch of biology may help to understand the expression of proteins in a biological

Food Proteomics
DOI: https://doi.org/10.1016/B978-0-323-90889-4.00012-9

system, including all the isoforms and modifications present, as well as the interactions among them, their structures, and higher-order complexes (Fox & Brodkorb, 2008). The development of the technologies involved in the characterization of proteomes in recent decades has helped to improve an arduous and complex task, reducing the time consumed and increasing the level of detail (Bendixen, 2013). This chapter is intended to offer an overview of the main proteomic approaches and the separation and identification techniques used in them.

Proteomics has the potential to improve and increase the performance of countless processes in different fields of human activity. This chapter discusses the applicability of proteomic techniques in food safety, quality assurance, and performance improvement of agricultural crops. Plant proteomic research is opening the doors to the development of a technology aimed to guarantee a controlled and adequate supply through the development and application of new practices committed to the environment, the conservation of biodiversity, and human health (Jorrín-Novo andValledor, 2013).

4.2 Definition and composition of vegetables

The definition of the vegetable can be somewhat ambiguous regarding the inclusion of certain foods. At first, any part of a plant consumed as food that it is not a fruit or seed is considered a vegetable. This meaning includes petioles, whole leaves, immature fruits, roots, and specialized structures, such as bulbs and tubers. However, this definition can be expanded and included as a vegetable ripe fruits that are consumed regularly during the main menu, without including dessert, such as tomato. However, melon, that is conventionally included in the dessert, is frequently accepted as a vegetable, despite being considered botanically, legally, and culinary as a fruit (Radovich, 2011). Another exception can also be made in the case of beans, where in many cases only the seed is consumed.

Compared to foods of animal origin, vegetables have a lower nutritional quality, but they contain a series of compounds of unquestionable relevance for health and that must be part of a balanced diet, such as vitamins, minerals (e.g., Ca, Fe, Cu, P, Zn, Cl, Na), digestible carbohydrates (e.g., starch, sucrose, glucose, fructose) and non-digestible carbohydrates (e.g., cellulose, hemicellulose, pectin, protides) (Caunii, Cuciureanu, Zakar, Tonea, & Giuchici, 2010). In addition, other secondary metabolites can be found, such as terpenes, polyphenols, alkaloids, and sulfur-containing compounds (Hounsome & Hounsome, 2011). The protein content of most vegetables is around 0.5%–1.5%, with the exception of green and dried legumes, which account for 5%–6% (peas and beans) and 20%–34% (lentils, dried beans, and soybeans) from plant proteins, respectively (Butnariu & Butu, 2015).

4.3 Cereals proteins. Content and classification

Cereals are plants belonging to the Gramineae family, whose grains are widely consumed worldwide. The grain or cereal seed is composed of the embryo, consisting of the embryonic axis and the scutellum, located at one end of the endosperm, which is the major grain tissue. This endosperm is divided into two clearly differentiated parts, the starchy endosperm, made up mainly of starch as its name indicates and proteins, and the aleurone layer, which surrounds the starchy endosperm, rich in proteins and lipids. Then, there are a serious of layers of tissue that wrap the endosperm and embryo. The innermost is called the nucellus, followed by the testa and the pericarp, the outermost (Evers, Blakeney, & O'Brien, 1999).

Cereal grains serve as food for both humans and livestock and provide a considerable amount of energy. Much of this energy (up to 389 kcal/100 g) is due to the high proportion of carbohydrates (primarily starch), which represent approximately 70% of the grain content. The next most important component of cereal grains is proteins, with an average content of 10%, considerably lower than carbohydrates. Finally, fat is the least of the macronutrients in the cereal grain, with values that range between 1.6% and 6.90% and constituted by triglycerides, phospholipids, and glycolipids (Tacer-Caba, Nilufer-Erdil, & Ai, 2015).

Cereal proteins are classified into four differentiated fractions according to their solubility in specific solvents (Wrigley, 2017). Albumin proteins are soluble in water, while globulins are subsequently extracted using a NaCl solution. Both proteins represent 20% of the total proteins of the cereal grain. The rest of grain proteins (80%) correspond to prolamines and glutelins. They are considered storage proteins and are insoluble in any of the aforementioned solvents. Prolamins are ethanol-soluble monomers and include gliadins, hordeins, zeins, avenins, and rhizins, which are found in wheat, barley, corn, oats, and rice, respectively. Regarding glutelins, they are generally polymerized and can be extracted using acidic solvents (Branlard & Bancel, 2007). Albumins and globulins are enzymes involved in poorly characterized plant functions. Their molecular mass is less than 30 kDa and the amino acid profile contains many lysine and glutamic acid residues. They are proteins related to the hardness of the grain, present in soft wheat but absent in durum wheat, for example. Then, prolamines are a heterogeneous group soluble in 70% aqueous alcohol or other alcohol mixtures. A well-known prolamin present in wheat grain is gliadin. This monomeric protein can have a

TABLE 4.1 Amino acid composition of some of the main cultivated vegetables and comparison with the daily requirements of amino acids stipulated by the WHO/FAO/UNU. (Iqbal, Khalil, Ateeq, & Sayyar Khan (2006); Kovalenko, Rippke, & Hurburgh (2006) WHO/FAO/UNU, 2007.

AA (mg/g DW)	[a]Chickpea	[a]Cowpea	[a]Lentil	[a]Green pea	[b]Soybean	[c]Adult AA requirements	
EAA						***AA***	***Quantity (mg/kg per day)***
Histidine	7.77	8.45	6.33	6.48	8.2	Histidine	10
Isoleucine	12.43	12.27	11.80	12.15	19.4	Isoleucine	20
Leucine	22.52	20.99	22.45	19.98	32.6	Leucine	39
Lysine	18.64	20.45	20.14	21.88	26.9	Lysine	30
Methionine	2.85	6	2.30	2.97	6.1	Methionine	10
						Methionine + cysteine	15
Phenylalanine	14.24	20.45	14.39	14.04	21.6	Phenylalanine + tyrosine	25
Threonine	8.03	10.36	10.07	10.26	16.2	Threonine	15
Tryptophan	2.33	1.91	2.01	2.16	5	Tryptophan	4
Valine	88.80	13.63	14.39	13.50	20.6	Valine	26
ΣEAA	177.61	114.50	103.88	103.43	156.6		
NEAA							
Alanine	12.87	11.45	12.09	14.04	17.9		
Arginine	21.49	20.45	22.45	19.44	31.7		
Asparagine	28.48	29.44	33.96	29.71	47.9		
Cysteine	1.55	1.36	2.59	4.86	7	Cysteine	4
Glutamine	44.79	46.89	61.87	47.26	76.6		
Glycine	9.58	10.36	10.36	12.15	17.7		
Proline	9.84	10.91	10.07	10.26	20.4		
Serine	9.58	12.27	14.96	13.77	19.2		
Tyrosine	7.25	8.18	92.08	9.99	15.3		
ΣNEAA	145.42	151.31	260.42	161.50	253.7		

[a]*Iqbal et al. (2006)*
[b]*Kovalenko et al. (2008)*
[c]*WHO/FAO/UNU (2007)*
WHO/FAO/UNU: World Health Organization/Food and Agriculture Organization/United Nations University; AA: amino acid; EAA: essential amino acid; NEAA: non-essential amino acid.

molecular mass between 30 and 75 kDa. Ω-gliadin is rich in glutamine, proline, and phenylalanine, representing 80% of the total amino acids, with the absence of cysteine (Table 4.1). On the other hand, γ-gliadin differs from α- and β-gliadins by the amounts of aspartic acid, proline, methionine, tyrosine, phenylalanine, and tryptophan. Finally, glutelins are proteins arranged to form polymers with a very high molecular weight, on the order of several million Daltons (Guerrieri & Cavaletto, 2018). These proteins can be separated into two subunits, one of high-molecular weight and the other of low-molecular weight. The first of the fractions is precipitated by using 60% aqueous 1-propanol heated to 60°C and dithioerythritol to reduce the disulfide bonds (Belitz, Grosch, & Schieberle, 2009).

4.4 Scope of vegetable and cereal proteins in agriculture and food

In today's first world society, there is a change in mentality in the way of eating that promotes healthier nutrition in which animal fats are being replaced with more heart-healthy and sustainable products. This is leading to an increase in the consumption of vegetable products causing a critical increase in the demand for plant proteins resulting in supply problems (Petrusan, Rawel, & Huschek, 2016). The cultivation of cereals and vegetables is a fundamental part of the diet in many parts of the world, providing most of the daily energy and being the livelihood and economic engine of many communities. For example, in many African countries, the access to animal protein sources is limited and population have to survive with a diet based on cereals (e.g., maize, rice, sorghum, millet), roots, and tubers (Temba, Njobeh, Adebo, Olugbile, & Kayitesi, 2016). Legumes are the second largest plant family in the world and a pillar for human nutrition, with a fundamental role in sustainable agriculture that together with cereal crops have integrated the nutritional base of many civilizations (Blair, Wu, Bhandari, Zhang, & Hao, 2016). Unlike cereals, whose consumption is more linked to obtaining energy due to their high carbohydrate content, legumes highlight, among other things, for their protein quantity. In fact, they can be used in baking to increase the protein quality of cereal flours (Swehata, Darwish, El-Nahry, Fatma, & Razek, 1988). Unfortunately, cereal-based diets do not meet nutritional requirements as they are deficient in the amino acid lysine (Sarwar, 2013). In contrast, legumes are deficient in methionine, cysteine (sulfur amino acids) and tryptophan (Sá, Moreno, & Carciofi, 2020). However, both protein sources may complement each other. In this sense, it has been found that the addition of legume flours to products made with cereal flours, such as cookies, may improve protein quality. Especially an interesting strategy in developing areas with a high rate of protein-energy malnutrition (Igbabul, Iorliam, & Umana, 2015).

The consumption of vegetables provides benefits, such as the maintenance of the hydration of the body and the regulation of metabolism. A series of nonprotein amino acids found in vegetables, including homoarginine, carnitine, citrulline, taurine, -aminobutyric, and -aminobutyric acids have a direct impact on health. -Aminobutyric acid, better known as GABA, is an inhibitory neurotransmitter in the central nervous system and in the retina and can be found in beans, spinach, potatoes, and kale. Carnitine is another bioactive compound found in peas, potatoes, and zucchini, which is involved in lipid metabolism in the heart and skeletal muscle. The well-known taurine, found in maize, lentils, and peanuts, has a role in the detoxification and stabilization of cell membranes. Despite all the beneficial effects reported by vegetable crops, deleterious effects have been found due to the presence of toxic compounds, such as the amino acids β-cyanoalanine and canavanine, which inhibit protein synthesis and the immune system (Hounsome, Hounsome, Tomos, & Edwards-Jones, 2008).

Therefore, both cereals and vegetable crops play an important role in human nutrition. They are a good source of protein and the food base for many communities. But not only plant crops are important in human nutrition, slaughter animals depend on certain crops for their maintenance and fattening. Cereals, such as corn, wheat, or barley can represent up to 60% of the daily energy ingested by livestock (Lee, Nam, & Kong, 2016). Most of the world's agricultural resources are focused on the animal production, using up to 70% of the arable land (Lal, 2016). For example, only 15% of the maize grown globally is intended for human consumption and most of the sorghum produced in the United States, the second-largest producer in the world, is used for animal feed (Awika, 2011).

4.5 Concept of proteomics and different approaches to proteome analysis

Proteomics is a word composed of the terms "protein" and "genomics" that defines an area of knowledge framed within the field of biology that came to light in the mid-1990s. However, the concept was promulgated a few years earlier by Norman G. Anderson and N. Leigh when back in 1979 they tried to unlock the genome using the two-dimensional gel electrophoresis (2-DGE) technique (Lawrance, Klopcic, & Wasinger, 2005). Wilkins et al. (1996) were the first to stablish the concept of proteomics or proteome analysis, as the separation, identification, and quantification of the set of proteins expressed by a genome, cell, or tissue. This definition includes all the isoforms and modifications present, as well as the

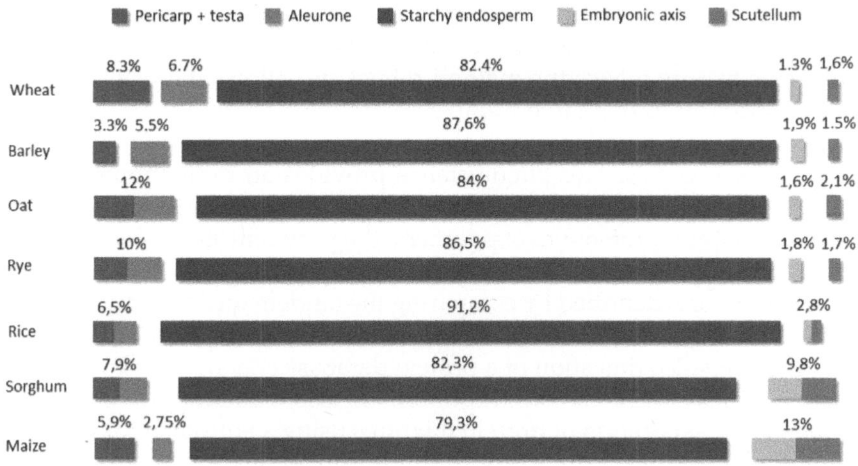

FIG. 4.1 Relation of the different tissues that compose the grain of some of the most consumed cereals. Data from Kent (1983).

interactions between them, their structures, and higher-order complexes (Tyers & Mann, 2003). The characterization of proteomes is an arduous and complex task given the vast variability of the forms in which proteins can appear. Unfortunately, at present there are no methods available that allow the complete analysis of proteomes in a single and simple event. However, technologies developed in recent decades have made it possible to exponentially improve speed, accuracy, and level of detail (Bendixen, 2013).

Today, mass spectrometry (MS) offers a clear advantage over other technologies for studying proteomes and its combination with a wide variety of separation methods makes it a very interesting tool. This analysis strategy along with an enzymatic digestion of the resolved proteins prior to their passage through MS is known as the bottom–up approach and has been and continues to be the main strategy for the characterization of proteins, in which they are first proteolytically purified or cleaved and the resulting products analyzed by MS. Recently, another strategy based on gas-phase fragmentation of intact protein ions or protein fragments of high molecular mass for mass spectrometric analysis is becoming more widespread within proteomics, known as the top–down approach. Fig. 4.1 briefly shows a flow chart with two summary examples of classic protocols for bottom–up and top–down approaches.

4.5.1 Bottom–up approach

The potential of a bottom–up proteomic analysis lies in working with peptides rather than proteins. Peptides can be separated better with reverse-phase liquid chromatography (RP-LC) than proteins, providing fragmentation patterns with fewer variations and therefore more predictable. This feature makes the method robust and allows the identification and quantification of a large number of proteins (Dupree et al., 2020).

When the test sample is composed of a mixture of proteins, it is called shotgun proteomics. This circumstance provides an indirect measure of proteins through the peptides obtained from the proteolytic digestion. In a classic shotgun proteomics experiment, the proteome is fractionated and subjected to LC with tandem mass spectrometry LC-MS/MS. The resulting peptides are identified by comparing the tandem mass spectra derived from their fragmentation with the theoretical tandem mass spectra generated from in silico digestion of a protein database (Zhang, Fonslow, Shan, Baek, & Yates, 2013). Gel electrophoresis is also a commonly used technique for the separation of protein mixtures using a bottom–up approach. By applying an electric current through the porous matrix of a gel at different speeds, the proteins are separated according to their size, and thus can be identified. This technique dates back to 1930 when Arne Tiselius developed the first electrophoretic apparatus. At first, the migration of molecules was very limited, but three decades later, a series of innovations modernized this technique, such as the sodium dodecyl sulphate (SDS) electrophoresis and the gradient gel electrophoresis (DGGE). In the early 1970s, the two most resolutive methods to date, isoelectric focusing (IEF) and discontinuous SDS-polyacrylamide gel electrophoresis (PAGE), were successfully implemented in the same separation experiment, which allowed the analysis of samples with complex matrices (Jorrin-Novo, Komatsu, Sanchez-Lucas, & Rodríguez de Francisco, 2019). The 2-DGE allows the analysis of hundreds of proteins in a single gel and with a single run, becoming a very popular separation technique in proteomics. In the IEF (first dimension), protein and peptide molecules are separated depending on their charge defined by the pKa values of the proton acceptance sites. This technique can provide very high resolving power, including the separation of protein post-translational modifications (PTMs) that alter charge. Modern IEF methods include fine polyacrylamide gels with an immobilized pH gradient (IPG) (Friedman, Hoving, & Westermeier, 2009). The pH gradients fixed in these gels remain stable for long periods of time at very high voltages, favoring the separation of proteins (Garfin, 2003). Once this first step in the resolution of a proteome is completed, the sample is resolved again, but this time based on their apparent molecular mass using SDS-PAGE (second dimension). In this step, the protein-SDS micelles migrate toward the anode of the electrophoretic chamber at a speed dependent on the molecular mass and to a lesser extent on the hydrophobicity of the molecules (Friedman et al., 2009). After this last separation, the proteins are stained for visualization. Universal staining methods include the use of coomassie brilliant blue, negatively charged metal salts, such as zinc, copper, or silver, and fluorescent and isotopic labels. On the other hand, there are specific staining methods such as immunochemical methods (e.g., western blotting) (Pomastowski & Buszewski, 2014).

Liquid phase separation for protein resolution of biological samples has advantages over 2-DGE separation, such as higher sensitivity, dynamic range, ease of automation, and speed. In LC, the use of separation mechanisms, such as size exclusion, reversed phase (RP), or ion exchange, allow the analysis of proteins of high acidity or basicity regardless of their molecular weight. However, the resolution is considerably lower than that of the 2-DGE technique (O'Donnell, Holland, Deeth, & Alewood, 2004). Experience of use has shown that high-performance liquid chromatography (HPLC) is an excellent alternative to two-dimensional polyacrylamide gel electrophoresis (2D-PAGE) for the resolution of proteomes due to its high separation power, reproducibility, and compatibility for connection to an electrospray ionization source (ESI) in MS equipment (Shi, Xiang, Horváth, & Wilkins, 2004).

A single-dimension LC separation is a basic working methodology in LC-MS-based proteomics and the RP is the most common way of separating peptides using packed or monolithic columns (Xie, Smith, & Shen, 2012). Packed columns are commonly used in HPLC, which consist of polymers or microparticulate porous silica arranged in a packed way, as the name suggests (Mistry & Grinberg, 2005). The size of these porous microparticles (a few microns) allows a high and effective separation of complex samples, which will be retained based on their chemical composition and structure (Unger, Skudas, & Schulte, 2008). In contrast, monolithic columns represent another increasingly used work concept to replace packed columns. Emerged in the late 1980s and early 1990s, these columns offer higher separation performance due to their silica skeletons and large pores (Buszewski, Szumski, & Sus, 2002). Structurally, they consist of a piece of a continuous porous material sealed to the wall of the tube that contains it, forcing the mobile phase to filter through the entire bed. The porous material usually has a bimodal distribution with macropores or also called throughpores that are interconnected in channels through which practically all of the mobile phase flows (Guiochon, 2007).

On the other hand, the protein separation can also be carried out using a multidimensional mechanism. Multidimensional chromatography coupled to MS is based on the combination of multiple dimensions of separation. This way of working allows obtaining higher separation and extended measured dynamic concentration range of peptides than single-dimensional chromatography, thus improving the identification and quantification capacity and reducing the effects of ion suppression in complex protein mixtures (Horvatovich, Hoekman, Govorukhina, & Bischoff, 2010). Strong-cation exchange (SCX) x RP-LC is one of the most used combinations in multidimensional LC systems. The sample is first fractionated according to the charge of the analytes (proteins/peptides) to subsequently being separated based on their hydrophobicity (Xie et al., 2012).

4.5.2 Top–down approach

Another way to characterize samples in proteomics is by analyzing intact proteins, without being subjected to enzymatic digestion. This analytic strategy known as the top–down approach is based on the gas-phase ionization of undigested proteins and their subsequent direct analysis by MS and tandem mass spectrometry (MS/MS) (Cunsolo, Muccilli, Saletti, & Foti, 2014). Some disadvantages of bottom–up approach, such as the lack of specificity of certain peptides for a single protein or the possible exclusion of considerable large regions due to loss of information about PTMs or variants of the sequence, have led to the emergence of this new analytical approach (Fig. 4.2). Direct introduction of the sample into the mass spectrometer instead of digested peptides increases sequence

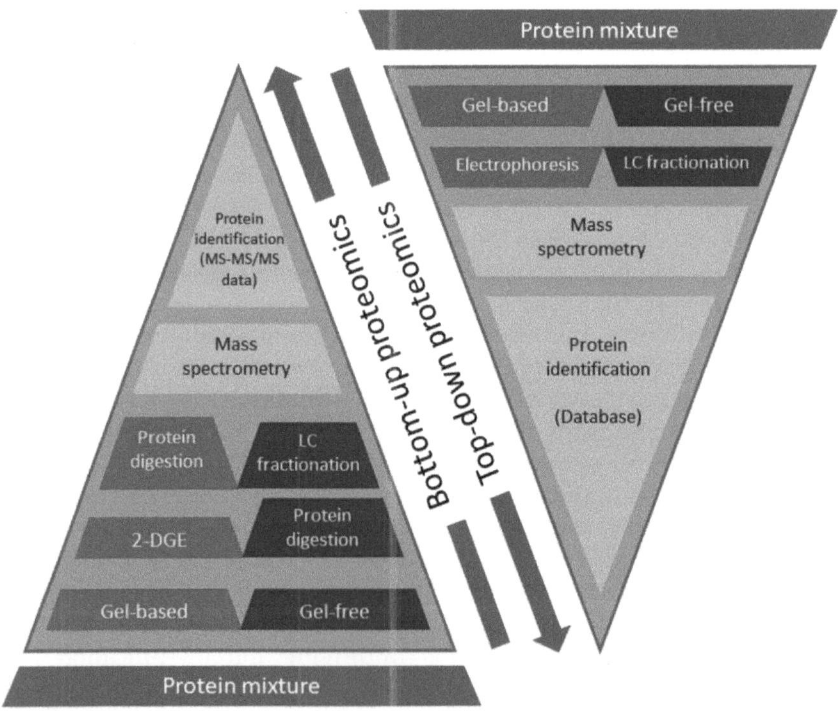

FIG. 4.2 Summary working diagram of the two strategies used in the proteomic analysis. The bottom–up and top–down approaches.

coverage, often being 100%, and complete identification of proteins and their isoforms, including PTMs (Catherman, Skinner, & Kelleher, 2014). Simple experiments based on this approach require several steps that begin with the careful preparation of the proteoforms of interest for ESI (includes a purification step) and their direct introduction into the equipment. To achieve greater separation, samples are often pre-fractionated based on their molecular mass or isoelectric point using various LC modes coupled online with ESI-MS. Generally, the cations generated, also called precursors, are collected in an ion trap, filtered with a quadrupole (Q) mass filter, or directly by ion optics into a high-resolution mass analyzer. These precursor ions are subsequent fragmented, often by collisional or electron-based dissociation, within the same spectrometer for identification and characterization (Savaryn, Toby, & Kelleher, 2016). This point in the analytical process is essential since it will directly influence the identification of proteoforms. Collision-induced dissociation (CID) and higher-energy collisional dissociation are the most robust methods for large-scale and high-throughput proteomic analysis under a top–down approach. The possible cleavage of labile PTMs by CID has limited its use for the localization and structural study of these compounds. However, the use of electron-based methods, such as electron transfer dissociation and electron capture dissociation generally preserve labile PTMs and are a very useful tool for complete characterization of biologically relevant proteins (Chen, Brown, Lin, & Ge, 2018).

4.6 Application of proteomics in the improvement of cereal and vegetable crops

4.6.1 Translational plant proteomics

The incessant increase in the demand for food as a consequence of the increase in the world population is putting more and more pressure on agricultural resources. Finding a way to reduce this environmental pressure while meeting demand is the short-medium term objective of the markets. In this regard, biotechnology plays a fundamental role in the generation of knowledge, but science and scientists must go beyond the simple generation of information, and in a joint effort with industry, translate laboratory data into solutions to real problems. The translational plant proteomics has this target through the development of new practices committed to the environment, the conservation of biodiversity, and human health (Jorrín-Novo & Valledor, 2013). Therefore, "translational proteomics" can be defined as the emerging subdiscipline within the field of proteomics in charge of translating the knowledge generated by basic science to solve problems related to human activity. In this case taken to the field of agronomy and food (Agrawal et al., 2012).

The following sections of this chapter discuss proteomic applications in the cereal and vegetable crops industry in search of qualitative improvement in areas, such as food safety, quality, authenticity, and crop diseases.

4.6.2 Food safety in cereal and vegetal crops

4.6.2.1 Identification of allergens

The plant tissues that we consume are composed of thousands of different proteins and some of these may cause an allergic response in atopic individuals, causing health problems that can even have a fatal outcome. Fortunately, only a small portion of foods contain proteins of this type, including plants, such as peanuts, nuts, wheat, and soybeans, being responsible for around 90% of adverse allergic manifestations (Alonso, Moro Moro, & Múgica García, 2015). These proteins belong to only a few families and superfamilies, including the cupin and prolamine superfamilies (Breiteneder & Radauer, 2004), and their intake leads to a variety of clinical symptoms that mainly affect the gastrointestinal and respiratory tracts, skin, and cardiovascular system (Shahali & Dadar, 2018). The significant level of allergens found in some of the most consumed vegetables, such as tomatoes, lettuce, asparagus, cabbage, celery, fennel, carrots, and artichokes, along with factors, such as the composition of food products, diversification of food sources and environmental practices, could explain the generalization of allergic events around the world in recent decades (Ahsan, Rao, Gruppuso, Ramratnam, & Salomon, 2016).

Early detection of the presence of allergens in plant products is essential for the protection of consumer health and is considered a primary issue for the food industry. Proteomics can help in the search and identification of these compounds and implement methods for their rapid and effective detection in plant materials. Gel-based analysis strategies, including 2-DGE and 2D immunoblotting coupled to MS, the use of HPLC-MS/MS, and the immunoglobulin E (IgE) binding assay of the trypsinized proteome, can provide valuable information on the presence, composition, and nature of food allergens. One of the most universal diet-related adverse reactions in nutrition is the gluten intolerance. This compound is the main storage protein in wheat grain and comprises a mixture of hundreds of proteins, mainly gliadin and glutenin, the intake of which causes the well-known celiac disease that affects to 1% of the Western population. On the other hand, there is what is known as a wheat allergy, which is an IgE-mediated reaction to the insoluble gliadins of this cereal (Biesiekierski, 2017). Importantly, celiac disease is different from wheat allergy. In the first, a specific protein (gluten) triggers a different type of abnormal immune system reaction. These immune problems are considered of special relevance due to the high consumption of wheat in countless prepared foods such as pasta, pastry products (cookies, muffins, or cakes), dairy and meat products.

For the detection of allergens by using proteomics, an analytical strategy is generally applied that includes 2D-PAGE of the protein extracts and subsequent staining of IgE-reactive spots by immunoblotting with sera from allergic individuals. Once IgE-binding proteins are enzymatically digested can be characterized by spectrometric techniques (Mamone, Picariello, Addeo, & Ferranti, 2011). Akagawa et al. (2007) found four allergens in wheat flour previously identified in the bibliography (serpin, R-amylase inhibitor, γ-gliadin, and low molecular weight glutenin) using matrix-assisted laser desorption/ionization (MALDI)-time-of-fligh (TOF)/TOF-MS for the analysis of peptides of IgE-binding proteins previously digested with trypsin and resolved from the flour protein mixture using IEF and lithium dodecyl sulfate-polyacrylamide gel electrophoresis (LDS-PAGE). The high resolving power of this electrophoretic separation technique achieved to identify nine subunits of low molecular weight glutenins as the most predominant IgE antigens, proving to be a suitable proteomic tool for the evaluation of the presence of wheat allergens in food. In a similar study, Šotkovský et al. (2008) also identified the allergen serpin in wheat flour along with 18 others, including α-amylase inhibitors, β-amylase, profilin, β-D-glucan exohydrolase, and 27K protein.

Several research articles have demonstrated the feasibility of proteomics for the detection of gluten contamination in foods. Fiedler, McGrath, Callahan, and Ross (2014) found two wheat peptide markers (W1-RPQQPYPQPQPQY and W4-LQLQPEPQPQLPY) from the /-gliadin protein at a concentration of 10 ppm of wheat flour in oatmeal using a proteomic approach based on 2-DGE gluten separation and subsequent LC-linear trap quadrupole (LQT) MS/MS analysis of the chymotrypsinized spots. By comprising gluten around 10% of the protein in wheat flour, it is possible to detect up to 1 ppm of wheat gluten in oatmeal. In the same line, Martínez-Esteso et al. (2016) used a multienzymatic digestion with lysC, trypsin, and chymotrypsin to cleave the protein fraction of a commercial mixture of flours that included the cereals wheat, rye, barley, oats, rice, and corn and the vegetable soybean. The authors identified 434 peptide sequences using ultraperformace liquid chromatography (UPLC) and TOF-MS, which were subsequently screened by UPLC-ESI-MS/MS based on selected reaction monitoring according to the presence of immunogenic and toxic sequences for those allergic individuals.

Proteomics can be a useful tool in the study of the presence of allergens in genetically modified organisms (GMOs). Introducing exogenous DNA sequences into a plant genome may lead to unwanted side effects, such as altered plantar metabolism, novel fusion proteins, or other pleiotropic effects, which may compromise product safety. This includes the possible appearance of new allergens or toxins (Zolla, Rinalducci, Antonioli, &

Righetti, 2008). Therefore, these products must be assessed in order to address the presence of potentially harmful compounds resulting from gene transformation. A typical example of a genetically modified vegetable crop is soybean. In order to increase the nutritional quality and harvest yield and to eliminate allergens and antinutrients, genetic modification is being implemented in this crop (Natarajan, Luthria, Bae, Lakshman, & Mitra, 2013). On the other hand, soybean is per se responsible of frequent allergic reactions, and for this reason, the analysis of the varieties reached by gene modification in search of new allergens in more than justified. For this purpose, a proteomic approach based on 2-DGE technology, followed by Western blotting with plasma of plant-sensitive individuals and a MS/MS analysis for the identification of protein-IgE bindings could provide valuable information for the evaluation of food safety in transgenic soybeans and other transgenic vegetable crops (Batista, Martins, Jenö, Ricardo, & Oliveira, 2007; Fonseca, Planchon, Renaut, Oliveira, & Batista, 2012). García-Molina et al. (2017) carried out a comparison between two genetically modified wheat lines for the lower expression of gliadin, responsible for allergies, and an unprocessed wild wheat for reference using 2-DGE, followed by RP-HPLC/nano(n)ESI-MS/MS. Products made with low-gliadin flour should always be evaluated since these lines still contain gluten proteins in their composition that triggers celiac disease. On the other hand, the transgenic lines analyzed contain a higher amounts of albumin and globulin proteins, associated with the wheat respiratory allergy or Bakers' asthma, and their consumption by allergy sufferers is not recommended.

The reality is that hardly any new allergens have been identified in transgenic varieties using comparative proteomics. Indeed, most studies in this regard reported similar protein profiles between GMOs and their natural counterparts. It should be noted that many modifications of the proteome are related to conventional reproduction and genotypic variation (Gong & Wang, 2013).

4.6.2.2 *Identification of pathogens*

The presence and diversity of microorganisms in crops is determined by different factors, such as its morphology and metabolic function, and other extrinsic to the crop, such as the type of production, agronomic practices, geographic location, and weather conditions (Ramos, Miller, Brandão, Teixeira, & Silva, 2013). Part of this microbiota may be pathogenic and cause public health problems if the appropriate measures are not taken. These microorganisms can reach crops in different ways. Irrigation water or that from surface runoff is a frequent source of biological contamination. Other routes for the arrival of pathogens to crops are the application of raw or poorly composted animal remains to the soil and the presence of wastewater and insects. On the other hand, a defective post-harvest handling (e.g., involuntary cuts, non-potable water) also opens the door to potentially dangerous microorganisms for health (Berger et al., 2010). Coliform bacteria are indicators of fecal contamination and are often found responsible for outbreaks in vegetal products (Buchholz et al., 2011; Grant et al., 2008; Mulchandani et al., 2021; Shin et al., 2016). Falomir, Gozalbo, and Rico (2010) found the presence of coliforms in half of the analyzed vegetables, collected directly from cultivated lands and from supermarkets and greengrocer's shops, identifying species that included the enterobacteria *Klebsiella pneumoniae, Klebsiella oxytoca, Serratia marcescens, Serratia rubidaea, Enterobacter cloacae, Kluyvera ascorbata,* and *Pantoea agglomerans,* and other species, such as *Acinetobacter baumannii* and *Stenotrophomonas maltophilia.* The Shigella species is another pathogen also belonging to this group of coliform bacteria, which together with *Salmonella* are frequently found in plant crops, such as scallions and parsley, and tomato, cantaloupe, and mamey, respectively (D'Alessandro & Zolla, 2012).

The early detection of these pathogenic microorganisms is essential to avoid the arrival of contaminated vegetables at the points of sale, causing infections and possible outbreaks. This early detection can be carried out by identifying specific proteins of the altering pathogens. These proteins can be used as fingerprints of each microorganism to establish differences between groups of pathogens, classifying them by species (Fratamico & Gunther, 2013). Proteomics offers the possibility to establish rapid and effective protocols for action in a global market with a fast and massive distribution of food products.

Genomic information based on the 16S ribosomal RNA gene sequence is used for microbial differentiation, since it is a highly conserved area of prokaryotic DNA and allows differentiations between microbial groups. However, this microbial differentiation methodology has limitations. Regardless of the applied technique, the genomic information obtained needs to be compared with a database of individually characterized strains. On the other hand, each technique used has a limit to define taxonomically the microorganism studied. That is, 16S rDNA sequencing is optimal for assigning the genus or species of a microorganism, but it is not suitable for finer typing (Welker, 2011).

The analysis of proteins contained in living cells can be considered an indirect analysis of the genome, since the sequence of each protein is encoded in the genome. Because microbial identification protocols must be routine, agile, and inexpensive, characterization of the entire proteome is not a feasible practice. The MALDI-TOF-MS technique has met the expectations of proteomic analysts by being able to detect individual proteins directly without a prior separation step (Beavis & Chait, 1990), and thus obtain a fingerprint that allows a taxonomic classification among closely related species. This analytical technique achieves a specificity of between 92 and 98% of correct species identification. A significantly higher result than those reached by commonly applied microbial identification tools such as 16S rARN sequencing (Böhme et al., 2013). MALDI-TOF-MS can identify microorganisms in a few minutes directly from colonies and reduces operational cost and diagnosis time, ensuring reliable and accurate results (Wieser, Schneider, Jung, & Schubert, 2012). This MS-based technique allowed the differentiation of *Escherichia coli* strains based on the variation of a single amino acid residue in the sequence of the biomarker protein called YahO, which resulted in a 1 Da difference in molecular weight. This distinction required the use of a tandem TOF mass analyzer for the analysis of the ions generated by MALDI, and the acquisition of MS/MS data of the intact proteins to detect this tiny difference between the protein profiles of *E. coli* O157:H7 from others *E.coli* non-O157:H7 (Fagerquist et al., 2010). Fagerquist et al. (2014) used the same proteomic approach for the analysis of 26 Shiga toxin-producing *Escherichia coli* strains for Shiga toxin 2 production, resulting in fast and highly specific identification.

Despite the benefits offered using MALDI-TOF-MS in microbial identification, the previous steps in the work protocols that include this spectrometric technique are time-consuming as they require the isolation of bacterial strains. On the other hand, the large number of overlapping signal ions precludes their use for mixed culture analysis (Böhme, Fernández-No, Calo-Mata, & Barros-Velázquez, 2017). Calo-Mata et al. (2016) used LC-ESI-MS/MS to identify the proteins cysteine synthase and cysteine proteases stophapain A and stophapain B, and 1-pyrroline-5-carboxylate dehydrogenase as potential biomarkers for *Staphylococcus aureus* differentiation. The sequence peptides AQKPVDNITQIIGGTPVVK, TESIPTGNNVTQLK, MTTYNEVDNLTK, YTINVSSFLSK, and KTGSPDYLLHFLEQKV were found specific for this bacterium and might be used to the fast detection of pathogens directly in food products without prior cultivation, which would be an advantage for the detection of those non-culturable microorganisms.

The invasion of crops by fungi can pose a serious health risk, especially the presence of mycotoxins in cereals and feed. These toxic compounds are secondary metabolites of fungi and their occurrence in agricultural foodstuff is influenced by factors, such as region, season, and growing, harvesting, and storage conditions. The intake of plant products contaminated with mycotoxins induces a series of harmful effects in humans and animals ranging from chronic disease and interference with reproductive efficiency to death (Dhumal & Salunkhe, 2019). Therefore, the presence of fungi in crops can be dangerous and proteomics, as in the case of bacteria, can help identify their presence. Wigmann, Behr, Vogel, and Niessen (2019) applied MALDI-TOF-MS to distinguish members belonging to the fungus species *Fusarium fujikuroi*, some of which are producers of mycotoxins. The target was to differentiate closely related microorganisms and to validate the efficacy of a standardized protocol by identifying field isolates that meet the morphological characteristics of the species. The authors created a database and demonstrated its discriminative power with 94.61% correct identifications to the species level. Similarly, Quéro et al. (2019) developed a database for the correct identification of filamentous fungi responsible for the decomposition of food, which is frequently involved in food spoilage and cause significant losses and substantial economic damage. MALDI-TOF-MS has once again become a key tool in the differentiation of fungal species. Specifically filamentous fungi. After applying a standardized extraction protocol, the authors acquired 6477 spectra of 618 strains of fungi belonging to 136 species. The performance of the database achieved was then evaluated by cross-validation, achieving approximately 95% correct identification to the species level. The database was also contrasted with external isolates belonging to 52 species, being able to identify 90% of the fungi to the species level.

4.6.3 Consumer protection against labeling fraud

At present, the labeling of food products is governed by strict laws. However, the incessant demand for food still causes fraud in the nutritional declaration of some products, favored by the weak regulation in certain countries, their ineffectiveness when applying the appropriate legislation, or the existence of poorly structured legal systems (Ehmke, Bonanno, Boys, & Smith, 2019). Sometimes, fraud can be accidental due to poor processing hygiene that can lead to product adulteration (Ayza & Belete, 2015). When a product is adulterated, its composition is modified to the detriment of the consumer, whose health is affected by being deprived of the necessary nutrients or even suffering from intoxication or allergy symptoms (Bansal, Singh, Mangal, Mangal, & Kumar, 2017). A common type of adulteration in the food industry is the replacement of expensive flours with cheaper ones. The wheat *Triticum aestivum* L., commonly known as common wheat, is sometimes used to replace durum wheat (*Triticum turgidum* ssp. *durum* L.), used in the production of pasta, and more expensive, in the elaboration of bakery products with the intention of lowering costs. In the discovery of these adulterations, a non-gel-based bottom–up approach using MS/MS has been reported as adequate, offering good selectivity and sensitivity. By multiple reaction monitoring (MRM)-based UPLC-ESI-MS/MS a specific marker was found for the selective identification of common wheat. This marker corresponded to a peptide derived from the tryptic digestion of the Pin-a protein, rich in cysteine and present in common wheat but not in durum (Russo et al., 2014). Bönick, Huschek, and Rawel (2017) carried out the same analytical approach focused on ensuring the authenticity of wheat, rye, and spelt flours in bread making. The specific peptide markers found for each cereal consisted of the amino acid sequences SQQQISQQPQQLPQQQQIPQQPQQF, QQHQIPQQPQQFPQQQQF, and QPHQPQQPYPQQ for wheat, ASIVVGIGGQ and SQQPGQIIPQQPQQPSPL for spelt, and LPQSHKQHVGQGAL, AQVQGIIQPQQL, and QQFPQQPQQSFPQQPQQPVPQQPL for rye. The addition of 0.5%–1% of these flours in bakery products was enough for their detection. On the other hand, the development of methodologies aimed at discovering a higher number of unique peptides would further facilitate the differentiation among closely related species.

The genetic transformation of crops is a well-established practice today and widely used for the introduction of desired traits in plants (Mall, Han, Tagliani, & Christensen, 2018). The artificial insertion of genes from another plant provides advantages that include the improvement in the useful life, higher yield, better quality, resistance to pests, tolerance to adverse climatic conditions, and ultimately to both biotic and abiotic stress situations (Jhansi Rani & Usha, 2013). Regulations around these GMOs are being implemented in more and more countries that trade these products. Laws have been established in the European Union (EU) that regulate aspects such as risk assessment, marketing, labeling, and traceability (Valdés, Simó, Ibáñez, & García-Cañas, 2013). Regarding the declaration of ingredients in GMO products, the EU regulation demands special labeling for those foods and feed in which transgenic ingredients account for more than 0.9% of the total (European Parliament, 2003). Despite all the benefits mentioned above, the production of these organisms might generate new proteins or the overproduction of existing ones with a negative impact on consumers' health, acting as toxins, antinutrients, or allergens. Comparing the proteomes of transgenic crops and control lines, the unwanted effects on the gene expression can be evaluated (Tan et al., 2019). In this regard, comparative proteomics is providing excellent results in evaluating GMOs at the protein level. Many studies in the last decade have successfully used a strategy based on protein differentiation and identification using 2-DGE and MS-based techniques in different cereals and plant crops, such as rice (Gong, Li, Yu, Wang, & Wang, 2012; Xue, Yang, Liu, & Xue, 2012), maize (Tan et al., 2017; Vidal, Barbosa, Jacob, & Arruda, 2015; Zolla et al., 2008), beans (Balsamo, Valentim-Neto, Mello, & Arisi, 2015), soybean (Barbosa, Arruda, Azevedo, & Arruda, 2012; Brandão, Barbosa, & Arruda, 2010), potato (Khalf et al., 2010), and tomato (Mora, Bramley, & Fraser, 2013).

4.6.4 Early detection and treatment of infected crops

Pathogens and pests negatively affect crop yields and the quality of agricultural production, reducing food safety and causing large economic losses (Savary et al., 2019). The yield loss in crops affected by pathogens ranges from 20% to 40%. In some important cereals, such as maize, barley, and rice, losses of around 12% due to pathogens have been recorded. About 12% of peanut crops are discarded for this reason, and the losses in wheat and cotton crops even reach 50% and 80%, respectively (Fang & Ramasamy, 2015). These figures offer a perspective on the seriousness of the problem and the imperative of developing methods for the early detection of these diseases, which include fungi, bacteria, viruses, and nematodes. In the fight for their survival, plants carry out a series of physiological, biochemical, cellular, and molecular modifications that lead to the generation of compounds that are responsible for reinforcing the walls of the cells that surround the infection and thus create a barrier that stops the expansion of the invading organism. Another strategy of plants to avoid the spread of infection is to activate mechanisms aimed at restricting their own growth. The plant can even develop a type of enhanced resistance to protect itself for a period ranging from several months to the full season (Fang et al., 2015). Therefore, both early detection and treatment of crop-related diseases are of utmost importance. Proteomics contributes with tools capable of identifying the genetic production of proteins responsible for the pathogenicity and virulence of infectious organisms. Knowledge of the mechanisms involved in infection would be useful in diagnosing diseases and protecting crops through the development of agrochemicals (Gonzalez-Fernandez & Jorrin-Novo, 2012). Different research studies in the last decade have contributed to the analysis of the changes produced at the protein level in the cells of infected plants (Table 4.2). Barnabas et al. (2016) investigated the proteome alterations that occurred in the meristem of a susceptible sugarcane cultivar infected with the fungus *Sporisorium scitamineum*. Differentially abundant proteins were identified by 2-DGE coupled with MALDI-TOF/TOF-MS, of which 53 proteins were involved in defense, stress, metabolism, protein folding, energy, and cell division. Similarly, Lin et al. (2020) studied the changes produced in sweet potato proteome as a result of infection with the *Fusarium fungus*, which is usually responsible for numerous agricultural losses. Applying a non-gel-based bottom–up proteomic approach using RP-HPLC followed by LC-MS/MS the authors obtained valuable data on the type of proteins involved in plant defense, such as those associated with signaling transduction, plant resistance, chitinase, and subtilisin-like protease.

On the other hand, knowledge of the essential pathogenicity/virulence factors for fungal infections is very important because it represents the targets that research works must attack in the struggle against these microorganisms (Acero et al., 2011). Manikandan, Harish, Karthikeyan, and Raguchander (2018) studied the possible mechanisms of virulence of fungal *Fusarium oxysporum* f.sp *lycopersici* in tomato. Analysis of 2-DGE revealed the expression of 17 differential proteins between two isolates of different virulence, of which 10 were showed to be involved in

TABLE 4.2 Proteomic approaches in the study of the interactions between pathogenic microorganisms and plants.

Microorganism	Plant	Purpose of the study	Techniques used	New and/or differentially expressed proteins identified in the plant	References
Bacteria					
Flavescence dorée phytoplasma	Grapevines	Identification of changes in the host proteome after infection	IEF/SDS-PAGE and MALDI-TOF/TOF MS/MS	33	Margaria & Palmano (2011)
Pseudomonas syringae	Tomato	Identification of changes in the proteome of a resistant and a susceptible cultivar	[a]iTRAQ HPLC-TOF MS/MS	477	Parker et al. (2013)
Herbaspirillum seropedicae	*Oryza sativa* ssp. *japonica* cv.Nipponbare	Identification of changes in the root host proteome after infection	IEF/SDS-PAGE and MALDI-TOF/TOF MS/MS	11	Alberton et al. (2013)
Herbaspirillum seropedicae	Maize	Identification of changes in the root host proteome after infection	IEF/SDS-PAGE and MALDI-TOF MS	18	Ferrari et al. (2014)
Xanthomonas oryzae pv. *oryzae*	Rice	Identification of changes in the leaf proteome of a resistant and a susceptible genotype	IEF/SDS-PAGE and MALDI-TOF/TOF MS-MS/MS	29	Kumar et al. (2015)
Ralstonia solanacearum	Potato	Identification of resistance genes and proteins in the host	IEF/SDS-PAGE and MALDI-TOF/TOF MS/MS	8	Park et al. (2016)

Xanthomonas oryzae pv. *oryzae*	Rice	Identification of changes in the leaf host proteome after infection	IEF/SDS-PAGE and MALDI-TOF/TOF MS	5	Sruthilaxmi and Babu (2020)
Acidovorax avenae subsp. avenae	Sugarcane	Identification of changes in the proteome of a resistant and a susceptible cultivar	IEF/SDS-PAGE and [a]iTRAQ UPLC-Q/Orbitrap MS/MS	36	Zhou et al. (2021)
Fungus					
Alternaria alternata	*Mentha arvensis*	Identification of changes in the leaf host proteome after infection	IEF/SDS-PAGE and MALDI TOT/TOF MS/MS	45	Sinha and Chattopad-hyay (2011)
Mycocentrospora acerina	Carrot	Identification of changes in the root proteome of an organic and a conventional crop	IEF/SDS-PAGE and MALDI TOT/TOF MS/MS	33	Louarn et al. (2012)
Colletotrichum fragariae	Strawberry	Identification of changes in the leaf proteome after infection	IEF/SDS-PAGE and MALDI TOT/TOF MS/MS	49	Fang et al. (2012)
Ganoderma boninense	Oil palm	Identification of changes in the leaf proteome after infection	IEF/SDS-PAGE and LC-[b]QTOF MS-MS/MS	51	Daim et al. (2015)
Blumeria graminis f. sp. *tritici*	Wheat	Identification of changes in the seedling leaf proteome of a resistant line after infection	[a]iTRAQ nHPLC-ESI-Q/Orbitrap MS-MS/MS	394	Fu et al. (2016)
Fusarium oxysporum f. sp. *ciceri* Race 1	Chickpea	Identification of changes in the root proteome of a resistant and a susceptible genotype	IEF/SDS-PAGE and MALDI-TOF/TOF MS-MS/MS	100	Chatterjee et al. (2014)

(Continued)

TABLE 4.2 Cont'd

Microorganism	Plant	Purpose of the study	Techniques used	New and/or differentially expressed proteins identified in the plant	References
Sporisorium scitamineum	Sugarcane	Identification of changes in the proteome of plant meristem after infection	IEF/SDS-PAGE and MALDI-TOF/TOF MS/MS	53	Barnabas et al. (2016)
Rhizoctonia solani	Cotton	Identification of changes in the proteome of a moderately resistant cultivar after infection	IEF/SDS-PAGE and [a]iTRAQ nHPLC-Obitrap MS/MS	174	Zhang et al. (2017)
Virus					
Maize rough dwarf disease	Maize	Identification of changes in the host proteome after infection	IEF/SDS-PAGE and MALDI-TOF/TOF MS/MS	91	Li, Xu, and Zhang (2011)
Papaya meleira virus	Papaya latex	Identification of changes in the latex host proteome after infection	IEF/SDS-PAGE and nHPLC-ESI MS/MS	160	Rodrigues et al. (2012)
Sugarcane mosaic virus	Maize	Identification of changes in the leaf proteome of a resistant and a susceptible ecotype	IEF/SDS-[c]DIGE and MALDI-TOF/TOF MS-MS/MS	93	Wu et al. (2013)
Rice stripe virus	Rice	Identification of changes in the leaf host proteome after infection	[a]iTRAQ HPLC-MS	681	Wang, Hajano, Ren, Lu, and Wang (2015)

Melon necrotic spot virus	Melon	Identification of changes in host phloem sap proteome after infection.	IEF/SDS-[c]DIGE, MALDI-TOF/TOF MS-MS/MS, nHPLC-nESI-Q/TOF MS/MS and IEF/SDS-PAGE	19	Serra-Soriano, Navarro, Genoves, and Pallás (2015)
Tomato yellow leaf curl virus	Tomato	Identification of changes in the leaf proteome of a resistant and a susceptible cultivar	IEF/SDS-PAGE and MALDI-TOF/TOF MS	86	Huang et al. (2016)
Potato virus Y	Potato	Identification of changes in the leaf host proteome after infection	nLC-Orbitrap [d]LTQ MS	21	Stare, Stare, Weckwerth, Wienkoop, and Gruden (2017)
Tomato yellow leaf curl China virus	Tobacco	Identification of changes in the leaf host proteome after infection	iTRAQ nLC-Q/Orbitrap MS/MS	272	Zhong et al. (2017)

pathogenicity, symptom expression and disease development, sporulation, growth, and higher penetration rate on tomato root tissue.

Another type of proteomic study are focused on the study of the defense mechanisms of varieties resistant to infections. Zhang et al. (2016) carried out a study of the differentially expressed proteins in different varieties of wheat resistant to the fungus *Aspergillus glaucus*. Four proteins associated with antifungal function were identified (peroxidase, endogenous alpha-amylase/subtilisin inhibitor, protein 1 xylanase inhibitor, and puroindoline B). Similarly, the over-accumulation of 15 proteins in a cucumber root resistant to the fungus *Fusarium oxysporum* f. sp. *cucumerium* after its infection helped to improve the understanding of the defense mechanisms against this microorganism. The presence of pathogen-resistant proteins in plants opens the door to the development of new antifungal agents or even the creation of new resistant transgenic varieties.

4.7 Conclusions

The consumption of vegetables represents a fundamental part of the diet and has been associated with a good health state due to their content in starch, fiber, vitamins, minerals, and amino acids, some of which are part of bioactive peptides with discovered healthy properties. Proteomics opens the door to a series of new technologies in the exploitation of plant resources that could have implications for food safety through a rapid and reliable detection of the presence of allergens or pathogenic microorganisms, the analysis and guarantee of the quality and authenticity of the product, or the early detection of crop diseases and the development of phytoremediation protocols, helping to reduce associated losses. With a disproportionate population growth rate and a continuously increasing demand for food, plant proteomics is presented as a discipline that could guarantee the future supply of agricultural products while safeguarding a balance with the environment. For this, it will be vital to promote the transfer of the knowledge generated to the business sector, which would translate into the creation of more reliable and faster methods of action that would increase the dynamism of the sector. In addition, this approach might create new job positions occupied by qualifying staff with the ability to generate knowledge exchange networks, contributing to the continuous improvement of proteomic methods.

Acknowledgment

Thanks to GAIN (Axencia Galega de Innovación) for supporting this publication (grant number IN607A2019/01).

References

Acero, F. J. F., Carbú, M., El-Akhal, M. R., Garrido, C., González-Rodríguez, V. E., & Cantoral, J. M. (2011). Development of proteomics-based fungicides: New strategies for environmentally friendly control of fungal plant diseases. *International Journal of Molecular Sciences, 12*(1), 795–816. https://doi.org/10.3390/ijms12010795.

Agrawal, G. K., Pedreschi, R., Barkla, B. J., Bindschedler, L. V., Cramer, R., Sarkar, A., et al. (2012). Translational plant proteomics: A perspective. *Journal of Proteomics, 75*(15), 4588–4601. https://doi.org/10.1016/j.jprot.2012.03.055.

Ahsan, N., Rao, R. S. P., Gruppuso, P. A., Ramratnam, B., & Salomon, A. R. (2016). Targeted proteomics: Current status and future perspectives for quantification of food allergens. *Journal of Proteomics, 143*, 15–23. https://doi.org/10.1016/j.jprot.2016.04.018.

Akagawa, M., Handoyo, T., Ishii, T., Kumazawa, S., Morita, N., & Suyama, K. (2007). Proteomic analysis of wheat flour allergens. *Journal of Agricultural and Food Chemistry, 55*(17), 6863–6870. https://doi.org/10.1021/jf070843a.

Alberton, D., Müller-Santos, M., Brusamarello-Santos, L. C. C., Valdameri, G., Cordeiro, F. A., Yates, M. G., et al. (2013). Comparative proteomics analysis of the rice roots colonized by *Herbaspirillum seropedicae* strain SmR1 reveals induction of the methionine recycling in the plant host. *Journal of Proteome Research, 12*(11), 4757–4768. https://doi.org/10.1021/pr400425f.

Alonso, M. A. T., Moro Moro, M., & Múgica García, M. V. (2015). Epidemiology of anaphylaxis. *Clinical and Experimental Allergy, 45*(6), 1027–1039. https://doi.org/10.1111/cea.12418.

Awika, J. M. (2011). Major cereal grains production and use around the world. In J. M. Awika, V. Piironen, & S. Bean (Eds.), *Advances in cereal science: Implications to food processing and health promotion* (pp. 1–13). Washington, DC (USA): American Chemical Society https://doi.org/10.1021/bk-2011-1089.

Ayza, A., & Belete, E. (2015). Food adulteration: Its challenges and impacts. *Food Science and Quality Mangement, 41*, 50–56.

Balsamo, G. M., Valentim-Neto, P. A., Mello, C. S., & Arisi, A. C. M. (2015). Comparative proteomic analysis of two varieties of genetically modified (GM) Embrapa 5.1 common bean (*Phaseolus vulgaris* L.) and their non-gm counterparts. *Journal of Agricultural and Food Chemistry, 63*(48), 10569–10577. https://doi.org/10.1021/acs.jafc.5b04659.

Bansal, S., Singh, A., Mangal, M., Mangal, A. K., & Kumar, S. (2017). Food adulteration: Sources, health risks, and detection methods. *Critical Reviews in Food Science and Nutrition, 57*(6), 1174–1189. https://doi.org/10.1080/10408398.2014.967834.

Barbosa, H. S., Arruda, S. C. C., Azevedo, R. A., & Arruda, M. A. Z. (2012). New insights on proteomics of transgenic soybean seeds: Evaluation of differential expressions of enzymes and proteins. *Analytical and Bioanalytical Chemistry, 402*(1), 299–314. https://doi.org/10.1007/s00216-011-5409-1.

Barnabas, L., Ashwin, N. M. R., Kaverinathan, K., Trentin, A. R., Pivato, M., Sundar, A. R., et al. (2016). Proteomic analysis of a compatible interaction between sugarcane and *Sporisorium scitamineum*. *Proteomics, 16*(7), 1111–1122. https://doi.org/10.1002/pmic.201500245.

Batista, R., Martins, I., Jenö, P., Ricardo, C. P., & Oliveira, M. M. (2007). A proteomic study to identify soya allergens—the human response to transgenic versus non-transgenic soya samples. *International Archives of Allergy and Immunology, 144*(1), 29–38. https://doi.org/10.1159/000102611.

Beavis, R. C., & Chait, B. T. (1990). Rapid, sensitive analysis of protein mixtures by mass spectrometry (sinapinic acid/laser desorption/apolipoproteins/milk). In , *Proceedings of the National Academy of Sciences of the United States of America* (*87*, 6873–6877.

Belitz, H.-D., Grosch, W., & Schieberle, P. (2009). Cereals and cereal products. In *Food chemistry* (pp. 670–745). New York, NY (USA): Springer. https://doi.org/10.1007/978-3-540-69934-7_16.

Bendixen, E. (2013). Understanding the proteome. In F. Toldrá, & L. M. L. Nollet (Eds.), *Proteomics in Food: Principles and Applications* (pp. 3–19). New York, NY (USA): Springer.

Berger, C. N., Sodha, S. V., Shaw, R. K., Griffin, P. M., Pink, D., Hand, P., et al. (2010). Fresh fruit and vegetables as vehicles for the transmission of human pathogens. *Environmental Microbiology, 12*(9), 2385–2397. https://doi.org/10.1111/j.1462-2920.2010.02297.x.

Biesiekierski, J. R. (2017). What is gluten? *Journal of Gastroenterology and Hepatology, 32*, 78–81. Australia. https://doi.org/10.1111/jgh.13703.

Blair, M. W., Wu, X., Bhandari, D., Zhang, X., & Hao, J. (2016). Role of legumes for and as horticultural crops in sustainable agriculture. In D. Nandwani (Ed.). *Organic farming for sustainable agriculture. Sustainable development and biodiversity* (*9* (pp. 185–211). New York, NY (USA): Springer https://doi.org/10.1007/978-3-319-26803-3_9.

Böhme, K., Fernández-No, I. C., Calo-Mata, P., & Barros-Velázquez, J. (2017). Proteomics of food spoilage pathogens. In M. L. Colgrave (Ed.), *Proteomics in food science: From farm to fork* (pp. 417–431). Amsterdam (Netherlands): Elsevier Inc https://doi.org/10.1016/B978-0-12-804007-2.00025-4.

Böhme, K., Fernández-No, I. C., Pazos, M., Gallardo, J. M., Barros-Velázquez, J., Cañas, B., et al. (2013). Identification and classification of seafood-borne pathogenic and spoilage bacteria: 16S rRNA sequencing versus MALDI-TOF MS fingerprinting. *Electrophoresis, 34*(6), 877–887. https://doi.org/10.1002/elps.201200532.

Bönick, J., Huschek, G., & Rawel, H. M. (2017). Determination of wheat, rye and spelt authenticity in bread by targeted peptide biomarkers. *Journal of Food Composition and Analysis, 58*, 82–91. https://doi.org/10.1016/j.jfca.2017.01.019.

Brandão, A. R., Barbosa, H. S., & Arruda, M. A. Z. (2010). Image analysis of two-dimensional gel electrophoresis for comparative proteomics of transgenic and non-transgenic soybean seeds. *Journal of Proteomics, 73*(8), 1433–1440. https://doi.org/10.1016/j.jprot.2010.01.009.

Branlard, G., & Bancel, E. (2007). Protein extraction from cereal seeds. In H. Thiellement, M. Zivy, C. Damerval, & V. Méchin (Eds.). *Plant proteomics* (*355* (pp. 15–25). Totowa, NJ (USA): Humana Press https://doi.org/10.1385/1-59745-227-0:15.

Breiteneder, H., & Radauer, C. (2004). A classification of plant food allergens. *Journal of Allergy and Clinical Immunology, 113*(5), 821–830. https://doi.org/10.1016/j.jaci.2004.01.779.

Buchholz, U., Bernard, H., Werber, D., Böhmer, M. M., Remschmidt, C., Wilking, H., et al. (2011). German outbreak of *Escherichia coli* O104:H4 associated with Sprouts. *New England Journal of Medicine, 365*(19), 1763–1770. https://doi.org/10.1056/nejmoa1106482.

Buszewski, B., Szumski, M., & Sus, S. (2002). Methacrylate-based monolithic columns for micro-HPLC and CEC. *Lc Gc Europe, 15*, 792–798.

Butnariu, M., & Butu, A. (2015). Chemical composition of vegetables and their products. In P. T. K. Cheung, & B. M. Metha (Eds.), *Handbook of food chemistry* (pp. 627–697). Berlin Heidelberg: Springer https://doi.org/10.1007/978-3-642-36605-5_17.

Calo-Mata, P., Carrera, M., Böhme, K., Caamaño-Antelo, S., Gallardo, J. M., Barros-Velázquez, J., et al. (2016). Novel peptide biomarker discovery for detection and identification of bacterial pathogens by LC-ESI-MS/MS. *Journal of Analytical & Bioanalytical Techniques, 7*(1). 10.4172/2155-9872.1000296.

Catherman, A. D., Skinner, O. S., & Kelleher, N. L. (2014). Top down proteomics: Facts and perspectives. *Biochemical and Biophysical Research Communications, 445*(4), 683–693. https://doi.org/10.1016/j.bbrc.2014.02.041.

Caunii, A., Cuciureanu, R., Zakar, A. M., Tonea, E., & Giuchici, C. (2010). Chemical composition of common leafy vegetables. Studia Universalis "Vasile Goldis". *Seria Stiintele Vietti, 20*(2), 45–48.

Chatterjee, M., Gupta, S., Bhar, A., Chakraborti, D., Basu, D., & Das, S. (2014). Analysis of root proteome unravels differential molecular responses during compatible and incompatible interaction between chickpea (*Cicer arietinum* L.) and *Fusarium oxysporum* f. sp. *ciceri* Race1 (Foc1). *Bmc Genomics [Electronic Resource], 15*(1), 1–19. 10.1186/1471-2164-15-949.

Chen, B., Brown, K. A., Lin, Z., & Ge, Y. (2018). Top-down proteomics: Ready for prime time? *Analytical Chemistry, 90*(1), 110–127. https://doi.org/10.1021/acs.analchem.7b04747.

Cunsolo, V., Muccilli, V., Saletti, R., & Foti, S. (2014). Mass spectrometry in food proteomics: A tutorial. *Journal of Mass Spectrometry, 49*(9), 768–784. https://doi.org/10.1002/jms.3374.

Daim, L. D. J., Ooi, T. E. K., Ithnin, N., Mohd Yusof, H., Kulaveerasingam, H., Majid, N. A., et al. (2015). Comparative proteomic analysis of oil palm leaves infected with *Ganoderma boninense* revealed changes in proteins involved in photosynthesis, carbohydrate metabolism, and immunity and defense. *Electrophoresis, 36*(15), 1699–1710. https://doi.org/10.1002/elps.201400608.

D'Alessandro, A., & Zolla, L. (2012). We are what we eat: Food safety and proteomics. *Journal of Proteome Research, 11*(1), 26–36. https://doi.org/10.1021/pr2008829.

Dhumal, S. S., & Salunkhe, D. K. (2019). Mycotoxins in foods. In A. T. Tu (Ed.), *Food poisoning* (pp. 291–334). Abingdon, Oxfordshire (UK): Routledge.

Dupree, E. J., Jayathirtha, M., Yorkey, H., Mihasan, M., Petre, B. A., & Darie, C. C. (2020). A critical review of bottom-up proteomics: The good, the bad, and the future of this field. *Proteomes, 8*(3), 1–26. https://doi.org/10.3390/proteomes8030014.

Ehmke, M. D., Bonanno, A., Boys, K., & Smith, T. G. (2019). Food fraud: Economic insights into the dark side of incentives. *Australian Journal of Agricultural and Resource Economics, 63*(4), 685–700. https://doi.org/10.1111/1467-8489.12346.

European Parliament, E. C. (2003). Regulation (EC) No 1830/2003 of September 2003 converning the traceability and labelling of genetically modified organisms and the traceability of food and feed products produced from genetically modified organisms and amending Directive 2001/18/EC. *Journal of European Union, L268*, 24–28.

Evers, A. D., Blakeney, A. B., & O'Brien, L. (1999). Cereal structure and composition. *Australian Journal of Agricultural Research, 50*(5), 629–650. https://doi.org/10.1071/AR98158.

Fagerquist, C. K., Garbus, B. R., Miller, W. G., Williams, K. E., Yee, E., Bates, A. H., et al. (2010). Rapid identification of protein biomarkers of *Escherichia coli* O157:H7 by matrix-assisted laser desorption Lonization-time-of-flight-time-of-flight mass spectrometry and top-down proteomics. *Analytical Chemistry, 82*(7), 2717–2725. https://doi.org/10.1021/ac902455d.

Fagerquist, C. K., Zaragoza, W. J., Sultan, O., Woo, N., Quiñones, B., Cooley, M. B., et al. (2014). Top-down proteomic identification of shiga toxin 2 subtypes from Shiga toxin-producing *Escherichia coli* by matrix-assisted laser desorption ionization-tandem time of flight mass spectrometry. *Applied and Environmental Microbiology, 80*(9), 2928–2940. https://doi.org/10.1128/AEM.04058-13.

Falomir, M. P., Gozalbo, D., & Rico, H. (2010). Coliform bacteria in fresh vegetables: From cultivated lands to consumers. In A. Méndez-Vilas (Ed.), *Current Research, Technology and Education Topics in Applied Microbiology and Microbial Biotechnology* (pp. 1175–1181). Badajoz, Extremadura (SPAIN): Formatex Research Center.

Fang, X., Chen, J., Dai, L., Ma, H., Zhang, H., Yang, J., et al. (2015). Proteomic dissection of plant responses to various pathogens. *Proteomics, 15*(9), 1525–1543. https://doi.org/10.1002/pmic.201400384.

Fang, X., Chen, W., Xin, Y., Zhang, H., Yan, C., Yu, H., et al. (2012). Proteomic analysis of strawberry leaves infected with *Colletotrichum fragariae*. *Journal of Proteomics, 75*(13), 4074–4090. https://doi.org/10.1016/j.jprot.2012.05.022.

Fang, Y., & Ramasamy, R. P. (2015). Current and prospective methods for plant disease detection. *Biosensors, 5*(3), 537–561. https://doi.org/10.3390/bios5030537.

Ferrari, C. S., Amaral, F. P., Bueno, J. C. F., Scariot, M. C., Valentim-Neto, P. A., & Arisi, A. C. M. (2014). Expressed Proteins of *Herbaspirillum seropedicae* in Maize (DKB240)

Roots-Bacteria Interaction Revealed Using Proteomics. *Applied Biochemistry and Biotechnology, 174*(6), 2267–2277. https://doi.org/10.1007/s12010-014-1197-3.

Fiedler, K. L., McGrath, S. C., Callahan, J. H., & Ross, M. M. (2014). Characterization of grain-specific peptide markers for the detection of gluten by mass spectrometry. *Journal of Agricultural and Food Chemistry, 62*(25), 5835–5844. https://doi.org/10.1021/jf500997j.

Fonseca, C., Planchon, S., Renaut, J., Oliveira, M. M., & Batista, R. (2012). Characterization of maize allergens - MON810 vs. its non-transgenic counterpart. *Journal of Proteomics, 75*(7), 2027–2037. https://doi.org/10.1016/j.jprot.2012.01.005.

Fox, P. F., & Brodkorb, A. (2008). The casein micelle: Historical aspects, current concepts and significance. *International Dairy Journal, 18*(7), 677–684. https://doi.org/10.1016/j.idairyj.2008.03.002.

Fratamico, P., & Gunther, N. W. (2013). *Advances in genomics and proteomics-based methods for the study of foodborne bacterial pathogens*. In J. Sofos (Ed.). *Advances in Microbial Food Safety* (*1* (pp. 462–497). Sawston, Cambridgeshire (UK): Woodhead Publishing https://doi.org/10.1533/9780857098740.6.462.

Friedman, D. B., Hoving, S., & Westermeier, R. (2009). Isoelectric focusing and two-dimensional gel electrophoresis. *Methods in Enzymology, 463*(C), 515–540. https://doi.org/10.1016/S0076-6879(09)63030-5.

Fu, Y., Zhang, H., Mandal, S. N., Wang, C., Chen, C., & Ji, W. (2016). Quantitative proteomics reveals the central changes of wheat in response to powdery mildew. *Journal of Proteomics, 130*, 108–119. https://doi.org/10.1016/j.jprot.2015.09.006.

García-Molina, M. D., Muccilli, V., Saletti, R., Foti, S., Masci, S., & Barro, F. (2017). Comparative proteomic analysis of two transgenic low-gliadin wheat lines and non-transgenic wheat control. *Journal of Proteomics, 165*, 102–112. https://doi.org/10.1016/j.jprot.2017.06.010.

Garfin, D. E. (2003). Two-dimensional gel electrophoresis: An overview. *TrAC—Trends in Analytical Chemistry, 22*(5), 263–272. https://doi.org/10.1016/S0165-9936(03)00506-5.

Gong, C. Y., Li, Q., Yu, H. T., Wang, Z., & Wang, T. (2012). Proteomics insight into the biological safety of transgenic modification of rice as compared with conventional genetic breeding and spontaneous genotypic variation. *Journal of Proteome Research, 11*(5), 3019–3029. https://doi.org/10.1021/pr300148w.

Gong, C. Y., & Wang, T. (2013). Proteomic evaluation of genetically modified crops: Current status and challenges. *Frontiers in Plant Science, 4*. doi:10.3389/fpls.2013.00041 MAR.

Gonzalez-Fernandez, R., & Jorrin-Novo, J. V. (2012). Contribution of proteomics to the study of plant pathogenic fungi. *Journal of Proteome Research, 11*(1), 3–16. https://doi.org/10.1021/pr200873p.

Grant, J., Wendelboe, A. M., Wendel, A., Jepson, B., Torres, P., Smelser, C., et al. (2008). Spinach-associated *Escherichia coli* O157:H7 Outbreak, Utah and New Mexico, 2006. *Emerging Infectious Diseases, 14*(10), 1633–1636. https://doi.org/10.3201/eid1410.071341.

Guerrieri, N., & Cavaletto, M. (2018). Cereals proteins. In R. Y. Yada (Ed.), *Proteins in Food Processing: Second Edition* (pp. 223–244). Amsterdam (Netherlands): Elsevier Inc https://doi.org/10.1016/B978-0-08-100722-8.00009-7.

Guiochon, G. (2007). Monolithic columns in high-performance liquid chromatography. *Journal of Chromatography A, 1168*(1–2), 101–168. https://doi.org/10.1016/j.chroma.2007.05.083.

Horvatovich, P., Hoekman, B., Govorukhina, N., & Bischoff, R. (2010). Multidimensional chromatography coupled to mass spectrometry in analysing complex proteomics samples. *Journal of Separation Science, 33*(10), 1421–1437. https://doi.org/10.1002/jssc.201000050.

Hounsome, N., & Hounsome, B. (2011). Biochemistry of vegetables: Major classes of primary (carbohydrates, amino acids, fatty acids, vitamins, and organic acids) and secondary metabolites (terpenoids, phenolics, alkaloids, and sulfur-containing compounds) in vegetables. In N. K. Sinha (Ed.), *Handbook of vegetables and vegetable processing*. (pp. 23–58). Hoboken, NJ (USA): Wiley-Blackwell. https://doi.org/10.1002/9780470958346.ch2.

Hounsome, N., Hounsome, B., Tomos, D., & Edwards-Jones, G. (2008). Plant metabolites and nutritional quality of vegetables. *Journal of Food Science, 73*(4), R48–R65. https://doi.org/10.1111/j.1750-3841.2008.00716.x.

Huang, Y., Ma, H. Y., Huang, W., Wang, F., Xu, Z. S., & Xiong, A. S. (2016). Comparative proteomic analysis provides novel insight into the interaction between resistant vs susceptible tomato cultivars and TYLCV infection. *BMC Plant Biology, 16*(1). doi:10.1186/s12870-016-0819-z.

Igbabul, B. D., Iorliam, B. M., & Umana, E. N. (2015). Physicochemical and sensory properties of cookies produced from composite flours of wheat, cocoyam and african yam beans. *Journal of Food Research, 4*(2), 150–158. https://doi.org/10.5539/jfr.v4n2p150.

Iqbal, A., Khalil, I. A., Ateeq, N., & Khan, M. S. (2006). Nutritional quality of important food legumes. *Food Chemistry, 97*(2), 331–335. doi:10.1016/j.foodchem.2005.05.011.

Jhansi Rani, S., & Usha, R. (2013). Transgenic plants: Types, benefits, public concerns and future. *Journal of Pharmacy Research, 6*(8), 879–883. https://doi.org/10.1016/j.jopr.2013.08.008.

Jorrin-Novo, J. V., Komatsu, S., Sanchez-Lucas, R., & Rodríguez de Francisco, L. E. (2019). Gel electrophoresis-based plant proteomics: Past, present, and future. Happy 10th anniversary. *Journal of Proteomics! Journal of Proteomics, 198*, 1–10. https://doi.org/10.1016/j.jprot.2018.08.016.

Jorrín-Novo, J., & Valledor, L. (2013). Translational proteomics special issue. *Journal of Proteomics, 93*, 1–4. https://doi.org/10.1016/j.jprot.2013.10.019.

Kaur, K. D., Jha, A., Sabikhi, L., & Singh, A. K. (2014). Significance of coarse cereals in health and nutrition: A review. *Journal of Food Science and Technology, 51*(8), 1429–1441. https://doi.org/10.1007/s13197-011-0612-9.

Kent, N. L. (1983). *Technology of cereals* (3rd ed.). Kidlington, Oxfordshire (UK): Pergamom Press.

Khalf, M., Goulet, C., Vorster, J., Brunelle, F., Anguenot, R., Fliss, I., et al. (2010). Tubers from potato lines expressing a tomato Kunitz protease inhibitor are substantially equivalent to parental and transgenic controls. *Plant Biotechnology Journal, 8*(2), 155–169. https://doi.org/10.1111/j.1467-7652.2009.00471.x.

Kovalenko, I. V., Rippke, G. R., & Hurburgh, C. R. (2006). Determination of amino acid composition of soybeans (*Glycine max*) by near-infrared spectroscopy. *Journal of Agricultural and Food Chemistry, 54*(10), 3485–3491. doi:10.1021/jf052570u.

Kumar, A., Bimolata, W., Kannan, M., Kirti, P. B., Qureshi, I. A., & Ghazi, I. A. (2015). Comparative proteomics reveals differential induction of both biotic and abiotic stress response associated proteins in rice during *Xanthomonas oryzae* pv. *oryzae* infection. *Functional and Integrative Genomics, 15*(4), 425–437. https://doi.org/10.1007/s10142-014-0431-y.

Lal, R. (2016). Feeding 11 billion on 0.5 billion hectare of area under cereal crops. *Food and Energy Security, 5*(4), 239–251. https://doi.org/10.1002/fes3.99.

Lawrance, I. C., Klopcic, B., & Wasinger, V. C. (2005). Proteomics: An overview proteomics: What does it mean? *Inflammatory Bowel Diseases, 11*(10), 927–936. http://www.nigms.nih.gov.

Lee, J., Nam, D. S., & Kong, C. (2016). Variability in nutrient composition of cereal grains from different origins. *SpringerPlus, 5*(1). https://doi.org/10.1186/s40064-016-2046-3.

Li, K., Xu, C., & Zhang, J. (2011). Proteome profile of maize (*Zea Mays* L.) leaf tissue at the flowering stage after long-term adjustment to *rice black-streaked dwarf virus* infection. *Gene, 485*(2), 106–113. https://doi.org/10.1016/j.gene.2011.06.016.

Lin, S. Q., Yang, Z. J., Huang, B. F., Bi, C. Y., Huang, X. F., Chen, G. T., et al. (2020). Comparative proteomic analysis of the sweetpotato provides insights into response mechanisms to *Fusarium oxysporum* f. sp. *batatas*. *Scientific Reports, 10*(1). doi:10.1038/s41598-020-78557-y.

Louarn, S., Nawrocki, A., Edelenbos, M., Jensen, D. F., Jensen, O. N., Collinge, D. B., et al. (2012). The influence of the fungal pathogen *Mycocentrospora acerina* on the proteome and polyacetylenes and 6-methoxymellein in organic and conventionally cultivated carrots (Daucus carota) during post harvest storage. *Journal of Proteomics, 75*(3), 962–977. https://doi.org/10.1016/j.jprot.2011.10.014.

Mall, T., Han, L., Tagliani, L., & Christensen, C. (2018). Transgenic crops: Status, potential, and challenges. In S. S. Gosal, & S. Hussain Wani (Eds.), *Biotechnologies of Crop Improvement, Volume 2: Transgenic Approaches* (pp. 451–485). New York, NY (USA): Springer International Publishing https://doi.org/10.1007/978-3-319-90650-8_16.

Mamone, G., Picariello, G., Addeo, F., & Ferranti, P. (2011). Proteomic analysis in allergy and intolerance to wheat products. *Expert Review of Proteomics, 8*(1), 95–115. https://doi.org/10.1586/epr.10.98.

Manikandan, R., Harish, S., Karthikeyan, G., & Raguchander, T. (2018). Comparative proteomic analysis of different isolates of *Fusarium oxysporum* f.sp. *lycopersici* to exploit the differentially expressed proteins responsible for virulence on tomato plants. *Frontiers in Microbiology, 9*. doi:10.3389/fmicb.2018.00420 MAR.

Margaria, P., & Palmano, S. (2011). Response of the *Vitis vinifera* L. cv. 'Nebbiolo' proteome to Flavescence dorée phytoplasma infection. *Proteomics, 11*(2), 212–224. doi:10.1002/pmic.201000409.

Martínez-Esteso, M. J., Nørgaard, J., Brohée, M., Haraszi, R., Maquet, A., & O'Connor, G. (2016). Defining the wheat gluten peptide fingerprint via a discovery and targeted proteomics approach. *Journal of Proteomics, 147*, 156–168. https://doi.org/10.1016/j.jprot.2016.03.015.

Mistry, K., & Grinberg, N. (2005). Application of monolithic columns in high performance liquid chromatography. *Journal of Liquid Chromatography and Related Technologies, 28*(7–8), 1055–1074. https://doi.org/10.1081/JLC-200052972.

Mora, L., Bramley, P. M., & Fraser, P. D. (2013). Development and optimisation of a label-free quantitative proteomic procedure and its application in the assessment of genetically modified tomato fruit. *Proteomics, 13*(12–13), 2016–2030. https://doi.org/10.1002/pmic.201200480.

Mulchandani, R., Brehmer, C., Butt, S., Vishram, B., Harrison, M., Marchant, E., et al. (2021). Outbreak of Shiga toxin-producing *Escherichia coli* O157 linked with consumption of a fast-food product containing imported cucumbers, United Kingdom, August 2020. *International Journal of Infectious Diseases*, 110, S62–S68. doi:10.1016/j.ijid.2021.04.001 In Press.

Natarajan, S., Luthria, D., Bae, H., Lakshman, D., & Mitra, A. (2013). Transgenic soybeans and soybean protein analysis: An overview. *Journal of Agricultural and Food Chemistry, 61*(48), 11736–11743. https://doi.org/10.1021/jf402148e.

O'Donnell, R., Holland, J. W., Deeth, H. C., & Alewood, P. (2004). Milk proteomics. *International Dairy Journal, 14*(12), 1013–1023. https://doi.org/10.1016/j.idairyj.2004.04.004.

Park, S., Gupta, R., Krishna, R., Kim, S. T., Lee, D. Y., Hwang, D. J., et al. (2016). Proteome analysis of disease resistance against *Ralstonia solanacearum* in potato cultivar CT206-10. *Plant Pathology Journal, 32*(1), 25–32. https://doi.org/10.5423/PPJ.OA.05.2015.0076.

Parker, J., Koh, J., Yoo, M. J., Zhu, N., Feole, M., Yi, S., et al. (2013). Quantitative proteomics of tomato defense against *Pseudomonas syringae* infection. *Proteomics, 13*(12–13), 1934–1946. https://doi.org/10.1002/pmic.201200402.

Petrusan, J.-I., Rawel, H., & Huschek, G. (2016). Protein-rich vegetal sources and trends in human nutrition: A review. *Current Topics in Peptide & Protein Research, 17*, 1–19.

Pomastowski, P., & Buszewski, B. (2014). Two-dimensional gel electrophoresis in the light of new developments. *TrAC - Trends in Analytical Chemistry, 53*, 167–177. https://doi.org/10.1016/j.trac.2013.09.010.

Quéro, L., Girard, V., Pawtowski, A., Tréguer, S., Weill, A., Arend, S., et al. (2019). Development and application of MALDI-TOF MS for identification of food spoilage fungi. *Food Microbiology, 81*, 76–88. https://doi.org/10.1016/j.fm.2018.05.001.

Radovich, T. J. K. (2011). Biology and classification of vegetables. In N. K. Sinha (Ed.), *Handbook of vegetables and vegetable processing* (pp. 3–22). Hoboken, NJ (USA): Blackwell Publishing.

Ramos, B., Miller, F. A., Brandão, T. R. S., Teixeira, P., & Silva, C. L. M. (2013). Fresh fruits and vegetables—an overview on applied methodologies to improve its quality and safety. *Innovative Food Science and Emerging Technologies, 20*, 1–15. https://doi.org/10.1016/j.ifset.2013.07.002.

Rodrigues, S. P., Ventura, J. A., Aguilar, C., Nakayasu, E. S., Choi, H. W., Sobreira, T. J. P., et al. (2012). Label-free quantitative proteomics reveals differentially regulated proteins in the latex of sticky diseased *Carica papaya* L. plants. *Journal of Proteomics, 75*(11), 3191–3198. https://doi.org/10.1016/j.jprot.2012.03.021.

Russo, R., Cusano, E., Perissi, A., Ferron, F., Severino, V., Parente, A., et al. (2014). Ultra-high performance liquid chromatography tandem mass spectrometry for the detection of durum wheat contamination or adulteration. *Journal of Mass Spectrometry, 49*(12), 1239–1246. https://doi.org/10.1002/jms.3451.

Sá, A. G. A., Moreno, Y. M. F., & Carciofi, B. A. M. (2020). Plant proteins as high-quality nutritional source for human diet. *Trends in Food Science and Technology, 97*, 170–184. https://doi.org/10.1016/j.tifs.2020.01.011.

Sarwar, H. (2013). The importance of cereals (Poaceae: Gramineae) nutrition in human health: A review. *Journal of Cereals and Oilseeds, 4*(3), 32–35. https://doi.org/10.5897/jco12.023.

Savary, S., Willocquet, L., Pethybridge, S. J., Esker, P., McRoberts, N., & Nelson, A. (2019). The global burden of pathogens and pests on major food crops. *Nature Ecology and Evolution, 3*(3), 430–439. https://doi.org/10.1038/s41559-018-0793-y.

Savaryn, J. P., Toby, T. K., & Kelleher, N. L. (2016). A researcher's guide to mass spectrometry-based proteomics. *Proteomics, 16*(18), 2435–2443. https://doi.org/10.1002/pmic.201600113.

Serra-Soriano, M., Navarro, J. A., Genoves, A., & Pallás, V. (2015). Comparative proteomic analysis of melon phloem exudates in response to viral infection. *Journal of Proteomics, 124*, 11–24. https://doi.org/10.1016/j.jprot.2015.04.008.

Shahali, Y., & Dadar, M. (2018). Plant food allergy: Influence of chemicals on plant allergens. *Food and Chemical Toxicology, 115*, 365–374. https://doi.org/10.1016/j.fct.2018.03.032.

Shi, Y., Xiang, R., Horváth, C., & Wilkins, J. A. (2004). The role of liquid chromatography in proteomics. *Journal of Chromatography A, 1053*(1–2), 27–36. https://doi.org/10.1016/j.chroma.2004.07.044.

Shin, J., Yoon, K. B., Jeon, D. Y., Oh, S. S., Oh, K. H., Chung, G. T., et al. (2016). Consecutive outbreaks of enterotoxigenic *Escherichia coli* O6 in schools in South Korea caused by contamination of fermented vegetable kimchi. *Foodborne Pathogens and Disease, 13*(10), 535–543. https://doi.org/10.1089/fpd.2016.2147.

Sinha, R., & Chattopadhyay, S. (2011). Changes in the leaf proteome profile of Mentha arvensis in response to *Alternaria alternata* infection. *Journal of Proteomics, 74*(3), 327–336. https://doi.org/10.1016/j.jprot.2010.11.009.

Šotkovský, P., Hubálek, M., Hernychová, L., Novák, P., Havranová, M., Šetinová, I., et al. (2008). Proteomic analysis of wheat proteins recognized by IgE antibodies of allergic patients. *Proteomics, 8*(8), 1677–1691. https://doi.org/10.1002/pmic.200700347.

Sruthilaxmi, C. B., & Babu, S. (2020). Proteome responses to individual pathogens and abiotic conditions in rice seedlings. *Phytopathology, 110*(7), 1326–1341. https://doi.org/10.1094/PHYTO-11-19-0425-R.

Stare, T., Stare, K., Weckwerth, W., Wienkoop, S., & Gruden, K. (2017). Comparison between proteome and transcriptome response in potato (*Solanum tuberosum* L.) leaves following potato virus Y (PVY) infection. *Proteomes, 5*(3). doi:10.3390/proteomes5030014.

Swehata, N. A., Darwish, N., El-Nahry, F., Fatma, A., & Razek, A. A. (1988). Supplementation of wheat flour with some local legumes. *Die Nahrung, 32*, 3–8.

Tacer-Caba, Z., Nilufer-Erdil, D., & Ai, Y. (2015). Chemical composition of cereals and their products. In P. C. K. Cheung, & B. M. Mehta (Eds.), *Handbook of Food Chemistry* (pp. 301–330). Berlin and Heidelberg: Springer https://doi.org/10.1007/978-3-642-36605-5_33.

Tan, Y., Tong, Z., Yang, Q., Sun, Y., Jin, X., Peng, C., et al. (2017). Proteomic analysis of phytase transgenic and non-transgenic maize seeds. *Scientific Reports, 7*(1). doi:10.1038/s41598-017-09557-8.

Tan, Y., Zhang, J., Sun, Y., Tong, Z., Peng, C., Chang, L., et al. (2019). Comparative proteomics of phytase-transgenic maize seeds indicates environmental influence is more important than that of gene insertion. *Scientific Reports, 9*(1). doi:10.1038/s41598-019-44748-5.

Temba, M. C., Njobeh, P. B., Adebo, O. A., Olugbile, A. O., & Kayitesi, E. (2016). The role of compositing cereals with legumes to alleviate protein energy malnutrition in Africa. *International Journal of Food Science and Technology, 51*(3), 543–554. https://doi.org/10.1111/ijfs.13035.

Tyers, M., & Mann, M. (2003). From genomics to proteomics. *Nature, 422*(6928), 193–197.

Unger, K. K., Skudas, R., & Schulte, M. M. (2008). Particle packed columns and monolithic columns in high-performance liquid chromatography-comparison and critical appraisal. *Journal of Chromatography A, 1184*(1–2), 393–415. https://doi.org/10.1016/j.chroma.2007.11.118.

Valdés, A., Simó, C., Ibáñez, C., & García-Cañas, V. (2013). Foodomics strategies for the analysis of transgenic foods. *TrAC—Trends in Analytical Chemistry, 52*, 2–15. https://doi.org/10.1016/j.trac.2013.05.023.

Vidal, N., Barbosa, H., Jacob, S., & Arruda, M. (2015). Comparative study of transgenic and non-transgenic maize (*Zea mays*) flours commercialized in Brazil, focussing on proteomic analyses. *Food Chemistry, 180*, 288–294. https://doi.org/10.1016/j.foodchem.2015.02.051.

Wang, B., Hajano, J. U. D., Ren, Y., Lu, C., & Wang, X. (2015). iTRAQ-based quantitative proteomics analysis of rice leaves infected by *Rice stripe virus* reveals several proteins involved in symptom formation. *Virology Journal, 12*(1). doi:10.1186/s12985-015-0328-y.

Welker, M. (2011). Proteomics for routine identification of microorganisms. *Proteomics, 11*(15), 3143–3153. https://doi.org/10.1002/pmic.201100049.

WHO/FAO/UNU. (2007). Protein and Amino Acid Requirements in Human Nutrition Report of a Joint WHO/FAO/UNU. Expert Consultation. WHO Technical Report Series no (935). Geneva (Switzerland): World Health Organization.

Wieser, A., Schneider, L., Jung, J., & Schubert, S. (2012). MALDI-TOF MS in microbiological diagnostics-identification of microorganisms and beyond (mini review). *Applied Microbiology and Biotechnology, 93*(3), 965–974. https://doi.org/10.1007/s00253-011-3783-4.

Wigmann, É. F., Behr, J., Vogel, R. F., & Niessen, L. (2019). MALDI-TOF MS fingerprinting for identification and differentiation of species within the *Fusarium fujikuroi* species complex. *Applied Microbiology and Biotechnology, 103*(13), 5323–5337. https://doi.org/10.1007/s00253-019-09794-z.

Wilkins, M. R., Pasquali, C., Appel, R. D., Ou, K., Golaz, O., Sanchez, J.-C., et al. (1996). From proteins to proteomes: Large scale protein identification by two-dimensional electrophoresis and arnino acid analysis. *Bio/Technology, 14*(1), 61–65. https://doi.org/10.1038/nbt0196-61.

Wrigley, C. (2017). Cereal-grain morphology and composition. In C. Wrigley, I. Batey, & D. Miskelly (Eds.), *Cereal Grains: Assessing and Managing Quality* (pp. 55–87). Amsterdam: Elsevier Inc. 2nd ed. https://doi.org/10.1016/B978-0-08-100719-8.00004-8.

Wu, L., Han, Z., Wang, S., Wang, X., Sun, A., Zu, X., et al. (2013). Comparative proteomic analysis of the plant-virus interaction in resistant and susceptible ecotypes of maize infected with sugarcane mosaic virus. *Journal of Proteomics, 89*, 124–140. https://doi.org/10.1016/j.jprot.2013.06.005.

Xie, F., Smith, R. D., & Shen, Y. (2012). Advanced proteomic liquid chromatography. *Journal of Chromatography A, 1261*, 78–90. https://doi.org/10.1016/j.chroma.2012.06.098.

Xue, K., Yang, J., Liu, B., & Xue, D. (2012). The integrated risk assessment of transgenic rice *Oryza sativa*: A comparative proteomics approach. *Food Chemistry, 135*(1), 314–318. https://doi.org/10.1016/j.foodchem.2012.04.042.

Zhang, M., Cheng, S. T., Wang, H. Y., Wu, J. H., Luo, Y. M., Wang, Q., et al. (2017). iTRAQ-based proteomic analysis of defence responses triggered by the necrotrophic pathogen *Rhizoctonia solani* in cotton. *Journal of Proteomics, 152*, 226–235. https://doi.org/10.1016/j.jprot.2016.11.011.

Zhang, S. B., Zhai, H. C., Lv, Y. Y., Cai, J. P., Wang, J. S., & Hu, Y. Sen (2016). Proteomic analysis reveals the fungal resistance of soft wheat during storage. *Journal of Stored Products Research, 69*, 195–198. https://doi.org/10.1016/j.jspr.2016.08.010.

Zhang, Y., Fonslow, B. R., Shan, B., Baek, M. C., & Yates, J. R. (2013). Protein analysis by shotgun/bottom-up proteomics. *Chemical Reviews, 113*(4), 2343–2394. https://doi.org/10.1021/cr3003533.

Zhong, X., Wang, Z. Q., Xiao, R., Wang, Y., Xie, Y., & Zhou, X. (2017). iTRAQ analysis of the tobacco leaf proteome reveals that RNA-directed DNA methylation (RdDM) has important roles in defense against geminivirus-betasatellite infection. *Journal of Proteomics, 152*, 88–101. https://doi.org/10.1016/j.jprot.2016.10.015.

Zhou, J. R., Sun, H. D., Ali, A., Rott, P. C., Javed, T., Fu, H. Y., et al. (2021). Quantitative proteomic analysis of the sugarcane defense responses incited by *Acidovorax avenae* subsp. *avenae* causing red stripe. *Industrial Crops and Products, 162*. doi:10.1016/j.indcrop.2021.113275.

Zolla, L., Rinalducci, S., Antonioli, P., & Righetti, P. G. (2008). Proteomics as a complementary tool for identifying unintended side effects occurring in transgenic maize seeds as a result of genetic modifications. *Journal of Proteome Research, 7*(5), 1850–1861. https://doi.org/10.1021/pr0705082.

CHAPTER

5

Proteomic advances in seafood and aquaculture

Robert Stryiński[a], *Elżbieta Łopieńska-Biernat*[a], *Mónica Carrera*[b]

[a]**Department of Biochemistry, Faculty of Biology and Biotechnology, University of Warmia and Mazury in Olsztyn (UWM), Michała Oczapowskiego 1a, Olsztyn, Poland**
[b]**Department of Food Technology, Spanish National Research Council, Marine Research Institute, Vigo, Spain**

5.1 Introduction

Global total production of aquatic organisms (fish, crustaceans, mollusks, and other aquatic animals, but excludes aquatic mammals, reptiles, seaweeds, and other aquatic plants) is estimated to have reached about 179 million tons in 2018 of which 82 million tones came from aquaculture production (FAO 2020). The increasing world population (United Nations; Department of Economic and Social Affairs (Population Division), 2019) has resulted in global food fish consumption increase at an average annual rate of 3.1% from 1961 to 2017, a rate higher than that of all other animal protein foods (meat, dairy, milk, etc.), which increased by 2.1% per year. Per capita food fish consumption grew from 9.0 kg in 1961 to 20.5 kg in 2018. All this translated into aquaculture enormous growth, accounted for 46% of the global aquatic organisms production and 52% of seafood for human consumption (FAO 2020). Moreover, the statistical model projects that the total fish supply and consumption will increase to 186 and 151 million tons in 2030, respectively (The World Bank, 2013).

Due to the feeding and management methods in aquaculture, farmed fish will always suffer losses due to infectious diseases, because they are more susceptible to various bacterial, virus, parasitic, and, to a lesser extent, fungal infections than wild fish (Ananda Raja & Jithendran, 2015;

Food Proteomics
DOI: https://doi.org/10.1016/B978-0-323-90889-4.00005-1

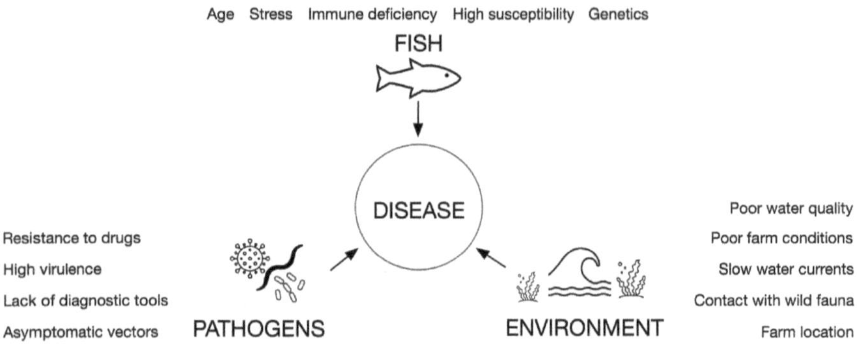

FIG. 5.1 Aquaculture disease triangle, with an indication of main parameters for evaluation of pathogen, host and environment interactions (based on Moreira et al., 2021; Sitjà-Bobadilla & Oidtmann, 2017).

Brugere, Onuigbo, & Morgan, 2017; Sharma, Shrivastav, Sahni, & Pandey, 2012; Waagbø, 2006). According to a group of experts, disease outbreaks in aquaculture are the result of a complex web of interactions in aquatic systems between farmed organism, environment, and possible pathogens, which pose a number of unique challenges to the health of farmed organisms (Freitas, Vaz-Pires, & Câmara, 2020; Hill, 2005; Iwama, 2007; Moreira et al., 2021) (Fig. 5.1).

In order to keep fish farms biologically balanced and to avoid disease outbreaks, and thus ensure quality for the consumer and help businesses to prosper, efforts have been made to integrate various aspects of science, such as pathogenesis, pathogen characterization, and description of its resistance to chemicals, description of hosts' molecular response to pathogens, characterization of host immune system responses, disease diagnosis, and disease biomarkers identification (Freitas et al., 2020; Oidtmann et al., 2014; Scarfe & Palić, 2020; Sitjà-Bobadilla & Oidtmann, 2017). Despite this, the growing need for fish products, and hence the steadily increasing number of fish farms along with the increasing pathogen virulence, demand still more rapid and effective tools for diagnosis and treatment (Burge et al., 2016; Gotesman, Menanteau-Ledouble, Saleh, Bergmann, & El-Matbouli, 2018; Lafferty, 2009; Parrington, 2002). Here, -omic technologies including genomics, metabolomics, and proteomics have become the "golden mean" and offers vast array of solutions used with a high positive impact on the aquaculture industry (Adams & Thompson, 2006; Denslow, Griffitt, & Martyniuk, 2012; Raposo de Magalhães et al., 2020).

In the last two decades, proteomics has become a powerful tool for describing dynamically changing biological systems. Proteomics methods were used for the identification and quantification of the protein composition of cells, sub-fractions of cells, or the medium or secretome surrounding cells, as well as tissue sections and whole organisms at a certain time, collectively termed "the proteome," but also to describe protein modifications and interactions (Gallardo, Ortea, & Carrera, 2013). Scientists distinguished two characteristic approaches using proteomic methods, i.e., discovery and targeted workflows (Aebersold, Bensimon, Collins, Ludwig, & Sabido, 2016; Borràs & Sabidó, 2017; Carrera, Piñeiro, & Martinez, 2020; Gallardo et al., 2013; Martinez, Sánchez-Alonso, Piñeiro, Careche, & Carrera, 2020; Stryiński, Łopieńska-Biernat, & Carrera, 2020). Discovery proteomics is applied to identify and characterize the proteins in the sample (e.g., global proteomes, cellular, subcellular, or excretory-secretory proteomes), while targeted proteomics is monitoring the previously identified protein biomarkers (single or multiple peptides) in analyzed samples. The techniques that proteomics offers are one of those most interesting new tools and approaches used for health management, epidemiology, disease research, and in food science (Ahmed et al., 2019; Cifuentes, 2013; Gallardo et al., 2013; Herrero, Simó, García-Cañas, Ibáñez, & Cifuentes, 2012; Moreira et al., 2021; Rodrigues, Schrama, Campos, Osório, & Freitas, 2012, 2016). It is therefore not surprising that proteomics applications have achieved recognition in the enormously fast-growing industry sector, like fishery and aquaculture.

In this chapter, some of the research regarding the use of proteomics to describe the physiology and pathology of aquatic organisms, its pathogens' characterization, and host–pathogen interactions, will be provided.

5.2 Proteomics of aquatic organisms and their pathogens

5.2.1 Proteomics of aquatic organisms to describe their physiology

Proteome characterization of aquatic organisms is useful for describing their physiology (Ahmed et al., 2019; Sanahuja & Ibarz, 2015). Proteome is dynamic which varies with tissue, physiological conditions, and environment where the organism lives. Global overview on the proteomic profile of specific tissues and secretions provides a complete information about physiological status of aquatic fauna. Recent studies describing protein biology of some aquatic species are presented in Table 5.1.

Proteomic methods were widely used in describing homeostasis of fish and other aquatic organisms. Therefore, research on the proteomic characteristics of skin mucus of healthy fish were performed with a great interest (Rajan et al., 2011). The liquid chromatography coupled with tandem mass spectrometry (LC-MS/MS) was applied on skin mucus of Atlantic cod (*Gadus* morhua) and lumpsucker (*Cyclopterus lumpus*) for characterization of the protein content of skin epidermis (Patel & Brinchmann, 2017; Rajan et al., 2011). Jurado et al. (2015) made a huge contribution to the description of epidermal mucus proteome. Using LC-MS/MS,

TABLE 5.1 Representation of some of the proteomic techniques applied to the characterization of fish physiology.

Species	Tissue/secretions	Proteomic method	References
Atlantic cod (*Gadus morhua*)	skin mucus	2-DE, LC-MS/MS	Rajan et al. (2011)
Common carp (*Cyprinus carpio*)	semen and blood plasma	1-DE SDS-PAGE LC-ESI-MS/MS, 2DE-DIGE LC-MS/MS	Dietrich et al. (2014)
European eel (*Anguilla anguilla*)	swim bladder	LC-MS/MS	Sialana, Schneebauer, Paunkov, Pelster, and Lubec (2018)
Fathead minnow (*Pimephales promelas*)	testes	LC-MS/MS	Martyniuk and Alvarez (2013)
Gilthead sea bream (*Sparus aurata*)	skin mucus	LC-MS/MS and 2-DE-PMF-MS/MS, 2-DE nano-LC-ESI-QTOF-MS/MS	Jurado et al. (2015), Sanahuja and Ibarz (2015)
Gilthead sea bream (*Sparus aurata*)	muscle	nano-LC-ESI-Q-TOF-MS/MS, nano-LC-MS/MS	Addis et al. (2010a), Piovesana et al. (2016)
Gilthead sea bream (*Sparus aurata*)	muscle	2-DE-DIGE MALDI-TOF/TOF-MS	Matos et al. (2013)
Jumbo squid (*Dosidicus gigas*)	skin	LC-MS/MS	Carrera et al. (2019a)
Lamprey (*Lampetra morii*)	buccal glands	nano-LC-MS/MS	Li et al. (2018a)
Lumpsucker (*Cyclopterus lumpus*)	skin mucus	2-DE, LC-MS/MS	Patel and Brinchmann (2017)
Rainbow trout (*Oncorhynchus mykiss*)	sperm	nano-LC-MS/MS	Nynca, Arnold, Fröhlich, Otte, and Ciereszko (2014a)
Rainbow trout (*Oncorhynchus mykiss*)	semen plasma	1-DE and LC-ESI-MS/MS	Nynca et al. (2014b)
Rainbow trout (*Oncorhynchus mykiss*)	ovarian fluid	1-DE LC-MS/MS	Nynca et al. (2015)

Rainbow trout (*Oncorhynchus mykiss*)	head kidney and spleen	nano-LC-MS/MS	Kumar, Hummel, Razzazi-Fazeli, and El-Matbouli (2018)
Scorpionfish (*Scorpaena plumieri*)	skin mucus	nano-LC-MS/MS	Borges et al. (2018)
Three-spined sticklebacks (*Gasterosteus aculeatus*)	gills	LC-MS/MS	Li et al. (2018b)
Tilapia fish (*Oreochromis spp.*)	buccal glands	2-DE MALDI-TOF/TOF	Iq and Shu-Chien (2011)
Zebrafish (*Danio rerio*)	blood plasma	LC-Q-TOF-MS/MS	Li, Tan, Lim, Lin, and Gong (2016)

Proteomic techniques abbreviations: 1/2-DE, one/two-dimensional electrophoresis; SDS-PAGE, sodium dodecyl sulphate-polyacrylamide gel electrophoresis; 2-DE DIGE, two-dimensional difference gel electrophoresis; 2-DE-PMF-MS/MS, two-dimensional gel electrophoresis peptide mass fingerprinting tandem mass spectrometry; iTRAQ, isobaric tag for relative and absolute quantitation; MALDI-TOF-MS, matrix-assisted laser desorption and ionization time-of-flight mass spectrometry; MALDI-TOF/TOF-MS, matrix-assisted laser desorption and ionization (Time-of-Flight)2 mass spectrometry; MALDI-TOF/TOF-MS/MS, matrix-assisted laser desorption and ionization (time-of-flight)2 tandem mass spectrometry; LC-MALDI-TOF/TOF-MS/MS, automated liquid chromatography matrix-assisted laser desorption and ionization (time-of-flight)2 tandem mass spectrometry; nano-LC-MS/MS, nano-scale liquid chromatography tandem mass spectrometry; LC-MS/MS, liquid chromatography tandem mass spectrometry; ESI MS/MS, electrospray ionization tandem mass spectrometry; LC-ESI-MS/MS, liquid chromatography electrospray ionization tandem mass spectrometry; LC-ESI-Q-TOF MS/MS, liquid chromatography electrospray ionization quadrupole time-of-flight tandem mass spectrometry; nano-LC-ESI-Q-TOF MS/MS, nano-scale liquid chromatography and electrospray ionization quadrupole time-of-flight tandem mass spectrometry; LC-ESI-MS/MS, liquid chromatography electrospray ionization tandem mass spectrometry; nano LC-ESI-MS/MS, nano-scale liquid chromatography electrospray ionization tandem mass spectrometry.

structural and metabolic proteins of gilthead sea bream were distinguished. However, Sanahuja and Ibarz (2015) during studies on the same species but using two-dimensional electrophoresis (2-DE) combined with nano liquid chromatography coupled to electrospray quadrupole time-of-flight mass spectrometer (2-DE nano-LC-ESI-QTOF-MS/MS) identified 62 proteins and classified into three main groups: structural, metabolic and protection related group. Recently, a skin of one of the largest cephalopods, Jumbo squid (*Dosidicus gigas*), was characterized (Carrera, Ezquerra-Brauer, & Aubourg, 2019a). A shotgun proteomics analysis (LC-MS/MS) identified in total 219 different proteins. The functional analysis revealed the presence of 11 different protein classes in the jumbo squid skin proteome, e.g., oxidoreductases, nucleic acid binding proteins, hydrolases, calcium-binding proteins, and transferases. Potential new valuable bioactive peptides such as antimicrobial, bioactive collagen peptides, antihypertensive and antitumoral peptides were predicted to be present in the jumbo squid skin proteome. The integration of the global proteomics results and the bioinformatics analysis of the jumbo squid skin proteome showed a comprehensive dataset and improved knowledge about this economically important fishery target (Carrera et al., 2019a).

Proteomics techniques were used for the identification and comprehensive characterization of protein content of healthy fish body fluids (including blood plasma and genital fluids) to understand the molecular mechanism of its function. In this field, group of Professor Andrzej Ciereszko from Polish Academy of Sciences made substantial progress. For understanding the reproductive mechanisms of common carp (*Cyprinus carpio*), a comparative proteome analysis of blood and seminal plasma was performed using one-dimensional sodium dodecyl sulphate-polyacrylamide gel electrophoresis (1-DE SDS-PAGE) and liquid chromatography electrospray ionization tandem mass spectrometry (LC-ESI-MS/MS) (Dietrich et al., 2014). Using this approach, they identified 137 proteins in carp seminal plasma and 88 proteins in carp blood plasma, most of which were newly identified in fish. Transferrin, serine proteinase inhibitors, apolipoproteins, and complement C3 were present in high abundance in carp seminal plasma. In carp blood plasma, besides these proteins, immunoglobulins, and macroglobulins were identified as major proteins (Dietrich et al., 2014). The semen plasma of another species, rainbow trout (*Oncorhynchus mykiss*), was also investigated with the use of the same method. 152 proteins were identified, where the most of them were, such as transferrin, apolipoproteins, complement C3, serum albumin, and hemopexin-, alpha-1-antiproteinase-, and precerebellin-like protein, and were recognized as acute-phase proteins (Nynca et al., 2014b). Another study regarding females unraveled the identity of several proteins not previously reported in the fish ovarian fluid. The proteome of trout ovarian fluid consists of diverse proteins participating in lipid binding and metabolism, carbohydrate and ion transport, innate immunity, maturation, and ovulation processes. What interesting, the most trout ovarian fluid proteins correspond to follicular fluid proteins of higher vertebrates, but 15% of the proteins were found to be different, such as those related to the immune system (precerebellin-like protein), proteolysis (myeloid cell lineage chitinase), or carbohydrate and lipid binding and metabolism (vitellogenins) (Nynca, Arnold, Fröhlich, & Ciereszko, 2015). Shotgun proteomics approach was also applied to testes of fathead minnow. Analysis revealed 1075 proteins including some significantly enriched to be involved in chromatin remodeling, DNA repair, or translation (Martyniuk & Alvarez, 2013).

A gill is a respiratory organ found in many aquatic organisms that extract dissolved oxygen from water and excretes carbon dioxide. This extremely important function prompted researchers to analyze the proteome of the gills from four populations of three-spined sticklebacks with different eco- and morphotypes (Li et al., 2018b). In the study, the population-specific gill proteome of four ecotypes of the same fish species identified by a data-independent acquisition (DIA) was described. The assay library, which consists of 1506 proteins, revealed unique and reproducible proteome signatures, which represent targets for testing molecular mechanisms of gill function. The gill assay library represents a model for other fish species (Li et al., 2018b).

To describe immunology of fish, the proteome profiles of lymphoid organs (head kidney and spleen) were also characterized, what will be discussed in the paragraph about host–pathogen interactions.

All above-mentioned results enhance the basic knowledge concerning protein composition of fish skin, genital fluids, reproductive and lymphoid organs, and their potential role in fish reproduction and immunology, as well as highly important respiratory organs–gills.

5.2.2 Proteomics of aquatic organisms in response to environmental stressors

Proteomic methods have found application in the context of determining changes that occur in aquatic organisms in response to physical and chemical environmental factors. Animal stress and welfare are important for producers, since growth and reproduction, the health status, and diseases susceptibility are strictly related to stress conditions (Conte, 2004; Huntingford et al., 2006). Nevertheless, the scientific efforts around the fish welfare are at an early stage compared with other animals farmed for human consumption (Huntingford et al., 2006).

However, proteomics has found its application here as well. The proteomic response of gilthead sea bream repeatedly handled and crowded at high stocking density revealed a differential expression in about 50%

of all detected proteins (Alves et al., 2010). Similarly, 2-DE MALDI-TOF/TOF MS proteomic approach identified a modulation of 27 proteins in muscles, and seventeen proteins in the blood plasma of the Atlantic salmon subjected to crowding (Veiseth-Kent, Grove, Færgestad, & Fjæra, 2010). Proteomics methods helped to understand the molecular adaptability of fish to live in different environments when temperature and/or salinity change. The long-term exposure to elevated temperature of Tasmanian Atlantic salmon showed quantitative changes in the protein expression (Nuez-Ortín, Carter, Nichols, Cooke, & Wilson, 2018). Many of the proteins modulated here (e.g., disulfide isomerases, peroxiredoxins, and glutathione transferases) were common to the thermal stress response. Such results are of increasing importance in high-temperature production areas. Similar study on *Takifugu fasciatus* liver, but at low temperatures, showed upregulation of the proteins involved in oxidative stress and mitochondrial enzymes as cold-tolerance response mechanism (Wen et al., 2019). Recently, Smith et al. performed a series of experiments on rainbow trout affected with another physical factor. Irradiating eggs at 48 h after fertilization, eyed eggs, yolk sac larvae, and first feeders with X-rays, revealed that at early life stages irradiation can cause sustained changes in the gill proteins, which play a key role in development but are also known to be anti-tumorigenic and anti-oxidant (Smith, Moccia, Mothersill, & Seymour, 2018a, 2018b). The varying water salinity also resulted in proteome changes in fish. The proteomic profile of the anterior and posterior intestine, and the intestinal fluid of Gulf toadfish acclimated in hypersaline seawater showed several differentially modulated proteins involved in important for growth and development processes, i.e., ion transport, digestion, nutrient absorption, and intestinal carbonate mineral precipitation (Schauer, Reddam, Xu, Wolfe, & Grosell, 2018). Ocean acidification and warming are both primarily caused by increased levels of atmospheric CO_2, and marine organisms are exposed to these two stressors simultaneously. A proteomics approach was used to assess changes that may occur from the exposure to these two environmental stressors (Bresolin de Souza, Jutfelt, Kling, Förlin, & Sturve, 2014). Gills and blood plasma of Atlantic halibut showed changes in protein abundance what indicates that moderate CO_2-driven acidification, alone and combined with high temperature, can elicit biochemical changes that may affect fish health (Bresolin de Souza et al., 2014).

Proteomic techniques have also had applications in identifying seafood exposure to potential emerging hazards, such as microplastic, the personal care product—triclosan, as well as pharmaceuticals and recreational drugs (see Martinez et al., 2020).

Plastic waste entering the oceans was calculated for 2010 at 4–12 million tons per year (Jambeck et al., 2015). Proteomic studies have been performed on a few aquatic species to assess the influence of the microplastic on health status (Green, Colgan, Thompson, & Carolan, 2019; Sussarellu et al., 2016). Although these studies were improvised in laboratory conditions, it is highly expected that the harmful effects of microplastics in the future will be recorded directly in the natural environment.

Ecotoxicoproteomics developed in the fishery and aquaculture industry identify proteins affected after exposure to contaminants and become a valuable approach for *in situ* biomonitoring to assess the health status of aquatic organisms (Gouveia et al., 2019). Triclosan is commonly used in several personal care products, textiles, and children's toys, which ecotoxicological impact is still largely unexplored (Bester, 2003; Singer, Müller, Tixier, & Pillonel, 2002). Proteomic approach, analyzing changes in protein expression profiles in gills of *Dreissena polymorpha* exposed for seven days to triclosan, showed effects on various biological processes, especially those involved in calcium-binding or stress response, what highlighted a pronounced cytogenotoxic effect of triclosan (Riva, Cristoni, & Binelli, 2012). Other chemical stressors influence, like pharmaceuticals and industrial contaminants, on aquatic organisms was also evaluated by proteomics methods. Analysis of gill tissues of three fish species: rainbow trout, common carp, and goldfish in response to copper exposure showed species-specific differences in metal-binding proteins (Eyckmans et al., 2012). Moreover, proteomics has also been applied to the identification of differentially regulated proteins in the gonads of the oyster *Crassostrea angulata* after $HgCl_2$ contamination (Zhang et al., 2013). Recently, the proteome dynamics of Atlantic cod plasma was evaluated by proteomic analysis after exposure to water contaminated with polycyclic aromatic hydrocarbons (PAH). In total, 369 proteins were identified and ranked according to their relative abundance. The levels of 12 proteins were found significantly altered in PAH exposed fish and were proposed as new biomarker candidates (Enerstvedt, Sydnes, & Pampanin, 2017a Enerstvedt, Sydnes, & Pampanin, 2017b; Enerstvedt, Sydnes, Larssen, & Pampanin, 2018.

Human and veterinary pharmaceuticals are being released into the environment in extremely large quantities. High stability of most of them results in accumulation in abiotic and biotic compartments of ecosystems, creating environmental problems, connected with presence in water and influence on aquatic organisms, especially vulnerable to contamination by pharmaceuticals (Breton & Boxall, 2003; European Environment Agency, 2010). The exposure of Atlantic salmon to the human medicinal drugs (acetaminophen also known as paracetamol, atenolol, and carbamazepine) induced notable changes in the proteome of their livers (Hampel, Alonso, Aparicio, Santos, & Leaver, 2015). Exposure to paracetamol, atenolol, and carbamazepine affected the expression of 11, 6, and 10 proteins, respectively. The authors point out that there is a lack of consistency between the results of proteomics and transcriptomics analysis, which indicates that current proteomics protocols may only detect changes in the most abundant and soluble proteins, resulting in an incomplete and biased picture (Hampel et al., 2015).

Recent studies characterizing proteomes affected by physical and chemical environmental factors of chosen aquatic species with an indication of the used proteomic approach are presented in Table 5.2. For more comprehensive information, the reader is advised to examine, cited here, recently published excellent reviews (Ahmed et al., 2019; Carrera et al., 2020; Martinez et al., 2020).

5.2.3 Proteomics of the pathogens of aquatic organisms

Diseases in aquatic organisms are caused by viruses, bacteria, fungi, and parasites (Chintagari, Hazard, Edwards, Jadeja, & Janes, 2018). Pathogen characterization is the main aspect of disease diagnosis and treatment. The classic approach to this topic, i.e., molecular methods and description of immunological mechanisms, but on a small scale and in a narrow scope, e.g., the genes expression or enzymes activity of one metabolic pathway, were usually used in this area (Burge et al., 2016). However, to understand exactly how pathogens affecting their hosts, and how they can cause such negative impacts on aquatic organisms, they need to be studied extensively. This is where the proteomic approach worked perfectly. The identification of proteins and the characterization of the proteomes of various pathogens were very helpful in describing their virulence.

Cyprinid herpesvirus 3 (CyHV-3) is the archetype of fish alloherpesviruses and the causative agent of major economic losses in common and koi carp worldwide (Vancsok et al., 2017). Using LC-MS/MS authors identified 16 virion transmembrane proteins of the CyHV-3, also demonstrating that eight of these proteins are essential for viral growth and two affect virulence *in vivo* (Vancsok et al., 2017). In another study, the virion-associated proteins of the spotted knifejaw iridovirus (SKIV) were characterized (Shuang et al., 2013). The spotted knifejaw (*Oplegnathus punctatus*) is an emerging economical aquaculture species in China in recent years (Huang et al., 2021). A total of 49 viral proteins and 33 cellular proteins were associated with the SKIV virions by LC-MS/MS, including 18 structural proteins that were detected by MALDI TOF/TOF-MS. One highly abundant protein from this group was identified as the virus-inducible stress protein (VISP) and further characterized as an envelope protein. However, knockdown of mVISP by siRNA method showed no effect in virion production (Shuang et al., 2013).

Studies on *Edwardsiella tarda*, a causative agent of hemorrhagic septicemia especially in turbot (*Scophthalmus maximus*), revealed proteins associated with virulent and avirulent strains (Srinivasa Rao, Yamada, Tan, &

TABLE 5.2 Representation of some of the proteomic techniques applied to aquatic organisms' characterization after exposure to physical or chemical environmental factors.

Species	Tissue/secretions	Stressor	Proteomic method	References
Atlantic cod (*Gadus morhua*)	Blood plasma	Polycyclic aromatic hydrocarbon	LC-MS/MS	Enerstvedt et al. (2018), Enerstvedt et al., 2017a Enerstvedt et al., 2017b
Atlantic halibut (*Hippoglossus hippoglossus*)	Gills and blood plasma	CO_2/temperature	nano-LC-MS/MS	Bresolin de Souza et al. (2014)
Atlantic salmon (*Salmo salar*)	Muscle and blood plasma	crowding	2-DE MALDI-TOF/TOF MS	Veiseth-Kent et al. (2010)
Atlantic salmon (*Salmo salar*)	Liver	Temperature	nano-LC-MS/MS	Nuez-Ortín et al. (2018)
Atlantic salmon (*Salmo salar*)	Liver	Human pharmaceuticals	2-DE DIGE MALDI-TOF/TOF-MS	Hampel et al. (2015)
Blue mussel (*Mytilus edulis*)	Hemolymph	Microplastic	LC-MS/MS	Green et al. (2019)
Fugu (*Takifugu fasciatus*)	Liver	Temperature	iTRAQ nano-LC-MS/MS	Wen et al. (2019)
Gilthead sea bream (*Sparus aurata*)	Liver	Repeated handling and crowding	2-DELC-MS/MS	Alves et al. (2010)
Gulf toadfish (*Opsanus beta*)	Intestine	Salinity	LC-MS/MS	Schauer et al. (2018)
Pacific oyster (*Crassostrea gigas*)	Oocytes	Microplastic	LC-MS/MS	Sussarellu et al. (2016)
Portuguese oyster (*Crassostrea angulate*)	Gonads	Mercury	2-DE MALDI-TOF/TOF-MS	Zhang et al. (2013)
Rainbow trout (*Oncorhynchus mykiss*)	Early life stages	X-rays	LC-ESI-Q-TOF-MS/MS	Smith et al. (2018a), 2018b
Several fish species	Gills	Copper	LC-MS/SM	Eyckmans et al. (2012)
Zebra mussel (*Dreissena polymorpha*)	Gills	Triclosan	2-DE MALDI-TOF/TOF-MS	Riva et al. (2012)

Proteomic techniques abbreviations: 1/2-DE, one/two-dimensional electrophoresis; 2-DE DIGE, two-dimensional difference gel electrophoresis; iTRAQ, isobaric tag for relative and absolute quantitation; MALDI-TOF/TOF-MS, matrix-assisted laser desorption and ionization (time-of-flight)2 Mass Spectrometry; nano-LC-MS/MS, nano-scale liquid chromatography tandem mass spectrometry; LC-MS/MS, liquid chromatography tandem mass spectrometry; ESI MS/MS, electrospray ionization tandem mass spectrometry; LC-ESI-MS/MS, liquid chromatography electrospray ionization tandem mass spectrometry; LC-ESI-Q-TOF MS/MS, liquid chromatography electrospray ionization quadrupole time-of-flight tandem mass spectrometry.

Leung, 2004; Tan, Rao, Zhang, Hew, & Leung, 2003). Moreover, using liquid chromatography electrospray ionization tandem mass spectrometry (LC-ESI-MS/MS) virulence determinants (e.g., iron-cofactored superoxide dismutase, heat shock protein 90, antigenic protein Et 46) of *E. tarda* were detected (Buján, Hernández-Haro, Monteoliva, Gil, & Magariños, 2015). A similar goal was achieved in the research on *Flavobacterium columnare* (Dumpala, Gülsoy, Lawrence, & Karsi, 2010). Columaris is one of the oldest known diseases among warm-water fish. Infections are the second leading cause of mortality in pond-raised catfish in the southeastern United States (Declercq, Haesebrouck, Van den Broeck, Bossier, & Decostere, 2013). Studies with the use of LC-ESI-MS/MS and matrix-assisted laser desorption and ionization coupled to time-of-flight/time-of-flight mass spectrometry (MALDI-TOF/TOF-MS), showed in the group of identified metabolic pathways, that the most significantly represented in obtained protein dataset, was folate biosynthesis, followed by the purine metabolism pathway (Dumpala et al., 2010). Moreover, 18 peptidases were identified in the protein list suggesting the breakdown of host proteins by *F. columnare* and usage of resultant amino acids as a major source of energy, carbon, and nitrogen. Similarly, glycosylphosphatidylinositol (GPI)-anchor biosynthesis and lipopolysaccharide (LPS) biosynthesis pathways were significantly represented in the dataset, which may play an important role in cell wall synthesis. Moreover, LPS biosynthesis may be one of the important virulence-related systems in *F. columnare* pathogenesis (Dumpala et al., 2010). Many more studies characterizing proteomes of the pathogenic bacteria of aquatic species with an indication of the used proteomic approach are presented in Table 5.3.

In the group of fungi, the *Saprolegnia parasitica* is one of the most severe fish pathogens. It is a causative agent of causes saprolegniasis, a disease characterized by visible white or gray patches of filamentous mycelium on the body or fins of freshwater fish (van West, 2006). Due to that, a proteomic analysis, using gel-based and isobaric tag for relative and absolute quantitation (iTRAQ) approaches, of four defined developmental stages of *S. parasitica* grown *in vitro*, i.e., the mycelium, primary cysts, secondary cysts, and germinated cysts, was performed to describe the groups of proteins linked to the different stages (Srivastava, Rezinciuc, & Bulone, 2018). Compared to the mycelium stage, most of the proteins in the different cyst stages showed similar enrichment analysis results, which were mainly related to energy metabolism, signal transduction, protein synthesis, and post-translational modifications. Candidate proteins potentially involved in both vegetative growth and infection processes were also identified. The differences between *S. parasitica* developmental stages' proteomes can potentially be exploited in saprolegniasis diagnosis and treatment strategies by targeting proteins specifically expressed at key infectious stages.

TABLE 5.3 Representation of some of the proteomic methods applied to fish pathogens for characterization and understanding pathogenesis.

Pathogen	Disease	Proteomic method	References
Viruses			
Cyprinid herpesvirus 3 (CyHV-3)	Koi herpesvirus (KHV) disease	SDS-PAGE LC-MS/MS	Vancsok et al. (2017)
Singapore grouper iridovirus (SGIV)	Systemic disease	SDS-PAGE MALDI-TOF-MS; 1-DE MALDI-TOF/TOF-MS/MS, LC-MALDI-TOF/TOF-MS/MS	Song et al. (2004), Zhou et al. (2011)
Spotted knifejaw iridovirus (SKIV)	Systemic disease	1-DE LC-MS/MS, MALDI-TOF/TOF-MS	Shuang et al., 2013)
Bacteria			
Aeromonas hydrophilla	Haemorrhagic septicaemia	2-DE SDS-PAGE	Poobalane et al. (2008)
Aeromonas hydrophilla	Haemorrhagic septicaemia	iTRAQ LC-MS/MS	Li et al. (2018c)
Aeromonas salmonicida	Furunculosis	2-DE LC-MS/MS, nano-LC-MS/MS	Ebanks, Goguen, McKinnon, Pinto, and Ross (2005); Menanteau-Ledouble, Kattlun, Nöbauer, and El-Matbouli (2014)
Edwardsiella ictaluri	Enteric septicaemia of catfish	2-DE LC-ESI-MS/MS and 2-DE MALDI-TOF/TOF-MS	Dumpala, Lawrence, and Karsi (2009)
Edwardsiella tarda	Haemorrhagic septicaemia	2-DE MALDI-TOF-MS	Kumar, Sharma, Rathore, Bisht, & Sengupta (2010), Srinivasa Rao et al. (2004), Tan et al. (2003)
Edwardsiella tarda	Haemorrhagic septicaemia	2-DE ESI MS/MS	Buján et al. (2015)

(Continued)

TABLE 5.3 Cont'd

Pathogen	Disease	Proteomic method	References
Edwardsiella tarda	Haemorrhagic septicaemia	2-DE MALDI-TOF/TOF, LC-ESI-MS/MS	Wang, Wang, Yang, and Zhang (2013)
Flavobacterium columnare	Columnaris disease	2-D LC-ESI-MS/MS and 2-DE MALDI TOF/TOF-MS	Dumpala et al. (2010)
Flavobacterium psychrophilum	Bacterial cold-water disease	2-DE LC-MS/MS	LaFrentz, LaPatra, Call, Wiens, and Cain (2009)
Flavobacterium psychrophilum	Bacterial cold-water disease	MALDI-TOF-MS	Pérez-Sancho et al. (2017)
Mycobacterium marinum	Mycobacteriosis	MALDI-TOF-MS	Kurokawa et al. (2013)
Photobacterium damselae	Pasteurellosis	MALDI-TOF-MS	Pérez-Sancho et al. (2016)
Photobacterium damselae subsp. *damselae*	Vibriosis	2-DE MALDI-TOF/TOF-MS	Puentes, Balado, Bermúdez-Crespo, Osorio, and Lemos (2017)
Photobacterium damselae subsp. *piscicida*	Pasteurellosis	2-DE nano-LC-ESI-Q-TOF MS/MS	Chuang, Tsai, Kuo, Chen, and Chen (2016)
Piscirickettsia salmonis	Salmonid rickettsial syndrome	MALDI-TOF-MS	López-Cortés et al. (2017)
Pseudomonas fluorescens	"Fin rot" disease	2-DE MALDI-TOF-MS, 2-DE MALDI-TOF/TOF-MS	Liu, Chi, and Sun (2015), Sun, Chi, and Sun (2016)
Several gram-negative species	–	MALDI-TOF-MS	Böhme et al. (2010)
Streptococcus agalactiae	Streptococcosis	nano-UPLC-HDMS^E	Tavares et al. (2018)
Streptococcus iniae	Streptococcosis	2-DE MALDI-TOF-MS	Zhang, Zhang, and Sun (2014)

Streptococcus iniae and *S. parauberis*	Streptococcosis	MALDI-TOF-MS	Yolanda, Clara, and Ysabel (2019)
Streptococcus parauberis	Streptococcosis	MALDI-TOF-MS	Kim et al. (2015)
Tenacibaculum sp.	Tenacibaculosis	MALDI-TOF-MS	Fernández-Álvarez, Torres-Corral, and Santos (2018), 2017)
Vagococcus salmoninarum	Cold-water streptococcosis	MALDI-TOF-MS	Torres-Corral and Santos (2019)
Vibrio parahaemolyticus	Vibriosis	2-DE LC-ESI-Q-TOF MS/MS	Shen, Kuo, Lin, Chow, and Chen (2010)
Vibrio salmonicida	Vibriosis or Hitra disease	nano-LC-ESI-MS/MS	Uttakleiv Ræder, Paulsen, Smalås, and Willassen (2007)
Yersinia ruckeri	Enteric redmouth disease	nano-LC-MS/MS	Kumar, Hummel, Welch, Razzazi-Fazeli, and El-Matbouli (2017), Kumar et al., (2016)
Yersinia ruckeri	Enteric redmouth disease	Gell free and 1-D nano-LC-ESI-MS/MS	Ormsby, Grahame, Burchmore, and Davies (2019)
Fungi			
Saprolegnia parasitica	Saprolegniasis	iTRAQ nano-LC-MS/MS	Srivastava et al. (2018)
Parasites			
Anisakis simplex s.s.	Anisakiasis	LC-MS/MS	Kochanowski et al. (2020b), Polak et al. (2020), Stryiński et al. (2019)
Ichtyophthirius multifiliis	Ichthyophthiriasis or White spot disease	iTRAQ LC-ESI-MS/MS	Yao et al. (2017)

(Continued)

TABLE 5.3 Cont'd

Pathogen	Disease	Proteomic method	References
Caligus rogercresseyi	Sea lice	MALD-TOF-MS	López-Cortés et al. (2017)
Contracaecum osculatum	Anisakidosis	LC-MS/MS	Kochanowski et al. (2020a)
Opisthorchis viverini and O. felineus	Opisthorchiasis	2-DE, LC-MS/MS	Boonmee et al. (2003), Mulvenna et al. (2010)
Pseudoterranova decipiens and P. krabbei	Pseudoterranovosis	LC-MS/MS	Carrera et al. (2016), Kochanowski et al. (2020a)

Proteomic techniques abbreviations: 1/2-DE, one/two-dimensional electrophoresis; SDS-PAGE, sodium dodecyl sulphate-polyacrylamide gel electrophoresis; iTRAQ, isobaric tag for relative and absolute quantitation; MALDI-TOF-MS, matrix-assisted laser desorption and ionization time-of-flight mass spectrometry; MALDI-TOF/TOF-MS, matrix-assisted laser desorption and ionization (Time-of-Flight)2 Mass Spectrometry; MALDI-TOF/TOF-MS/MS, matrix-assisted laser desorption and ionization (Time-of-Flight)2 tandem Mass Spectrometry; LC-MALDI-TOF/TOF-MS/MS, automated liquid chromatography matrix-assisted laser desorption and ionization (Time-of-Flight)2 tandem mass spectrometry; nano-LC-MS/MS, nano-scale liquid chromatography tandem mass spectrometry; LC-MS/MS, liquid chromatography tandem mass spectrometry; ESI MS/MS, electrospray ionization tandem mass spectrometry; LC-ESI-MS/MS, liquid chromatography electrospray ionization tandem mass spectrometry; LC-ESI-Q-TOF MS/MS, liquid chromatography electrospray ionization quadrupole time-of-flight tandem mass spectrometry; nano-LC-ESI-Q-TOF MS/MS, nano-scale liquid chromatography and electrospray ionization quadrupole time-of-flight tandem mass spectrometry; LC-ESI-MS/MS, liquid chromatography electrospray ionization tandem mass spectrometry; nano LC-ESI-MS/MS, nano-scale liquid chromatography electrospray ionization tandem mass spectrometry, nano UPLC-HDMSE, ultra-performance liquid chromatography with high definition tandem mass spectrometry.

In the group of fish pathogens, there is a significant place for parasites. Many parasitic species develop in aquatic organisms which later land on our tables. However, foodborne parasites are considered neglect in relation to bacteria and viruses that can be transmitted to human by eating food products made of infected organisms (Trevisan, Torgerson, & Robertson, 2019). Moreover, parasitic species becoming recognized as serious pathogens weakening the condition of farmed fish or reducing the suitability for consumption of many aquatic organisms, on which different stages of various parasitic species develop. The proteomic techniques were also applied for characterization of the life stages biology of parasitic species.

The *Opisthorchis viverrini*–endemic countries of Cambodia, the Lao People's Democratic Republic, Thailand, and Vietnam have hugely increased freshwater fish production, e.g., in Vietnam, freshwater fish production increased from 41,750 tons in 1962 to 390,000 tons 40 years later. The development cycle of *O. viverrini* is complex. After the egg is eaten by the appropriate Bithynia snail (first intermediate host), it begins to transform initially into ciliated miracidium, then into sporocyst, redia, and invasive cercaria (Wykoff, Harinasuta, Juttijudata, & Winn, 1965). The expelled, swimmable cercaria attacks the body of a freshwater fish of the Cyprinidae family (second intermediate host), where it encapsulates the muscles to form metacercariae. In 1965, nine fish hosts of *O. viverrini* were known. Up to 2002, 15 species of fish from seven genera of the family Cyprinidae were known to serve as second intermediate host. Further research by Rim et al. (2008) showed an additional five host species, many of which are commonly known in aquaculture (i.e., *Barbonymus gonionotus*, or *Systomus rubripinnis*). Methods for the prevention of *O. viverrini* in aquaculture fishponds were proposed by Khamboonruang et al. (1997). However, the increasing demand for fish meat, larger and larger breeding tanks and more stringent animal production requirements, as well as consumer requirements, make molecular methods, including proteomics, applicable in research on the prevention of *O. viverrini*.

A few proteomic studies were performed on *O. viverrini* to discover potential candidates for diagnostic biomarkers or new drug development. Comparative 2-DE analysis was used to highlight proteins that are significantly modulated between the juvenile and adult form of the parasite and are thought to be important for survival and pathogenesis. Compared with the 1-week-old juvenile fluke, 35 protein spots in four-week-old adults were differentially regulated (Boonmee, Imtawil, Wongkham, & Wongkham, 2003). Moreover, Mulvenna et al. (2010) using LC-MS/MS characterized 300 proteins from the *O. viverrini* excretory-secretory products. In addition, more than 160 tegumental proteins were identified using sequential solubilization of isolated teguments, and the several proteins functions are still unknown (Mulvenna et al., 2010).

The comparative proteomic strategy was also applied to two developmental stages of *Ichtyophthirius multifiliis*, a parasitic ciliate, which name literally translates as "the fish louse with many children" (Yao et al., 2017). The parasite can infect most freshwater fish species and shows very low host specificity. It penetrates gill epithelia, skin and fins of the fish host and resides as a feeding stage (the trophont) inside the epidermis (Jørgensen, 2017). When the organism gets into a large fish culture facility, it is difficult to control due to its fast-reproductive cycle. If not controlled, the infection may lead to 100% mortality in the tank (Heinecke & Buchmann, 2009). A total of 2300 proteins were identified in the two developmental stages, of which 1520 proteins were differentially expressed. Among them, 84 proteins were uniquely expressed in the theronts–infectious stage, while 656 proteins were expressed only in trophonts. This study provided a repository of proteins that could help to conduct future studies for the identification of potential drug targets and vaccines candidates for the control of this devastating fish pathogen (Yao et al., 2017).

Currently, parasitic infections of fish species intended for consumption, by a large part of society, are often associated with infections caused by nematodes of the Anisakidae family. In 1988, a standardized nomenclature recommended three different terms: (1) anisakidosis caused by any members of the family Anisakidae, (2) anisakiasis caused by members of the genus *Anisakis*, and (3) pseudoterranovosis caused by members of the genus *Pseudoterranova* (Audicana & Kennedy, 2008; Klimpel & Palm, 2011). The first case of a human infection with *Anisakis* sp. was reported in the Netherlands (Van Thiel, 1962). It is well established that one of the most important first intermediate hosts in the subfamily Anisakinae life cycle are the crustaceans (Smith & Wootten, 1978). Infected crustaceans such as krill (e.g., Euphasiids) are eaten by fish or squid and become a source of the infective 3rd stage larvae for humans, which become an accidental host by eating infected fish or squids.

Almost three decades ago, the most common fish and shellfish sources of *A. simplex* were the spotted chub mackerel (*Scomber japanicus*) and Japanese flying squid (*Todarodes pacificus*) (Nagasawa & Moravec, 1995). However, just over a decade later, Abollo and colleagues concluded that most species of cephalopods and fish can potentially harbor these marine parasitic nematodes (200 fish and 25 cephalopods species have been identified as hosts for *Anisakis* spp.) (Abollo, Gestal, & Pascual, 2001). The *Anisakis* larvae are found in diverse species of fish, crustacean, and cephalopods of commercial importance as herring (*Clupea harengus*), hake (*Merluccius merluccius*), anchovy (*Engraulis encrasicolus*), mackerel (*Scomber scombrus*), cod (*Gadus morhua*), sardines (*Sardina pilchardus*), and salmon (*Salmo salar*). Other intermediate host reported for *Anisakis* include squid, octopus-molluscs, and one-eyed crayfish (Aibinu, Smooker, & Lopata, 2019).

Due to that, scientists studied those parasitic nematodes extensively. The comparative proteomic profiling of *A. simplex*, *P. decipiens*, and *C. osculatum* L3 larvae using LC-MS/MS was performed (Kochanowski et al., 2020a). Recently, for the first time, the global proteome of the third and fourth stage larvae of *A. simplex* was analyzed using quantitative proteomics based on tandem mass tag (TMT) (Stryiński et al., 2019). In addition, using MALDI-TOF/TOF analysis, differentially expressed proteins for *A. simplex* s.s., *A. pegreffii*, and their hybrid were described, and potential new allergens were identified (Arcos et al., 2014). Moreover, the response to the invasive larvae of *A. simplex* s.s. to ivermectin (anthelminthic drug) was also evaluated using TMT-based methodology (Polak, Łopieńska-Biernat, Stryiński, Mateos, & Carrera, 2020). The *P. krabbei* proteome was characterized in a study performed by Carrera, Gallardo, Pascual, González, and Medina (2016). Much more proteomic techniques were used to ensure the safety of food products that could be affected by Anisakid nematodes, which will be revived closer in subsection six of this chapter.

Proteomic studies on Anisakid nematodes were also reviewed meticulously and accurately by D'Amelio, Lombardo, Pizzarelli, Bellini, and Cavallero (2020), which the reader is advised to examine. Studies describing proteomes of the pathogens of chosen aquatic species with an indication of the used proteomic approach are presented in Table 5.3.

5.2.4 Proteomics of host–pathogen interactions

The influence of the pathogen on the host, and vice versa, begins when bacteria, viruses, fungi, or parasites with their presence cause reactions that change the metabolism of the host, and when the host organism responds by defense mechanisms. This interaction implies gene expression and protein synthesis on both sides. Weakened defense mechanisms or the infectious process can lead to the death of the host or pathogen (Sen, Nayak, & De, 2016). The host–pathogen interactions are extremely complex and can be established at multiple levels, ranging from molecular, cellular, physiological, to populations and ecosystems (Näpflin et al., 2019). The knowledge on which genes or proteins are involved in host–pathogens interactions during infection has been easy to obtain through the use of high-throughput technologies, like next generation sequencing of RNA or mass spectrometry to obtain comparative proteomics data, however, until now the proteomic approach has been less frequently chosen (Moreira et al., 2021).

In this section, selected studies applied to fish host for characterization of the proteomic dynamics as a response against various pathogens, will be discussed in more detail. In most of these studies, the proteomics was used to understand the immune response of fish to the infecting pathogen, and to describe these interactions on the molecular level. Several proteomic studies applied on fish organs and tissues during certain infections to describe associated molecular processes are summarized in Table 5.4.

Fish, like all other living organisms, can be infected by virus representatives from most of the virus families (Kim & Leong, 1999). The pancreas disease (PD) affecting Atlantic salmon and rainbow trout is a major burden in European salmonid aquaculture and causes major economic losses every year. The disease is caused by Salmon Pancreas disease virus, also referred to as salmonid alphavirus (SAV), which belongs to the genus *Alphavirus* within the family Togaviridae. The proteomic study of salmon showed 72 proteins response to the alpha virus subtype 3 (SAV3) infection (Braceland et al., 2013). The proteins were identified by 2-DE, and peptide MS/MS fingerprinting. For example, creatine kinase, enolase, and malate dehydrogenase serum concentrations were shown to correlate with PD whose etiological agent is SAV3. Those proteins may act as biomarkers of the PD and may be used to monitor disease progression and enhance understanding of its pathology (Braceland et al., 2013).

Spring viremia of carp (SVC) is a viral disease that can cause significant mortality in several carp species including the common carp (*Cyprinus carpio*). These species are bred as food fish in many countries, and koi carp are selectively bred for the ornamental fish industry. The disease has been a problem in Europe, the Middle East, and Russia, but it was first reported in koi and feral carp in the United States in 2002. The World Organization for Animal Health (OIE) lists it as a reportable disease. Diagnosing SVC in fish farmed in the United States may result in the isolation and depopulation of infected fish (Petty, Francis-Floyd, & Yanong, 2012). The zebrafish as a model organism was used to better understand spring viremia of carp virus (SVCV) pathogenesis (Medina-Gali et al., 2019). After LC-MS/MS analysis a total of 3062 proteins were identified of which, 137, 63, and 31 proteins were enriched in blood samples harvested at 1-, 2-, and 5-days post-SVCV infection, respectively. What is interesting, a number of up-regulated plasma proteins in SVCV-infected samples increased rapidly and dropped after 1 day post-infection. Moreover, evaluation of the list of the up-regulated proteins during SVCV infection showed the highest fold change for the proteins of the vitellogenin family (Vtg) and the grass carp reovirus-induced gene (Gig) proteins (Medina-Gali et al., 2019). This was the first proteomic study revealed the involvement of vitellogenins in adult fish response to viral infection and proposed those proteins as a novel biomarker of viral infections in fish. The authors also highlight the role of Gig proteins as important factors in antiviral response in fish (Medina-Gali et al., 2019).

Bacteria are a wide group of organisms which are pathogenic for fish and other aquatic species. An important example is *Yersinia ruckeri*, the causative agent of enteric red mouth disease (ERM) in rainbow trout,

TABLE 5.4 Representation of some of the proteomic methods applied to fish host for characterization of the proteomic dynamics and metabolic changes as a response against pathogens.

Host	Pathogen	Organ/tissue	Proteomic method	References
Viruses				
Atlantic salmon (*Salmo salar*)	Salmonid alphavirus subtype 3	Blood and serum	2-DE nano-LC-ESI-MS/MS	Braceland et al. (2013)
Crucian carp (Carassius carassius)	Cyprinid herpesvirus 2	Head kidney	nano-LC-MS/MS	Liu et al. (2018)
Rainbow trout (Oncorhynchus mykiss)	Viral septicemia hemorrhagic virus	Red blood cells from blood and head kidney	LC-ESI-MS/MS	Nombela et al., 2019
Turbot (*Scophthalmus maximus*)	Megalocyti virus	Spleen	2-DE MALDI-TOF-MS/MS	Zhang et al. (2013)
Zebrafish (*Danio rerio*)	Infectious spleen and kidney necrosis virus (ISKNV)	Spleen	2-DE MALDI-TOF/TOF-MS/MS	Xiong et al. (2011)
Zebrafish (Danio rerio)	Rhabdovirus	Fins	2-DE DIGE MALDI-TOF/TOF-MS	Encinas et al. (2010)
Zebrafish (Danio rerio)	Spring viremia of carp virus	Plasma	LC-MS/MS	Medina-Gali et al. (2019)
Bacteria				
Atlantic cod (Gadus morhua)	Vibrio anguillarum	Mucus	2-DE LSI-ESI-MS/MS	Rajan, Lokesh, Kiron, and Brinchmann (2013)
Atlantic salmon (*Salmo salar*)	*Piscirickettsia salmonis*	Extracellular vesicles	LC-MS/MS	Lagos et al. (2017)
Channel catfish (Ictalurus punctatus)	Edwardsiella ictaluri	Head kidney	MALDI-TOF-MS/MS	Booth and Bilodeau-Bourgeois (2009)
Common carp (Cyprinus carpio)	Aeromonas hydrophila	Intestinal mucosa	LC-MS/MS	Di et al. (2017))

(Continued)

TABLE 5.4 Cont'd

Host	Pathogen	Organ/tissue	Proteomic method	References
Gilthead seabream (Sparus aurata)	Moraxella sp.	Kidney	2-DE MALDI-TOF-MS, nano-LC-ESI-Q-TOF-MS/MS	Addis et al. (2010b))
Olive flounder (Paralichthys olivaceus)	Streptococcus parauberis	Kidney	nano-LC-MS/MS	Cha et al. (2012))
Rainbow trout (Oncorhynchus mykiss)	Aeromonas salmonicida	Liver	LC-MS/MS	Causey et al. (2018))
Rainbow trout (Oncorhynchus mykiss)	Aeromonas salmonicida	Spleen	iTRAQ nano-LC-MS/MS	Long et al. (2015))
Rainbow trout (Oncorhynchus mykiss)	Yersinia ruckeri	Intestine	Micro-LC-ESI-QTOF-MS/MS	Kumar et al. (2019))
Zebrafish (Danio rerio)	Aeromonas hydrophila	Gills	iTRAQ LC-MS/MS	Lü et al. (2014))
Parasites				
Atlantic salmon (Salmo salar)	Lepeophtheirus salmonis	Mucus	LC-MS/MS	Easy and Ross (2009), Provan et al. (2013))
Atlantic salmon (Salmo salar)	Neoparamoeba perurans	Gill	2-DE LC-MS/MS	Marcos-López et al. (2017))
Atlantic salmon (Salmo salar)	Neoparamoeba perurans	Gill and skin mucus	2-DE nano-LC-ESI-MS/MS	Valdenegro-Vega et al. (2014))
Common carp (Cyprinus carpio)	Ichthyophthirius multifiliis	Mucus	nano-LC-QTOF-MS/MS	Saleh et al. (2019))
Gilthead seabream (Sparus aurata)	Amyloodinium ocellatum	Plasma	2-DE DIGE LC-MS/MS	Moreira et al. (2017)

Proteomic techniques abbreviations: 1/2-DE, one/two-dimensional electrophoresis; iTRAQ, isobaric tag for relative and absolute quantitation; MALDI-TOF-MS, matrix-assisted laser desorption and ionization time-of-flight mass spectrometry; MALDI-TOF/TOF-MS, matrix-assisted laser desorption and ionization (Time-of-Flight)2 Mass Spectrometry; MALDI-TOF/TOF-MS/MS, matrix-assisted laser desorption and ionization (Time-of-Flight)2 tandem mass spectrometry; LC-MALDI-TOF/TOF-MS/MS, automated liquid chromatography matrix-assisted laser desorption and ionization (Time-of-Flight)2 tandem mass spectrometry; nano-LC-MS/MS, nano-scale liquid chromatography tandem mass spectrometry; LC-MS/MS, liquid chromatography tandem mass spectrometry; ESI MS/MS, electrospray ionization tandem mass spectrometry; LC-ESI-MS/MS, liquid chromatography electrospray ionization tandem mass spectrometry; LC-ESI-Q-TOF MS/MS, liquid chromatography electrospray ionization quadrupole time-of-flight tandem mass spectrometry; micro/nano-LC-ESI-Q-TOF MS/MS, micro-/nano-scale liquid chromatography and electrospray ionization quadrupole time-of-flight tandem mass spectrometry; LC-ESI-MS/MS, liquid chromatography electrospray ionization tandem mass spectrometry; nano LC-ESI-MS/MS, nano-scale liquid chromatography electrospray ionization tandem mass spectrometry.

resulting in heavy commercial loses (Furones, Rodgers, & Munn, 1993; Novoslavskij et al., 2016). The proteomic study of the rainbow trout intestine (Kumar, Hummel, Razzazi-Fazeli, & El-Matbouli, 2019) was performed due to the fact, that in fish, the gastrointestinal tract is a multifunctional organ, which functions not only for nutrient absorption but in which important mechanisms of innate defense occur to maintain tissue homeostasis (Salinas & Parra, 2015). Therefore, Kumar et al. (2019) aimed to evaluate the possible effects of *Y. ruckeri* exposure on the modulation of the proteome profile in the posterior intestine of rainbow trout using microflow high-performance liquid chromatography electrospray-ionization quadrupole time-of-flight tandem mass spectrometry (micro-LC-ESI-QTOF-MS/MS). Intestinal endopeptidase, exopeptidase, and the proteins involved in antioxidant defense processes were upregulated upon *Y. ruckeri* exposure, while those involved in lipid metabolism, actin filament, and translation processes were downregulated. The characterized pathways might be useful in understanding the gut defense mechanisms of rainbow trout and further research work in this direction.

Parasites, as previously mentioned, are a very important group of organisms causing fish pathologies. Protozoan parasites, *I. multifiliis*, cause important economic losses to the aquaculture sector. In the skin mucus of common carp, under exposure to the parasite, LC-ESI-MS/MS revealed the differential regulation of some immune-related signal transduction proteins (Saleh et al., 2019). The analysis revealed increased abundance values of epithelial chloride channel protein, galactose-specific lectin nattection, high choriolytic enzyme 1 (nephrosin), lysozyme C, granulin and protein-glutamine gamma-glutamyltransferase 2 in *I. multifiliis*-exposed carp skin mucus. A variety of lectins and several serpins with protease inhibitor activity were identified as related to lectin pathway activation and proteolytic regulation, indicating that these proteins support the carp innate immune system and the preventive properties of skin mucus (Saleh et al., 2019). The studies on *Neoparamoeba perurans*, causative agent of the most significant infectious disease for Atlantic salmon in aquaculture—amoebic gill disease, showed upregulation of 23 proteins in the host gills, and may contribute to the ongoing understanding of the mechanisms and markers of the disease (Marcos-López et al., 2017). Moreover, the disease induce excessive mucus production from gills, as well as skin, which was also the subject of the studies performed by Valdenegro-Vega et al. (2014). The study results demonstrated, except 52 differentially regulated proteins in gills, also 42 skin mucus proteins in *N. perurans*-affected fish, such as C-reactive protein, apolipoprotein 1, granulin, cathepsin, angiogenin-1. In the studies on other parasite affecting Atlantic salmon i.e., on sea lice (*Lepeophtheirus salmonis*), the LC-MS/MS analysis gave similar results to previously mentioned studies on fish parasites affecting gills and skin, identifying intracellular proteins, calmodulin, actin, and hemopexin and plasma proteins, such as apolipoproteins, lectin, plasminogen and transferrin in the skin mucus of salmon (Easy & Ross, 2009).

Recently, scientists have looked closer to the immune-related proteins identified in the extracellular vesicles (EVs), a class of membrane-bound organelle secreted by a variety of cell types. EVs are now recognized as representing a form of intercellular communication involving horizontal transfer of biologically active cargo, including proteins, messenger RNA (mRNA), and non-coding RNA (Lawson, Kovacs, Finding, Ulfelder, & Luis-Fuentes, 2017). Due to that, the isolation and characterization of EVs have become more common. Lagos et al. (2017) performed proteomic analysis of serum EVs from Atlantic salmon infected with *Piscirickettsia salmonis*, acausative agent of salmonid rickettsial septicemia, which is chronic and often fatal disease in salmonid (Saavedra et al., 2017). Mass spectrometry identified 180 proteins in serum EVs from both infected and non-infected fish. Interestingly, 35 unique proteins were identified in serum EVs isolated from the fish infected with *P. salmonis*. The study was the first to show that EVs carrying specific host proteins could represent a mechanism by which stimulatory molecules can be released from infected cells to promote an immune response in fish (Lagos et al., 2017).

The importance of EV-mediated interactions between aquatic organisms and their hosts seems to be the one of the most promising sources of detailed mechanisms of pathogens infection and host defense, as well as biomarkers of pathologies caused by viruses, bacteria, or parasites. Nevertheless, such approach requires additional studies but seems to be the most advanced in current biology. We also believe that the studies on EVs are one of the most promising and interesting, which can boost future knowledge, not only of fish pathogenesis, but also will lead to gain new information about many more biological systems.

5.2.5 Proteomics of food products from aquatic organisms

Except health and disease management, nutritional management, studies related to reproduction, as well as studies on aquatic toxicology and environment and climate change, the proteomics became useful in fishery and aquaculture for food products authentication, quality assessment, and allergen detection (Nissa, Pinto, Parkar, Goswami, & Srivastava, 2021). Safety and quality are very important not only for the consumers but also for food industry during the global fish-food-based consumption enhancement.

The proteomics techniques were applied in food science for the identification purposes of commercial fish species in food products, which helps in the seafood authentication (Carrera, Cañas, & Gallardo, 2006, 2007, 2009, 2011, 2013; Mazzeo et al., 2008; Ortea, Cañas, & Gallardo, 2011;

Pepe et al., 2010; Wulff, Nielsen, Deelder, Jessen, & Palmblad, 2013). The monitoring of the species-specific proteins or peptides helps in distinguish seafood from closely related commercial fish. Accurate identification of the species of fish used in the preparation of the food product is a key factor in controlling frauds, especially when dealing with fillets and minced fish. Species identification is done in target detection phase of a protein or peptide marker, chosen during discovery phase of proteomic approach (Martinez et al., 2020).

To wider explore the discussed issues, the reader is directed to read the literature examples presented in Table 5.5. For more comprehensive

TABLE 5.5 Representation of the proteomic methods applied for authentication and quality assessment of aquatic-based food products.

Species	Organ/tissue	Proteomic method	References
15 fish species	Muscle	Gel-free LC-ESI-MS/MS	Carrera et al. (2019b)
Peled (Coregonus peled)	Muscle	2-DE MALDI-TOF-MS/MS	Deng, Lei, Yu, Lu, and Zhang (2019)
10 fish species	Different cell lines	2-DE	Goswami, Dubey, Yadav, Sharma, and Lakra (2016)
Atlantic salmon (*Salmo salar*)	Kidney	1-DE MS/MS	Peñaranda, Jensen, Tollersrud, Bruun, and Jørgensen (2019)
Mozambique tilapia (*Oreochromis mossambicus*)	Brain and lip epithelium	LC-MS/MS	Gardell, Qin, Rice, Li, and Kültz (2014)
Rainbow trout (Oncorhynchus mykiss) and three more teleosts	Different cell lines	2-DE	Wagg & Lee, 2005
22 different species	Muscle	LC-MS/MS	Wulff et al. (2013)
7 species of Decapoda shrimps	Muscle	MALDI-TOF MS and ESI-MS/MS	Ortea, Cañas, and Gallardo (2009)
Large and small yellow croaker (*Larimichthys crocea* and *L. polyactis*) and four tonguefishes from the genus *Cynoglossus*	Skin	MALDI-TOF MS	Bi et al. (2019)

Proteomic techniques abbreviations: 1/2-DE, one/two-dimensional electrophoresis; MALDI-TOF-MS, matrix-assisted laser desorption and ionization time-of-flight mass spectrometry; MALDI-TOF-MS/MS, matrix-assisted laser desorption and ionization time-of-flight tandem mass spectrometry; LC-MS/MS, liquid chromatography tandem mass spectrometry; ESI MS/MS, electrospray ionization tandem mass spectrometry; LC-ESI-MS/MS, liquid chromatography electrospray ionization tandem mass spectrometry.

information, the reader is also advised to examine recently published excellent reviews (Gallardo et al., 2013; Martinez et al., 2020; Nissa et al., 2021; Ortea, O'Connor, & Maquet, 2016).

Moreover, the proteomics methods could be helpful in quality assessment and allergen detection. A great achievement of proteomics was identifying, in a group of physiological proteins, parvalbumin as a protein with high allergenic potential and developing its detection methods in food products (Carrera, Cañas, & Gallardo, 2012; Kobayashi et al., 2016; Piovesana et al., 2016; Sun et al., 2019).

The discovery of potential allergens, based on proteomics of previously described fish pathogens, has prompted scientists to use a targeted approach and develop methodologies for detection of those pathogens in foods. Lately, the method for detection of *A. simplex* allergens in fresh, frozen, and cooked fish meat was proposed with the use of immunoglobulin G (IgG) immunoblotting (Rodríguez-Mahillo, González-Muñoz, de las Heras, Tejada, & Moneo, 2010). The proteome characterization of Anisakids nematodes supported methods for the fast monitoring of these parasites. Fæste, Plassen, Løvberg, Moen, and Egaas (2015) have detected by ELISA, immunostaining, and MS, proteins of *A. simplex* in farmed salmon intended for use in sushi and other fish products on the Norwegian market. The same group proposed two more methodologies for *A. simplex* protein detection in fish (Fæste et al., 2016). Both proposed protocols were based on multiple reaction monitoring (MRM)-MS/MS, which is applied to quantify previously identified target peptides by measuring specific precursor-to-product ion transitions. Both of approaches, the label-free semi-quantitative nLC-nESI-Orbitrap-MS/MS and the heavy peptide-applying absolute-quantitative (AQUA) LC-TripleQ-MS/MS use unique reporter peptides derived from Anisakids hemoglobin and SXP/RAL-2 protein as analytes (Fæste et al., 2016).

Recently, the analysis performed by Carrera et al. (2016) showed possible practical use of peptide biomarkers in food industry. The discovery phase was based on the isolation of heat-stable proteins of *A. simplex*, *P. krabbei*, and *P. decipiens*, then the monitoring of several peptide biomarkers by parallel reaction monitoring (PRM) mass spectrometry in a linear ion trap mass spectrometer. The target detection step showed the same proportional relationships between the proposed peptide biomarkers that spiked in hake protein extracts, like those of the buffer diluted sample, which confirms the effectiveness of the PRM method in real fish samples. This method can quickly detect Anisakids in less than 2 h, and if it is made a part of a control protocol defined by food safety authorities, it may facilitate testing and thus increase consumer safety (Carrera et al., 2016).

5.3 Conclusions

To date, the use of proteomics in description of aquatic organisms' biology, to improve fishery and aquaculture has been limited. Due to the fact, that total fish production has seen important increases in all the continents in the last few decades, more advanced and powerful technologies, like proteomic approaches, will be applied more extensively in this field. As presented in this chapter, proteomics is a very promising tool for research on aquatic organisms' health, its pathogen diagnosis and treatment, as well as, it has potential to provide quality control methodologies in the food sector.

The implementation of proteomics to the fishery and aquaculture sector to study the impact of pathogens or farming conditions on food quality and safety might be helpful for achieving the goals and objectives of the Blue Revolution, which is now a critical part of the global food system. Further progress in aquatic organisms' proteomics, will lead to new valuable and important discoveries benefiting human and animal health.

Funding

This work was supported by the "Development Program of the University of Warmia and Mazury in Olsztyn," co-financed by the European Union under the European Social Fund from the Operational Program "Knowledge Education Development". Robert Stryiński is a recipient of a scholarship from the program "Interdisciplinary Doctoral Studies in Biology and Biotechnology" (POWR.03.05.00-00-Z310/17), which is funded by the European Social Fund. This work was also supported by the GAIN-Xunta de Galicia Project (IN607D 2017/01) and the Spanish AEI/EU-FEDER PID2019-103845RB-C21 project.

References

Abollo, E., Gestal, C., & Pascual, S. (2001). Anisakis infestation in marine fish and cephalopods from Galician waters: An updated perspective. *Parasitology Research, 87*, 492–499.

Adams, A., & Thompson, K. D. (2006). Biotechnology offers revolution to fish health management. *Trends in Biotechnology, 24*, 201–205.

Addis, M. F., Cappuccinelli, R., Tedde, V., Pagnozzi, D., Porcu, M. C., Bonaglini, E., et al. (2010a). Proteomic analysis of muscle tissue from gilthead sea bream (Sparus aurata, L.) farmed in offshore floating cages. *Aquaculture, 309*, 245–252.

Addis, M. F., Cappuccinelli, R., Tedde, V., Pagnozzi, D., Viale, I., Meloni, M., et al. (2010b). Influence of Moraxella sp. colonization on the kidney proteome of farmed gilthead sea breams (Sparus aurata, L.). *Proteome Science, 8*(1), , 1–8. https://doi.org/10.1186/1477-5956-8-50.

Aebersold, R., Bensimon, A., Collins, B. C., Ludwig, C., & Sabido, E. (2016). Applications and developments in targeted proteomics: From SRM to DIA/SWATH. *Proteomics, 16*, 2065–2067.

Ahmed, F., Kumar, G., Soliman, F. M., Adly, M. A., Soliman, H. A. M., El-Matbouli, M., et al. (2019). Proteomics for understanding pathogenesis, immune modulation and host pathogen interactions in aquaculture. *Comp. Biochem. Physiol. Part D Genomics Proteomics, 32*, 1–15, Article 100625.

Aibinu, I. E., Smooker, P. M., & Lopata, A. L. (2019). Anisakis nematodes in fish and shellfish- from infection to allergies. *Int. J. Parasitol. Parasites Wildl., 9*, 384–393.

Alves, R. N., Cordeiro, O., Silva, T. S., Richard, N., de Vareilles, M., Marino, G., et al. (2010). Metabolic molecular indicators of chronic stress in gilthead seabream (Sparus aurata) using comparative proteomics. *Aquaculture, 299*, 57–66.

Ananda Raja, R., & Jithendran, K. P. (2015). *Aquaculture disease diagnosis and health management*. In *Advances in Marine and Brackishwater Aquaculture* (pp. 247–255). New Delhi: Springer India.

Arcos, S. C., Ciordia, S., Roberston, L., Zapico, I., Jiménez-Ruiz, Y., Gonzalez-Muñoz, M., et al. (2014). Proteomic profiling and characterization of differential allergens in the nematodes Anisakis simplex sensu stricto and A. pegreffii. *Proteomics, 14*, 1547–1568.

Audicana, M. T., & Kennedy, M. W. (2008). Anisakis simplex: From obscure infectious worm to inducer of immune hypersensitivity. *Clinical Microbiology Reviews, 21*, 360–379.

Bester, K. (2003). Triclosan in a sewage treatment process—balances and monitoring data. *Water Research, 37*, 3891–3896.

Bi, H., Zhong, C., Shao, M., Wang, C., Yi, J., Qiao, L., et al. (2019). Differentiation and authentication of fishes at the species level through analysis of fish skin by matrix-assisted laser desorption/ionization time-of-flight mass spectrometry. *Rapid Communications in Mass Spectrometry, 33*, 1336–1343.

Böhme, K., Fernández-No, I. C., Barros-Velázquez, J., Gallardo, J. M., Calo-Mata, P., & Cañas, B. (2010). Species differentiation of seafood spoilage and pathogenic gram-negative bacteria by MALDI-TOF mass fingerprinting. *Journal of Proteome Research, 9*, 3169–3183.

Boonmee, S., Imtawil, K., Wongkham, C., & Wongkham, S. (2003). Comparative proteomic analysis of juvenile and adult liver fluke, Opisthorchis viverrini. *Acta Tropica, 88*, 233–238.

Booth, N. J., & Bilodeau-Bourgeois, A. L. (2009). Proteomic analysis of head kidney tissue from high and low susceptibility families of channel catfish following challenge with Edwardsiella ictaluri. *Fish & Shellfish Immunology, 26*, 193–196.

Borges, M. H., Andrich, F., Lemos, P. H., Soares, T. G., Menezes, T. N., Campos, F. V., et al. (2018). Combined proteomic and functional analysis reveals rich sources of protein diversity in skin mucus and venom from the Scorpaena plumieri fish. *Journal of Proteomics, 187*, 200–211.

Borràs, E., & Sabidó, E. (2017). What is targeted proteomics? A concise revision of targeted acquisition and targeted data analysis in mass spectrometry. *Proteomics, 17*, 1–8, Article 1700180.

Braceland, M., Bickerdike, R., Tinsley, J., Cockerill, D., Mcloughlin, M. F., Graham, D. A., et al. (2013). The serum proteome of Atlantic salmon, Salmo salar, during pancreas disease (PD) following infection with salmonid alphavirus subtype 3 (SAV3). *Journal of Proteomics, 94*, 423–436.

Bresolin de Souza, K., Jutfelt, F., Kling, P., Förlin, L., & Sturve, J. (2014). Effects of increased CO2 on fish gill and plasma proteome. *Plos One, 9*, 1–9, Article e102901.

Breton, R., & Boxall, A. (2003). Pharmaceuticals and personal care products in the environment: Regulatory drivers and research needs. *Qsar & Combinatorial Science, 22*, 399–409.

Brugere, C., Onuigbo, D. M., & Morgan, K. L. (2017). People matter in animal disease surveillance: Challenges and opportunities for the aquaculture sector. *Aquaculture, 467*, 158–169.

Buján, N., Hernández-Haro, C., Monteoliva, L., Gil, C., & Magariños, B. (2015). Comparative proteomic study of Edwardsiella tarda strains with different degrees of virulence. *Journal of Proteomics, 127*, 310–320.

Burge, C. A., Friedman, C. S., Getchell, R., House, M., Lafferty, K. D., Mydlarz, L. D., et al. (2016). Complementary approaches to diagnosing marine diseases: A union of the modern and the classic. *Philos. Trans. R. Soc. B Biol. Sci., 371*, 1–11, Article 20150207.

Carrera, M., Barros, L., Cañas, B., & Gallardo, J. M. (2009). Discrimination of South African commercial fish species (Merluccius capensis and Merluccius paradoxus) by LC-MS/MS analysis of the protein aldolase. *Journal of Aquatic Food Product Technology, 18*, 67–78.

Carrera, M., Cañas, B., & Gallardo, J. M. (2012). Rapid direct detection of the major fish allergen, parvalbumin, by selected MS/MS ion monitoring mass spectrometry. *Journal of Proteomics, 75*, 3211–3220.

Carrera, M., Cañas, B., & Gallardo, J. M. (2013). The sarcoplasmic fish proteome: Pathways, metabolic networks and potential bioactive peptides for nutritional inferences. *Journal of Proteomics, 78*, 211–220.

Carrera, M., Cañas, B., López-Ferrer, D., Piñeiro, C., Vázquez, J., & Gallardo, J. M. (2011). Fast monitoring of species-specific peptide biomarkers using high-intensity-focused-ultrasound-assisted tryptic digestion and selected MS/MS ion monitoring. *Analytical Chemistry, 83*, 5688–5695.

Carrera, M., Cañas, B., Piñeiro, C., Vázquez, J., & Gallardo, J. M. (2006). Identification of commercial hake and grenadier species by proteomic analysis of the parvalbumin fraction. *Proteomics, 6*, 5278–5287.

Carrera, M., Cañas, B., Piñeiro, C., Vázquez, J., & Gallardo, J. M. (2007). De novo mass spectrometry sequencing and characterization of species-specific peptides from nucleoside diphosphate kinase B for the classification of commercial fish species belonging to the family merlucciidae. *Journal of Proteome Research, 6*, 3070–3080.

Carrera, M., Ezquerra-Brauer, J. M., & Aubourg, S. P. (2019a). Characterization of the Jumbo Squid (Dosidicus gigas) skin by-product by shotgun proteomics and protein-based bioinformatics. *Mar. Drugs, 18*, 1–27.

Carrera, M., Gallardo, J. M., Pascual, S., González, Á. F., & Medina, I. (2016). Protein biomarker discovery and fast monitoring for the identification and detection of Anisakids by parallel reaction monitoring (PRM) mass spectrometry. *Journal of Proteomics, 142*, 130–137.

Carrera, M., González-Fernández, Á., Magadán, S., Mateos, J., Pedrós, L., Medina, I., et al. (2019b). Molecular characterization of B-cell epitopes for the major fish allergen, parvalbumin, by shotgun proteomics, protein-based bioinformatics and IgE-reactive approaches. *Journal of Proteomics, 200*, 123–133.

Carrera, M., Piñeiro, C., & Martinez, I. (2020). Proteomic strategies to evaluate the impact of farming conditions on food quality and safety in aquaculture products. *Foods, 9*, 1–17, Article 1050.

Causey, D. R., Pohl, M. A. N., Stead, D. A., Martin, S. A. M., Secombes, C. J., & Macqueen, D. J. (2018). High-throughput proteomic profiling of the fish liver following bacterial infection. *Bmc Genomics [Electronic Resource], 19*, 1–17, Article 719.

Cha, I.-S., Kwon, J., Park, S.-H., Nho, S.-W., Jang, H.-B., Park, S.-B., et al. (2012). Kidney proteome responses in the teleost fish Paralichthys olivaceus indicate a putative immune response against Streptococcus parauberis. *Journal of Proteomics, 75*, 5166–5175.

Chintagari, S., Hazard, N., Edwards, G., Jadeja, R., & Janes, M. (2018). Risks associated with fish and seafood. *Microbiol. Spectr., 5*, 1–16.

Chuang, M.-Y., Tsai, W.-C., Kuo, T.-Y., Chen, H.-M., & Chen, W.-J. (2016). Comparative proteome analysis reveals proteins involved in salt adaptation in Photobacterium damselae subsp. piscicida. *Journal of Basic Microbiology, 56*, 1234–1243.

Cifuentes, A. (2013). *Foodomics: Advanced Mass Spectrometry in Modern Food Science and Nutrition* (pp. 1–580). Hoboken, New Jersey: Wiley.

Conte, F. (2004). Stress and the welfare of cultured fish. *Appl. Anim. Behav. Sci., 86*, 205–223.

D'Amelio, S., Lombardo, F., Pizzarelli, A., Bellini, I., & Cavallero, S. (2020). Advances in Omic studies drive discoveries in the biology of anisakid nematodes. *Genes (Basel), 11*, 1–18, Article 801.

Declercq, A. M., Haesebrouck, F., Van den Broeck, W., Bossier, P., & Decostere, A. (2013). Columnaris disease in fish: A review with emphasis on bacterium-host interactions. *Veterinary Research, 44*, 1–17, Article 27.

Deng, X., Lei, Y., Yu, Y., Lu, S., & Zhang, J. (2019). The discovery of proteins associated with freshness of coregonus peled muscle during refrigerated storage. *Journal of Food Science, 84*, 1266–1272.

Denslow, N. D., Griffitt, R. J., & Martyniuk, C. J. (2012). Advancing the Omics in aquatic toxicology: SETAC North America 31st Annual Meeting. *Ecotoxicol. Environ. Saf., 76*, 1–2.

Di, G., Li, H., Zhang, C., Zhao, Y., Zhou, C., Naeem, S., et al. (2017). Label-free proteomic analysis of intestinal mucosa proteins in common carp (Cyprinus carpio) infected with Aeromonas hydrophila. *Fish & Shellfish Immunology, 66*, 11–25.

Dietrich, M. A., Arnold, G. J., Nynca, J., Fröhlich, T., Otte, K., & Ciereszko, A. (2014). Characterization of carp seminal plasma proteome in relation to blood plasma. *Journal of Proteomics, 98*, 218–232.

Dumpala, P. R., Gülsoy, N., Lawrence, M. L., & Karsi, A. (2010). Proteomic analysis of the fish pathogen Flavobacterium columnare. *Proteome Science, 8*, 1–11, Article 26.

Dumpala, P. R., Lawrence, M. L., & Karsi, A. (2009). Proteome analysis of Edwardsiella ictaluri. *Proteomics, 9*, 1353–1363.

Easy, R. H., & Ross, N. W. (2009). Changes in Atlantic salmon (Salmo salar) epidermal mucus protein composition profiles following infection with sea lice (Lepeophtheirus salmonis). *Comp. Biochem. Physiol. Part D Genomics Proteomics, 4*, 159–167.

Ebanks, R., Goguen, M., McKinnon, S., Pinto, D., & Ross, N. (2005). Identification of the major outer membrane proteins of Aeromonas salmonicida. *Diseases of Aquatic Organisms, 68*, 29–38.

Encinas, P., Rodriguez-Milla, M. A., Novoa, B., Estepa, A., Figueras, A., & Coll, J. (2010). Zebrafish fin immune responses during high mortality infections with viral haemorrhagic septicemia rhabdovirus. A proteomic and transcriptomic approach. *Bmc Genomics [Electronic Resource], 11*, 1–16, Article 518.

Enerstvedt, K. S., Sydnes, M. O., Larssen, E., & Pampanin, D. M. (2018). Screening for protein adducts of naphthalene and chrysene in plasma of exposed Atlantic cod (Gadus morhua). *Chemosphere, 200*, 67–79.

Enerstvedt, K. S., Sydnes, M. O., & Pampanin, D. M. (2017a). Study of the plasma proteome of Atlantic cod (Gadus morhua): Effect of exposure to two PAHs and their corresponding diols. *Chemosphere, 183*, 294–304.

Enerstvedt, K. S., Sydnes, M. O., & Pampanin, D. M. (2017b). Identification of an albumin-like protein in plasma of Atlantic cod (Gadus morhua) and its biomarker potential for PAH contamination. *Heliyon, 3*, 1–13.

European Environment Agency (2010). Pharmaceuticals in the environment. EEA Technical report No 1/2010, Luxembourg.

Eyckmans, M., Benoot, D., Van Raemdonck, G. A. A., Zegels, G., Van Ostade, X. W. M., Witters, E., et al. (2012). Comparative proteomics of copper exposure and toxicity in rainbow trout, common carp and gibel carp. *Comp. Biochem. Physiol. Part D Genomics Proteomics, 7,* 220–232.

Fæste, C. K., Moen, A., Schniedewind, B., Haug Anonsen, J., Klawitter, J., & Christians, U. (2016). Development of liquid chromatography-tandem mass spectrometry methods for the quantitation of Anisakis simplex proteins in fish. *Journal of Chromatography A, 1432,* 58–72.

Fæste, C. K., Plassen, C., Løvberg, K. E., Moen, A., & Egaas, E. (2015). Detection of proteins from the fish parasite Anisakis simplex in Norwegian farmed salmon and processed fish products. *Food Analytical Methods, 8,* 1390–1402.

The State of World Fisheries and Aquaculture. Sustainability in action. (2020). (pp. 1–244). Rome, Italy: FAO. https://doi.org/10.4060/ca9229en.

Fernández-Álvarez, C., Torres-Corral, Y., Saltos-Rosero, N., & Santos, Y. (2017). MALDI-TOF mass spectrometry for rapid differentiation of Tenacibaculum species pathogenic for fish. *Applied Microbiology and Biotechnology, 101,* 5377–5390.

Fernández-Álvarez, C., Torres-Corral, Y., & Santos, Y. (2018). Use of ribosomal proteins as biomarkers for identification of Flavobacterium psychrophilum by MALDI-TOF mass spectrometry. *Journal of Proteomics, 170,* 59–69.

Freitas, J., Vaz-Pires, P., & Câmara, J. S. (2020). From aquaculture production to consumption: Freshness, safety, traceability and authentication, the four pillars of quality. *Aquaculture, 518,* 1–17, Article 734857.

Furones, M. D., Rodgers, C. J., & Munn, C. B. (1993). Yersinia ruckeri, the causal agent of enteric redmouth disease (ERM) in fish. *Annu. Rev. Fish Dis., 3,* 105–125.

Gallardo, J. M., Ortea, I., & Carrera, M. (2013). Proteomics and its applications for food authentication and food-technology research. *TrAC—Trends Anal. Chem., 52,* 135–141.

Gardell, A. M., Qin, Q., Rice, R. H., Li, J., & Kültz, D. (2014). Derivation and osmotolerance characterization of three immortalized tilapia (Oreochromis mossambicus) cell lines. *Plos One, 9,* 1–14, Article e95919.

Goswami, M., Dubey, A., Yadav, K., Sharma, B., & Lakra, W. (2016). Identification of fish cell lines using 2-D electrophoresis based protein expression signatures. *Current Proteomics, 12,* 245–252.

Gotesman, M., Menanteau-Ledouble, S., Saleh, M., Bergmann, S. M., & El-Matbouli, M. (2018). A new age in AquaMedicine: Unconventional approach in studying aquatic diseases. *BMC Vet. Res., 14,* 1–9, 178.

Gouveia, D., Almunia, C., Cogne, Y., Pible, O., Degli-Esposti, D., Salvador, A., et al. (2019). Ecotoxicoproteomics: A decade of progress in our understanding of anthropogenic impact on the environment. *Journal of Proteomics, 198,* 66–77.

Green, D. S., Colgan, T. J., Thompson, R. C., & Carolan, J. C. (2019). Exposure to microplastics reduces attachment strength and alters the haemolymph proteome of blue mussels (Mytilus edulis). *Environmental Pollution, 246,* 423–434.

Hampel, M., Alonso, E., Aparicio, I., Santos, J. L., & Leaver, M. (2015). Hepatic proteome analysis of atlantic salmon (Salmo salar) after exposure to environmental concentrations of human pharmaceuticals. *Molecular & Cellular Proteomics, 14,* 371–381.

Heinecke, R. D., & Buchmann, K. (2009). Control of Ichthyophthirius multifiliis using a combination of water filtration and sodium percarbonate: Dose-response studies. *Aquaculture, 288,* 32–35.

Herrero, M., Simó, C., García-Cañas, V., Ibáñez, E., & Cifuentes, A. (2012). Foodomics: MS-based strategies in modern food science and nutrition. *Mass Spectrometry Reviews, 31,* 49–69.

Hill, B. J. (2005). The need for effective disease control in international aquaculture. *Dev. Biol. (Basel)., 121,* 3–12.

Huang, X., Wei, J., Zheng, Q., Zhang, Y., Zhu, W., Liu, J., et al. (2021). Isolation, identification and genomic analysis of an ISKNV-type megalocytivirus from spotted knifejaw (*Oplegnathus punctatus*). *Aquaculture, 532,* 1–7, Article 736032.

Huntingford, F. A., Adams, C., Braithwaite, V. A., Kadri, S., Pottinger, T. G., Sandoe, P., et al. (2006). Current issues in fish welfare. *Journal of Fish Biology, 68*, 332–372.

Iq, K. C., & Shu-Chien, A. C. (2011). Proteomics of buccal cavity mucus in female Tilapia Fish (*Oreochromis spp.*): A comparison between parental and non-parental fish. *Plos One, 6*, 1–7, Article e18555.

Iwama, G. (2007). The welfare of fish. *Diseases of Aquatic Organisms, 75*, 155–158.

Jambeck, J. R., Geyer, R., Wilcox, C., Siegler, T. R., Perryman, M., Andrady, A., et al. (2015). Plastic waste inputs from land into the ocean. *Science, 80*-(347), 768–771.

Jurado, J., Fuentes-Almagro, C. A., Guardiola, F. A., Cuesta, A., Esteban, M.Á., & Prieto-Álamo, M.-J. (2015). Proteomic profile of the skin mucus of farmed gilthead seabream (Sparus aurata). *Journal of Proteomics, 120*, 21–34.

Khamboonruang, C., Keawvichit, R., Wongworapat, K., Suwanrangsi, S., Hongpromyart, M., Sukhawat, K., et al. (1997). Application of hazard analysis critical control point (HACCP) as a possible control measure for Opisthorchis viverrini infection in cultured carp (Puntius gonionotus). *Southeast Asian Journal of Tropical Medicine and Public Health, 28*, 65–72. Suppl 1.

Kim, C. H., & Leong, J.-A. (1999). Fish viruses. In *Encyclopedia of virology* (2nd ed., pp. 558–568). Cambridge, MA, United States: Academic Press.

Kim, S. W., Jang, H. Bin, Lee, J. S., Im, S. P., Lazarte, J. M. S., Seo, J. P., et al. (2015). Comparison of proteome typing and serotyping of Streptococcus parauberis isolates from olive flounder (Paralichthys olivaceus). *Journal of Microbiological Methods, 118*, 168–172.

Klimpel, S., & Palm, H. W. (2011). Anisakid Nematode (Ascaridoidea) life cycles and distribution: Increasing zoonotic potential in the time of climate change? In H. Mehlhorn (Ed.), *Progress in parasitology. Parasitology Research Monographs* (pp. 201–222). Berlin and Heidelberg: Springer.

Kobayashi, Y., Yang, T., Yu, C.-T., Ume, C., Kubota, H., Shimakura, K., et al. (2016). Quantification of major allergen parvalbumin in 22 species of fish by SDS–PAGE. *Food Chemistry, 194*, 345–353.

Kochanowski, M., Dąbrowska, J., Różycki, M., Karamon, J., Sroka, J., & Cencek, T. (2020a). Proteomic profiling reveals new insights into the allergomes of anisakis simplex, pseudoterranova decipiens, and contracaecum osculatum. *Journal of Parasitology, 106*(5), 572–588.

Kochanowski, M., Różycki, M., Dąbrowska, J., Bełcik, A., Karamon, J., Sroka, J., et al. (2020b). Proteomic and bioinformatic investigations of heat-treated anisakis simplex third-stage larvae. *Biomolecules, 10*, 1–35.

Kumar, G., Hummel, K., Ahrens, M., Menanteau-Ledouble, S., Welch, T. J., Eisenacher, M., et al. (2016). Shotgun proteomic analysis of Yersinia ruckeri strains under normal and iron-limited conditions. *Veterinary Research, 47*, 1–13, Article 100.

Kumar, G., Hummel, K., Razzazi-Fazeli, E., & El-Matbouli, M. (2018). Proteome profiles of head kidney and spleen of rainbow trout (Oncorhynchus Mykiss). *Proteomics, 18*, 1–4, Article 1800101.

Kumar, G., Hummel, K., Razzazi-Fazeli, E., & El-Matbouli, M. (2019). Modulation of posterior intestinal mucosal proteome in rainbow trout (Oncorhynchus mykiss) after Yersinia ruckeri infection. *Veterinary Research, 50*, 1–14, Article 54.

Kumar, G., Hummel, K., Welch, T. J., Razzazi-Fazeli, E., & El-Matbouli, M. (2017). Global proteomic profiling of Yersinia ruckeri strains. *Veterinary Research, 48*, 1–11.

Kumar, G., Sharma, P., Rathore, G., Bisht, D., & Sengupta, U. (2010). Proteomic analysis of outer membrane proteins of *Edwardsiella tarda*. *Journal of Applied Microbiology* 108, 2214–2221, Article 55.

Kurokawa, S., Kabayama, J., Fukuyasu, T., Hwang, S. D., Park, C.-I., Park, S.-B., et al. (2013). Bacterial classification of fish-pathogenic mycobacterium species by multigene phylogenetic analyses and MALDI biotyper identification system. *Marine Biotechnology, 15*, 340–348.

Lafferty, K. D. (2009). The ecology of climate change and infectious diseases. *Ecology, 90*, 888–900.

LaFrentz, B., LaPatra, S., Call, D., Wiens, G., & Cain, K. (2009). Proteomic analysis of Flavobacterium psychrophilum cultured in vivo and in iron-limited media. *Diseases of Aquatic Organisms, 87*, 171–182.

Lagos, L., Tandberg, J., Kashulin-Bekkelund, A., Colquhoun, D., Sørum, H., & Winther-Larsen, H. (2017). Isolation and characterization of serum extracellular vesicles (EVs) from atlantic salmon infected with *Piscirickettsia salmonis*. *Proteomes, 5*(4), 1–13, Article 34.

Lawson, C., Kovacs, D., Finding, E., Ulfelder, E., & Luis-Fuentes, V. (2017). Extracellular vesicles: Evolutionarily conserved mediators of intercellular communication. *The Yale Journal of Biology and Medicine, 90*, 481–491.

Li, B., Gou, M., Han, J., Yuan, X., Li, Y., Li, T., et al. (2018a). Proteomic analysis of buccal gland secretion from fasting and feeding lampreys (Lampetra morii). *Proteome Science, 16*, 1–9.

Li, C., Tan, X. F., Lim, T. K., Lin, Q., & Gong, Z. (2016). Comprehensive and quantitative proteomic analyses of zebrafish plasma reveals conserved protein profiles between genders and between zebrafish and human. *Science Reports, 6*, 1–15, Article 24329.

Li, J., Levitan, B., Gomez-Jimenez, S., & Kültz, D. (2018b). Development of a gill assay library for ecological proteomics of threespine sticklebacks (Gasterosteus aculeatus). *Molecular & Cellular Proteomics, 17*, 2146–2163.

Li, W., Ali, F., Cai, Q., Yao, Z., Sun, L., Lin, W., et al. (2018c). Reprint of: Quantitative proteomic analysis reveals that chemotaxis is involved in chlortetracycline resistance of Aeromonas hydrophila. *Journal of Proteomics, 180*, 138–146.

Liu, L., Chi, H., & Sun, L. (2015). Pseudomonas fluorescens: Identification of Fur-regulated proteins and evaluation of their contribution to pathogenesis. *Diseases of Aquatic Organisms, 115*, 67–80.

Liu, M., Wu, T., Li, S., Wei, P., Yan, Y., Gu, W., et al. (2018). Combined transcriptomic/proteomic analysis of crucian carp Carassius auratus gibelio in cyprinid herpesvirus 2 infection. *Fish & Shellfish Immunology, 82*, 386–399.

Long, M., Zhao, J., Li, T., Tafalla, C., Zhang, Q., Wang, X., et al. (2015). Transcriptomic and proteomic analyses of splenic immune mechanisms of rainbow trout (Oncorhynchus mykiss) infected by Aeromonas salmonicida subsp. salmonicida. *Journal of Proteomics, 122*, 41–54.

López-Cortés, X. A., Nachtigall, F. M., Olate, V. R., Araya, M., Oyanedel, S., Diaz, V., et al. (2017). Fast detection of pathogens in salmon farming industry. *Aquaculture, 470*, 17–24.

Lü, A., Hu, X., Wang, Y., Shen, X., Li, X., Zhu, A., et al. (2014). iTRAQ analysis of gill proteins from the zebrafish (Danio rerio) infected with Aeromonas hydrophila. *Fish & Shellfish Immunology, 36*, 229–239.

Marcos-López, M., Rodger, H. D., O'Connor, I., Braceland, M., Burchmore, R. J. S., Eckersall, P. D., et al. (2017). A proteomic approach to assess the host response in gills of farmed Atlantic salmon Salmo salar L. affected by amoebic gill disease. *Aquaculture, 470*, 1–10.

Martinez, I., Sánchez-Alonso, I., Piñeiro, C., Careche, M., & Carrera, M. (2020). Protein signatures to trace seafood contamination and processing. *Foods, 9*, 1–23, Article 1751.

Martyniuk, C. J., & Alvarez, S. (2013). Proteome analysis of the fathead minnow (Pimephales promelas) reproductive testes. *Journal of Proteomics, 79*, 28–42.

Matos, E., Silva, T. S., Wulff, T., Valente, L. M. P., Sousa, V., Sampaio, E., et al. (2013). Influence of supplemental maslinic acid (olive-derived triterpene) on the post-mortem muscle properties and quality traits of gilthead seabream. *Aquaculture, 396–399*, 146–155.

Mazzeo, M. F., Giulio, B. De, Guerriero, G., Ciarcia, G., Malorni, A., Russo, G. L., et al. (2008). Fish authentication by MALDI-TOF mass spectrometry. *Journal of Agricultural and Food Chemistry, 56*, 11071–11076.

Medina-Gali, R., Belló-Pérez, M., Ciordia, S., Mena, M. C., Coll, J., Novoa, B., et al. (2019). Plasma proteomic analysis of zebrafish following spring viremia of carp virus infection. *Fish & Shellfish Immunology, 86*, 892–899.

Menanteau-Ledouble, S., Kattlun, J., Nöbauer, K., & El-Matbouli, M. (2014). Protein expression and transcription profiles of three strains of *Aeromonas salmonicida* ssp. salmonicida under normal and iron-limited culture conditions. *Proteome Science, 12*, 1–9, Article 29.

Moreira, M., Schrama, D., Farinha, A. P., Cerqueira, M., Raposo de Magalhães, C., Carrilho, R., et al. (2021). Fish pathology research and diagnosis in aquaculture of farmed fish; a proteomics perspective. *Animals, 11*, 1–25, Article 125.

Moreira, M., Schrama, D., Soares, F., Wulff, T., Pousão-Ferreira, P., & Rodrigues, P. (2017). Physiological responses of reared sea bream (Sparus aurata Linnaeus, 1758) to an Amyloodinium ocellatum outbreak. *Journal of Fish Diseases, 40*, 1545–1560.

Mulvenna, J., Sripa, B., Brindley, P. J., Gorman, J., Jones, M. K., Colgrave, M. L., et al. (2010). The secreted and surface proteomes of the adult stage of the carcinogenic human liver fluke Opisthorchis viverrini. *Proteomics, 10*, 1063–1078.

Nagasawa, K., & Moravec, F. (1995). Larval Anisakid nematodes of Japanese common squid (*Todarodes pacificus*) from the sea of Japan. *Journal of Parasitology, 81*, 69–75.

Näpflin, K., O'Connor, E. A., Becks, L., Bensch, S., Ellis, V. A., Hafer-Hahmann, N., et al. (2019). Genomics of host-pathogen interactions: Challenges and opportunities across ecological and spatiotemporal scales. *PeerJ, 7*, 1–37, Article e8013.

Nissa, M. U., Pinto, N., Parkar, H., Goswami, M., & Srivastava, S. (2021). Proteomics in fisheries and aquaculture: An approach for food security. *Food Control, 127*, 108–125.

Nombela, I., Lopez-Lorigados, M., Salvador-Mira, M. E., Puente-Marin, S., Chico, V., Ciordia, S., et al. (2019). Integrated transcriptomic and proteomic analysis of red blood cells from rainbow trout challenged with VHSV point towards novel immunomodulant targets. *Vaccines, 7*, 1–24, Article 63.

Novoslavskij, A., Terentjeva, M., Eizenberga, I., Valciņa, O., Bartkevičs, V., & Bērziņš, A. (2016). Major foodborne pathogens in fish and fish products: A review. *Ann. Microbiol., 66*, 1–15.

Nuez-Ortín, W. G., Carter, C. G., Nichols, P. D., Cooke, I. R., & Wilson, R. (2018). Liver proteome response of pre-harvest Atlantic salmon following exposure to elevated temperature. *BMC Genomics, 19*, 1–13, Article 133.

Nynca, J., Arnold, G. J., Fröhlich, T., & Ciereszko, A. (2015). Shotgun proteomics of rainbow trout ovarian fluid. *Reproduction, Fertility, and Development, 27*(3), 504–512.

Nynca, J., Arnold, G. J., Fröhlich, T., Otte, K., & Ciereszko, A. (2014a). Proteomic identification of rainbow trout sperm proteins. *Proteomics, 14*, 1569–1573.

Nynca, J., Arnold, G. J., Fröhlich, T., Otte, K., Flenkenthaler, F., & Ciereszko, A. (2014b). Proteomic identification of rainbow trout seminal plasma proteins. *Proteomics, 14*, 133–140.

Oidtmann, B. C., Peeler, E. J., Thrush, M. A., Cameron, A. R., Reese, R. A., Pearce, F. M., et al. (2014). Expert consultation on risk factors for introduction of infectious pathogens into fish farms. *Preventive Veterinary Medicine, 115*, 238–254.

Ormsby, M. J., Grahame, E., Burchmore, R., & Davies, R. L. (2019). Comparative bioinformatic and proteomic approaches to evaluate the outer membrane proteome of the fish pathogen Yersinia ruckeri. *Journal of Proteomics, 199*, 135–147.

Ortea, I., Cañas, B., & Gallardo, J. M. (2009). Mass Spectrometry characterization of species-specific peptides from arginine kinase for the identification of commercially relevant shrimp species. *Journal of Proteome Research, 8*, 5356–5362.

Ortea, I., Cañas, B., & Gallardo, J. M. (2011). Selected tandem mass spectrometry ion monitoring for the fast identification of seafood species. *Journal of Chromatography A, 1218*, 4445–4451.

Ortea, I., O'Connor, G., & Maquet, A. (2016). Review on proteomics for food authentication. *Journal of Proteomics, 147*, 212–225.

Parrington, J. (2002). Use of emerging genomic and proteomic technologies in fish physiology. *Aquat. Living Resour., 15*, 193–196.

Patel, D. M., & Brinchmann, M. F. (2017). Skin mucus proteins of lumpsucker (Cyclopterus lumpus). *Biochem. Biophys. Reports, 9*, 217–225.

Peñaranda, M. M. D., Jensen, I., Tollersrud, L. G., Bruun, J.-A., & Jørgensen, J. B. (2019). Profiling the Atlantic Salmon IgM+ B Cell surface proteome: Novel information on teleost fish B cell protein repertoire and identification of potential B cell markers. *Frontiers in Immunology, 10*, 1–19, Article 37.

Pepe, T., Ceruso, M., Carpentieri, A., Ventrone, I., Amoresano, A., & Anastasio, A. (2010). Proteomics analysis for the identification of three species of Thunnus. *Veterinary Research Communications, 34*, 153–155.

Pérez-Sancho, M., Vela Ana, I., Awad, M., Kostrzewa, M., Domínguez, L., & Fernández-Garayzábal, J. F. (2016). Differentiation of Photobacterium damselae subspecies using matrix-assisted laser-desorption/ionization time-of-flight mass spectrometry (MALDI-TOF MS) in fish isolates. *Aquaculture, 464*, 159–164.

Pérez-Sancho, M., Vela, A. I., Wiklund, T., Kostrzewa, M., Domínguez, L., & Fernández-Garayzábal, J. F. (2017). Differentiation of Flavobacterium psychrophilum from Flavobacterium psychrophilum -like species by MALDI-TOF mass spectrometry. *Research in Veterinary Science, 115*, 345–352.

Petty, B. D., Francis-Floyd, R., & Yanong, R. P. E. (2012). Spring viremia of carp. UF/IFAS Extension, 1–6, VM-142.

Piovesana, S., Capriotti, A. L., Caruso, G., Cavaliere, C., La Barbera, G., Zenezini Chiozzi, R., et al. (2016). Labeling and label free shotgun proteomics approaches to characterize muscle tissue from farmed and wild gilthead sea bream (Sparus aurata). *Journal of Chromatography A, 1428*, 193–201.

Polak, I., Łopieńska-Biernat, E., Stryiński, R., Mateos, J., & Carrera, M. (2020). Comparative proteomics analysis of *Anisakis simplex* s.s.—evaluation of the response of invasive larvae to Ivermectin. *Genes (Basel), 11*, 1–15, Article 710.

Poobalane, S., Thompson, K. D., Diab, A., Ardó, L., Jeney, G., & Adams, A. (2008). Protein expression by Aeromonas hydrophila during growth in vitro and in vivo. *Microbial Pathogenesis, 45*, 60–69.

Provan, F., Jensen, L. B., Uleberg, K. E., Larssen, E., Rajalahti, T., Mullins, J., et al. (2013). Proteomic analysis of epidermal mucus from sea lice-infected Atlantic salmon, Salmo salar L. *Journal of Fish Diseases, 36*, 311–321.

Puentes, B., Balado, M., Bermúdez-Crespo, J., Osorio, C. R., & Lemos, M. L. (2017). A proteomic analysis of the iron response of Photobacterium damselae subsp. damselae reveals metabolic adaptations to iron levels changes and novel potential virulence factors. *Veterinary Microbiology, 201*, 257–264.

Rajan, B., Fernandes, J. M. O., Caipang, C. M. A., Kiron, V., Rombout, J. H. W. M., & Brinchmann, M. F. (2011). Proteome reference map of the skin mucus of Atlantic cod (Gadus morhua) revealing immune competent molecules. *Fish & Shellfish Immunology, 31*, 224–231.

Rajan, B., Lokesh, J., Kiron, V., & Brinchmann, M. F. (2013). Differentially expressed proteins in the skin mucus of Atlantic cod (*Gadus morhua*) upon natural infection with *Vibrio anguillarum. BMC Veterinary Research, 9*, 1–11, Article 103.

Raposo de Magalhães, C. S. F., Cerqueira, M. A. C., Schrama, D., Moreira, M. J. V., Boonanuntanasarn, S., & Rodrigues, P. M. L. (2020). A proteomics and other Omics approach in the context of farmed fish welfare and biomarker discovery. *Rev. Aquac., 12,* 122–144.

Rim, H.-J., Sohn, W.-M., Yong, T.-S., Eom, K. S., Chai, J.-Y., Min, D.-Y., et al. (2008). Fishborne trematode metacercariae detected in freshwater fish from vientiane municipality and savannakhet province. *Lao PDR. Korean J. Parasitol., 46*(4), 253–260.

Riva, C., Cristoni, S., & Binelli, A. (2012). Effects of triclosan in the freshwater mussel Dreissena polymorpha: A proteomic investigation. *Aquat. Toxicol, 118–119,* 62–71.

Rodrigues, P. M., Schrama, D., Campos, A., Osório, H., & Freitas, M. (2016). *Applications of proteomics in aquaculture.* In *Agricultural Proteomics Volume 1* (pp. 175–209). Cham: Springer International Publishing.

Rodrigues, P. M., Silva, T. S., Dias, J., & Jessen, F. (2012). PROTEOMICS in aquaculture: Applications and trends. *Journal of Proteomics, 75,* 4325–4345.

Rodríguez-Mahillo, A. I., González-Muñoz, M., de las Heras, C., Tejada, M., & Moneo, I. (2010). Quantification of Anisakis simplex allergens in fresh, long-term frozen, and cooked fish muscle. *Foodborne Pathog. Dis., 7,* 967–973.

Saavedra, J., Hernandez, N., Osses, A., Castillo, A., Cancino, A., Grothusen, H., et al. (2017). Prevalence, geographic distribution and phenotypic differences of Piscirickettsia salmonis EM-90-like isolates. *Journal of Fish Diseases, 40,* 1055–1063.

Saleh, M., Kumar, G., Abdel-Baki, A.-A. S., Dkhil, M. A., El-Matbouli, M., & Al-Quraishy, S. (2019). Quantitative proteomic profiling of immune responses to Ichthyophthirius multifiliis in common carp skin mucus. *Fish & Shellfish Immunology, 84,* 834–842.

Salinas, I., & Parra, D. (2015). *Fish mucosal immunity: Intestine.* In *Mucosal Health in Aquaculture* (pp. 135–170). Cambridge, MA, United States: Academic Press.

Sanahuja, I., & Ibarz, A. (2015). Skin mucus proteome of gilthead sea bream: A non-invasive method to screen for welfare indicators. *Fish & Shellfish Immunology, 46,* 426–435.

Scarfe, A. D., & Palić, D. (2020). Aquaculture biosecurity: Practical approach to prevent, control, and eradicate diseases. In *Aquaculture Health Management* (pp. 75–116). Amsterdam: Elsevier.

Schauer, K. L., Reddam, A., Xu, E. G., Wolfe, L. M., & Grosell, M. (2018). Interrogation of the Gulf toadfish intestinal proteome response to hypersalinity exposure provides insights into osmoregulatory mechanisms and regulation of carbonate mineral precipitation. *Comp. Biochem. Physiol. Part D Genomics Proteomics, 27,* 66–76.

Sen, R., Nayak, L., & De, R. K. (2016). A review on host–pathogen interactions: Classification and prediction. *European Journal of Clinical Microbiology & Infectious Diseases, 35,* 1581–1599.

Sharma, M., Shrivastav, A. B., Sahni, Y. P., & Pandey, G. (2012). Overwiews of the treatment and control of the common fish diseases. *Int. Res. J. Pharm., 3,* 123–127.

Shen, C.-J., Kuo, T.-Y., Lin, C.-C., Chow, L.-P., & Chen, W.-J. (2010). Proteomic identification of membrane proteins regulating antimicrobial peptide resistance in Vibrio parahaemolyticus. *Journal of Applied Microbiology, 108,* 1398–1407.

Shuang, F., Luo, Y., Xiong, X., Weng, S., Li, Y., He, J., et al. (2013). Virions proteins of an RSIV-type megalocytivirus from spotted knifejaw Oplegnathus punctatus (SKIV-ZJ07). *Virology, 437,* 89–99.

Sialana, F. J., Schneebauer, G., Paunkov, A., Pelster, B., & Lubec, G. (2018). Proteomic studies on the swim bladder of the European Eel (Anguilla anguilla). *Proteomics, 18,* 1–5, Article 1700445.

Singer, H., Müller, S., Tixier, C., & Pillonel, L. (2002). Triclosan: Occurrence and fate of a widely used biocide in the aquatic environment: Field measurements in wastewater treatment plants, surface waters, and lake sediments. *Environmental Science & Technology, 36,* 4998–5004.

Sitjà-Bobadilla, A., & Oidtmann, B. (2017). Integrated pathogen management strategies in fish farming. In D. Jeney (Ed.), *Fish Diseases* (pp. 119–144). Cambridge, MA: Academic Press.

Smith, J. W., & Wootten, R. (1978). *Anisakis and Anisakiasis*. In *Advances in Parasitology* (pp. 93–163). London: Academic Press Inc.

Smith, R. W., Moccia, R. D., Mothersill, C. E., & Seymour, C. B. (2018a). Irradiation of rainbow trout at early life stages results in a proteomic legacy in adult gills. Part B; the effect of a second radiation dose, after one year, on the proteomic responses in the irradiated and non-irradiated bystander fish. *Environmental Research, 163*, 307–313.

Smith, R. W., Moccia, R. D., Seymour, C. B., & Mothersill, C. E. (2018b). Irradiation of rainbow trout at early life stages results in a proteomic legacy in adult gills. Part A; proteomic responses in the irradiated fish and in non-irradiated bystander fish. *Environmental Research, 163*, 297–306.

Song, W. J., Qin, Q. W., Qiu, J., Huang, C. H., Wang, F., & Hew, C. L. (2004). Functional genomics analysis of singapore grouper iridovirus: Complete sequence determination and proteomic analysis. *Journal of Virology, 78*, 12576–12590.

Srinivasa Rao, P. S., Yamada, Y., Tan, Y. P., & Leung, K. Y. (2004). Use of proteomics to identify novel virulence determinants that are required for Edwardsiella tarda pathogenesis. *Molecular Microbiology, 53*, 573–586.

Srivastava, V., Rezinciuc, S., & Bulone, V. (2018). Quantitative proteomic analysis of four developmental stages of *Saprolegnia parasitica*. *Frontiers in Microbiology, 8*, 1–13, Article 2658.

Stryiński, R., Łopieńska-Biernat, E., & Carrera, M. (2020). Proteomic insights into the biology of the most important foodborne parasites in Europe. *Foods, 9*(10), 1–34, Article 1403.

Stryiński, R., Mateos, J., Pascual, S., González, Á. F., Gallardo, J. M., Łopieńska-Biernat, E., et al. (2019). Proteome profiling of L3 and L4 Anisakis simplex development stages by TMT-based quantitative proteomics. *Journal of Proteomics, 201*, 1–11.

Sun, L., Lin, H., Li, Z., Sun, W., Wang, J., Wu, H., et al. (2019). Development of a method for the quantification of fish major allergen parvalbumin in food matrix via liquid chromatography-tandem mass spectrometry with multiple reaction monitoring. *Food Chemistry, 276*, 358–365.

Sun, Y.-Y., Chi, H., & Sun, L. (2016). *Pseudomonas fluorescens* filamentous hemagglutinin, an iron-regulated protein, is an important virulence factor that modulates bacterial pathogenicity. *Frontiers in Microbiology, 7*, 1–11, Article 1320.

Sussarellu, R., Suquet, M., Thomas, Y., Lambert, C., Fabioux, C., Pernet, M. E. J., et al. (2016). Oyster reproduction is affected by exposure to polystyrene microplastics. *Proceedings of the National Academy of Sciences, 113*, 2430–2435.

Tan, Y. P., Rao, P. S. S., Zhang, Y. L., Hew, C. L., & Leung, K. Y. (2003). *Identification of virulence genes in bacterial fish pathogens: A genomic and proteomic approach*. In *Aquatic genomics* (pp. 340–351). Tokyo: Springer.

Tavares, G. C., Carvalho, A. F., Pereira, F. L., Rezende, C. P., Azevedo, V. A. C., Leal, C. A. G., et al. (2018). Transcriptome and Proteome of Fish-Pathogenic *Streptococcus agalactiae* Are Modulated by Temperature. *Frontiers in Microbiology, 9*, 1–21, Article 2639.

The World Bank. (2013). *FISH TO 2030—prospects for fisheries and aquaculture*. Washington, DC: The World Bank.

Torres-Corral, Y., & Santos, Y. (2019). Identification and typing of Vagococcus salmoninarum using genomic and proteomic techniques. *Journal of Fish Diseases, 42*, 597–612.

Trevisan, C., Torgerson, P. R., & Robertson, L. J. (2019). Foodborne Parasites in Europe: Present status and future trends. *Trends in Parasitology, 35*, 695–703.

United Nations;(2019). Department of economic and social affairs (Population Division), World Population Prospects 2019: Highlights.

Uttakleiv Ræder, I. L., Paulsen, S. M., Smalås, A. O., & Willassen, N. P. (2007). Effect of fish skin mucus on the soluble proteome of Vibrio salmonicida analysed by 2-D gel electrophoresis and tandem mass spectrometry. *Microbial Pathogenesis, 42*, 36–45.

Valdenegro-Vega, V. A., Crosbie, P., Bridle, A., Leef, M., Wilson, R., & Nowak, B. F. (2014). Differentially expressed proteins in gill and skin mucus of Atlantic salmon (Salmo salar) affected by amoebic gill disease. *Fish & Shellfish Immunology, 40*, 69–77.

Van Thiel, P. H. (1962). Anisakiasis. *Parasitology, 52*, 16–17.

van West, P. (2006). Saprolegnia parasitica, an oomycete pathogen with a fishy appetite: New challenges for an old problem. *Mycologist, 20*, 99–104.

Vancsok, C., Peñaranda, M. M. D., Raj, V. S., Leroy, B., Jazowiecka-Rakus, J., Boutier, M., et al. (2017). Proteomic and functional analyses of the virion transmembrane proteome of Cyprinid Herpesvirus 3. *Journal of Virology* 91(21), 1–20, Article e01209–17.

Veiseth-Kent, E., Grove, H., Færgestad, E. M., & Fjæra, S. O. (2010). Changes in muscle and blood plasma proteomes of Atlantic salmon (Salmo salar) induced by crowding. *Aquaculture, 309*, 272–279.

von Jørgensen, L. G. (2017). The fish parasite Ichthyophthirius multifiliis—Host immunology, vaccines and novel treatments. *Fish & Shellfish Immunology, 67*, 586–595.

Waagbø, R. (2006). Feeding and disease resistance in fish. In R. Mosenthin, J. Zentek, & T. Żebrowska (Eds.), *Biology of growing animals* (pp. 387–415). Amsterdam: Elsevier.

Wagg, S. K., & Lee, L. E. J. (2005). A proteomics approach to identifying fish cell lines. *Proteomics, 5*, 4236–4244.

Wang, Y., Wang, Q., Yang, M., & Zhang, Y. (2013). Proteomic analysis of a twin-arginine translocation-deficient mutant unravel its functions involved in stress adaptation and virulence in fish pathogen Edwardsiella tarda. *Fems Microbiology Letters, 343*, 145–155.

Wen, X., Zhang, X., Hu, Y., Xu, J., Wang, T., & Yin, S. (2019). iTRAQ-based quantitative proteomic analysis of Takifugu fasciatus liver in response to low-temperature stress. *Journal of Proteomics, 201*, 27–36.

Wulff, T., Nielsen, M. E., Deelder, A. M., Jessen, F., & Palmblad, M. (2013). Authentication of fish products by large-scale comparison of tandem mass spectra. *Journal of Proteome Research, 12*, 5253–5259.

Wykoff, D. E., Harinasuta, C., Juttijudata, P., & Winn, M. M. (1965). *Opisthorchis viverrini* in Thailand: The life cycle and comparison with *O. felineus*. *Journal of Parasitology, 51*, 207–214.

Xiong, X.-P., Dong, C.-F., Xu, X., Weng, S.-P., Liu, Z.-Y., & He, J.-G. (2011). Proteomic analysis of zebrafish (Danio rerio) infected with infectious spleen and kidney necrosis virus. *Developmental and Comparative Immunology, 35*, 431–440.

Yao, J.-Y., Xu, Y., Yuan, X.-M., Yin, W.-L., Yang, G., Lin, L.-Y., et al. (2017). Proteomic analysis of differentially expressed proteins in the two developmental stages of Ichthyophthirius multifiliis. *Parasitology Research, 116*, 637–646.

Yolanda, T.-C., Clara, F.-Á., & Ysabel, S. (2019). Proteomic and molecular fingerprinting for identification and tracking of fish pathogenic Streptococcus. *Aquaculture, 498*, 322–334.

Zhang, B., Zhang, J., & Sun, L. (2014). *Streptococcus iniae* SF1: Complete genome sequence, proteomic profile, and immunoprotective antigens. *Plos One, 9*, 1–14, Article e91324.

Zhang, J., Hu, Y., Xiao, Z., & Sun, L. (2013). Megalocytivirus-induced proteins of turbot (Scophthalmus maximus): Identification and antiviral potential. *Journal of Proteomics, 91*, 430–443.

Zhang, Q.-H., Huang, L., Zhang, Y., Ke, C.-H., & Huang, H.-Q. (2013). Proteomic approach for identifying gonad differential proteins in the oyster (Crassostrea angulata) following food-chain contamination with HgCl2. *Journal of Proteomics, 94*, 37–53.

Zhou, S., Wan, Q., Huang, Y., Huang, X., Cao, J., Ye, L., et al. (2011). Proteomic analysis of Singapore grouper iridovirus envelope proteins and characterization of a novel envelope protein VP088. *Proteomics, 11*, 2236–2248.

CHAPTER

6

Proteomics advances in beef production

Mohammed Gagaoua, Yao Zhu

Food Quality and Sensory Science Department, Teagasc Food Research Centre, Ashtown, Dublin, Ireland

6.1 Introduction

Foodomics is the term used to designate the OMICs-based approaches developed in the last two decades for an in-depth characterization and control of several food products including muscle foods (Munekata, Pateiro, López-Pedrouso, Gagaoua, & Lorenzo, 2021; Valdés et al., 2022). Very briefly, foodomics can be described as the discipline that combines food and nutrition studies through the implementation of OMICs technologies to improve the quality and safety of food products for consumer's health and well-being, and for increasing our knowledge on the underlying pathways and mechanisms behind the development of technological and sensory qualities of foods. It can be further described as the approach allowing comprehensive and high-throughput information of composition, safety, nutritional value, etc., by using OMICs-based approaches such as genomics, transcriptomics, proteomics, metabolomics, lipidomics, etc. The huge amount of data yielded by these OMICs approaches are mostly analyzed with advanced statistics and chemometrics (Gagaoua, Monteils, Couvreur, & Picard, 2019a) or bioinformatics (Gagaoua et al., 2021a, 2021b) to decipher the underlying pathways or to produce new knowledge about food quality variation or defects, including in muscle foods. In the field of meat science, OMICs-based approaches were efficiently implemented to study in untargeted manner the biological mechanisms in *post-mortem* muscle or fresh meat and processed meat products, hence advancing our knowledge in regards of meat quality variation and the origin of certain meat quality defects (Gagaoua, 2021; Gagaoua et al., 2021a; Munekata et al., 2021). For example, to develop new

Food Proteomics
DOI:https://doi.org/10.1016/B978-0-323-90889-4.00008-7

insights about the molecular mechanisms of novel biomarkers as well as in exploring the unknowns, foodomics was successful in analyzing the biochemical pathways underpinning the conversion of muscle into the meat (Gagaoua, Troy, & Mullen, 2021c; Ouali et al., 2013; Purslow, Gagaoua, & Warner, 2021). Therefore, foodomics is an inevitable trend in the development of biomarkers-based studies for a better understanding and management of meat quality variation (D'Alessandro & Zolla, 2013).

Among the foodomics approaches used in food sciences and more specifically in meat science, proteomics has been acclaimed by the majority of researchers for several reasons. First, the analysis of the complex proteome has been boosted in recent years by the development and improvement of high-resolving separation systems, along with the even more accurate high-resolution tandem mass spectrometers. It was then extensively used for such analytical potential as an effective quantitative analysis method through the large-scale and systematic characterization of the global (whole proteins) or partial analysis of the proteome (subproteome) of the muscle tissue or fluids such as plasma (Picard, Gagaoua, & Hollung, 2017; Zhu et al., 2021a). Second, proteomics allows a better understanding of the mechanisms underpinning the phenome (phenotypes of interest in animal production including meat quality traits), mainly because the proteome is the main constituent of muscle tissue and it has important roles in the biochemical changes occurring during the transformation of muscle into meat and structural alterations in early *post-mortem* muscle and during meat aging. Third, proteomics was successfully applied to propose explanatory mechanisms at the origin of the variability of several beef eating quality traits (Gagaoua, Claudia Terlouw, Boudjellal, & Picard, 2015a, 2017a; Hollung, Veiseth, Jia, Færgestad, & Hildrum, 2007; Purslow et al., 2021; Verónica et al., 2021).

The analysis of the proteome depends on five important steps including (1) protein separation, (2) detection and identification of proteins of interest, (3) in-depth characterization of the proteins, (4) quantification and functional characterization using statistical and bioinformatics tools, and finally the study of the (5) interactions among the proteins through interactomics and biological interpretation in the context of the defined scientific question or experimental design (Gagaoua, 2021). The separation step is mainly performed at the protein and/or peptide level, typically based on one- or two-dimensional gel electrophoresis (1-DE or 2-DE gel-based approach) and/or liquid chromatography (LC gel-free approach). The subsequent analysis that is detection, is then performed by means of mass spectrometry (MS) using ionization methods such as matrix-assisted laser desorption/ionization (MALDI) and electrospray ionization (ESI). It is worth noting that two main, but different MS-based proteomics workflows, can be followed depending on whether the MS analysis is performed on the peptide fragments (bottom–up) or on intact proteins (top–down). In the later approach, the intact proteins are directly infused to a high-resolution mass spectrometer (HRMS) and the intact mass and further fragmentation inside the mass spectrometer are acquired using MALDI-time-of-flight/time-of-flight (TOF/TOF), ESI-ion trap (IT), ESI-quadrupole (Q)/TOF, Orbitrap and HRMS instrumentation Fourier transform ion cyclotron resonance (FTICR) MS (Álvarez-Rivera, Valdés, León, & Cifuentes, 2021).

The large number of papers and reviews concerning the application of proteomics in food sciences and meat science (Fig. 6.1) highlights its

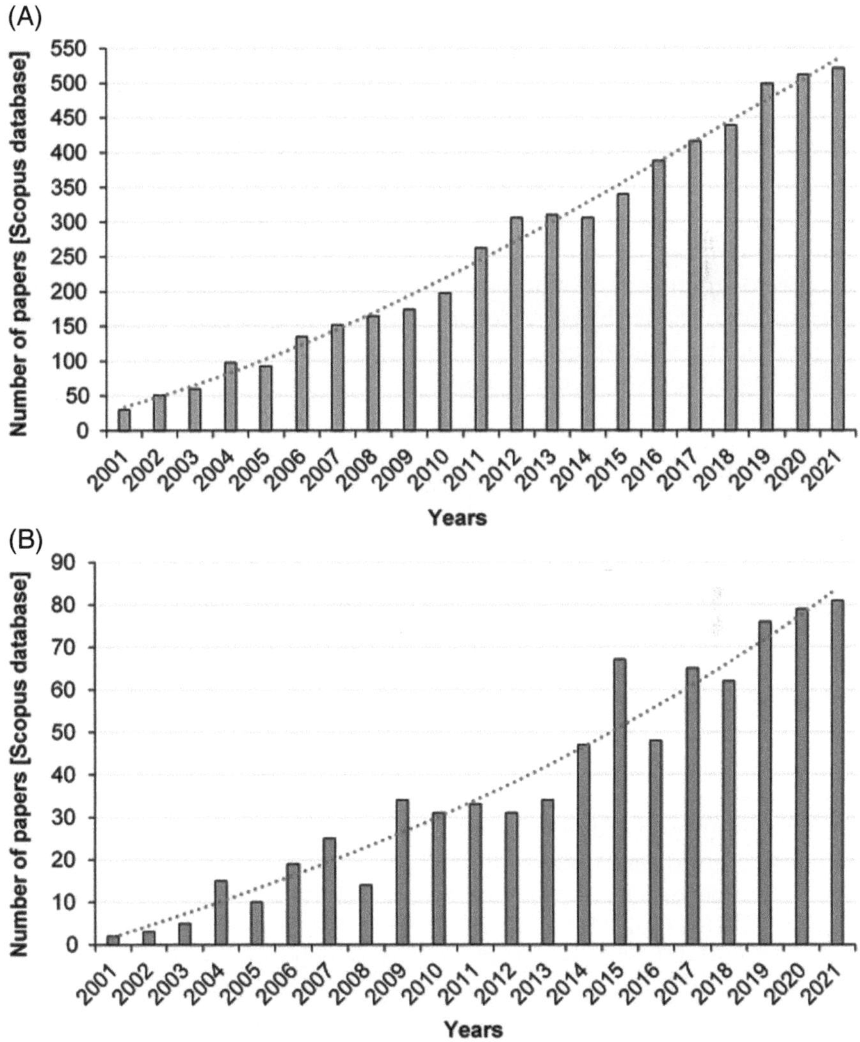

FIG. 6.1 Number of the published studies over the past 20 years in the field of proteomics and (A) food or (B) meat using the key words "proteom* & food," or "proteom* & meat," respectively, in Scopus database (accessed November 1, 2021).

importance as illustrated by the studies conducted to characterize the cellular and molecular mechanisms behind most meat quality traits from all species, muscle growth and development in the context of livestock production, or biological traits and diseases in farm animals using different biological sources (tissues or biofluids). Indeed, proteomics by means of gel-based and gel-free approaches coupled with MS has emerged as a well-defined strategy to study meat quality traits which, through its in-depth characterization of the whole bovine muscle proteome, allowed to provide substantial data on the biochemical mechanisms underpinning important beef quality traits (López-Pedrouso, Lorenzo, Gagaoua, & Franco, 2020; Picard & Gagaoua, 2017; Picard et al., 2017). Proteomics in beef research was also used to investigate other important production traits and factors related to cattle production such as age, breed, gender or sex, animal types, the effects of rearing practices or production systems, pre-slaughter stress, animal performances through for example proxies to feed efficiency, etc. These factors can be other sources of beef quality variation as they were evidenced in several studies to impact the muscle proteome among other biological sources. Thus, this chapter deals with the current advances and studies in beef proteomics research and attempts to summarize the findings evidenced using this high-throughput analytical method to strengthen our knowledge behind the unknowns and underlying biological mechanisms.

6.2 Proteomics to investigate cattle breeding, animal performances, and rearing practices

The global meat production is significantly increasing and expected to be around 200 million tons by 2050 to meet the expectations of a growing global population (Hunter, Smith, Schipanski, Atwood, & Mortensen, 2017). The main way to achieve such goal is to make stronger beef production using genetically superior animals. The physiology of growth in meat animals is an orchestrated process with a fine coordination of cell proliferation and increase in cell size in organs, skeletal muscle, bone, and adipose tissue from birth to market weight (Moloney & McGee, 2017). Specifically, growth characteristics likely body weight and feed efficiency are the two important traits evaluated at different ages of cattle and used as selection criteria to improve cattle characteristics (Gagaoua, Picard, & Monteils, 2018a; Guidolin et al., 2012). The other important trait routinely evaluated for beef cattle is carcass weight, which reflects life-time growth rate. Bovine skeletal muscle is a tissue of significant value in carcass weight and to the beef industry and global economy. In the case of dairy progeny destined for beef production, carcass weight accounts around ~70% of the monetary value of the total beef merit index (Keane, Dunne,

Kenny, & Berry, 2011). Breeding value for carcass weight is estimated using an animal prediction model and expressed as expected progeny difference (EPD_{CWT}) for carcass weight (Campion, Keane, Kenny, & Berry, 2009). Carcass weight can be increased by the use of sires of higher genetic merit for growth, that is mainly a consequence of skeletal muscle mass (Beline et al., 2021; Keane et al., 2011; Picard & Gagaoua, 2020a). Several studies reported that sire EPD_{CWT} had a positive association with progeny carcass weight (Crews, Pollak, & Quaas, 2004), while Keane & Diskin (2007) reported that the progeny of Limousin sires with high growth rate (high EPD_{CWT}) had heavier carcasses and a higher kill-out proportion than the progeny of sires selected for low growth rate. Thus, it is important to understand the detailed effects of genetic merit on animal performance as differences in genetic potential may require differences in feeding and management practices over production periods to ensure efficient use and exploitation of the animals. Rearing practices were also evidenced to impact the biochemical and structural properties of skeletal muscles, which consequently impact final meat quality traits and carcass profitability (Gagaoua, Picard, & Monteils, 2019b; Kim et al., 2018). Proteomics allowed to decipher of several of these changes and mechanisms and proposed in numerous of them putative biomarkers that can be used as management tools following the pipeline of validation.

6.2.1 Age at slaughter, genetic merit for carcass weight, growth rate, and path of cattle

Keady et al. (2013) have applied a proteomics approach to investigate the effects of both sire breed and EPDcwt on the proteome of *M. longissimus thoracis* (LT, a muscle contributing to beef prime-cuts with a high economic value compared to other muscles)from two breeds, Belgian Blue (BB) × Holstein Friesian and Aberdeen Angus (AA) × Holstein Friesian using 2-DE followed by MS. Multiple proteins with different abundances were identified between the two breeds and found to belong to several pathways including glycolysis (glycogen phosphorylase (PYGM), phosphoglycerate mutase (PGM1)) and the citric acid cycle (aconitate hydratase, ACO2, and 2-oxoglutarate dehydrogenase, OGDH). Expression of proteins involved in muscle structure, such as myosin light chain isoforms and troponins I (TNNI2) and T (TNNT3), were also different in the muscles of the two breeds breed. Furthermore, heat shock protein β-1 (known as HSPB1 or Hsp27) and peroxiredoxin 6 (PRDX6) had increased abundance in muscle of AA-sired relative to BB-sired steers (Keady et al., 2013). These findings were the first that enlightened us about the genetic influences controlling muscle growth and fat accumulation and possibility of using such proteomics knowledge for future breeding programs to increase lean tissue gain of beef cattle.

A study that focused on growth paths of the Alentejana bulls, investigated the differences in the proteome of LT muscle of continuous growth (CG) to the discontinuous growth (DG) animals using label-free quantitative proteomics (Almeida et al., 2017). From a list of quantifiable proteins (n = 510), 26 proteins belonging to muscle structure and contractility, energy and lipid metabolism, as well as cell signaling, proliferation, and responses to stress were found to differ significantly in their abundance between the two groups (25 were shown to have higher abundances in the CG group and only one protein; Myosin-binding protein H (MYBPH) had higher abundance in the DG group). In this study, a high number (n = 12) of the changing proteins were from the structure, development, morphogenesis, and muscle contractility. For example, Troponins T (slow, TNNT1) and I (TNNI1) and Myozenin-2 (MYOZ2) were more abundant in the CG group. Such results can be related to a higher muscular development, as a consequence of the younger age and putatively higher metabolic rates of the animals in the CG group. Other proteins from energy metabolism pathways: Phosphoglycerate mutase (PGM1), l-lactate dehydrogenase B chain (LDHB), Fructose-bisphosphate aldolase C (ALDOC), ADP-ribosylarginine hydrolase-like protein 1 (ADPRHL1) and ATP-dependent (S)-NAD(P)H-hydrate dehydratase isoform X1 (NAXD) and Carbonic Anhydrase 3 (CA3) and the 14-3-3 protein zeta/delta (YWHAZ) were further proposed as putative biomarkers for a more intensive growth path. Myosin binding protein H (MYBPH) known as a robust negative biomarker of beef tenderness whatever the gender (Gagaoua et al., 2021a) was more abundant in the DG animals, suggesting it to be associated with DG path as a consequence of feed restriction that reduced metabolic activity and delayed (muscle) growth. In support of these findings, MYBPH evaluated by Dot-Blot was in a previous study negatively related to EUROP conformation scores, ribeye, M. *Longissimus thoracis*, and hot carcass weights (Gagaoua, Monteils, & Picard, 2018c). Located in the A-band of the myofibrils, MYBPH is closely related to the thick filaments explaining the role it would play in muscularity development as previously evidenced under a marked (grass fed) rearing practice effect (Gagaoua, Monteils, Couvreur, & Picard, 2017b). Moreover, this protein has been previously associated using a proteomics approach to reduced muscle contractility due to potassium depletion (Thongboonkerd et al., 2006). The study by Almeida and co-workers further evidenced that some of the differences in the muscle proteome could be a direct consequence of the growth path, visible in the changes in marbling and meat color, that were higher in the CG animals (Almeida et al., 2017). In fact, carbonic anhydrase 3 (CA3) and myosin light chain (MYL6B) were previously identified as markers of higher marbling (Ceciliani, Lecchi, Bazile, & Bonnet, 2018) and tenderness or color of meat (Gagaoua, 2021) were suggested as markers of a more intensive growth path. Another label-free quantitative proteomics was performed to quantify the variations in protein abundance between cattle undergoing dietary restriction and subsequent compensatory growth in comparison to cattle that were continuously fed or fed a restricted diet (Mullins et al., 2020). The authors revealed 39 changing proteins belonging mainly to oxidative phosphorylation, cellular binding processes, and mitochondrial function. The study further evidenced the up regulation of three other pathways involved in nucleotide biosynthesis in addition to the validation of functional genes at the proteome level contributing to the expression of compensatory growth (Mullins et al., 2020).

The age at slaughter is an important consideration for the consumers where some of them have a preference for meat obtained from animals slaughtered at light weight or youthful age because it is believed that they yield tender meat (Vergara, Molina, & Gallego, 1999). Increasing age seems to be favorable for certain beef quality traits likely juiciness and flavor (due to more marbling), but not recommended for tenderness due to the development of connective tissue despite an attenuation of this effect when the amounts of intramuscular fat (IMF) are higher (Gagaoua, Picard, Soulat, & Monteils, 2018b; Listrat et al., 2020a, 2020b). Proteomics studies have allowed a better understanding of beef quality differences among breeds with different ages at slaughter (Romao, He, McAllister, & Guan, 2014; Silva et al., 2019; Zhao, Basu, Dodson, Basarb, & Guan, 2010). For example, Romao et al. (2014) investigated in eight British-Continental crossbred steers at 12 and 15 months of age the proteomic changes in subcutaneous adipose tissue of animals fed a high-grain diet in an effort to understand the molecular mechanisms of fat development during feedlot production. It resulted that 123 proteins of 627 differed in their abundance between the two ages from which 20.2% of the proteins were linked to cellular growth and proliferation of adipose tissue. Seventeen proteins from the lipid metabolism pathway were differentially expressed between the two groups which were annexin A1 (ANXA1), EH-domain containing 1 (EHD1), apolipoprotein C-III (APOC3), renin binding protein (RENBP), retinol-binding protein 4, plasma (RBP4), myosin, heavy chain 9, non-muscle (MYH9), acyl-CoA synthetase medium-chain family member 1 (ACSM1), apolipoprotein H (beta-2-glycoprotein I; APOH), lamin A/C (LMNA), gelsolin (GSN), coagulation factor II (thrombin; F2), mitogen-activated protein kinase 1 (MAP2K1), succinate dehydrogenase complex, subunit A, flavoprotein (Fp; SDHA), protein disulfide isomerase family A, member 3 (PDIA3), orosomucoid 1 (ORM1), serpin peptidase inhibitor, clade C (antithrombin), member 1 (SERPINC1), and serpin peptidase inhibitor, clade G (C1 inhibitor), member 1 (SERPING1). These findings revealed a coordinated balance between subcutaneous fat mass and the cellular abundance of lipogenic proteins to control the rate of fat deposition in growing beef cattle; hence expanding the current knowledge how the proteome of bovine adipose tissue is regulated during growth, which might help the development in the future of new strategies to manipulate adiposity in beef cattle in a manner that improves meat quality and animal productivity.

Another proteomics study investigated the fattening stages (early, 12 months *versus* late, 27 months) in Hanwoo steers (Kim et al., 2009). Using a 2-DE and MS, eight proteins were found to differ significantly between the two groups in LT muscle, among which Zinc finger and SCAN domain-containing protein 31 (ZSCAN31) and myosin light chain polypeptide 6 (MYL6) were highly expressed, further validated by western-blotting, in late-fattening stage, and two energy metabolism enzymes, triosephosphate isomerase 1 (TPI1) and succinate dehydrogenase (SDH) were more abundant in the early fattening stage. Slaughter weight also affects the carcass properties and meat quality traits, which mainly concern the muscle structure properties (Sañudo et al., 2004). Accordingly, Gulyás *et al.* studied the differences in the LT muscle proteome profiles of Charolais bulls slaughtered at 500 kg and 700 kg live weight (Gulyás et al., 2015) using two-dimensional difference gel electrophoresis (2D-DIGE), which is a modified form of 2-DE electrophoresis that allows the simultaneous comparison on the same gel of two or three protein sample extracts (Di Luca, Hamill, Mullen, & Elia, 2018). The study identified several energy metabolism proteins, mainly from glycolysis as main discriminators including ENO1, ENO3, and TPI1, and one structural protein (ACTA1) for being highly expressed in bulls of 700 kg live weight group while HSPB1 encoding for HSP27 was more abundant in 500 kg Charolais bulls.

Feed efficiency of growing animals at the farm is another important aspect of cattle production mainly measured by means of residual feed intake (RFI). Indeed, energy expenditure, and efficiency of energy conversion into body weight gain are of paramount importance for livestock due to the increasing demand for animal protein and sustainable use of resources (Baldassini et al., 2018). Efficient animals (low RFI) eat less than what is predicted based on their weights and growth rate, while inefficient animals (high RFI) eat more than predicted. The genetic and molecular mechanisms associated with the feed efficiency of animals were investigated in several studies (Abo-Ismail et al., 2018; Alexandre et al., 2014; Santana et al., 2014). However, the relationships between gene expression and the abundances of proteins are not very strong. Therefore, proteomics studies were applied (1) to decipher further molecular signatures, (2) to identify potential protein biomarkers, and (3) to strengthen the current knowledge. Researchers observed that several muscle proteins, including from mitochondria, are associated with feed efficiency (Ramos & Kerley, 2013). Liver proteomics can also unravel the metabolic pathways related to feed efficiency in beef cattle (Fonseca et al., 2019). Regarding metabolic pathways in skeletal muscle, proteomic studies have revealed that mitochondrial energy is lower for high feed efficiency (Baldassini et al., 2018). Integrated OMICs including metabolomics and proteomics of hepatic tissue allowed further insights into the mechanisms of different feed efficiency with high or low dietary forage levels in Holstein heifers (Zhang et al., 2019). This study revealed that 29 metabolites and 60 proteins were significantly different between the liver samples of the two groups. Enhanced hepatic ability of fatty acid oxidation, amino acid metabolism, pentose phosphate pathway, gluconeogenesis, ketone bodies synthesis, and oxidative stress defense was found in high forage consumed heifers, which were all related to energy utilization in ruminants and might be responsible for the reduced feed efficiency. Another study conducted by Carvalho and co-workers used a 2-DE and mass spectrometry ESI-MS approach to reveal the differences in key proteins associated with energy metabolism in skeletal muscle of divergent RFI young Nellore bulls using LT muscle sampled after slaughter (Carvalho et al., 2019). It resulted that YWHAE (14–3-3 protein epsilon) was more abundant in skeletal muscle of RFI high cattle (RFI-high) and HSPB1 was more abundant in RFI-low animals. These results suggest that differences in energy expenditure in the skeletal muscle may contribute to their difference in feed efficiency. The rate of oxygen uptake by muscle mitochondria and respiratory chain protein concentrations was also shown to differ in another study between high and low RFI animals (Davis, Brooks, & Kerley, 2016). For a comprehensive overview of the use of other OMICs methods including metabolomics, transcriptomics, and epigenomics to investigate in a comprehensive manner feed efficiency in cattle with a focus on the impact of prenatal nutrition on physiological phenotypes in calves, the reader may refer to this very recent review (Foroutan, Wishart, & Fitzsimmons, 2021).

6.2.2 Sex/gender

The possibility to control the sex of cattle is a topic of great interest due to the role that sex counts in the variation of meat quality and other phenotypes of interest. Overall, for a given slaughter weight, age, and fattening duration in the same breed, bulls have lower IMF contents than heifers (Harper & Pethick, 2004). These observations indicate that sex hormones may affect both hyperplasia and hypertrophy of intramuscular adipocytes in cattle. Thus, controlling the ratio between males and females entails direct returns in the livestock industry allowing a better management of food production, faster genetic selection, animal welfare improvement, and a decrease of environmental impact (De Canio et al., 2014). In fact, castration as a major tool in beef cattle management is, for example, used to reduce aggressive behavior and improve management ease of the animals. Understanding the sex effects and their manipulation has then huge importance for the industry, due to several sex-related traits, like milking, herd replacement, and growth rates as well as the differences in meat quality outcomes such as marbling, color, and texture of the produced meat. For example, castration is known to affect meat yield (reduced muscle mass/strength) and quality (increased amount of subcutaneous fat), and usually, bulls are heavier and leaner than steers owing to the anabolic effect of testosterone on skeletal muscle tissue, while steers had greater carcass fat content (Moreira et al., 2017). However, divergences might exist in the literature and there is no unique consensus on such comparisons due to the multiple factors underlying the relationships and differences in the genotypes. On another hand, the sensory qualities of meat from steers were found to be higher (mainly in terms of overall liking) than bulls while heifers had intermediate levels (Gagaoua et al., 2016). A recent proteomics study investigated such divergences for a better understanding of the mechanisms and found that several pathways are changing for instance between bulls and steers of Nellore breed (*Bos taurus indicus*) (Silva et al., 2019). The authors showed that castration upregulated the abundances of certain glycolytic enzymes likely fructose-bisphosphate aldolase (ALDOA), glyceraldehyde-3-phosphate dehydrogenase (GAPDH) and TPI1, hence indicating an increase in skeletal muscle glucose metabolism in steers while the mitochondrial ATP synthase subunit beta (ATP5B), involved in oxidative phosphorylation, was down-regulated. Based on these results, one can speculate that the greater glycolysis along with reduced oxidative phosphorylation in the muscles of steers suggests a greater synthesis of glycerol-3-phosphate and fatty acids from glucose, in agreement with the high IMF content in steers loins. The authors further evidenced by means of phosphoproteomics differences in the abundances of muscle structure proteins such as troponin T fast isoform (TNNT3), which was upregulated by castration, while the slow isoform (TNNT1) tended to decrease, compared to bulls. Moreover, this study revealed an upregulation of creatine kinase M-type (CKM) in the muscle of steers, an enzyme related to ATP regeneration by creatine phosphate, which is a metabolism more used by fast-twitch muscles (Picard & Gagaoua, 2020b), and inversely myosin light chain 1 isoform (MYL1) was more abundant in the muscle of bulls.

Another in-depth study compared the protein profiles associated with testosterone-related differences in adipogenesis (adipose tissue) and myogenesis (muscle) between steers and bulls (Zhang et al., 2012). Four proteins were different in their abundance in skeletal muscle, where ankyrin repeat domain-containing protein 1 (ANKRD1), myosin light chain 1 (MLC1), and heat shock protein beta 1 (HSPB1) were up-regulated in bulls and cofilin 2 (CFL2) was down-regulated, compared to bulls. On another hand, the levels of L-lactate dehydrogenase B chain (LDHB) and Transaldolase (TALDO1) from the adipose tissue were lower in bulls compared to steers. The greater abundance of ANKRD1 and HSPB1 in bulls can be a consequence of higher muscle hypertrophy, while the greater abundance of CFL2 in steers would be negatively related to muscle hypertrophy due to actin filaments depolymerisation (Kammoun et al., 2016; Picard & Gagaoua, 2020b).

Guillemin and co-workers measured by means of the dot-blot technique (Gagaoua et al., 2015a), the abundances of 24 protein biomarkers in LT and ST muscles of bulls and steers (Guillemin, Bonnet, Jurie, & Picard, 2011a). The authors confirmed the previous suggestions by validating the higher abundances in steers of fast skeletal muscle MYL1, myosin heavy chain IIx and the glycolytic enzymes enolase-1 (ENO1) and enolase-3 (ENO3). Steers were found also to have greater abundances of m- and μ-calpains. In terms of heat shock proteins (HSPs), bulls had higher abundance of αB-crystallin (CRYAB) and HSPB1, two small HSPs and chaperones related to early *post-mortem* muscle conversion into meat and programmed cell death onset (Ouali et al., 2013; Picard & Gagaoua, 2017) but lower abundances of the large HSP70 members.

The abundances of 20 targeted proteins, previously identified as putative biomarkers of beef marbling and tenderness, were also quantified by the Reverse Phase Protein Array (RPPA), a quantitative microformat Dot-Blot approach (Gagaoua, Bonnet, Ellies-Oury, De Koning, & Picard, 2018d) and compared in terms of their relative abundance between cows and steers of the Protected Designation Origin (PDO) Maine Anjou breed among five different muscles: LT, *Semimembranosus* (SM), *Rectus abdominis* (RA), *Triceps brachii* (TB) and *Semitendinosus* (ST) (Picard, Gagaoua, Al Jammas, & Bonnet, 2019). The study showed that the muscles of cows comparatively to steers differed by the abundances of 8 proteins belonging to energy metabolism, contraction, and cellular stress pathways: HSP20, PRDX6, PGK1, ALDOA, TPI1, MDH1, MyHC-IIX (fast glycolytic isoform), and TNNT1 (slow isoform). It is worth noting that the gender effect was muscle type-dependent with greater abundances in ST, RA, and LT muscles. The few dissimilarities were mainly explained by differences in sex hormones between the two genders in agreement with the knowledge of the effect of estrogens on skeletal muscle properties (Enns & Tiidus, 2010). Further, the fattening duration was found to impact certain proteins among the five muscles of PDO Maine Anjou cows (Picard et al., 2019). Thus, the abundances of 12 protein biomarkers differed significantly and mainly in TB muscle as the abundance of six proteins was modified. For TB muscle, the abundances of MLC1F, Glycogen phosphorylase (PYGB), PRDX6, and Four and a half LIM domains protein 1 (FHL1) decreased when fattening duration increased whereas the abundance of HSP70-1A and TTN (titin) increased. PYGB was also modified in LT and ST muscles (negatively) but not in RA and SM muscles. HSP70-1A was modified in RA but inversely in comparison with TB muscle.

ENO3 was inversely correlated with fattening duration in LT (positively) and SM (negatively). These variations seemed to be the result of the types of diet and animal activity in the farm that was independent of the slaughter weight and age in agreement to previous studies that found the influencing role of fattening duration of cattle (Gagaoua et al., 2017b, 2018b, 2018c).

A recent integrative proteomics study of 12 independent experiments conducted in the same laboratory using two-dimensional electrophoresis (2-DE) associated with MS (MALDI-TOF) of different types of cattle: young bulls, steers or cows from beef breeds (Charolais, Limousin, Blond d'Aquitaine), hardy breed (Salers), or mixed breed (PDO Maine-Anjou known also as Rouge des Près) revealed that HSPA1B (Heat shock 70 kDa protein 1B) and TPI1 are common protein biomarkers whatever the gender (Picard & Gagaoua, 2020a). The abundance of HSPA1B was always lower in tender meat for the three genders. The abundance of TPI1 was lower in tender meat of young bulls and higher in tender meat of steers and cows. As stated above, such differences could be explained mainly by differences in sex hormones between the genders (Enns & Tiidus, 2010). On another hand, two proteins: TNNT3 (Troponin T, fast skeletal muscle) and YWHAE were differentially abundant in males but inversely in steers and young bulls. In young bulls, TNNT3 was lower abundant in tender meat, while in steers it was higher. YWHAE is a member of a large and highly conserved family functioning as a dimer in diverse biological processes such as metabolism, protein trafficking, apoptosis, signal transduction, cell cycle regulation, and potassium channel regulation (Fu, Subramanian, & Masters, 2000). YWHAE was reported as an interesting biomarker of beef tenderness in several studies as evidenced in the recent integromics meta-analysis study (Gagaoua et al., 2021a). Among steers and cows, MSRA (Mitochondrial peptide methionine sulfoxide reductase) and HSPA9 (HSP70-GRP75) were the only significant proteins significantly different in groups of divergent tenderness. MSRA was higher abundant and GRP75 was lower abundant in groups of high tenderness. Finally, five proteins were common to cows and young bulls: HSPB1 and MYBPH in the same direction, and ENO1, MYH1, and PGM1 in opposite directions. In cows, PGM1 and ENO1 were higher abundant in tender group and MYH1 was lower abundant, the opposite was observed in young bulls. ENO3 was higher abundant in tender LT muscle only of cows (Picard & Gagaoua, 2020a).

6.2.3 Breed or genotype

In addition to the above, breed or genotype is another major well-known factor affecting both carcass properties and the final beef quality traits (Gagaoua et al., 2016; Picard et al., 2014). For example, early maturing Anglo-Saxon breeds such as AA and Japanese black cattle are characterized with high degree of fatness compared to the late-maturing breeds such as French beef breeds or double-muscled cattle known to have high muscle yield and low-fat cover scores (Christensen et al., 2011; Gotoh et al., 2009). Foodomics studies allowed huge advances to understand breed effect and to decipher the main consequences on the phenotypes or variations among breeds (Rodríguez-Vázquez et al., 2020). In the first studies where genomics was applied, significant knowledge on valuable genome-wide data for a very high number of markers and data on the inter-breed genomic diversity were encouraging, hence allowing identification of quantitative trait loci underlying inter-breed variations in carcass and beef quality traits (Ramayo-Caldas, Renand, Ballester, Saintilan, & Rocha, 2016; Silva et al., 2017). This boosted animal/meat scientists to concentrate further efforts on the phenotypes of interest by investigating the potential of comparative proteomics as an essential bridge from genome and transcriptome investigations to the production of functional proteins, to provide superimposed data to genetic studies and unveil the complex molecular pathways underpinning the variation in meat and carcass quality properties among cattle breeds (Gagaoua, Terlouw, Richardson, Hocquette, & Picard, 2019c; Ohsaki et al., 2007; Rodrigues et al., 2017; Timperio et al., 2009).

Gagaoua and co-workers investigated the differences in the protein abundances of AA and Limousin (LIM) young bulls (Gagaoua, Terlouw, & Picard, 2017c). The two breeds were selected on the basis of their divergences in physiological maturation rates (early *versus* late) and IMF content and perceived meat quality (Picard et al., 2014). From the primary list of 21 proteins quantified by Dot-Blot, 12 proteins differed significantly between AA and LIM bulls. Among the proteins, four structural proteins were found as good discriminators of AA and LIM namely F-actin-capping protein subunit beta (CAPZB) and all the isoforms of myosin heavy chains MyHC-I (MYH7), -IIa (MYH2), and -IIx/b (MYH1) evaluated by an adapted SDS-PAGE electrophoresis. The percentage of MYH1 was highest in LIM and the other proteins were highest in AA bulls. CAPZB is one of the most abundant barbed-end-capping proteins known to anchor the barbed ends of sarcomeric actin to the Z-disc (Maiti & Bamburg, 2013). This protein was reported in several proteomics studies as a biomarker of both beef tenderness and color (Gagaoua et al., 2020a, 2021a, 2021b) and having in Large White and Meishan pigs different levels of expression (Xu et al., 2009). More interestingly, Gagaoua *et al.* reported in the AA and LIM study differences in five of the seven HSPs measured proteins (Gagaoua et al., 2017c). The small HSPs proteins (CRYAB, HSPB6, and HSPB1) and the co-chaperone HSP40 were more abundant in AA breed, while those of the large and inducible HSP70 proteins were not different unless the non-inducible Hsp70-Grp75 (HSPA9) that was higher in LIM compared to AA breed. These interesting divergences in the abundance of HSPs can be related to differences in the proportions of myosin heavy chains fibers known for having different levels of HSPs, namely the abundances of small HSPs that are higher in oxidative fibers. Worthy to mention that slow oxidative activity in LT muscle of AA breed compared LIM, would contain more small HSPs (Folkesson et al., 2013). A companion study to this experimental design where Blonde d'Aquitaine breed was further included, allowed to strengthen our knowledge on the differences in the abundances of proteins among breeds confirming also that biomarkers of beef tenderness are not only breed-dependent but they are also depending on the country origin of the panelists and end-point cooking temperature of the meat (Gagaoua et al., 2019c). Other proteomics findings using 2-DE gels on the LT muscle of French breeds (Charolais, Limousin, and Blond d'Aquitaine) allowed the confirmation of an important role of proteins from muscle structure, energy metabolisms, and from the HSP family but with certain relationships with beef tenderness for being breed-dependent (Chaze et al., 2013).

An Italian group used 24 young bulls of breeds with different purposes, being Friesian (dairy), Romagnola x Podolian crossbreed (beef), and Podolian (rustic) to investigate by means of 2-DE the proteolytic patterns of myofibrillar proteins at 1, 7, 14, and 21 days aging (Marino et al., 2013). It resulted that Podolian cattle had lower marbling content and tougher beef at 1 and 7 days of aging and more important protein changes at 21days of aging that is mainly related to protein fragments of myosin heavy chain fibers. The authors found also several spots of proteins in Podolian that were identified as troponins, which can be explained by the tenderization of meat recently confirmed by a high-throughput proteomics study on heifers and steers with different glycolysing rates (Gagaoua et al., 2021c).

In the context of molecular breeding programs and authentication of beef origin, a recent Spanish study investigated the quantitative proteomic distances based on 2-DE gels among the three major cattle breeds in the Spanish meat industry that account around 78% of total beef production, these being Asturiana de los Valles (AV), Retinta (RE), and Rubia Gallega (RG) (Rodríguez-Vázquez et al., 2020). Estimates of qualitative and quantitative proteomic distances computed from 2-DE profiles using LT muscle provided opposite information about the overall proteomic differentiation among the three breeds. Therefore, eighteen differentially abundant proteins (myofibrillar and sarcoplasmic) identified by tandem MS allowed to build and explain the proteomics distances among the three breeds. The proteins belong to different biological pathways mainly from the contractile and muscle structure (MYBPH, ACTA1, ANK2, TNNT1, MYL6B, and MYLPF), followed by energy metabolism (CKM, PRDX3, PDHB, and YWHAE), nervous system development (TMEM233) and regulation of cell cycle (CCNG1). Among these proteins, MYLPF (Myosin regulatory light chain 2, skeletal muscle isoform) that is a structural protein was a major player in causing differences in tenderness among the three breeds. Interestingly, several of these proteins were previously proposed as robust biomarkers of beef tenderness (Gagaoua et al., 2021a), color (Gagaoua et al., 2020a), or to play a role in the development or dark-cutting beef condition, known also as DFD (Dark, Firm, Dry) meat (Gagaoua et al., 2021b).

The muscle proteomes of Japanese Black known for its high IMF content, tenderness, and flavor and Holstein cattle were compared using 2-DE gels (Ohsaki et al., 2007). It resulted that the LT muscle of Holstein had greater abundance of skeletal muscle MYLPF, ANKRD2 (Ankyrin repeat domain-containing protein 2), and OGN (Mimecan). The data were interpreted for MYLPF by the greater percentages of fast-twitch fibers (higher expression of MYLPF fast-twitch fibers), and for OGN by its role in collagen fibrillogenesis and connective tissue. In fact, this protein was recently identified by our group as a negative biomarker of beef tenderness in Irish young Limousin beef (Zhu et al., 2021b). OGN belongs to a secreted family of small leucine-rich proteoglycans located in the extracellular matrix (Hu et al., 2005). Therefore, the greater abundance of OGN in tougher beef samples (breeds) may be related to a higher abundance of connective tissue content (Listrat et al., 2020c).

A recent study combined proteomics and phosphoproteomics to compare Nellore (*Bos indicus*) and Angus (*Bos taurus*) breeds (Rodrigues et al., 2017). The authors found that Nellore breed had greater levels of Tropomyosin alpha-1 chain (TPM1), TNNT3, myosin light chain-1 fragment (MYL1), malate dehydrogenase 1 (MDH1), α-enolase (ENO1), and 78 kDa glucose-regulated protein (HSPA5), while myosin light chain 3 (MYL3), prohibitin (PHB), and two large HSP70 proteins HSPA9 and HSPA6 to be more abundant in Angus LT muscle. In terms of the phosphorylated proteins, Nellore breed had higher phosphorylation of MYL1, MYLPF, α-actin (ACTA1), TPI1, and YWHAE, while Angus had greater phosphorylation of PGM1 and TNNT3. Overall, the proteins involved in the contraction and muscle organization, myofilaments expressed in fast or slow-twitch fibers and HSPs localized in sarcoplasmic reticulum or mitochondria and involved in cell flux of calcium and apoptosis might be associated with differences in beef quality such as tenderness, color, and pH decline between Angus and Nellore breeds.

Another in-depth study applied proteomics by means of 2-DE and MS to understand changes in the LT muscle of Nellore breed genotyped for the CAPN4751 (calpain-1) and UOGCAST (calpastatin) polymorphisms and phenotyped for shear force at different aging times (1, 7, and 14 days) (Rosa et al., 2018). The results indicated differentially abundant proteins for the effect of interaction among molecular markers, including several structural proteins (myosin, actin, and troponin) such as myosins isoforms (MYL2, MYLPF, and MYH6), ACTA1, CAPZB, and TNNT3, and energy metabolism proteins (TPI1, CKM, ENO3, UQCRC1, TRIM72) and small heat shock proteins likely HSP27.

In line with the negative control role that the mutation MSTN (myostatin) plays on the proliferation of fast-twitch glycolytic muscle fibers, a proteomics analysis on ST muscle of double-muscled Belgian Blue showed greater differences in the abundance of proteins of the fast glycolytic type in favor of homozygote cattle comparatively to their controls, while the heterozygotes being intermediate (Bouley et al., 2005). This comprehensive study allowed to decipher certain of the mechanisms behind muscle hypertrophy and proposing some biomarkers among which MYBPH, also known as a biomarker of beef tenderness (Gagaoua et al., 2021a) and beef color (Gagaoua et al., 2020a).

6.2.4 Muscle types

The development of comparative and targeted proteomics allowed in the last years a better understanding of the key role of muscle types in meat quality variations (Beline et al., 2021; Gagaoua et al., 2015a, 2017a, 2018c; Picard & Gagaoua, 2020a; Picard et al., 2014, 2018). Integromics studies permitted further comparisons on the published proteome studies among different muscles, hence proposing robust biomarkers as can be exemplified by the identification of ENO3 for beef color whatever the muscle (Gagaoua et al., 2020a, 2021d). The analysis of the proteome of LT and ST muscle allowed to build the first networks of beef tenderness of Charolais bulls and steers (Guillemin et al., 2011b) that were followed some years after by original correlation networks on young bulls of three breeds (AA, Limousin, and Blond d'Aquitaine) (Gagaoua et al., 2015a). The tenderness biomarkers were further compared using several multivariate and regression analyses (Picard et al., 2014). In this study, we found that certain proteins such as m-calpain, CRAYB (a small HSPs), Peroxiredoxin 6 (PRDX6) and DJ-1 (PARK7, involved in oxidative stress), MYL1, MYH2, and MYBPH (involved in structure and contraction) were related to tenderness in the same direction in LT and ST muscles. However, some relationships were muscle-specific (Picard et al., 2014). Later on, comparisons between LT and RA muscles were reported for PDO Maine Anjou breed (Gagaoua, Couvreur, Le Bec, Aminot, & Picard, 2017a). Among the quantified biomarkers, 18 of 23 proteins quantified by Dot-Blot and belonging to six different and interconnected biological functions: heat shock proteins; muscle fiber; energy metabolism; proteolysis; oxidative resistance, and apoptosis were statistically different. Very recently, both proteomic and peptidomic approaches on New Zealand-raised Angus steers using 4 muscles (LT, ST, PM, and *Infraspinatus*) revealed molecular signatures distinguishing the muscles (Clerens et al., 2016). Further, Wu, Fu, Therkildsen, Li, and Dai (2015) reported differentially abundant sarcoplasmic proteins in LT, PM, and ST muscles from Luxi yellow cattle during *post-mortem* storage for 0, 5, 10, and 15 days. *Rectus abdominis* was evidenced in a comprehensive RPPA proteomics study among five muscles to have molecular peculiarities (Picard et al., 2018). In RA muscle, the proteins of fast glycolytic type were less abundant while those of the slow contractile type (ALDH1A1 and TNNT1), and small HSPs (HSPB6, HSPB1, and CRYAB) known to be more abundant in slow fibers (Liu & Steinacker, 2001; Picard et al., 2018), were highly abundant in RA in agreement with its biochemical characteristics (Gagaoua et al., 2017a). The differences in the proteome of muscles are overall in relation with the differences in their contractile and metabolic properties.

6.3 Proteomics to investigate beef quality and impact of post-slaughter effects: a focus on electrical stimulation and aging

The meat industry must strive to consistently provide high-quality meat in order to meet the expectations of discerning consumers. The supply of meat that is wholesome, safe, nutritionally beneficial, and of high quality ensures continued consumption of meat and meat products (Santos et al., 2021; Troy & Kerry, 2010). Meat quality can be defined by the traits which consumers find desirable and include sensory traits and confidence in the perceived safety, health, and welfare status of the animals and the production system (Becker, 2000). Many purchasing decisions are based upon the appearance of the meat in terms of color, texture, amount, and distribution of fat and absence of excess water in a packaged product. Once meat is cooked, meat quality is largely determined by the attributes of tenderness, flavor, and juiciness (Gagaoua et al., 2018b). These three attributes have also been identified as the most important determinants of sensory enjoyment by the consumer (O'Quinn, Legako, Brooks, & Miller, 2018). However, tenderness is still the primary sensory attribute that has huge impact on consumer acceptability for which several studies were conducted to control its variability and understand the underlying molecular signatures (Gagaoua et al., 2021a). On another hand, the modern beef industry currently lacks objective and fast methods to estimate the beef quality potential (mainly tenderness) of live animals or online in slaughterhouses, paying specific attention on sensory quality traits. In order to predict and best manage beef tenderness, there was a consensus that it is useful to first reveal the gene and protein networks involved in the development of this desired quality phenotype before moving toward the development of simple and fast tools. These targeted tools would allow managing the marketing of a meat quality trait of interest according to its potential or to develop genetic selection schemes.

6.3.1 Proteomics and electrical stimulation of carcasses

Electrical stimulation (ES) of a carcass in the early *post-mortem* period causes an increase in glycolysis that results in a rapid pH decline during the early development of *rigor mortis* (Simmons et al., 2008). ES is known to improve meat tenderness as early rigor development can be used to prevent cold shortening (Gursansky, O'Halloran, Egan, & Devine, 2010). ES has also been reported to enhance the proteolysis of myofibrillar and cytoskeletal proteins including structural proteins such as desmin, titin, troponin-T, and nebulin (Ferguson, Jiang, Hearnshaw, Rymill, & Thompson, 2000). The degradation patterns of structural proteins have been well documented and it is generally accepted that degradation of myofibrillar proteins contributes significantly to the development of meat tenderness (Hopkins & Thompson, 2001; Hwang, Devine, & Hopkins, 2003). Although a contribution of ES to protein degradation has been postulated, to our knowledge, there have been few proteomic studies addressing this issue. The earlier study by Bjarnadóttir and co-workers have shown that electrically stimulated muscle (low voltage) of Norwegian Red young bulls had lower abundance of enzymes related to energy metabolism namely ENO1 and CKM than unstimulated muscle, that can be a consequence of higher glycolysis and energy metabolism (Bjarnadóttir, Hollung, Faergestad, & Veiseth-Kent, 2010). Moreover, several protein changes in terms of the abundance were observed for the myofibrillar and cytoskeletal proteins including TNNT3, actin (ACTA1), and desmin (DES), in agreement with the role these proteins play in meat tenderization during aging (Gagaoua et al., 2021c). The authors found also higher quantities of the small HSPs proteins such as HSP27 and CRYAB in the muscle that had been ES indicating an increase in stress levels and suggested that this indicates a change in the stabilization of myofibrillar proteins partly due to the enhanced tenderization. Accordingly, the study by Contreras-Castillo, Lomiwes, Wu, Frost, and Farouk (2016) determined the effect of ES and pHu on shear force, myofibrillar protein degradation, and small HSPs levels in LT muscle. They confirmed the previous findings by showing that the protein levels of αβ-crystallin (CRYAB) and HSP20 (HSPB6) were significantly impacted with higher concentrations of these small HSPs protein members in ES muscles at later aging time-points compared to non-ES muscles. Using 2D-DIGE, Kim et al. (2013) determined the influence of low voltage ES on the tenderness development of beef round muscles. The results showed that ES did not influence proteolysis and tenderness, but did alter the predominance of metabolic proteins in the soluble fraction of muscle such as GAPDH, ALDOA, phosphoglycerate kinase-1 (PGK1), and PKM. The authors reported other protein changes namely in the abundance of Annexin A5 (ANXA5) that increased in the ES muscles in agreement with the role it plays in calcium homeostasis and possible activation of calpain systems (Purslow et al., 2021). Li et al. (2012)

investigated also the response of sarcoplasmic proteins in bovine LT muscle to low-voltage ES after carcass dressing. The proteomics analysis showed in line it previous observtions that ES resulted in decreased phosphorylation of CKM and other energy metabolism enzymes likely ALDOA, ENO3, and pyruvate kinase (PKM).

6.3.2 Proteomics and meat aging

After critical early biochemical events such as apoptosis, pH decline, and *rigor mortis*, the tenderization period known also as maturation commences leading to improvements of both the texture and sensory aspects of meat (Ouali et al., 2013). Aging consists of a number of biochemical processes, not fully elucidated, that are thought to be related to the breakdown of muscle proteins namely those of structure by endogenous proteolytic systems leading to the appearance of new protein fragments (Gagaoua et al., 2021c). In fact, two main effects can be observed: a progressive increase in tenderness, as the muscle structure is disintegrated, and a release and accumulation of peptides and free amino acids. Such molecules have an impact on color (Gagaoua et al., 2020a) and other quality attributes, but more importantly as precursors of flavor (Kim et al., 2018). Proteomics allowed to report the main changes that occur in *post-mortem* muscle and aged meat under several conditions. For example, changes in the abundances of TNNT3 and myosin fragments were evidenced by Polati et al., 2012. Very recently, we further confirmed that the maturation process involves interconnected molecular pathways in a pH-dependent manner leading for instance to the concomitant appearance of two major proteolytic fragments at 110 and 30 kDa based on 1DE electrophoresis (Gagaoua et al., 2021c). These two protein bands appearing during aging increase in their intensity as a function of *post-mortem* time in a pH decline-dependent manner. LC-MS/MS analysis yielded 22 unique proteins for the 110 kDa fragment and 13 for the 30 kDa, with four common proteins related to both actin and fibrinogen complex. The Gene Ontology analysis revealed that a myriad of biological pathways is influential with many of them related to proteins involved primarily in muscle contraction and structure. Other pathways were apoptotic mitochondrial changes, calcium, and ion transport energy metabolism, etc. Interestingly, most of the proteins composing these two fragments among others that appear or disappear during the tenderization process have been so far identified as biomarkers of beef tenderness (Gagaoua et al., 2021a). The changes in the abundance of the proteins as a consequence of pH support previous findings that identified for instance in Nellore cattle histone H2AJ (H2AFJ), alpha-actinin-2 (ACTN2), UTP-glucose-1-phosphate uridylyltransferase (UGP2), and voltage-dependent anion-selective channel protein 3 (VDAC3) to present differential abundances in addition of significant correlations with some meat quality attributes, such as color and tenderness.

Another recent study used Tandem mass tag labeling to examine the proteome profile changes between beef LL and PM muscle during the early *post-mortem* period at 45 min, 12 h, and 36 h (Zhai et al., 2020). A total of 71 proteins differed significantly among the muscles (PM *versus* LL at 45 min, 12 h, and 36 h). These proteins belong to different pathways including muscle contraction, oxygen transport, tricarboxylic acid cycle, oxidative phosphorylation, and ATP-related transport, fatty acid degradation, NADPH regeneration, chaperone activity, calcium signaling, as well as degradation of the extracellular matrix. At early *post-mortem*, it seemed that more abundant anti-apoptotic proteins in LT muscle could cause high metabolic stability, enhanced autophagy, and delayed apoptosis, while overabundant metabolic enzymes and pro-apoptotic proteins in PM muscle could accelerate the generation of reactive oxygen species and cell death onset. For a comprehensive reading on the proteomics studies in terms of the consequences of proteolysis on *post-mortem* muscle and meat aging, we suggest to the reader this proteomic perspective from the point of view of each single protein (Lana & Zolla, 2016).

6.4 Brief overview on proteomics of meat quality traits and discovery of biomarkers: a focus on beef tenderness and color

One arena of research that intensively used proteomics with particular interest is in the discovery of meat quality biomarkers of beef tenderness (Gagaoua, Bonnet, & Picard, 2020b, 2021a; Picard & Gagaoua, 2020a; Zhu et al., 2021a) as well as other meat quality traits such as color (Gagaoua et al., 2020a; Nair, Li, Beach, Rentfrow, & Suman, 2018), water-holding capacity (Di Luca, Hamill, Mullen, Slavov, & Elia, 2016; Tao et al., 2021), and pH decline (Gagaoua et al., 2015b; Poleti et al., 2018), etc. In addition, proteomics was used to investigate the dynamic changes and modifications occurring in *post-mortem* muscle proteome (Gagaoua et al., 2021c; Zhou, Pan, Cao, & Zhou, 2021) and most importantly for the identification of putative protein biomarkers (Gagaoua, 2021; Gagaoua et al., 2021a; Ouali et al., 2013). In addition, proteomics was used to investigate the dynamic changes and modifications occurring in *post-mortem* muscle proteome as a consequence of processing methods on the carcasses or meat cuts.

6.4.1 Beef tenderness proteomics and putative biomarkers evidenced by integromics

As introduced above, several studies have recognized that the meat tenderizing process involves myriad pathways, such as the degradation of structural proteins, energy metabolism pathways, response to stress and oxidative stress, apoptosis, and signaling pathways (Ouali et al., 2013; Picard & Gagaoua, 2017) as confirmed recently by the integromics meta-analysis study of Gagaoua and co-workers (Gagaoua et al., 2021a). In our quest for beef tenderness biomarkers, more than 100 proteomics studies were conducted in the last 20 years for several objectives. It is beyond the scope of this chapter to review the entire range of proteomics studies on beef tenderness quality. However, it is important to mention the recent ground-breaking integromics meta-analysis performed by our group on 28 eligible beef tenderness proteomics experiments (22 papers) from the literature. This analysis allowed to propose the first comprehensive list of 124 biomarkers from which 64 were found in a minimum of two studies, allowing then to shortlist for future validation a robust panel of 33 biomarkers that were identified in at least four independent experiments (Gagaoua et al., 2021a). Twelve proteins were from muscle structure and contraction pathway: ACTA1, MYH1, MYL1, TNNT3, MYH7, MYBPH, FHL1, MYLPF, TNNT1, CAPZB, TNNC1, and TNNI2). Nine proteins were from energy metabolism pathway: CKM, ENO3, ENO1, GAPDH, PGM1, TPI1, MDH1, ALDOA, and PKM. Eight proteins were from response to stress pathway: HSPB1, HSPB6, HSPA1A, CRYAB, HSPA8, HSPA1B, HSPA9, and YWHAE. Three proteins were from the oxidative stress pathway: PARK7, PRDX6, and SOD1 and one protein (CA3) from other pathways and involved in cell detoxification.

Very briefly, this integromics meta-analysis allowed to identify in a robust manner the importance of the changing integrity of muscle contractile and structure proteins, energy metabolism enzymes, response to stress, and oxidative stress proteins in the determination of beef tenderness, in that order of importance. Moreover, protein networks analysis delivered a functional annotation of the 124 proteins from LT muscle and provided key insights into the interconnectedness among various pathways and processes in the muscle which are pivotal in producing high-quality beef. The meta-analysis study further revealed the importance of mitochondria, thus suggesting to conduct more studies in the future to understand the specific role of mitochondrial metabolism post-mortem on tenderness development, and how this is related to differences between muscles with different proportions of fiber types (Picard & Gagaoua, 2020b), where mitochondrial metabolism differs.

On another hand, similar integromics meta-analysis of published proteomics studies were recently conducted by our group to create new biomarkers repertoires for beef color as discussed on the following subsection (Gagaoua et al., 2020a) and the quality defect of dark-cutting beef (Gagaoua et al., 2021b). From our data integration experience, it seemed that meta-analysis and data-mining of published proteomics studies are very useful tools to gather and compare meat-eating quality OMICs studies conducted under various platforms (Boudon et al., 2020; Gagaoua et al., 2020a; Gagaoua et al., 2021a) or even from a single laboratory using the same proteomics laboratory platform (Picard & Gagaoua, 2020a) in the frame of integrated animal systems biology. These analyses allowed also to shortlist protein biomarkers for validation along with broadening our knowledge on the underlying mechanisms compared to traditional biochemical studies.

6.4.2 Beef color proteomics and putative biomarkers evidenced by integromics

Lean beef color is critical to fresh meat marketability as it often influences consumer purchasing decisions (indicator of freshness and wholesomeness) and attractiveness at the point of sale. Historically, the role of muscle proteins in meat color have been identified including the major role of muscle fiber type (Picard & Gagaoua, 2020b), glycolysis and sarcoplasmic proteins (Nair et al., 2018), oxidation, and myofibrillar structure (Gagaoua et al., 2017c; Purslow et al., 2021). Proteomics has only recently been used to study the muscle proteome associated with beef color stability and/or variation (Joseph, Suman, Rentfrow, Li, & Beach, 2012), allowing new insights about the underlying mechanisms. Indeed, proteomics was mainly used to investigate the biochemical basis of pre- and postharvest aspects affecting color and to identify predictive candidate protein biomarkers for color stability (Suman, Hunt, Nair, & Rentfrow, 2014). Currently, there exist around 13 proteomics studies that aimed to study beef color with the goal of identifying biomarkers that are correlated with one of the color parameters (included in the meta-analysis by (Gagaoua et al., 2020a)). Using an integromics data-mining approach, these studies were gathered in a unique repertoire of 79 protein biomarkers for 5 different muscles with a major number of them (n = 59) from LT muscle (Gagaoua et al., 2020a). This protein list was also subjected to bioinformatics and pathway process enrichments, therefore allowing to confirm several biological pathways previously known to be involved in meat color development, including energy metabolism (mainly glycolysis), contractile and associated proteins, proteolysis, chaperones and heat shock proteins, oxidative stress and cell redox homeostasis, and binding, cofactor and transport proteins, including signaling or apoptosis.

The analysis of the pathways governing color stability showed that similar mechanisms are shared with tenderness, however glycolysis and other associated energy metabolism functional pathways are predominant for color while muscle structure and contractile proteins were for beef tenderness (Gagaoua et al., 2020a; Gagaoua et al., 2021a). For beef color proteomics and with a cut-off level ≥3 studies that identified a given protein, 27 potential biomarkers were shortlisted as robust from which ENO3, PRDX), HSPB1, PGM1, superoxide dismutase (SOD1), and μ-calpain (CAPN1) were consistently reported (cut-off level ≥5) by multiple studies as being differentially expressed and having a pivotal role in beef color determination. Apart from the role of energy metabolism in determining meat color, the study highlighted the potential of integromics to reveal the importance of oxidative stress, cell redox and contraction pathways, and particularly their interactions, in beef color. In fact, the changes in muscle structure and contractile proteins were proposed to have their role through light scattering from structural elements and the paleness of the meat surface. The oxidative/redox proteins were proposed to have a role in the onset of oxidation *post-mortem*, hence impacting beef color, and importantly, color stability during storage and retail display.

6.5 Conclusions

This chapter provided an overview of the applications of proteomics in the field of beef research with a focus on production systems, animal performances and productivity, the impact of post-slaughter treatments such as ES and aging on the variation of the muscle proteome, and consequences on meat quality. The current integromics proteomics studies were further exemplified with a focus on beef tenderness and color. At the era of big data and data analytics, and in a time when OMICs data is easier than ever to collect, an important limitations on fully elucidating molecular signatures of relevance relates to the 'shape' of data that is commonly generated. Within proteomics datasets, and OMICs data in general, the number of data recorded for each sample being vastly greater than the number of samples/phenotypes in a traditional (conventional) study, and this challenges statistical inference. Therefore, for the future, multi-OMICs approaches and sophisticated statistical and bioinformatics would allow further development in the topic livestock phenomics and multi-OMICs.

Acknowledgments

Dr. Mohammed Gagaoua acknowledges the funding support of the Marie Sklodowska-Curie grant agreement No. 713654 under the project number MF20180029.

References

Abo-Ismail, M. K., Lansink, N., Akanno, E., Karisa, B. K., Crowley, J. J., Moore, S. S., et al. (2018). Development and validation of a small SNP panel for feed efficiency in beef cattle. *Journal of Animal Science, 96*(2), 375–397.

Alexandre, P. A., Gomes, R. C., Santana, M. H., Silva, S. L., Leme, P. R., Mudadu, M. A., et al. (2014). Bovine NR1I3 gene polymorphisms and its association with feed efficiency traits in Nellore cattle. *Meta Gene, 2*, 206–217.

Almeida, A. M., Nanni, P., Ferreira, A. M., Fortes, C., Grossmann, J., Bessa, R. J., et al. (2017). The longissimus thoracis muscle proteome in Alentejana bulls as affected by growth path. *Journal of Proteomics, 152*, 206–215.

Álvarez-Rivera, G., Valdés, A., León, C., & Cifuentes, A. (2021). Chapter 1:Foodomics—fundamentals, state of the art and future trends. *Foodomics: Omic Strategies and Applications in Food Science*. Croydon, UK: The Royal Society of Biochemistry, 1–53. https://doi.org/10.1039/9781839163005-00001.

Baldassini, W. A., Bonilha, S. F. M., Branco, R. H., Vieira, J. C. S., Padilha, P. M., & Lanna, D. P. D. (2018). Proteomic investigation of liver from beef cattle (Bos indicus) divergently ranked on *residual* feed intake. *Molecular Biology Reports, 45*(6), 2765–2773. https://doi.org/10.1007/s11033-018-4341-2.

Becker, T. (2000). Consumer perception of fresh meat quality: A framework for analysis. *British Food Journal, 102*(3), 158–176. https://doi.org/10.1108/00070700010371707.

Beline, M., Gómez, J. F. M., Antonelo, D. S., Silva, J., Buarque, V. L. M., Cônsolo, N. R. B., et al. (2021). Muscle fiber type, postmortem metabolism, and meat quality of Nellore cattle with different post-weaning growth potential. *Livestock Science, 244*, 104348.

Bjarnadóttir, S. G., Hollung, K., Faergestad, E. M., & Veiseth-Kent, E. (2010). Proteome changes in bovine longissimus thoracis muscle during the first 48 h postmortem: Shifts in energy status and myofibrillar stability. *Journal of Agricultural and Food Chemistry, 58*(12), 7408–7414. https://doi.org/10.1021/jf100697h.

Boudon, S., Ounaissi, D., Viala, D., Monteils, V., Picard, B., & Cassar-Malek, I. (2020). Label free shotgun proteomics for the identification of protein biomarkers for beef tenderness in muscle and plasma of heifers. *Journal of Proteomics, 217*, 103685. https://doi.org/10.1016/j.jprot.2020.103685.

Bouley, J., Meunier, B., Chambon, C., Smet, S. D., Hocquette, J. F., & Picard, B. (2005). Proteomic analysis of bovine skeletal muscle hypertrophy. *Proteomics, 5*(2), 490–500. https://doi.org/10.1002/pmic.200400925.

Campion, B., Keane, M. G., Kenny, D. A., & Berry, D. P. (2009). Evaluation of estimated genetic merit for carcass weight in beef cattle: Live weights, feed intake, body measurements, skeletal and muscular scores, and carcass characteristics. *Livestock Science, 126*(1–3), 87–99. https://doi.org/10.1016/j.livsci.2009.06.004.

Carvalho, E. B., Gionbelli, M. P., Rodrigues, R. T., Bonilha, S. F., Newbold, C. J., Guimarães, S. E., et al. (2019). Differentially expressed mRNAs, proteins and miRNAs associated to energy metabolism in skeletal muscle of beef cattle identified for low and high residual feed intake. *BMC Genomics, 20*(1), 1–12. https://doi.org/10.1186/s12864-019-5890-z.

Ceciliani, F., Lecchi, C., Bazile, J., & Bonnet, M. (2018). Proteomics Research in the Adipose Tissue. In A. M. de Almeida, D. Eckersall, & I. Miller (Eds.), *Proteomics in domestic animals: From farm to systems biology* (pp. 233–254). Switzerland: Springer International Publishing https://doi.org/10.1007/978-3-319-69682-9_12.

Chaze, T., Hocquette, J. F., Meunier, B., Renand, G., Jurie, C., Chambon, C., et al. (2013). Biological markers for meat tenderness of the three main french beef breeds using 2-DE and MS approach. In *Proteomics in Foods* (pp. 127–146). Boston, MA: Springer.

Christensen, M., Ertbjerg, P., Failla, S., Sañudo, C., Richardson, R. I., Nute, G. R., et al. (2011). Relationship between collagen characteristics, lipid content and raw and cooked texture of meat from young bulls of fifteen European breeds. *Meat Science, 87*(1), 61–65.

Clerens, S., Thomas, A., Gathercole, J., Plowman, J. E., Yu, T. Y., Grosvenor, A. J., et al. (2016). Proteomic and peptidomic differences and similarities between four muscle types from New Zealand raised Angus steers. *Meat Science, 121*, 53–63.

Contreras-Castillo, C. J., Lomiwes, D., Wu, G., Frost, D., & Farouk, M. M. (2016). The effect of electrical stimulation on post mortem myofibrillar protein degradation and small heat shock protein kinetics in bull beef. *Meat Science, 113*, 65–72. https://doi.org/10.1016/j.meatsci.2015.11.012.

Crews, D. H., Pollak, E. J., & Quaas, R. L. (2004). Evaluation of Simmental carcass EPD estimated using live and carcass data. *Journal of Animal Science, 82*(3), 661–667. https://doi.org/10.2527/2004.823661x.

D'Alessandro, A., & Zolla, L. (2013). Foodomics to investigate meat tenderness. *TrAC Trends in Analytical Chemistry, 52*, 47–53. https://doi.org/10.1016/j.trac.2013.05.017.

Davis, M. P., Brooks, M. A., & Kerley, M. S. (2016). Relationship between residual feed intake and lymphocyte mitochondrial complex protein concentration and ratio in crossbred steers12. *Journal of Animal Science, 94*(4), 1587–1591. https://doi.org/10.2527/jas.2015-9843.

De Canio, M., Soggiu, A., Piras, C., Bonizzi, L., Galli, A., Urbani, A., et al. (2014). Differential protein profile in sexed bovine semen: Shotgun proteomics investigation. *Molecular BioSystems, 10*(6), 1264–1271.

Di Luca, A., Hamill, R. M., Mullen, A. M., Slavov, N., & Elia, G. (2016). Comparative proteomic profiling of divergent phenotypes for water holding capacity across the post mortem ageing period in porcine muscle exudate. *Plos One, 11*(3), e0150605. https://doi.org/10.1371/journal.pone.0150605.

Di Luca, A., Hamill, R., Mullen, A. M., & Elia, G. (2018). DIGE Analysis of Animal Tissues. In K. Ohlendieck (Ed.), *Difference gel electrophoresis: Methods and protocols* (pp. 137–152). New York, U.S.A: Springer https://doi.org/10.1007/978-1-4939-7268-5_12.

Enns, D. L., & Tiidus, P. M. (2010). The influence of estrogen on skeletal muscle. *Sports Medicine, 40*(1), 41–58. https://doi.org/10.2165/11319760-000000000-00000.

Ferguson, D. M., Jiang, S.-T., Hearnshaw, H., Rymill, S. R., & Thompson, J. M. (2000). Effect of electrical stimulation on protease activity and tenderness of M. longissimus from cattle with different proportions of Bos indicus content. *Meat Science, 55*(3), 265–272. https://doi.org/10.1016/S0309-1740(99)00131-X.

Folkesson, M., Mackey, A. L., Langberg, H., Oskarsson, E., Piehl-Aulin, K., Henriksson, J., et al. (2013). The expression of heat shock protein in human skeletal muscle: Effects of muscle fibre phenotype and training background. *Acta Physiologica, 209*(1), 26–33.

Fonseca, L. D., Eler, J. P., Pereira, M. A., Rosa, A. F., Alexandre, P. A., Moncau, C. T., et al. (2019). Liver proteomics unravel the metabolic pathways related to Feed Efficiency in beef cattle. *Scientific Reports, 9*(1), 1–11.

Foroutan, A., Wishart, D. S., & Fitzsimmons, C. (2021). Exploring biological impacts of prenatal nutrition and selection for residual feed intake on beef cattle using Omics technologies: A review. *Frontiers in Genetics, 12*, 2074. https://doi.org/10.3389/fgene.2021.720268.

Fu, H., Subramanian, R. R., & Masters, S. C. (2000). 14-3-3 proteins: Structure, function, and regulation. *Annual Review of Pharmacology and Toxicology, 40*, 617–647. https://doi.org/10.1146/annurev.pharmtox.40.1.617.

Gagaoua, M. (2021). The path from protein profiling to biomarkers: The potential of proteomics and data integration in beef quality research. IOP Conference Series: Earth and Environmental Science, *854*(1), 012029. https://doi.org/10.1088/1755-1315/854/1/012029

Gagaoua, M., Bonnet, M., Ellies-Oury, M.-P., De Koning, L., & Picard, B. (2018d). Reverse phase protein arrays for the identification/validation of biomarkers of beef texture and their use for early classification of carcasses. *Food Chemistry, 250*, 245–252. https://doi.org/10.1016/j.foodchem.2018.01.070.

Gagaoua, M., Bonnet, M., & Picard, B. (2020b). Protein array-based approach to evaluate biomarkers of beef tenderness and marbling in cows: Understanding of the underlying mechanisms and prediction. *Foods, 9*(9), 1180. https://doi.org/10.3390/foods9091180.

Gagaoua, M., Claudia Terlouw, E. M., Boudjellal, A., & Picard, B. (2015a). Coherent correlation networks among protein biomarkers of beef tenderness: What they reveal. *Journal of Proteomics, 128*, 365–374. https://doi.org/10.1016/j.jprot.2015.08.022.

Gagaoua, M., Couvreur, S., Le Bec, G., Aminot, G., & Picard, B. (2017a). Associations among protein biomarkers and pH and color traits in longissimus thoracis and rectus abdominis muscles in protected designation of origin Maine-Anjou Cull Cows. *Journal of Agricultural and Food Chemistry, 65*(17), 3569–3580. https://doi.org/10.1021/acs.jafc.7b00434.

Gagaoua, M., Hughes, J., Terlouw, E. C., Warner, R. D., Purslow, P. P., Lorenzo, J. M., et al. (2020a). Proteomic biomarkers of beef colour. *Trends in Food Science & Technology*, 101, 234–252.

Gagaoua, M., Monteils, V., Couvreur, S., & Picard, B. (2017b). Identification of biomarkers associated with the rearing practices, carcass characteristics, and beef quality: An integrative approach. *Journal of Agricultural and Food Chemistry, 65*(37), 8264–8278. https://doi.org/10.1021/acs.jafc.7b03239.

Gagaoua, M., Monteils, V., Couvreur, S., & Picard, B. (2019a). Beef tenderness prediction by a combination of statistical methods: Chemometrics and supervised learning to manage integrative farm-to-meat continuum data. *Foods, 8*(7), 274. https://doi.org/10.3390/foods8070274.

Gagaoua, M., Monteils, V., & Picard, B. (2018c). Data from the farmgate-to-meat continuum including omics-based biomarkers to better understand the variability of beef tenderness: An integromics approach. *Journal of Agricultural and Food Chemistry, 66*(51), 13552–13563. https://doi.org/10.1021/acs.jafc.8b05744.

Gagaoua, M., Picard, B., & Monteils, V. (2018a). Associations among animal, carcass, muscle characteristics, and fresh meat color traits in Charolais cattle. *Meat Science, 140*, 145–156. https://doi.org/10.1016/j.meatsci.2018.03.004.

Gagaoua, M., Picard, B., & Monteils, V. (2019b). Assessment of cattle inter-individual cluster variability: The potential of continuum data from the farm-to-fork for ultimate beef tenderness management. *Journal of the Science of Food and Agriculture, 99*(8), 4129–4141. https://doi.org/10.1002/jsfa.9643.

Gagaoua, M., Picard, B., Soulat, J., & Monteils, V. (2018b). Clustering of sensory eating qualities of beef: Consistencies and differences within carcass, muscle, animal characteristics and rearing factors. *Livestock Science, 214*, 245–258. https://doi.org/10.1016/j.livsci.2018.06.011.

Gagaoua, M., Terlouw, E. M. C., Micol, D., Hocquette, J. F., Moloney, A. P., Nuernberg, K., et al. (2016). Sensory quality of meat from eight different types of cattle in relation with their biochemical characteristics. *Journal of Integrative Agriculture, 15*(7), 1550–1563.

Gagaoua, M., Terlouw, C., & Picard, B. (2021d). Contribution of proteomics to discovery of putative biomarkers for the study of beef color. INRA Productions Animales, *34*. doi:10.20870/productions-animales.2021.34.3.4899

Gagaoua, M., Terlouw, C., Richardson, I., Hocquette, J.-F., & Picard, B. (2019c). The associations between proteomic biomarkers and beef tenderness depend on the end-point cooking temperature, the country origin of the panelists and breed. *Meat Science, 157*, 107871. https://doi.org/10.1016/j.meatsci.2019.06.007.

Gagaoua, M., Terlouw, E. M. C., Micol, D., Boudjellal, A., Hocquette, J.-F., & Picard, B. (2015b). Understanding early post-mortem biochemical processes underlying meat color and pH decline in the longissimus thoracis muscle of Young Blond d'Aquitaine bulls using protein biomarkers. *Journal of Agricultural and Food Chemistry, 63*(30), 6799–6809. https://doi.org/10.1021/acs.jafc.5b02615.

Gagaoua, M., Terlouw, E. C., Mullen, A. M., Franco, D., Warner, R. D., Lorenzo, J. M., et al. (2021a). Molecular signatures of beef tenderness: Underlying mechanisms based on integromics of protein biomarkers from multi-platform proteomics studies. *Meat Science, 172*, 108311.

Gagaoua, M., Terlouw, E. M. C., & Picard, B. (2017c). The study of protein biomarkers to understand the biochemical processes underlying beef color development in young bulls. *Meat Science, 134*, 18–27. https://doi.org/10.1016/j.meatsci.2017.07.014.

Gagaoua, M., Troy, D., & Mullen, A. M. (2021c). The extent and rate of the appearance of the major 110 and 30 kDa proteolytic fragments during post-mortem aging of beef depend on the glycolysing rate of the muscle and aging time: An LC–MS/MS approach to decipher their proteome and associated pathways. *Journal of Agricultural and Food Chemistry, 69*(1), 602–614. https://doi.org/10.1021/acs.jafc.0c06485.

Gagaoua, M., Warner, R. D., Purslow, P., Ramanathan, R., Mullen, A. M., López-Pedrouso, M., et al. (2021b). Dark-cutting beef: A brief review and an integromics meta-analysis at the proteome level to decipher the underlying pathways. *Meat Science, 181*, 108611.

Gotoh, T., Albrecht, E., Teuscher, F., Kawabata, K., Sakashita, K., Iwamoto, H., et al. (2009). Differences in muscle and fat accretion in Japanese Black and European cattle. *Meat Science, 82*(3), 300–308.

Guidolin, D. G. F., Buzanskas, M. E., Ramos, S. B., Venturini, G. C., Lôbo, R. B., Paz, C. C. P., et al. (2012). Genotype–environment interaction for post-weaning traits in Nellore beef cattle. *Animal Production Science, 52*(11), 975–980.

Guillemin, N., Bonnet, M., Jurie, C., & Picard, B. (2011a). Functional analysis of beef tenderness. *Journal of Proteomics, 75*(2), 352–365. https://doi.org/10.1016/j.jprot.2011.07.026.

Guillemin, N., Jurie, C., Cassar-Malek, I., Hocquette, J.-F., Renand, G., & Picard, B. (2011b). Variations in the abundance of 24 protein biomarkers of beef tenderness according to muscle and animal type. *Animal: An International Journal of Animal Bioscience, 5*(6), 885–894. https://doi.org/10.1017/S1751731110002612.

Gulyás, G., Czeglédi, L., Béri, B., Harangi, S., Csősz, E., Szabó, Z., et al. (2015). Proteomic analysis of skeletal muscle at different live weights in Charolais bulls. *Acta Alimentaria, 44*(1), 132–138.

Gursansky, B., O'Halloran, J. M., Egan, A., & Devine, C. E. (2010). Tenderness enhancement of beef from Bos indicus and Bos taurus cattle following electrical stimulation. *Meat Science, 86*(3), 635–641. https://doi.org/10.1016/j.meatsci.2010.05.002.

Harper, G. S., & Pethick, D. W. (2004). How might marbling begin? *Australian Journal of Experimental Agriculture, 44*(7), 653–662. https://doi.org/10.1071/ea02114.

Hollung, K., Veiseth, E., Jia, X., Færgestad, E. M., & Hildrum, K. I. (2007). Application of proteomics to understand the molecular mechanisms behind meat quality. *Meat Science, 77*(1), 97–104. https://doi.org/10.1016/j.meatsci.2007.03.018.

Hopkins, D. L., & Thompson, J. M. (2001). The relationship between tenderness, proteolysis, muscle contraction and dissociation of actomyosin. *Meat Science, 57*(1), 1–12. https://doi.org/10.1016/S0309-1740(00)00065-6.

Hu, S. M., Li, F., Yu, H. M., Li, R. Y., Ma, Q. Y., Ye, T. J., et al. (2005). The mimecan gene expressed in human pituitary and regulated by pituitary transcription factor-1 as a marker for diagnosing pituitary tumors. *The Journal of Clinical Endocrinology & Metabolism, 90*(12), 6657–6664. https://doi.org/10.1210/jc.2005-0322.

Hunter, M. C., Smith, R. G., Schipanski, M. E., Atwood, L. W., & Mortensen, D. A. (2017). Agriculture in 2050: Recalibrating targets for sustainable intensification. *Bioscience, 67*(4), 386–391. https://doi.org/10.1093/biosci/bix010.

Hwang, I. H., Devine, C. E., & Hopkins, D. L. (2003). The biochemical and physical effects of electrical stimulation on beef and sheep meat tenderness. *Meat Science, 65*(2), 677–691. https://doi.org/10.1016/S0309-1740(02)00271-1.

Joseph, P., Suman, S. P., Rentfrow, G., Li, S., & Beach, C. M. (2012). Proteomics of muscle-specific beef color stability. *Journal of Agricultural and Food Chemistry, 60*(12), 3196–3203. https://doi.org/10.1021/jf204188v.

Kammoun, M., Picard, B., Astruc, T., Gagaoua, M., Aubert, D., Bonnet, M., et al. (2016). The invalidation of HspB1 gene in mouse alters the ultrastructural phenotype of muscles. *Plos One, 11*(8), e0158644.

Keady, S. M., Kenny, D. A., Ohlendieck, K., Doyle, S., Keane, M. G., & Waters, S. M. (2013). Proteomic profiling of bovine M. longissimus lumborum from crossbred aberdeen angus and belgian blue sired steers varying in genetic merit for carcass weight. *Journal of Animal Science, 91*(2), 654–665. https://doi.org/10.2527/jas.2012-5850.

Keane, M. G., Dunne, P. G., Kenny, D. A., & Berry, D. P. (2011). Effects of genetic merit for carcass weight, breed type and slaughter weight on performance and carcass traits of beef × dairy steers. *Animal, 5*(2), 182–194. https://doi.org/10.1017/S1751731110001758.

Keane, M. G., Diskin, M. G. (2011). Performance and carcass traits of progeny of Limousin sires differing in genetic merit. *Irish Journal of Agricultural and Food Research*, 63–76.

Kim, N. K., Lee, S. H., Cho, Y. M., Son, E. S., Kim, K. Y., Lee, C. S., et al. (2009). Proteome analysis of the m. Longissimus dorsi between fattening stages in Hanwoo steer. *BMB Reports, 42*(7), 433–438.

Kim, Y. H. B., Lonergan, S. M., Grubbs, J. K., Cruzen, S. M., Fritchen, A. N., Della Malva, A., et al. (2013). Effect of low voltage electrical stimulation on protein and quality changes in bovine muscles during postmortem aging. *Meat Science, 94*(3), 289–296.

Kim, Y. H. B., Ma, D., Setyabrata, D., Farouk, M. M., Lonergan, S. M., Huff-Lonergan, E., et al. (2018). Understanding postmortem biochemical processes and post-harvest aging factors to develop novel smart-aging strategies. *Meat Science, 144*, 74–90.

Lana, A., & Zolla, L. (2016). Proteolysis in meat tenderization from the point of view of each single protein: A proteomic perspective. *Journal of Proteomics, 147*, 85–97. https://doi.org/10.1016/j.jprot.2016.02.011.

Li, C. B., Li, J., Zhou, G. H., Lametsch, R., Ertbjerg, P., Brüggemann, D. A., et al. (2012). Electrical stimulation affects metabolic enzyme phosphorylation, protease activation, and meat tenderization in beef. *Journal of Animal Science, 90*(5), 1638–1649.

Listrat, A., Gagaoua, M., Andueza, D., Gruffat, D., Normand, J., Mairesse, G., et al. (2020b). What are the drivers of beef sensory quality using metadata of intramuscular connective tissue, fatty acids and muscle fiber characteristics? *Livestock Science, 240*, 104209.

Listrat, A., Gagaoua, M., Normand, J., Andueza, D. J., Gruffat, D., Mairesse, G., et al. (2020c). Are there consistent relationships between major connective tissue components, intramuscular fat content and muscle fibre types in cattle muscle? *Animal, 14*(6), 1204–1212.

Listrat, A., Gagaoua, M., Normand, J., Gruffat, D., Andueza, D., Mairesse, G., et al. (2020a). Contribution of connective tissue components, muscle fibres and marbling to beef tenderness variability in longissimus thoracis, rectus abdominis, semimembranosus and semitendinosus muscles. *Journal of the Science of Food and Agriculture, 100*(6), 2502–2511.

Liu, Y., & Steinacker, J. M. (2001). Changes in skeletal muscle heat shock proteins: Pathological significance. *Frontiers in Bioscience: A Journal and Virtual Library, 6*, D12–D25. https://doi.org/10.2741/liu.

López-Pedrouso, M., Lorenzo, J. M., Gagaoua, M., & Franco, D. (2020). Application of proteomic technologies to assess the quality of raw pork and pork products: An overview from farm-to-fork. *Biology, 9*(11), 393.

Maiti, S., & Bamburg, J. R. (2013). Actin-capping and -severing proteins. In W. J. Lennarz, & M. D. Lane (Eds.), *Encyclopedia of biological chemistry* (2nd ed.) (pp. 18–26). London, U.K.: Academic Press. https://doi.org/10.1016/B978-0-12-378630-2.00415-1.

Marino, R., Albenzio, M., della Malva, A., Santillo, A., Loizzo, P., & Sevi, A. (2013). Proteolytic pattern of myofibrillar protein and meat tenderness as affected by breed and aging time. *Meat Science, 95*(2), 281–287. https://doi.org/10.1016/j.meatsci.2013.04.009.

Medeiros de Oliveira Silva, R., Bonvino Stafuzza, N., de Oliveira Fragomeni, B., Miguel Ferreira de Camargo, G., Matos Ceacero, T., Noely dos Santos Gonçalves Cyrillo, J., et al. (2017). Genome-wide association study for carcass traits in an experimental nelore cattle population. *Plos One, 12*(1), e0169860.

Moloney, A. P., & McGee, M. (2017). Chapter 2—factors influencing the growth of meat animals. In F. Toldra´ (Ed.), *Lawrie´s Meat Science* (Eighth Edition) (pp. 19–47). Duxford, U.K.: Woodhead Publishing. https://doi.org/10.1016/B978-0-08-100694-8.00002-9.

Moreira, A. D., Siqueira, G. R., Lage, J. F., Benatti, J. M. B., Moretti, M. H., Miguel, G. Z., et al. (2017). Castration methods in crossbred cattle raised on tropical pasture. *Animal Production Science, 58*(7), 1307–1315.

Mullins, Y., Keogh, K., Kenny, D. A., Kelly, A., O' Boyle, P., & Waters, S. M. (2020). Label-free quantitative proteomic analysis of M. longissimus dorsi from cattle during dietary restriction and subsequent compensatory growth. *Scientific Reports, 10*(1), 2613. https://doi.org/10.1038/s41598-020-59412-6.

Munekata, P. E., Pateiro, M., López-Pedrouso, M., Gagaoua, M., & Lorenzo, J. M. (2021). Foodomics in meat quality. Current Opinion in Food Science, *38*, 79–85. https://doi.org/10.1016/j.cofs.2020.10.003

Nair, M. N., Li, S., Beach, C. M., Rentfrow, G., & Suman, S. P. (2018). Changes in the sarcoplasmic proteome of beef muscles with differential color stability during postmortem aging. Meat and Muscle Biology, 2(1), Article 1. https://doi.org/10.22175/mmb2017.07.0037

Ohsaki, H., Okada, M., Sasazaki, S., Hinenoya, T., Sawa, T., Iwanaga, S., et al. (2007). Proteomic comparison between Japanese black and holstein cattle by two-dimensional gel electrophoresis and identification of proteins. *Asian-Australasian Journal of Animal Sciences, 20*(5), 638–644. doi: 10.5713/ajas.2007.638.

O'Quinn, T. G., Legako, J. F., Brooks, J. C., & Miller, M. F. (2018). Evaluation of the contribution of tenderness, juiciness, and flavor to the overall consumer beef eating experience. *Translational Animal Science, 2*(1), 26–36. https://doi.org/10.1093/tas/txx008.

Ouali, A., Gagaoua, M., Boudida, Y., Becila, S., Boudjellal, A., Herrera-Mendez, C. H., et al. (2013). Biomarkers of meat tenderness: Present knowledge and perspectives in regards to our current understanding of the mechanisms involved. *Meat Science, 95*(4), 854–870.

Picard, B., & Gagaoua, M. (2017). Chapter 11—Proteomic investigations of beef tenderness. In M. L. Colgrave (Ed.), *Proteomics in food science* (pp. 177–197). London, United Kingdom: Academic Press. https://doi.org/10.1016/B978-0-12-804007-2.00011-4.

Picard, B., & Gagaoua, M. (2020a). Meta-proteomics for the discovery of protein biomarkers of beef tenderness: An overview of integrated studies. *Food Research International, 127*, 108739. https://doi.org/10.1016/j.foodres.2019.108739.

Picard, B., & Gagaoua, M. (2020b). Muscle fiber properties in cattle and their relationships with meat qualities: An overview. *Journal of Agricultural and Food Chemistry*, 68(22), 6021–6039. https://doi.org/10.1021/acs.jafc.0c02086.

Picard, B., Gagaoua, M., Al Jammas, M., & Bonnet, M. (2019). Beef tenderness and intramuscular fat proteomic biomarkers: Effect of gender and rearing practices. *Journal of Proteomics, 200*, 1–10. https://doi.org/10.1016/j.jprot.2019.03.010.

Picard, B., Gagaoua, M., Al-Jammas, M., De Koning, L., Valais, A., & Bonnet, M. (2018). Beef tenderness and intramuscular fat proteomic biomarkers: Muscle type effect. Peer J, *6*, e4891. https://doi.org/10.7717/peerj.4891

Picard, B., Gagaoua, M., & Hollung, K. (2017). Chapter 12—Gene and protein expression as a tool to explain/predict meat (and Fish) quality. In P. P. Purslow (Ed.), *New Aspects of Meat Quality* (pp. 321–354). Duxford, United Kingdom: Woodhead Publishing. https://doi.org/10.1016/B978-0-08-100593-4.00013-8.

Picard, B., Gagaoua, M., Micol, D., Cassar-Malek, I., Hocquette, J.-F., & Terlouw, C. E. M. (2014). Inverse relationships between biomarkers and beef tenderness according to contractile and metabolic properties of the muscle. *Journal of Agricultural and Food Chemistry, 62*(40), 9808–9818. https://doi.org/10.1021/jf501528s.

Polati, R., Menini, M., Robotti, E., Millioni, R., Marengo, E., Novelli, E., et al. (2012). Proteomic changes involved in tenderization of bovine Longissimus dorsi muscle during prolonged ageing. *Food Chemistry, 135*(3), 2052–2069.

Poleti, M. D., Moncau, C. T., Silva-Vignato, B., Rosa, A. F., Lobo, A. R., Cataldi, T. R., et al. (2018). Label-free quantitative proteomic analysis reveals muscle contraction and metabolism proteins linked to ultimate pH in bovine skeletal muscle. *Meat Science, 145*, 209–219.

Purslow, P. P., Gagaoua, M., & Warner, R. D. (2021). Insights on meat quality from combining traditional studies and proteomics. *Meat Science, 174*, 108423. https://doi.org/10.1016/j.meatsci.2020.108423.

Ramayo-Caldas, Y., Renand, G., Ballester, M., Saintilan, R., & Rocha, D. (2016). Multi-breed and multi-trait co-association analysis of meat tenderness and other meat quality traits in three French beef cattle breeds. *Genetics Selection Evolution, 48*(1), 37. https://doi.org/10.1186/s12711-016-0216-y.

Ramos, M. H., & Kerley, M. S. (2013). Mitochondrial complex I protein differs among residual feed intake phenotype in beef cattle. *Journal of Animal Science, 91*(7), 3299–3304. https://doi.org/10.2527/jas.2012-5589.

Rodrigues, R. T.D.S., Chizzotti, M. L., Vital, C. E., Baracat-Pereira, M. C., Barros, E., et al. (2017). Differences in beef quality between Angus (Bos taurus taurus) and Nellore (Bos taurus indicus) Cattle through a Proteomic and Phosphoproteomic Approach. *Plos One, 12*(1), e0170294.

Rodríguez-Vázquez, R., Mato, A., López-Pedrouso, M., Franco, D., Sentandreu, M. A., & Zapata, C. (2020). Measuring quantitative proteomic distance between Spanish beef breeds. *Food Chemistry, 315*, 126293. https://doi.org/10.1016/j.foodchem.2020.126293.

Romao, J. M., He, M. L., McAllister, T. A., & Guan, L. L. (2014). Effect of age on bovine subcutaneous fat proteome: Molecular mechanisms of physiological variations during beef cattle growth1. *Journal of Animal Science, 92*(8), 3316–3327. https://doi.org/10.2527/jas.2013-7423.

Rosa, A. F., Moncau, C. T., Poleti, M. D., Fonseca, L. D., Balieiro, J. C., Silva, S. L., et al. (2018). Proteome changes of beef in Nellore cattle with different genotypes for tenderness. *Meat Science, 138*, 1–9.

Santana, M. H., Utsunomiya, Y. T., Neves, H. H., Gomes, R. C., Garcia, J. F., Fukumasu, H., et al. (2014). Genome-wide association analysis of feed intake and residual feed intake in Nellore cattle. *BMC Genetics, 15*(1), 1–8.

Santos, D., Monteiro, M. J., Voss, H.-P., Komora, N., Teixeira, P., & Pintado, M. (2021). The most important attributes of beef sensory quality and production variables that can affect it: A review. *Livestock Science, 250*, 104573. https://doi.org/10.1016/j.livsci.2021.104573.

Sañudo, C., Macie, E. S., Olleta, J. L., Villarroel, M., Panea, B., & Albertí, P. (2004). The effects of slaughter weight, breed type and ageing time on beef meat quality using two different texture devices. *Meat Science*, 66(4), 925–932. https://doi.org/10.1016/j.meatsci.2003.08.005.

Silva, L. H. P., Rodrigues, R. T. S., Assis, D. E. F., Benedeti, P. D. B., Duarte, M. S., & Chizzotti, M. L. (2019). Explaining meat quality of bulls and steers by differential proteome and phosphoproteome analysis of skeletal muscle. Journal of Proteomics, 199, 51–66. https://doi.org/10.1016/j.jprot.2019.03.004.

Simmons, N. J., Daly, C. C., Cummings, T. L., Morgan, S. K., Johnson, N. V., & Lombard, A. (2008). Reassessing the principles of electrical stimulation. *Meat Science, 80*(1), 110–122. https://doi.org/10.1016/j.meatsci.2008.05.006.

Suman, S. P., Hunt, M. C., Nair, M. N., & Rentfrow, G. (2014). Improving beef color stability: Practical strategies and underlying mechanisms. *Meat Science, 98*(3), 490–504. https://doi.org/10.1016/j.meatsci.2014.06.032.

Tao, Y., Ma, L., Li, D., Tian, Y., Liu, J., & Liu, D. (2021). Proteomics analysis to investigate the effect of oxidized protein on meat color and water holding capacity in Tan mutton under low temperature storage. LWT, *146*, 111429. https://doi.org/10.1016/j.lwt.2021.111429

Thongboonkerd, V., Kanlaya, R., Sinchaikul, S., Parichatikanond, P., Chen, S.-T., & Malasit, P. (2006). Proteomic identification of altered proteins in skeletal muscle during chronic potassium depletion: Implications for hypokalemic myopathy. *Journal of Proteome Research, 5*(12), 3326–3335. https://doi.org/10.1021/pr060136h.

Timperio, A. M., D'Alessandro, A., Pariset, L., D'amici, G. M., Valentini, A., & Zolla, L. (2009). Comparative proteomics and transcriptomics analyses of livers from two different Bos taurus breeds: "Chianina and Holstein Friesian". *Journal of Proteomics* 73(2), 309–322. doi:10.1016/j.jprot.2009.09.015.

Troy, D. J., & Kerry, J. P. (2010). Consumer perception and the role of science in the meat industry. *Meat Science, 86*(1), 214–226. https://doi.org/10.1016/j.meatsci.2010.05.009.

Valdés, A., Álvarez-Rivera, G., Socas-Rodríguez, B., Herrero, M., Ibáñez, E., & Cifuentes, A. (2022). Foodomics: Analytical opportunities and challenges. *Analytical Chemistry,* 94(1), 366–381. doi:10.1021/acs.analchem.1c04678.

Vergara, H., Molina, A., & Gallego, L. (1999). Influence of sex and slaughter weight on carcass and meat quality in light and medium weight lambs produced in intensive systems. *Meat Science, 52*(2), 221–226. https://doi.org/10.1016/s0309-1740(98)00171-5.

Verónica, S., Laura, G.-B., Yolanda, D., Fernando, D., María, J. G.-E., Ana, C.-M., et al. (2021). New Insights on the Impact of Cattle Handling on Post-Mortem Myofibrillar Muscle Proteome and Meat Tenderization. Foods, 10(12), 3115. doi:10.3390/foods10123115.

Wu, W., Fu, Y., Therkildsen, M., Li, X.-M., & Dai, R.-T. (2015). Molecular understanding of meat quality through application of proteomics. *Food Reviews International, 31*(1), 13–28. https://doi.org/10.1080/87559129.2014.961073.

Xu, Y. J., Jin, M. L., Wang, L. J., Zhang, A. D., Zuo, B., Xu, D. Q., et al. (2009). Differential proteome analysis of porcine skeletal muscles between Meishan and Large White. *Journal of Animal Science, 87*(8), 2519–2527.

Zhai, C., Djimsa, B. A., Prenni, J. E., Woerner, D. R., Belk, K. E., & Nair, M. N. (2020). Tandem mass tag labeling to characterize muscle-specific proteome changes in beef during early postmortem period. *Journal of Proteomics, 222*, 103794. https://doi.org/10.1016/j.jprot.2020.103794.

Zhang, J., Shi, H., Li, S., Cao, Z., Yang, H., & Wang, Y. (2019). Integrative hepatic metabolomics and proteomics reveal insights into the mechanism of different feed efficiency with high or low dietary forage levels in Holstein heifers. *Journal of Proteomics, 194*, 1–13. https://doi.org/10.1016/j.jprot.2018.12.026.

Zhang, Q., Lee, H. G., Han, J. A., Kang, S. K., Lee, N. K., Baik, M., et al. (2012). Differentially expressed proteins associated with myogenesis and adipogenesis in skeletal muscle and adipose tissue between bulls and steers. *Molecular Biology Reports, 39*(2), 953–960.

Zhao, Y. M., Basu, U., Dodson, M. V., Basarb, J. A., & Guan, L. L. (2010). Proteome differences associated with fat accumulation in bovine subcutaneous adipose tissues. *Proteome Science, 8*, 14. https://doi.org/10.1186/1477-5956-8-14.

Zhou, C.-Y., Pan, D.-D., Cao, J.-X., & Zhou, G.-H. (2021). A comprehensive review on molecular mechanism of defective dry-cured ham with excessive pastiness, adhesiveness, and bitterness by proteomics insights. *Comprehensive Reviews in Food Science and Food Safety, 20*(4), 3838–3857. https://doi.org/10.1111/1541-4337.12779.

Zhu, Y., Gagaoua, M., Mullen, A. M., Kelly, A. L., Sweeney, T., Cafferky, J., et al. (2021b). A proteomic study for the discovery of beef tenderness biomarkers and prediction of Warner–Bratzler shear force measured on longissimus thoracis muscles of Young Limousin-Sired Bulls. *Foods, 10*(5), 952.

Zhu, Y., Gagaoua, M., Mullen, A. M., Viala, D., Rai, D. K., Kelly, A. L., et al. (2021a). Shotgun proteomics for the preliminary identification of biomarkers of beef sensory tenderness, juiciness and chewiness from plasma and muscle of young Limousin-sired bulls. *Meat Science, 176*, 108488.

CHAPTER

7

Proteomic advances in poultry science

Xue Zhang[a], *Surendranath Suman*[b], *M. Wes Schilling*[a]

[a]Department of Food Science, Nutrition and Health, Promotion, Mississippi State University, MS, United States
[b]Department of Animal and Food Sciences, University of Kentucky, Lexington, KY, United States

7.1 Introduction

The chicken genome has 20,000–23,000 protein-coding genes. There are several millions of protein species, however, due to the presence of these gene codes, mutations in the coding sequence, the variation in the starting site for translation of mRNA into proteins, and post-translational modifications (Timp & Timp, 2020). These are all the key elements that determine protein structure and chemistry in a living cell. Proteins are the functional units that make the biology work, while the proteome is a complete set of proteins that are characterized regarding their localization, interactions, post-translational modifications, and turnover at a given time point and for a given condition (Petracci and Cavani, 2012; Aslam, Basit, Nisar, Khurshid, & Rasool, 2017). A full picture of the proteome provides information about the cellular activities, mechanisms for signaling between cells and tissues, and causes of diseases and defects such as microbial infections in the chicken gut and muscle myopathy in broilers and turkeys. Two-dimensional gel electrophoresis (2-DE) in combination with mass spectrometry (MS) is widely used for proteomic studies. Since proteins are the end products of genes, differences in protein abundance can reflect differences in gene expression. Specific proteomic signatures that are differentially expressed provide clues on biochemical processes and poultry quality and allow for the determination of biological markers that are indicators of poultry quality defects (van de Wiel & Zhang, 2007).

Food Proteomics
DOI: https://doi.org/10.1016/B978-0-323-90889-4.00010-5

Proteomic studies in poultry are limited compared to that in mammals, and mainly include proteomics in egg components and allergens, muscle development that is related to growth performance, infectious diseases, and meat defect such as pale, soft, and exudative (PSE) and woody breast (WB) meat. These studies contribute to the understanding of cellular processes and how they are regulated. However, challenges remain in the field of proteomics. It becomes complicated when identifying protein isoforms since the structure and chemistry of proteins can be modified post-translationally by the addition of chemical groups, polypeptides, complex molecules, directly modification of amino acids, or proteolysis (Spoel, 2018). This is also partially due to the sensitivity of MS and the availability of existing protein database. Additionally, linking proteomics to DNA/RNA sequencing on a single-cell basis is still a challenge although this will ultimately provide an in-depth understanding of cellular state especially with respect to disease or a meat quality defect.

Poultry proteomics has received research attention since the early twenty-first century. Remignon, Molette, Babile, and Fernandez (2006) analyzed the proteomes from fast-glycolyzing and normal glycolyzing turkeys and suggested that the differences in proteomes were indicative of differences in meat quality. This study highlighted the potential of proteomic methods to understand the biochemical basis for color, water-holding capacity, and texture of broiler meat. Molette, Rémignon, and Babilé (2005) studied the proteomic basis of BUT9 turkeys and reported differences in fast and normal glycolyzing breast muscles and their relationship to meat quality. Researchers used 2DE and isoelectric focusing to determine that there were differences in three protein spots between normal and fast glycolyzing birds, and the proteins were identified as myosin heavy chain, actin fragments, and glyceraldehyde-3 phosphate dehydrogenase (GAPDH). In another study, a proteomic approach was used for the identification of chicken meat that was mechanically recovered and hand deboned. Results suggested that the amount of hemoglobin can be used as a marker to differentiate mechanically recovered chicken meat from deboned chicken meat (Surowiec, Koistinen, Fraser, & Bramley, 2011). Proteomic characterization of the sarcoplasmic proteins in the *pectoralis major* and *supracoracoideus* breast muscles was conducted for two different chicken genotypes, which included Ross 708 commercial broilers and Leghorn chicks, Hyline W-36. Results suggested that glycogen phosphorylase, enolase, elongation factor 1, creatine kinase, fructose-bisphosphate aldolase, and GAPDH were different in the two strains during breast muscle growth (Zapata, Reddish, Miller, Lilburn, & Wick, 2012). These are some of the first proteomic studies on poultry that provided a foundation for some of the more recent research that is presented in the subsequent text.

7.2 Egg proteomics

An egg is composed of egg yolk, egg white, the vitelline membrane, chalaza, eggshell membranes, and the eggshell. Proteomics has been used to evaluate egg component protein profiles, egg allergenicity, authentication, egg quality, and safety. Egg allergenicity is one of the most common food allergies in both adults and children. Foods including eggs and their byproducts should be properly labeled in accordance with Food and Drug Administration regulations. In the last decade, proteomic techniques have been applied with success to allergen detection in egg white, egg yolk, and liquid and powdered egg products. The most common allergens include the egg white proteins ovomucoid, ovalbumin, ovotransferrin, and lysozyme as well as the egg yolk proteins vitellogenin, α-livetin, and IgY (Monaci, De Angelis, Montemurro, & Pilolli, 2018). The allergic reactions of human to eggs are interspecific. Therefore, comparative proteomics of eggs from different avian egg species is necessary to characterize the unique protein profiles of each species to ensure food safety. Different 2DE gel patterns were observed in egg yolk proteins from chicken, duck, and quail eggs, which could be used as a unique fingerprint for authentication (Liu, Qiu, Gao, & Ma, 2018). Protein isomers, including serum albumin, ovalbumin, and vitellogenin-related, were detected in these three poultry species (Liu et al., 2018), which may explain the different stabilities and antigenic activities of the same protein from different species. Similarly, a proteomic study of egg white proteins revealed that the quail egg white proteome pattern on 2DE gels is closely related to that of chicken eggs, while the duck egg white protein pattern is more distinct (Hu et al., 2016). The diversity in egg yolk and egg white proteome among species might be associated with the changes in protein functions to adapt to the living environment or reproduction demand.

A deeper understanding of posttranslational modifications is required to investigate the functional properties of proteins and the relationship between biological activities and variability in species. Protein glycosylation is a ubiquitous posttranslational modification that modulates protein properties, thereby influencing bioactivities within a system. In chicken egg yolk, 217 N-glycosylation sites and 86 glycoproteins were identified. Among them, 22 were proteases and protease inhibitors, indicating the presence of a proteinase/inhibitor regulation system; 15 were proteins related to the complement and immune systems, which provide protection against potential threats during hatching (Geng et al., 2018). This study revealed structural information about chicken egg yolk glycoproteins, which aid in the understanding of the underlying mechanism of embryo development as well as changes in egg yolk functional characteristics during storage and processing. A higher N-glycosylation level was found in duck egg yolk than that of chicken, which may associate with a stronger resistance to the influenza virus disease of the duck. A large set of corresponding glycoproteins between chicken and duck egg yolks were detected with similar molecular functions, which suggested slower changes in avian egg yolk proteins than that of egg white proteins over time (Meng et al., 2019, 2020). N-glycoproteomes between chicken and duck egg white proteins also differed. The different N-glycoproteins were mainly involved in molecular functions and biological processes, including allergenicities and antibacterial abilities (Meng et al., 2019). N-glycosylation of egg proteins is tissue specific. Mucin-5B was the most heavily N-glycosylated protein in the chicken egg vitelline membrane (CEVM; with 28 N-glycosites), in egg white (with 15 N-glycosites) and chalaza (with 34 N-glycosites) but was absent in egg yolk and the eggshell matrix (Xiao et al., 2020). In eggshells, 547 proteins, 182 N-glycoproteins, and 69 phosphoproteins were identified. These eggshell matrix proteins were enriched in the regulation of enzyme activity, phosphoproteins were enriched in brain development and cell adhesion, and glycoproteins were enriched in the immune process and hydrolase activity of glycosidic bonds (Yang et al., 2020). The glycoproteins enriched in the immune process may enhance the embryos' protection from foreign substances. Studies on egg proteins and their posttranslational modifications provide not only a comprehensive understanding of protein profiles but also insights into the functions of each egg component and underlying mechanisms.

7.3 Growth performance

Proteomics has been used as an effective method to study the molecular basis of muscle growth in poultry. Skeletal muscle is a major determinant of muscle mass and therefore meat production, while embryonic myogenesis is a critical process to increase muscle mass. Ouyang et al. (2017) used isobaric tags for relative and absolute quantification (iTRAQ) to study the protein expression during embryonic skeletal muscle growth in leg muscle samples of female Xinghua chicken. The differentially expressed proteins during embryonic development of skeletal muscle were mainly involved in collagen fibril organization, cytoplasmic translation, epithelial cell differentiation, muscle contraction, and protein folding, which contribute to the development of embryonic muscle in chicken.

The protein biomarkers associated with the growth rate were investigated in chicken and quails using 2-dimensional gel electrophoresis and MS analysis. Phongpa-Ngan, Grider, Mulligan, Aggrey, and Wicker (2011) separated broilers into slow and fast growth rate groups and identified biomarkers from water-soluble protein and crude myofibrillar protein extracts of breast muscle in both groups. For water-soluble proteins, serum albumin precursor, creatine kinase M type, and protein DJ-1 were overexpressed in slow-growth broilers, and ubiquitin was overexpressed in fast-growth broilers. For crude myofibrillar proteins, actin, keratin, tropomyosin, and a few actin fragments were overexpressed in fast-growth broilers and myosin heavy chain was overexpressed in slow-growth broilers. These results showed that contractile proteins are important for chicken muscle growth and related to meat quality since muscle fibers are mainly characterized by contractile and metabolic properties. In Japanese quails, 24 proteins were differentially expressed in breast muscle between low growth and high growth quails (Nasirifar, Rezvannejad, Maleki, & Sami, 2021). Alpha-tropomyosin, myosin-1B, myosin-2, keratin type II cytoskeletal 2 epidermal, alpha-enolase isoform X1, 3-mercaptopyruvate sulphurtransferase, cytochrome b-c1 complex subunit 1, mitochondrial, hypothetical protein, myosin heavy chain skeletal muscle, 26S protease regulatory subunit 6B were overexpressed in the fast growth group, while serum albumin was overexpressed in the slow-growing group. The authors concluded that increased growth rate may cause disturbances in metabolism in most organs. However, the bird can adapt to the changes, which is evident due to the increased expression in those protein biomarkers. The knowledge on protein biomarkers associated with growth rate can be used to assist in the selection of genetic biomarkers in the poultry industries.

Feed efficiency (FE), the ability of an animal to convent feed into body mass, is one of the most important traits in poultry production. Proteomic data has been used to understand the cellular basis of FE in poultry. Kong et al. (2016) studied the protein profiles of breast muscle from broilers with high or low FE phenotypes using in-gel trypsin digestion and tandem MS. One hundred and fifty-two proteins were differentially expressed between the high and low FE phenotypes. Mitochondria plays an important role in the phenotypic expression of FE. Mitochondrial expression was greater in the high FE phenotype than the low FE phenotype. Mitochondrial dysfunction and oxidative phosphorylation pathways were upregulated in high FE birds as indicated by Ingenuity Pathway Analysis. Among them, upregulated insulin signaling and mTOR (mamFfimalian target of rapamycin) pathways may be potential biomarker candidates for high FE selection. However, inhibition of MAP4K4 (Mitogen-Activated Protein Kinase Kinase Kinase Kinase 4) activity may contribute to muscle growth in the high FE phenotype. Two physiological processes, site-specific defects in electron transport chain and proton leak, might contribute to the mitochondrial inefficiency and the overall expression of a low FE phenotype (Bottje & Kong, 2013). The growing condition, such as heat stress, may also affect mitochondrial function. He et al. (2019) investigated the proteomic differences in duck from birds exposed to heat stress using a tandem mass tag proteomic approach. The differentially expressed proteins included

178 down-regulated and 34 up-regulated proteins, revealing that heat stress affects the oxidation status of duck muscle due to the suppression of mitochondrial functions and respiratory chains. It was also found that sarcoplasmic/endoplasmic reticulum calcium ATPases, Mn-superoxide dismutase, heat shock protein family B member 7, methyltransferase like 21C, myosin-binding protein C, and carbonic anhydrase 3 play important roles in defensive mechanisms against heat stress.

7.4 Meat quality attributes

Proteomic technologies have been applied to characterize biochemical mechanisms governing meat quality traits, such as color, tenderness, and water-holding capacity (Gagaoua et al., 2021; Gobert, Sayd, Gatellier, & Santé-Lhoutellier, 2014; Purslow, Gagaoua, & Warner, 2021). While extensive investigations examined the role of proteome in the quality of red meats, a proteomic approach in poultry meat quality research is still in its early stages. In addition, proteomic investigations on meat from poultry species were focused primarily on tenderness and water-holding capacity (Cao, Xiao, Ge, & Wu, 2020).

Mekchay, Teltathum, Nakasathien, and Pongpaichan (2010) evaluated the proteome of breast muscles (with high- and low-shear force values) from Thai native chicken and commercial broiler chicken to determine biomarker proteins for tenderness. Proteomic data indicated that five proteins were up- and down-regulated in association with shear force values. While pyruvate kinase, phosphoglycerate mutase, and triosephosphate isomerase were associated with increased shear force in native Thai birds, two unidentified proteins were associated with low shear force in commercial broilers. These results suggested that glycolytic enzymes play a major role in the development of meat quality and that the abundance of these proteins could potentially be used as functional biomarkers for poultry meat quality.

With the goals to identify proteome biomarkers for water-holding capacity in broiler meat, Phongpa-Ngan et al. (2011) examined the proteome of chicken breast muscles that were categorized by high and low water-holding capacity. Several protein spots were differentially expressed in breast muscles exhibiting high and low water-holding capacities. Heat shock protein and pyruvate kinase muscle isozyme were more abundant in high water-holding capacity muscles. Interestingly, triosephosphate isomerase appeared in three spots (possibly due to post-translational modifications); while two spots had increased expression in high water-holding capacity samples, one spot was increased expression in low water-holding capacity muscles. These findings suggest that the aforementioned proteins could be employed as biomarkers for water-holding capacity in broiler meat and that protein post-translational modifications influence meat quality.

Meat from Korean native chickens differs in flavor from the meat from commercial broilers. To elucidate the contributions of muscle proteins and peptides to flavor in meat from Korean native chickens and commercial broilers, Liu et al. (2012) evaluated the proteome of breast and thigh meat from the two breeds. While seven protein spots exhibited differential abundance in breast meat, nine spots were differentially abundant in thigh meat from the two breeds. Phosphoglucomutase 1, phosphoglycerate mutase, ATP synthase subunit beta (mitochondrial), myosin heavy chain, heat shock protein B1, cytochrome c reductase, glyoxylase 1, and DNA methyltransferase 3B were overexpressed in Korean native chickens. The authors concluded that these overabundant proteins could contribute to increased degradation of muscle proteins during post-mortem aging leading to increased flavor in Korean native chickens.

The proteome basis of improved carcass traits and meat quality in chickens supplemented with the probiotic *Enterococcus faecium* has been examined (Zheng et al., 2014). The results indicated that probiotic supplementation improved color and water-holding capacity of breast muscle, and 22 proteins were differentially abundant as a result of probiotic feeding. The differentially abundant proteins included enzymes involved in energy metabolism, cytoskeletal proteins, and chaperones, which are major influencers of meat quality traits. The increased pH in *E. facecium* supplemented muscle might be related to the downregulation of beta-enolase and pyruvate kinase muscle isozyme. While the enhance expression of heat shock protein 70 and GAPDH are correlated with the improved water-holding capacity and meat tenderness of broilers fed *E. facecium*. These findings suggested that dietary supplementation of probiotic influences the expression of muscle proteins, which in turn reflect in an improvement in meat quality.

To elucidate the mechanisms of and identify protein markers associated with water-holding capacity, proteome profiles of geese pectoralis major muscle from low and high drip loss groups were evaluated using label-free MS (Zhang, Wang, Xu, & Xu, 2019a). Twenty-one differentially abundant proteins (structural proteins, metabolic enzymes, antioxidant enzymes, and stress response proteins) were identified in high and low drip loss groups. Similar to Zheng et al. (2014), heat shock protein 70 and 90 were associated with the water-holding capacity of goose meat and they were enriched in the low-drip group. Metabolic enzymes including glycogen synthase, GAPDH and ATP synthase subunit were more abundant in the high drip group with a pH of 5.8, which was not different from the pH of low drip group of 6.1. Another category of proteins is structural proteins and the changes in structural proteins influence the drip formation. In this study, myosin regulatory light chain 2 and keratin were more abundant but desmin and cofilin were less abundant in low drip samples.

Lipid oxidation is the major cause of flavor deterioration in poultry meat. Zhang et al. (2021a) examined the proteome profile of duck pectoralis major muscle, using isobaric tags for relative and absolute quantification and parallel reaction monitoring, to determine protein biomarkers for lipid oxidation. Proteome analyses of pectoralis muscle from low and high lipid oxidation groups identified more than 300 differentially abundant proteins. The differentially abundant proteins included proteins involved in lipid metabolism, lipolysis, stress response, oxidative respiratory chain, and redox regulation, which could contribute to variations in free radical generation and lipid oxidation. Adipocyte-like fatty acid-binding protein and apolipoprotein C-III were more abundant in the high lipid oxidation group; these two proteins are involved in lipid binding, transport, and metabolism. Cytosolic phospholipase A2, which cleaves phospholipids to release free fatty acids, was less abundant in the high lipid oxidation group. Lipid oxidation is positively associated with the expression of stress proteins including phosphoprotein 1, heat shock protein 90-alpha, and alpha-crystallin B chain. The stress proteins were expressed in response to stress to regulated cellular homeostasis, cell death and survival.

7.5 Meat quality defects

The poultry industry produces high-quality meat protein in a cost-effective manner through selective breeding and optimized nutrition. This contributes to increased meat yield to meet consumers' growing demand for poultry products. However, the production of fast-growing birds that increases yields also contributes to an increased incidence of decreased meat quality. Poultry meat defect includes PSE, white striping (WS), WB, spaghetti, and green muscle in broilers and turkeys. Gonzalez (2020), and Baldi, Soglia, and Petracci (2020), Zampiga et al. (2019) summarized breast muscle abnormalities in broilers and turkeys, macroscopic traits, potential causes, and their impact on meat quality. Affected muscle is characterized by abnormal cell structure and protein functions. Therefore, proteomics research has been conducted to characterize poultry meat defects (Table 7.1).

7.5.1 Pale, soft, and exudative meat

PSE-like meat is characterized by light or pale color (increased lightness values), lower pH, soft texture, and reduced water-holding capacity and cooking yield. PSE-like meat is caused by the combination of genetics, antemortem stress, such as heat stress, and processing conditions. The combination of low pH and high temperature of the breast meat results in protein profile changes, protein denaturation, and partial loss of protein functionality. Normal and PSE meat differ in protein expression

TABLE 7.1 Recent studies on the poultry meat defects using proteomic approaches.

Meat defect	Studied factors	Animal protein source	Proteomic approach	Main findings	References
PSE	Stress	Male Hubbard × Cobb 500 broiler breast samples	2-DE and liquid chromatography-electrospray ionization-tandem Mass Spectrometry (LC-ESI-MS/MS)	Over-abundant of proteins involved in glycolytic, muscle contraction, proteolytic, ATP regeneration, energy metabolism, and CO2 hydration in PSE-like breast meat	Desai et al. (2016)
White striping and woody breast	Meat myopathy	High breast yielding male broiler breast samples	LC-MS/MS	Upregulated pathways indicate an increased protein synthesis in breast muscle with severe myopathies	Kuttappan et al. (2017)
Woody breast	Meat myopathy	ROSS 708 broiler breast samples	2-DE and LC-MS/MS	The differentially expressed proteins suggested an increased oxidative stress in woody breast. The glycolytic enzymes were under-expressed in WB meat	Cai et al. (2018)
Woody breast	Genetics, diet, postmortem	Five genetically different commercial broiler strain breast samples	2-DE and LC-MS/MS	WB and NB had different proteome profiles, and the difference was dependent on genetic strain, diet, and postmortem time	Zhang et al. (2020a)
Woody breast	Postmortem	Male broiler breast samples	2-DE and LC-MS/MS	There are differences in the postmortem metabolism of NB and WB. These changes also indicate a faster and more oxidative muscle to meat conversion process for WB	Zhang et al. (2020b)
Woody breast	Genetics	Five genetically different commercial broiler strain breast samples	2-DE and LC-MS/MS	There were differences in the protein expression changes in different broiler strains and the breast phenotypes (NB and WB). Strain 5 birds are genetically more prone to develop WB when birds were fed the control diet	Zhang et al. (2021b)

(Desai et al., 2016). Glycolytic enzymes, including phosphoglycerate kinase, beta-enolase, and pyruvate kinase M type were overabundant in PSE meat, were positively correlated with lightness, and negatively correlated with pH at 24 h postmortem. Other glycolytic enzymes including alpha-enolase, ATP-dependent 6-phosphofructokinase (PFK), fructose 1,6-bisphosphatase, and phosphoglycerate mutase-1 were overabundant in normal meat (Table 7.1). The abundance of PFK and phosphoglycerate mutase-1 were negatively correlated with lightness and positively correlated with pH at 24 h postmortem. Myofibrillar proteins that were differentially expressed were overabundant in PSE meat, including myosin heavy chain, malate dehydrogenase, creatine kinase M-type, carbonic anhydrase 2, and proteasome subunit alpha. These differentially expressed proteins could potentially be used as PSE biomarkers in broiler production. PSE-like meat in turkeys has been well studied. Malila et al. (2014) compared the transcriptome profiles of PSE turkey breast meat and identified differential expression of 494 loci. The differentially expressed genes suggested abnormal calcium homeostasis and signaling pathways that regulate actin cytoskeleton structure, carbohydrate metabolism, and energy production in PSE turkey breast samples. Studies on protein profiles of turkey PSE meat are needed to characterize dynamic activity that happens in a cell and post-translational modifications of proteins to determine how to lessen the production incidence of PSE meat.

7.5.2 Woody breast, white striping, and spaghetti

WB is characterized by bulging and pale areas of Pectoralis major muscle (Sihvo, Immonen, & Puolanne, 2014). WS is characterized by the presence of white striations parallel to the direction of muscle fibers (Kuttappan et al., 2017), while spaghetti is the separation of muscle fiber bundles (Baldi et al., 2018). Although broiler WB, WS, and spaghetti meat are macroscopically different, they exhibit similar histological alteration microscopically (Soglia, Mazzoni, & Petracci, 2019). The common histological features include the loss of normal muscle architecture, abnormal fibers, proliferation of perimysial and endomysial connective tissue, and infiltration of fat and inflammatory cells in muscle tissue. Although these myopathies have been extensively studied in broilers, limited information is available for the potential abnormalities in turkey. Zampiga et al. (2019) tested the two genetic lines of turkey hybrids and noticed that WS was present in almost all breast samples that were evaluated. The occurrence of spaghetti meat-like conditions was negligible, while the WB condition was not evaluated. Similarly, Mudalal (2019) also found that the incidence of moderate and severe WS in turkey breast was 61.3%. The prevalence of WS in turkey breast was relatively high and the affected breast was characterized with high fat and low protein contents, which negatively affects the quality traits of turkey breast meat, including water-holding capacity, texture, and ability to use the protein in processed meats.

In the past, our research group and others studied the proteomic profiles of normal and WB meat from commercially available broiler strains (Cai et al., 2018; Kuttappan et al., 2017; Zhang et al., 2020a, b, 2021b) using 2-DE and LC-MS/MS analysis (Table 7.1). Kuttappan et al. (2017) compared the proteomic profiles of normal (no or few minor myopathic lesions) and severe (with severe myopathic changes) muscle tissue using a hybrid LTQ-QrbitrapXL mass spectrometer (Table 7.1). The 141 differentially expressed proteins (Table 7.2) suggested down-regulation of glycolysis and gluconeogenesis pathways and up-regulation of eukaryotic initiation factor-2 (eIF-2) signaling, mechanistic target of rapamycin (mTOR) signaling, and eIF4 and $p70^{S6K}$ signaling. These differences in protein expressions indicated that the myopathy is associated with up-regulated protein metabolism and protein translation, which might be attributed to the higher growth rate seen in these birds. Research done by our group utilized 2-DE and LC-MS/MS and focused on the differentially expressed proteins between normal and myopathy-affected breast tissue.

TABLE 7.2 Protein biomarkers (>1.5 fold change in protein expression) of woody breast and white striping meat in broilers.

Factor	Over-expressed in affected tissue	Under-expressed in affected tissue
Woody breast and white striping (Kuttappan et al., 2017)	cadherin 13, calreticulin, crystallin alpha B, desmin, eukaryotic translation initiation factor 3 subunit A, eukaryotic translation initiation factor 3 subunit L, heat shock protein 90 kDa beta family member 1, hexokinase 1, prolyl 4-hydroxylase subunit beta, protein disulfide isomerase family A member 3, receptor for activated C kinase 1, ribosomal protein, tyrosine 3-monooxygenase, vimentin	bisphosphoglycerate mutase, dual specificity phosphatase 3, lactate dehydrogenase A, malic enzyme 1, myosin light chain 1, myosin light chain kinase 2, phosphoglycerate mutase 1, phosphoglycerate kinase 1, phosphoglucomutase 1, phosphorylase, glycogen (brain), protein phosphatase 1 regulatory subunit 3A pyruvate kinase (muscle), triosephosphate isomerase 1
Woody breast (Cai et al., 2018)	14–3-3 protein gamma, carbonic anhydrase III, myosin regulatory light chain 2, phosphoglycerate mutase 1, protein deglycase DJ-1, serum albumin	creatine kinase M-type triosephosphate isomerase

(Continued)

TABLE 7.2 Cont'd

Factor	Over-expressed in affected tissue	Under-expressed in affected tissue
Woody breast (Zhang et al., 2020a)	annexin A2, annexin A5, apolipoprotein A-1, cofilin-2 (muscle isoform), creatine kinase M-type, desmin, elongation factor 1-alpha, elongation factor Tu (mitochondrial), four and a half LIM domains protein 1, heat shock protein beta-1, L-lactate dehydrogenase B chain, phosphofructokinase, phospho-glycerate mutase-1, phosphoglucomutase-1 isoform X1, peptidyl-prolyl cis-trans isomerase B, serum albumin, T-complex protein 1 subunit gamma, ovotransferrin, troponin I (fast skeletal muscle), troponin T (fast skeletal muscle), ubiquitin carboxyl-terminal hydrolase	carbonic anhydrase 2 beta-enolase glycerol-3-phosphate dehydrogenase 1

Cai et al. (2018) found that six proteins were overabundant in WB meat samples and two proteins were overabundant in NB meat (Table 7.2). Differences between WB and normal breast proteomes revealed that there was increased oxidative stress and decreased glycolytic activity in WB meat. When we compared the proteome differences between NB and WB of different genetic broiler strains that were fed with standard and amino acid-reduced diets at different postmortem time points (Table 7.1; Zhang et al., 2020a). Twenty-one of these proteins were overexpressed in WB meat and three proteins were underexpressed in WB meat (Table 7.2). The differentially expressed proteins in each comparison were consistently upregulated or downregulated in WB tissue although the different protein profiles were noticed for each comparison. Strains 2 and 5 had more protein profile differences between WB and NB meat than strains 1, 3, and 4, which potentially indicates a stronger genetic component for strains 2 and 5 with respect to WB formation. The proteins that were overexpressed in WB are involved in carbohydrate metabolism, oxidative stress, cytoskeleton structure, and transport and signaling.

The conversion of muscle to meat postmortem is a critical step in the development of meat quality traits. We characterized the biochemical changes during rigor mortis for NB and WB (Table 7.1; Zhang et al., 2020b). Twenty out of twenty-four proteins changed exclusively in either NB or WB, indicating differences in post-mortem metabolism. Reduced glycolytic potential and a shorter extension of glycolysis were seen in post-mortem WB as glycolytic proteins declined faster in WB than in NB. The WB condition affected the post-mortem metabolism, as indicated by changes in structural proteins, metabolic proteins, stress-related proteins, transport proteins as well as the degree of post-translational modifications due to the differences in antemortem conditions in NB and WB. These differences are indicative of a faster and more oxidative muscle to meat conversion process for WB.

Broiler genetics have a major impact on the growth rate of broiler chicks, feed conversion ratios, abdominal fat pad weight, carcass weight, and carcass composition. We also observed that genetic variation causes phenotypic differences in five strains of broilers that are commonly used in the broiler industry and produce NB and WB (Table 7.1, Zhang et al., 2021b). The proteomes of broiler breast muscle act as a molecular link between the genome and phenotypic characteristics. Strain 5 breast proteomes were most different from other strains in both NB and WB groups. Strain 5 birds are genetically more prone to develop WB when birds were fed the control diet due to the evidence of increased apoptosis and protein synthesis, more intense contraction, and high oxidative stress in comparison to other strains. In the future, the genome sequencing of broiler strains will facilitate the application of proteomics to detect biomarkers for meat defects and monitoring the effectiveness of corrective strategies. In summary, proteomic analyses indicated that the WB myopathy in broilers is related to genetics, nutrition, and rigor mortis (Fig. 7.1). The differentially expressed proteins in WB are potential indicators of hypertrophic muscle cells, impaired protein synthesis and skeletal muscle regeneration, cell apoptosis, and increased cellular and oxidative stress.

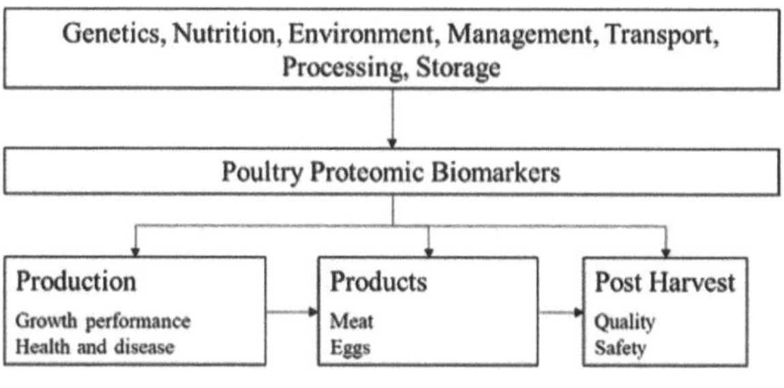

FIG 7.1 Major application areas in poultry proteomics.

7.6 Infectious disease

Proteomic studies have also been applied to determine the proteins in the samples of interest and to determine the protein expression changes under pathological conditions or challenges. Differentially expressed proteins can be used to characterize the bird's response to viral or bacterial infections and biomarkers of bird's health status. Infectious bronchitis coronavirus (IBV), Newcastle disease virus (NDV), and avian influenza virus (AIV) are major avian respiratory viruses, which result in morbidity and mortality in poultry. The proteome changes in the tracheal tissues of the chickens after IBV, NDV, and AIV H9 infections were investigated using 2D-DIGE and followed by MALDI-TOF/TOF-MS analysis (Sun et al., 2014). The proteomic analysis indicated that IBV, NDV, and AIV H9 induced similar core host responses. The differential expressions of proteins, such as annexin proteins (ANXA1, ANXA2) and a heat shock protein (HSPB1), in infected chickens that were induced by different viral infections may be able to help elucidate the specific pathogenetic mechanism of IBV, NDV, and AIV H9 viruses.

Salmonella enterica serovar Enteritidis affects human health through contamination of eggs and egg products. A shotgun proteomics analysis was used to determine the proteome changes in the liver and blood serum of S. *Enteritidis* infected chickens. In the liver, proteome changes indicated decreased carbon metabolism, amino acid metabolism, glycolysis, citric cycle, oxidative phosphorylation, and fatty acid metabolism, and increased immune system function related pathways, such as immunoglobulin expression and TNF (tumor necrosis factor) signaling. The abundance of some known acute phase proteins (serum amyloid A, avidin, lysozyme G2, leukocyte cell-derived chemotaxin-2) were increased after Salmonella enteritidis infection. In addition, the increase in pantetheinase VNN1 may explain the decrease in fatty acid metabolism in the liver. Another bacteria, *Campylobacter jejuni*, a main reservoir for human infections, has also been studied for its protein profiles in poultry. After challenging broilers with *C. jejuni*, actin and actin-associated cytoskeletal proteins were more abundant in the small intestine; however, an actin-associated anti-apoptotic protein villin-1 was less abundant, indicating that the differential expression of cytoskeletal proteins was attributed to the increased rate of apoptosis in challenged birds (O'Reilly et al., 2017). Proteins involved in energy metabolism, glycolysis, and stress (heat shock proteins 70, 710, and 108) decreased over time in the intestine of birds that were challenged with *C. jejuni*. Proteomic studies provide information on the pathophysiology of various infection processes induced by viruses or pathogens and the protein biomarkers identified in pathogenic conditions in poultry for early diagnosis and therapeutic targets.

7.7 Future directions

In the post-genomic era, the applications of proteomics in food science are exponentially growing and represent a cornerstone to interpret and improve food quality. Proteomic tools, such as MS, gel electrophoresis, and bioinformatic databases, offer high-throughput and highly sensitive analytical techniques to analyze protein-based concerns in food quality and provide robust data to meaningfully associate with food quality traits with proteins. Characterizing the functional association between proteins and quality traits will enable developing suitable strategies to process and retail poultry products that are rich in high-quality animal proteins. Future research could integrate other omic disciplines (such as genomics, transcriptomics, metabolomics, lipidomics, and metagenomics) with proteomics to systematically elucidate the biological basis behind the complex traits in order to understand and improve the nutrition, meat quality and disease resistance in poultry. This integrative omic approach can, in turn, lead to minimizing food waste, improving food security, and providing animal proteins to the growing global population.

7.8 Conclusions

Proteomic methods have many applications in poultry research and in the poultry industry. First, proteomics can be used to differentiate the proteome between eggs proteins that do and do not exhibit allergenicity. Second, proteomics can be used to explain differences in variability among muscle proteins of different avian species. Third, it is useful in exploring potential biomarkers that can be linked to meat quality defects. In order to work toward minimizing these defects, biomarkers need to be linked to genetics, nutrition, and the environment to explain the factors that lead to poultry meat quality defects. In addition, instrumental meat quality and sensory characterization are also necessary to fully elucidate the cause of meat defects and the effect on the meat product. Future research needs to be linked to other omic methodologies, quality determinations, sensory data, and the microbiome to work toward improving animal health and performance as well as eliminating quality defects from meat production.

References

Aslam, B., Basit, M., Nisar, M. A., Khurshid, M., & Rasool, M. H. (2017). Proteomics: Technologies and their applications. *Journal of Chromatographic Science, 55*(2), 182–196. https://doi.org/10.1093/chromsci/bmw167.

Baldi, G., Soglia, F., Mazzoni, M., Sirri, F., Canonico, L., Babini, E., et al. (2018). Implications of white striping and spaghetti meat abnormalities on meat quality and histological features in broilers. *Animal, 12*(1), 164–173. https://doi.org/10.1017/S1751731117001069.

Baldi, G., Soglia, F., & Petracci, M. (2020). Current status of poultry meat abnormalities. *Meat and Muscle Biology, 4*(2), 1–7. https://doi.org/10.22175/mmb.9503.

Bottje, W., & Kong, B. W. (2013). Cell biology symposium: Feed efficiency: Mitochondrial function to global gene expression. *Journal of Animal Science, 91*(4), 1582–1593. https://doi.org/10.2527/jas.2012-5787.

Cai, K., Shao, W., Chen, X., Campbell, Y. L., Nair, M. N., Suman, S. P., et al. (2018). Meat quality traits and proteome profile of woody broiler breast (pectoralis major) meat. *Poultry Science, 97*(1), 337–346. https://doi.org/10.3382/ps/pex284.

Cao, C., Xiao, Z., Ge, C., & Wu, Y. (2020). Application and research progress of proteomics in chicken meat quality and identification: A review. *Food Reviews International, 38*, 1–22. https://doi.org/10.1080/87559129.2020.1733594.

Desai, M. A., Jackson, V., Zhai, W., Suman, S. P., Nair, M. N., Beach, C. M., et al. (2016). Proteome basis of pale, soft, and exudative-like (PSE-like) broiler breast (Pectoralis major) meat. *Poultry Science, 95*(11), 2696–2706. https://doi.org/10.3382/ps/pew213.

Gagaoua, M., Terlouw, E. M. C., Mullen, A. M., Franco, D., Warner, R. D., Lorenzo, J. M., et al. (2021). Molecular signatures of beef tenderness: Underlying mechanisms based on integromics of protein biomarkers from multi-platform proteomics studies. *Meat Science, 172*, 108311. doi:10.1016/j.meatsci.2020.108311.

Geng, F., Xie, Y., Wang, J., Majumder, K., Qiu, N., & Ma, M. (2018). N -Glycoproteomic analysis of chicken egg yolk. *Journal of Agricultural and Food Chemistry, 66*(43), 11510–11516. https://doi.org/10.1021/acs.jafc.8b04492.

Gobert, M., Sayd, T., Gatellier, P., & Santé-Lhoutellier, V. (2014). Application to proteomics to understand and modify meat quality. *Meat Science, 98*(3), 539–543. https://doi.org/10.1016/j.meatsci.2014.06.035.

Gonzalez, J. M. (2020). Poultry and pork muscle defects and meat quality - consequences, causes, and management. *Journal of Animal Science, 98*(9), SKAA263. doi:10.1093/JAS/SKAA263.

He, J., Xia, C., He, Y., Pan, D., Cao, J., Sun, Y., et al. (2019). Proteomic responses to oxidative damage in meat from ducks exposed to heat stress. *Food Chemistry, 295*, 129–137. https://doi.org/10.1016/j.foodchem.2019.05.073.

Hu, S., Qiu, N., Liu, Y., Zhao, H., Gao, D., Song, R., et al. (2016). Identification and comparative proteomic study of quail and duck egg white protein using 2-dimensional gel electrophoresis and matrix-assisted laser desorption/ionization time-of-flight tandem mass spectrometry analysis. *Poultry Science, 95*(5), 1137–1144. https://doi.org/10.3382/ps/pew033.

Kong, B. W., Lassiter, K., Piekarski-Welsher, A., Dridi, S., Reverter-Gomez, A., Hudson, N. J., et al. (2016). Proteomics of breast muscle tissue associated with the phenotypic expression of feed efficiency within a pedigree male broiler line: I. Highlight on mitochondria. *Plos One, 11*(5), 0155679. https://doi.org/10.1371/journal.pone.0155679.

Kuttappan, V. A., Bottje, W., Ramnathan, R., Hartson, S. D., Coon, C. N., Kong, B. W., et al. (2017). Proteomic analysis reveals changes in carbohydrate and protein metabolism associated with broiler breast myopathy. *Poultry Science, 96*(8), 2992–2999. https://doi.org/10.3382/ps/pex069.

Liu, X. D., Jayasena, D. D., Jung, Y., Jung, S., Kang, B. S., Heo, K. N., et al. (2012). Differential proteome analysis of breast and thigh muscles between Korean native chickens and commercial broilers. *Asian-Australasian Journal of Animal Sciences, 25*(6), 895–902. https://doi.org/10.5713/ajas.2011.11374.

Liu, Y., Qiu, N., Gao, D., & Ma, M. (2018). Comparative proteomic analysis of chicken, duck, and quail egg yolks. *International Journal of Food Properties, 21*(1), 1311–1321. https://doi.org/10.1080/10942912.2018.1489836.

Malila, Y., Carr, K. M., Ernst, C. W., Velleman, S. G., Reed, K. M., & Strasburg, G. M. (2014). Deep transcriptome sequencing reveals differences in global gene expression between normal and pale, soft, and exudative Turkey meat. *Journal of Animal Science, 92*(3), 1250–1260. https://doi.org/10.2527/jas.2013-7293.

Mekchay, S., Teltathum, T., Nakasathien, S., & Pongpaichan, P. (2010). Proteomic analysis of tenderness trait in thai native and commercial broiler chicken muscles. *The Journal of Poultry Science, 47*(1), 8–12. https://doi.org/10.2141/jpsa.009033.

Meng, Y., Qiu, N., Geng, F., Huo, Y., Sun, H., & Keast, R. (2019). Identification of the duck egg white N-glycoproteome and insight into the course of biological evolution. *Journal of Agricultural and Food Chemistry, 67*(35), 9950–9957. https://doi.org/10.1021/acs.jafc.9b03059.

Meng, Y., Qiu, N., Geng, F., Keast, R., Li, B., & Zheng, X. (2020). N-glycoproteomic analysis of duck egg yolk proteins: Implications for biofunctions and evolution. *International Journal of Biological Macromolecules, 151*, 19–26. https://doi.org/10.1016/j.ijbiomac.2020.02.038.

Molette, C., Rémignon, H., & Babilé, R. (2005). Modification of glycolyzing enzymes lowers meat quality of turkey. *Poultry Science, 84*(1), 119–127. https://doi.org/10.1093/ps/84.1.119.

Monaci, L., De Angelis, E., Montemurro, N., & Pilolli, R. (2018). Comprehensive overview and recent advances in proteomics MS based methods for food allergens analysis. *TrAC—Trends in Analytical Chemistry, 106*, 21–36. https://doi.org/10.1016/j.trac.2018.06.016.

Mudalal, S. (2019). Incidence of white striping and its effect on the quality traits of raw and processed Turkey breast meat. *Food Science of Animal Resources, 39*(3), 410–417. https://doi.org/10.5851/kosfa.2019.e35.

Nasirifar, E., Rezvannejad, E., Maleki, M., & Sami, M. (2021). The impact of differential expression of the pectoral muscle proteome in two groups of Japanese quail with different growth rates on meat quality. *British Poultry Science, 62*(1), 38–45. https://doi.org/10.1080/00071668.2020.1812526.

O'Reilly, E. L., Burchmore, R. J., Sparks, N. H., & Eckersall, P. D. (2017). The effect of microbial challenge on the intestinal proteome of broiler chickens. *Proteome Science, 15*(1), 1–17. https://doi.org/10.1186/s12953-017-0118-0.

Ouyang, H., Wang, Z., Chen, X., Yu, J., Li, Z., & Nie, Q. (2017). Proteomic analysis of chicken skeletal muscle during embryonic development. *Frontiers in Physiology, 8*(MAY), 1–11. doi:10.3389/fphys.2017.00281.

Petracci, M., Cavani, C. (2012). Muscle growth and poultry meat quality issues. *Nutrients, 4*(1), 1–12.

Phongpa-Ngan, P., Grider, A., Mulligan, J. H., Aggrey, S. E., & Wicker, L. (2011). Proteomic analysis and differential expression in protein extracted from chicken with a varying growth rate and water-holding capacity. *Journal of Agricultural and Food Chemistry, 59*(24), 13181–13187. https://doi.org/10.1021/jf202622n.

Purslow, P. P., Gagaoua, M., & Warner, R. D. (2021). Insights on meat quality from combining traditional studies and proteomics. *Meat Science, 174*, 108243. doi:10.1016/j.meatsci.2020.108423.

Remignon, H., Molette, C., Babile, R., & Fernandez, X. (2006). Current advances in proteomic analysis and its use for the resolution of poultry meat quality problems. *World's Poultry Science Journal, 62*(1), 123–191. https://doi.org/10.1079/WPS200589.

Sihvo, H. K., Immonen, K., & Puolanne, E. (2014). Myodegeneration with fibrosis and regeneration in the pectoralis major muscle of broilers. *Veterinary Pathology, 51*(3), 619–623. https://doi.org/10.1177/0300985813497488.

Soglia, F., Mazzoni, M., & Petracci, M. (2019). Spotlight on avian pathology: Current growth-related breast meat abnormalities in broilers. *Avian Pathology, 48*(1), 1–3. https://doi.org/10.1080/03079457.2018.1508821.

Spoel, S. H. (2018). Orchestrating the proteome with post-translational modifications. *Journal of Experimental Botany, 69*(19), 4499–4503. https://doi.org/10.1093/jxb/ery295.

Sun, J., Han, Z., Shao, Y., Cao, Z., Kong, X., & Liu, S. (2014). Comparative proteome analysis of tracheal tissues in response to infectious bronchitis coronavirus, Newcastle disease virus, and avian influenza virus H9 subtype virus infection. *Proteomics, 14*(11), 1403–1423. https://doi.org/10.1002/pmic.201300404.

Surowiec, I., Koistinen, K. M., Fraser, P. D., & Bramley, P. M. (2011). Proteomic approach for the detection of chicken mechanically recovered meat. *Meat Science, 89*(2), 233–237. https://doi.org/10.1016/j.meatsci.2011.04.004.

Timp, W., & Timp, G. (2020). Beyond mass spectrometry, the next step in proteomics. *Science Advances, 6*(2), eaax8978. https://doi.org/10.1126/sciadv.aax8978.

van de Wiel, D. F. M., & Zhang, W. L. (2007). Identification of pork quality parameters by proteomics. *Meat Science, 77*(1), 46–54. https://doi.org/10.1016/j.meatsci.2007.04.017.

Xiao, J., Wang, J., Cheng, L., Gao, S., Li, S., Qiu, N., et al. (2020). A puzzle piece of protein N-glycosylation in chicken egg: N-glycoproteome of chicken egg vitelline membrane. *International Journal of Biological Macromolecules, 164*, 3125–3132. https://doi.org/10.1016/j.ijbiomac.2020.08.193.

Yang, R., Geng, F., Huang, X., Qiu, N., Li, S., Teng, H., et al. (2020). Integrated proteomic, phosphoproteomic and N-glycoproteomic analyses of chicken eggshell matrix. *Food Chemistry, 330*, Article 127167. https://doi.org/10.1016/j.foodchem.2020.127167.

Zampiga, M., Tavaniello, S., Soglia, F., Petracci, M., Mazzoni, M., Maiorano, G., et al. (2019). Comparison of 2 commercial Turkey hybrids: Productivity, occurrence of breast myopathies, and meat quality properties. *Poultry Science, 98*(5), 2305–2315. https://doi.org/10.3382/ps/pey607.

Zapata, I., Reddish, J. M., Miller, M. A., Lilburn, M. S., & Wick, M. (2012). Comparative proteomic characterization of the sarcoplasmic proteins in the pectoralis major and supracoracoideus breast muscles in 2 chicken genotypes. *Poultry Science, 91*(7), 1654–1659. https://doi.org/10.3382/ps.2011-02029.

Zhang, M., Wang, D., Xu, X., & Xu, W. (2019). Comparative proteomic analysis of proteins associated with water holding capacity in goose muscles. *Food Research International, 116*, 354–361. https://doi.org/10.1016/j.foodres.2018.08.048.

Zhang, M., Wang, D., Xu, X., Xu, W., & Zhou, G. (2021a). iTRAQ-based proteomic analysis of duck muscle related to lipid oxidation. *Poultry Science, 100*(4), Article 101029. https://doi.org/10.1016/j.psj.2021.101029.

Zhang, X., Antonelo, D., Hendrix, J., To, V., Campbell, Y., Von Staden, M., et al. (2020). Proteomic characterization of normal and woody breast meat from broilers of five genetic strains. *Meat and Muscle Biology, 4*(1), 1–17. doi:10.22175/mmb.8759.

Zhang, X., To, K. V., Jarvis, T. R., Campbell, Y. L., Hendrix, J. D., Suman, S. P., et al. (2021b). Broiler genetics influences proteome profiles of normal and woody breast muscle. *Poultry Science, 100*(4), 1–14. doi:10.1016/j.psj.2021.01.017.

Zhang, X., Zhai, W., Li, S., Suman, S. P., Chen, J., Zhu, H., et al. (2020). Early postmortem proteome changes in normal and woody broiler breast muscles. *Journal of Agricultural and Food Chemistry, 68*(39), 11000–11010. https://doi.org/10.1021/acs.jafc.0c03200.

Zheng, A., Luo, J., Meng, K., Li, J., Zhang, S., Li, K., et al. (2014). Proteome changes underpin improved meat quality and yield of chickens (Gallus gallus) fed the probiotic Enterococcus faecium. *Bmc Genomics [Electronic Resource], 15*(1), 1–14. doi:10.1186/1471-2164-15-1167.

CHAPTER

8

Current trends in proteomic development towards milk and dairy products

Anand Raj Dhanapal[a], *Baskar Venkidasamy*[b], *Muthu Thiruvengadam*[c], *Maksim Rebezov*[d,f,i], *Natalya Fedoseeva*[e], *Mohammad Ali Shariati*[f], *Ruben Agregán*[g], *Jose M. Lorenzo*[g,h]

[a]Department of Biotechnology, Karpagam Academy of Higher Education (Deemed to be University), Coimbatore, Tamil Nadu, India
[b]Department of Biotechnology, Sri Shakthi Institute of Engineering and Technology, Coimbatore, Tamil Nadu, India
[c]Department of Applied Bioscience, College of Life and Environmental Science, Konkuk University, Seoul, Republic of Korea
[d]V. M. Gorbatov Federal Research Center for Food Systems of Russian Academy of Sciences, 26 Talalikhina St., Moscow, Russian Federation
[e]Russian State Agrarian Correspondence University, Department of Zootechnii, Production and Processing of Livestock Products, 50 Shosse Entuziastov, Balashikha, Russian Federation
[f]K.G. Razumovsky Moscow State University of Technologies and Management (The First Cossack University), 73, Zemlyanoy Val St., Moscow, Russian Federation
[g]Centro Tecnológico de la Carne de Galicia, Ourense, Spain
[h]University of Vigo, Spain
[i]Faculty of Biotechnology and Food Engineering, Ural State Agricultural University, 42 Karl Liebknecht str., Yekaterinburg, Russian Federation

Food Proteomics
DOI: https://doi.org/10.1016/B978-0-323-90889-4.00002-6

8.1 Introduction

Vast number of studies on milk proteins was carried out continuously for over 75 years. Till date, knowledge toward this complex protein system has incrementally progressed in recent decades that largely concurring through technology advances (Anagnostopoulos et al., 2016). Even though there are an excess of in-depth analyzing methods fora number of milk protein components, there raises many questionnaires on concern toward milk protein expression, structure, and modifications. Henceforward, detailed analysis on protein interactions the modification causing with disease and dispensation are also poorly understood. Overall, proteomic associated technologies proven potential to facilitate further detailed understanding of milk proteins. Proteomics approach lets for the detection, identification, and characterization of milk proteins and peptides in a detailed and in-depth manner. Most importantly, proteomic techniques facilitate the analysis on large numbers of milk proteins as well as dairy products simultaneously (Arena, Renzone, D'Ambrosio, Salzano, & Scaloni, 2017).

With the application of proteomic tools leads to the global then dynamic view of proteins and peptides which are expressed by bacteria. Therefore, with the currently available bacterial genomes for homology searches, it is possible to screen the large number of proteins expressed by microorganisms in various fermented foods through such advanced techniques (Manso, Léonil, Jan, & Gagnaire, 2005). Furthermore, proteomic tool objective to investigate toward protein heterogeneity in protein-rich foods (H. McKenzie, 1970). Milk is considered as valuable source of proteins along with other nutrients for human consumptionas of different animal species. Henceforward, dynamic range of proteomic methods delivers different nutritional profiles between animal species (Soggiu, Roncada, & Piras, 2018).

In general, milk can be consumed in plain or with other renovated developments like fermentation, ripening, and many others. Henceforward, the development and advantage of proteomics to study proteins along with safety, quality, and revolution practices need clear and deep understanding (Abd El-Salam, 2014). The large-scale study on proteins through proteomic technologies could evidence about different protein profiling with exploring unique characteristics of milk, nutrients, lactation stage, as well as relative health status of the consumers (Gagnaire, Jardin, Jan, & Lortal, 2009). More extensively, proteomics is related to the assessment of safety along with quality of milk and milk-derived dairy products (Agregán et al., 2021). Henceforward, we aim to summarize the methods toward sample preparation as well as investigation through recent applications of proteomics to make them available and applicable for research and analytical purposes.

Overall, this chapter is not a review, still a brief overview of recent research techniques that enlighten the effort toward proteins and polysaccharides interaction. Furthermore, to complex between milk proteins and polysaccharides with application of such advanced techniques to dairy industry.

Conversely, the application of proteomics varies according to the samples with origins sources such as vegetable, microbes, and animal. Vast variety of nutritionally relevant proteins have identified and depicted (Le, Deeth, & Larsen, 2017). Proteomic tools perhaps permit such interactions between proteins then polysaccharides to characterize quality of the milk products. Proteomic studies on milk products allow the separation of major proteins, such as caseins and whey proteins, which also regulate the relationship between structure and properties (Tolstoguzov, 1997). Heterogeneity and several molecular forms of milk proteins exist as important challenge that elucidated by means of detailed proteomic study. Protein techniques with high resolution also explore the large number of low abundance proteins in dairy products (Schmitt, Sanchez, Desobry-Banon, & Hardy, 1998). The main aspect concerning with proteomic techniques is to explore the ability to adaptation of milk toward different environmental stress factors.

Furthermore, a new type of food biopolymer termed inter-biopolymer complex, in which their functional properties vary from those macromolecular reactants can be observed (Doublier, Garnier, Renard, & Sanchez, 2000). With multifaceted by other biopolymers, the functional properties of individual biopolymers usually enhanced (Dickinson, 1998), which leads to increased protein expression at a particular point in time. Recent investigations provide static information as well as overall view on cell machinery toward the interaction of milk protein dynamic process (Agregán et al., 2021; Bhattacharya, Salcedo, Robinson, Henrick, & Barile, 2019). Elucidation of interaction on whey proteins or caseins with polysaccharides at low pH thus to form soluble or insoluble complex. Next, to explore biological functions toward dairy industrial applications requires genome expression (D'Auria et al., 2018; Weinbreck, Nieuwenhuijse, Robijn, & de Kruif, 2004).

Henceforward, we provide the application of proteomic tools toward the study of milk proteins expression employed in dairy industries, as well as the recent advances in the application of advanced techniques to milk proteins. In precise, the emergence proteomic techniques with its time-consuming procedures perhaps provide the vast number of unidentified proteins with high quantification on milk proteins and dairy products.

8.2 Milk proteins

Milk is assessed as liquid secreted by female of living species on meeting the complete nutritional requirement primarily (Kinsella & Morr, 1984). It remains to be foremost requirements for energy (supplied by lipids and lactose in addition to proteins), essential amino acids and amino groups for the biosynthesis of non-essential amino acids (both supplied by proteins), and essential fatty acids, vitamins, and inorganic elements (Lönnerdal, 2003). It is characterized to be the essential nutritional requirement of the neonate that depends on birth till growth for energy requirements. Milk is a food full of nutrients with an optimal balance but variable in proportion according to factors intrinsic to the animal, such as the species or the breed, and extrinsic, such as the lactation period or the diet (Guetouache, Guessas, & Medjekal, 2014). Most of the milk is water (87%) and the remaining 13% are solids, with a very wide range of compounds. Specifically, more than 100 different components have been identified only in cow's milk (Miller, Jarvis, & McBean, 2007). The milk of the most commercially exploited species, such as cow, goat, sheep, or buffalo, and others such as yak, horse, and pig have been quite well characterized, even those of mice and rats of experimentation and human milk, concluding that the species is a factor that considerably affects the concentration of the main macronutrients, with variation ranges of 2%–55% in lipids, 1%–20% in protein, and 0%–10% in lactose. In addition, other constituents, such as minerals, vitamins, hormones, or enzymes can be found to a lesser extent, most at trace levels (Fox, 2008).

Milk proteins are a heterogeneous mixture of proteins that are divided into two large fractions, the casein fraction, and the whey protein fraction. The Swedish scientist Olav Hammarsten identified these two fractions back in 1880 through a simple experiment. If the milk was acidified to a pH of 4.6 (isoelectric point of caseins) and heated to about 30%, he observed that most of the proteins present precipitated, specifically about 80%, which corresponds to the casein fraction. The remaining 20% of the proteins that remained soluble under these pH and temperature conditions were designated as serum or whey proteins, also known as non-casein nitrogen (Fox, Uniacke-Lowe., McSweeney., & O'Mahony, 2015). Fig. 8.1 displays an overview of milk proteins and their different fractions.

Domesticated sheep, goats, and cattle are characterized by their use in use toward milk production since almost 8000 BC (McKenzie, 1967). However, toward the industrial point of view, the evolution of milk protein profiling revealed a wide range of different and attractive products appearance displaying an important milk feature (Pihlanto, 2006). Such products development allows the exploitation of broad and narrow properties of milk proteins that attracted the attention of researchers. Lawrance, Klopcic, and Wasinger (2005) stated the milk proteomic research eased by the nature of proteins, which was isolated from milk free from other constituents. Up till now, milk proteins characterization is probably the best toward all other food proteins (Boland & Singh, 2019). A method described for the preparation of protein from milk by acid precipitation

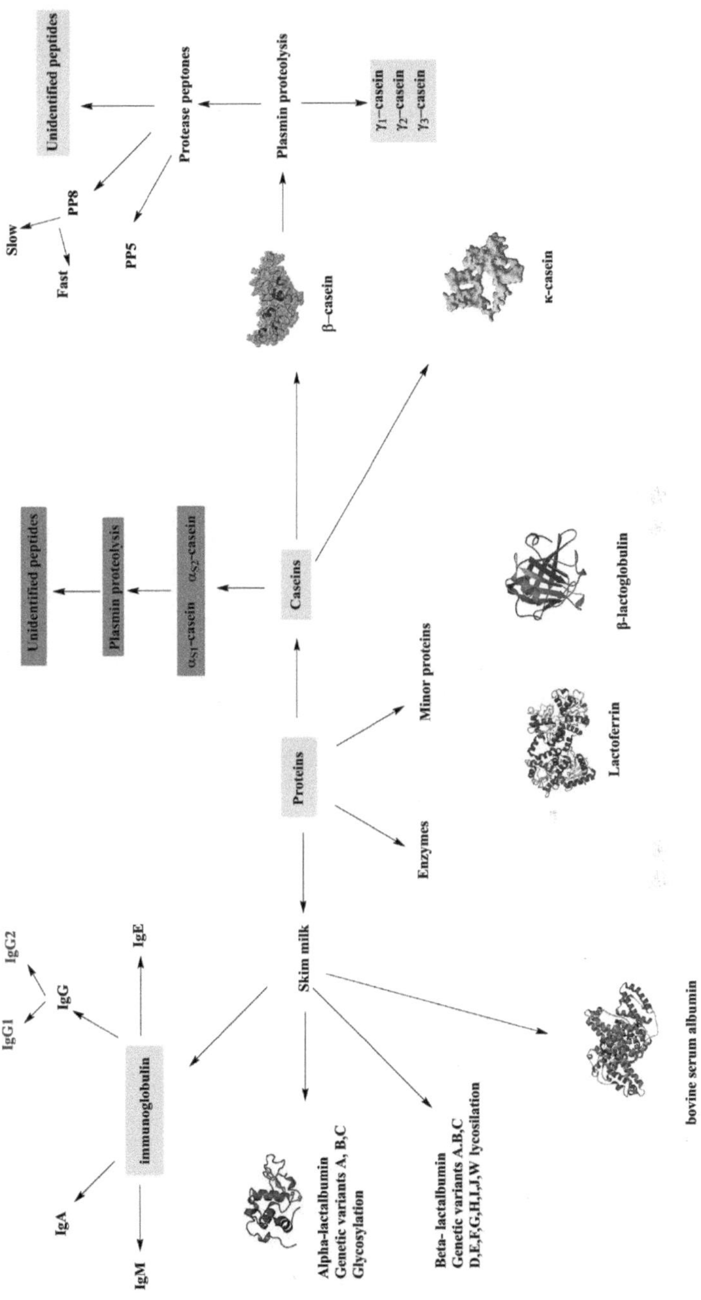

FIG. 8.1 Scheme of various sections of milk proteins.

was in 1838 by J.G. Mulder, who coined the term "protein." In food fermentation and production, there was a growing trend toward widespread use of proteins, it became desirable to produce milk with increased casein levels and also to understand the relationships between the physical properties of proteins with their behavior in food systems (Fox et al., 2015). Due to their specific functional attributes for a wide range of milk-derived protein preparations (e.g., dry milk, milk proteins, caseins, whey proteins, and lactalbumin) were used in foods (Wong, Camirand, Pavlath, Parris, & Friedman, 1996).

In this chapter, we discussed some of the recent applications toward analyzing the properties of the milk protein components with high-resolution techniques. Thus, we reviewed the characters of milk proteins in recent decades with the advance in proteomic study. Detailed analysis of various methods was employed on determining the functional properties along with the variability in composition, as well as preparations. Although, it is difficult to compare data and/or reconcile differences in published information. Henceforth, the desirability of developing standard methods is emphasized on increased quality of the final products.

8.3 Milk proteomics

Characterization of milk proteins came forth continuously for the last 5 decades. However, in-depth knowledge on proteomics to investigate the complex protein includes caseins and whey proteins in recent decades. Advance in proteomics and associated technologies perhaps be the key characteristics to facilitate further improvements on our knowledge toward cheese manufacturing process in industries. Proteomics allows not only for the detection, identification, and characterization of milk proteins, but more importantly, it aids in depicting variants in milk genetic and glycosylation patterns simultaneously (Roncada et al., 2012). In this chapter, we condense recent applications of proteomics toward low abundance proteins/peptides, also with some hormones and enzymes, highlighting pre-fraction methods to bioinformatics analysis. In particular, emphasis is placed on the efficacy of two-dimensional gel electrophoresis (2DE) in combination with various mass spectrometry (MS) techniques for the detailed characterization of such proteins.

Investigation of major proteins, including α_{s1}-, α_{s2}-, β-, and κ-casein, along with whey proteins, such as α-lactalbumin, β-lactoglobulin, and bovine serum albumin, perhaps it is the remarkable feature of the proteomic tool (Pan, Brentnall, & Chen, 2015). The casein polymorphisms and the separation of large number of low abundant proteins are fundamental characteristics specifically considered in the field of proteomics (Giacometti & Buretić-Tomljanović, 2017). Less abundance proteins in the milk are classified by great heterogeneity with the presence of several isoforms difference in ruminants. Moreover, milk containing the high number of low abundance proteins, such as lactoferrin, immunoglobulins, glycoproteins, hormones, and enzymes are identified and presented with proteomic approach (Abd El-Salam, 2014). In addition, advances in proteomic analysis of milk from farm animals are fundamental to understand the milk proteomes in detail (Table 8.1). Focusing on the challenge on fractioning the different proteins while studying milk from farm animals, proteomic studies possibly explore benefits in livestock production.

Characterization of buttermilk and dairy products through proteomic approach facilitates the understanding of lipid secretion and synthesis in detail. In addition, molecular exploration with milk as a source of proteins associated with membranes provided understandings toward signaling and secretory pathways in detail (Low et al., 2013). Additionally, studying the protein fractions of the cytoplasm may help to understand the biological state of milk. Thus, the discovery of many undeclared proteins hereafter could be used in mapping the milk proteome in all species (Reinhardt, Lippolis, Nonnecke, & Sacco, 2012).

Detailed knowledge on secretion of milk fat remains incomplete (Lu, Liu, Gao, Lv, & Yu, 2017), and micro and macro heterogeneity along with different degree of phosphorylation must be explored as soon as possible (Chen, Chen, & Hsieh, 2016). Proteomic approach shows great heterogeneity owing to post-translational modification (PTM) with different patterns on amino acid residues. With an experimental procedure, initially proteins were pre-fractionated by centrifugation following 2DE (Jin et al., 2021) which revealed the host defense proteins in milk protein families (Sui et al., 2014). Analysis of milk during mastitis and colostrum formation were resolved by 2DE, exploring many peptides detected by MS technique.

8.3.1 Concept of proteomics

The stable foundation for research into the human health and disease since application of genomics was documented. Later, proteomics development was the next consistent step (Hagenstein et al., 2003). In 1995, the term proteomics was first made up from a combination of "protein" and "genomics." The concept to characterize the human proteome was first proposed by Norman G. Anderson and N. Leigh Anderson in a submission entitled the "Human Proteins Index Project" in 1979. Since then, protein identification through employing the newly developed technique of two-dimensional polyacrylamide gel electrophoresis (2D-PAGE) would unlock the genome. Henceforth, the understanding of proteomics has evolved much faster than expected (Birgit et al., 2002).

Synthetic and natural peptides separation are presented in proposed separation models toward experimental analysis (Baggerman, Vierstraete, De Loof, & Schoofs, 2005), in addition to chain connectivity of amino acid

TABLE 8.1 List of milk proteins identified in recent proteomic studies using advanced techniques.

Milk protein/peptide and PTMs identified	Technique	References
7 Phosphorylated forms of αs1-casein, 8 αs2-casein, and 5 β-casein	SDS-PAGE, IEF+IPG, nUPLC-nESI MS/MS	Li et al. (2012)
α_{S1}-Casein 1–17, α_{S1}-casein 1–22, α_{S1}-casein 1–36, α_{S2}-casein 182–207, α_{S2}-casein 189–207, α_{S2}-casein 198–207 α_{S2}-casein 200–207, β-casein 1–11, β-casein 191–209, β-casein 192–209, β-casein 193–209, β-casein 196–209, and β-casein 200–209	MALDI-TOF MS	Ebner, Baum, and Pischetsrieder (2016)
Caseinomacropeptides	SDS-PAGE, 2DE, LC-ESI MS and LC-ESI MS/MS	Le et al. (2016)
Peptides from hydrolysis of sodium caseinate	UPLC, SDS-PAGE, LC-ESI-QTOF MS	McGrath, Kinsella, Huppertz, McSweeney, and Kelly (2016)
CSA, lactoferrin, lipoprotein lipase, lactotransferrin, α-1-antiproteinase, serpin A3-7, serpin A3-8, β-1,4-galactosyltransferase 1, peptidoglycan recognition protein 1, ribonuclease 4, angiogenin-1, histone H4, L-amino-acidoxidase, sulfhydryloxidase, and ADP-ribosylation factor 1	HPLC-QTOF, SDS-PAGE, nLC-nESI MS/MS	Felfoul, Jardin, Gaucheron, Attia, and Ayadi (2017)
GlyCAM1, CD59 molecule and lactadherin, phospholipid scramblase 2, secreted phosphoprotein 1, guanine nucleotide-binding protein, annexin A4, selenium-binding protein 1, and keratin 10	SDS-PAGE, iTRAQ nLC-ESI MS/MS	Ji, Li, Ma, and Li (2017)
Enolase 1, Epidermal growth factor receptor, Glyceraldehyde-3-phosphate dehydrogenase, Hexokinase 3, Inhibitor of carbonic anhydrase-like, and Serotransferrin	nLC-MS/MS	Sun, Wang, Sun, and Guo (2019)
Apolipoprotein B, prothrombin, tetraspanin, neural EGF-like 2, thymosin beta 4, Perilipin, cathepsin B, cathepsin Z, perilipin, zinc-alpha-2-glyco-protein, thrombospondin 4, follistatin like 1, and cartilageacidic protein 1	LC-MS/MS, LC-PRM MS	Li et al. (2020)

residues with their interaction on surface. Later, on revealing the proposed model, prediction was accurate on peptide separations with matching amino acid contents and different sequence units within the chain (Nedelkov, Kiernan, Niederkofler, Tubbs, & Nelson, 2006). The variances in the sequence allow variation of amino acid residues with their isomers in the chain besides the occurrence of terminal groups. The employed separation model toward predicting the retention times of peptides equipped the enzymatic hydrolysis proteins in E. coli (Wright & Van Eyk, 2017). Furthermore, the general model accurately clarifies the array on experimental data toward such peptide separation existence the first successful attempt to relate the chain sequence to the retention volume. Fig. 8.2 portrays a possible workflow of milk and its product analysis.

8.3.2 Analytical techniques

Mostly, "unbiased" and "biased" methods are employed for protein analysis in detail (Gilchrist et al., 2006). Such methods classically include separation of proteins by 2D-PAGE, through which proteins are separated based on their mass and electrical charge (Cho, Shan, Winsor, & Diamandis, 2007). Liquid chromatography (LC) technique is based on the flow of mixed proteins through a column packed with porous beads of specific binding properties. In addition, both methods are monitored using MS for protein or peptide identification (Schneider & Riedel, 2010). On the other hand, "biased" methods included the antibody-based methods, on which identified protein of interest is directed with highly specific antibodies (Zhang, Wu, Stenoien, & Paša-Tolić, 2014).

In Caenorhabditis elegans sketching network maps on protein–protein interaction high-throughput version of yeast 2-hybrid technology was employed along with the second newly developed LUminescence-based Mammalian IntERactome (LUMIER). Henceforward, LUMIER technology facilitates the systematical mapping such protein-protein interactions in mammalian cells (Borgo et al., 2017; Wang, Tai, & Chen, 2008). Next, this technology was applied to the TGFβ pathway, which made it possible on identifying the previously unknown physical interaction between the TGFβ type I receptor and Occludin.

8.3.2.1 Characterization techniques

Until the date, 2DE serving as dominant routine to separate and isolate interest proteins (Balkir, Kemahlioglu, & Yucel, 2021; O'Donnell, Holland, Deeth, & Alewood, 2004), based on orthogonal separation of proteins through isoelectric point besides molecular weight, that resolving the related proteins with improved forms of assumed protein. Studies on PTMs rely on taking place selection with a precise analysis toward detecting intact protein modification in the gel through 2DE-based methods

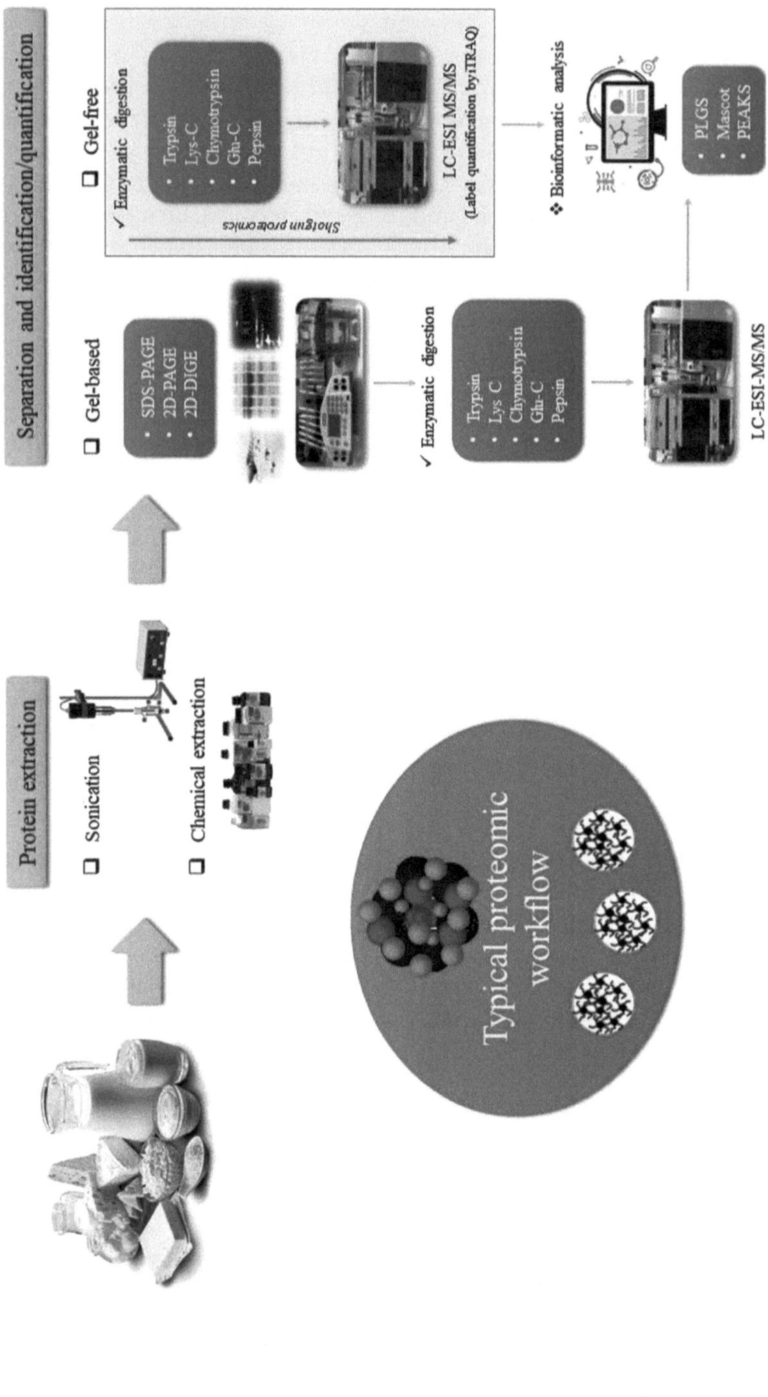

FIG. 8.2 Scheme of proteomic workflow (adapted from Soggiu et al., 2018).

after blotting to membranes (Zhao & Jensen, 2009). In addition, the enrichment of modified proteins through various newly emerged methods before electrophoresis and MS analysis given added value in the food biotechnology field.

Generally, post-translationally modified proteins along with peptides serve to increase the relative abundance of modified polypeptides by means of enrichment through affinity-based methods in sample. The level of intact protein and peptide in a given sample (Chandramouli & Qian, 2009) was explored later with application of such protein methods. To study the cell signaling, antibodies such as anti-phosphotyrosine are extensively cast-off for immune precipitation with intact proteins via MS-driven proteomic strategies (Aslam, Basit, Nisar, Khurshid, & Rasool, 2017). The enrichment of phosphoproteins to reveal novel kinase substrates was possibly facilitated by Anti-pSer/pThr antibodies (Cufaro et al., 2019). Overall, through the development as well as availability with broader range of specific anti-phosphor amino acid antibodies for such immune precipitation will allow a detailed analysis on phosphor-proteome dynamics toward dairy products in the near future.

Mass spectrometers largely solved the difficulties of the proteomics community from their increased sensitivity. In addition, with higher mass accuracy and resolving power improved sensitivity displays the prediction of effective fragmentation on peptides achieved through MS/MS (Savaryn, Catherman, Thomas, Abecassis, & Kelleher, 2013). For several years, electrospray ionization (ESI) MS/MS was employed toward peptide sequencing for generating the diagnostics. PTM-specific reporter ions otherwise MS/MS in precursor ion scans, besides neutral loss scans, stand as efficient technique toward detection of specific modified peptides, including selection on phosphor-peptides and glycopeptides (Peng et al., 2003). However, mass spectrometer hardware and software are required for further advances and enhancements for preparing the samples. On employing these methods of scanning has its application toward the analysis of the highly complex peptide mixtures in proteomics experiments in milk besides their products (Yates III, Gilchrist, Howell, & Bergeron, 2005).

Well ahead, the emergence of matrix-assisted laser desorption/ionization (MALDI)-time-of-flight (TOF) MS/MS instruments proved its efficiency toward post-translationally modified peptides sequencing (e.g., phosphopeptides). Liquid–liquid extraction is a simple and rapid technique to separate the hydrophobic as well as hydrophilic peptides on previously MALDI MS and MS/MS via in-situ (Otto, Becher, & Schmidt, 2014). Elusive hydrophilic peptides, such as phosphor peptides and glycopeptides, may be detected through graphite column by MS-specific (Abdelhamid & Wu, 2015). Capillary chromatography in addition to electrophoresis had interfaced to MALDI MS/MS in various ways to apply in the milk and dairy products analysis. Moreover, MALDI MS and

MS/MS separation are effective when decoupled with time, as this approach allows the revisit of previously analyzed fractions, facilitating in-depth analysis of crude peptide mixtures (Aslam et al., 2017).

8.3.3 Proteomics on characterization of milk and dairy products

The field proteomics by way of that encompass the identification and characterization of peptides, their quantification, discovery of markers, and analysis employing high throughput techniques toward milk and dairy products (Arena et al., 2017). These aspects require specialized expertise in sample preparation technique, besides separate instrumentation handling. On the other hand, analysis of the data generated is crucial for successful research on buttermilk and dairy products (Agregán et al., 2021). Meanwhile, the standard principles of MS-based protein structure analysis gained prominence in the field of peptidomics with fluctuating bioinformatics characteristics.

Milk and dairy products contain more metabolically active microorganisms that secrete extracellular proteins with the discharging intracellular enzymes. Investigation on dietary milk proteins catabolizing in digestive system as well as all over the body employing peptidomics were explored (Giacometti & Buretić-Tomljanović, 2017). The milk *proteome* being extremely complex with adaptable proteolytic processes, the presence of numerous variants in the genetics of milk proteins can be explored (Sánchez-Rivera, Martínez-Maqueda, Cruz-Huerta, Miralles, & Recio, 2014). Also, the study of these products from fermented milk, such as yogurt and cheese, showed their enormous nutritional value as well as health assistance in addition to economic importance (Minkiewicz et al., 2008).

The markers for product originality can be employed through some peptides as well as other product degradation (Barbé et al., 2014). Differences in the peptidic profile of products such as cheeses can be due to variables such as raw material, maturation, and coagulation, through which it is possible to evaluate the quality and characterize artisan cheeses such as "Coalho" from the Brazilian region, Emmentaler cheese (Agyei, Tsopmo, & Udenigwe, 2018) and Gouda cheeses.

Thus, peptides generated on milk protein *hydrolysis* are mainly identified as caseins and α-lactalbumin. Another multi-analyte multiple reaction monitoring (MRM) method on detecting of species-built with species-specific peptides was successfully applied for viewing high specificity to different compositions of cheeses (Piovesana et al., 2015). In addition, high-performance liquid chromatography (HPLC) followed By Edman sequencing peptides released from bovine, goat, and sheep's milk and dairy products were employed and documented (Izquierdo-González et al., 2019). Table 8.2 summarized the techniques used in milk and dairy products.

TABLE 8.2 Advanced proteomic techniques applied to milk and dairy products.

Technique	Application	References
nHPLC-LIT-Orbitrap MS/MS	Exhaustive peptidomic analysis of commercial donkey milk to provide a complete description of nutritional potential	Piovesana et al. (2015)
SDS-PAGE, LC-nESI-LTQ-Orbitrap MS/MS	Characterization of the milk fat globule membrane proteome	Nguyen et al. (2017)
nLC-MS/MS, SDS-PAGE	Characterization of the proteins associated with casein micelles in Holstein and Jersey cows, buffaloes, yaks, goats, camels, and horses	Wang et al. (2017)
nLC-MS/MS	Mapping of the N-glycoproteome of whey from Holstein and Jersey cows, buffaloes, yaks, goats, camels, and horses	Yang et al. (2017b)
iTRAQ nLC-MS/MS	Exploration of protein composition in human and bovine calostrum and mature milk	Yang et al. (2017a)
nLC-nESI-IT MS/MS	Detection of adulteration processes of specific milks in mixtures	Nardiello, Natale, Palermo, Quinto, and Centonze (2018)
Urea-PAGE, micro LC-ESI-tripleTOF MS/MS	Evaluation of the proteomic profile of Fiore Sardo cheese and possible differences among the use of raw and heat-treated milk in its production	Guglielmetti et al. (2019)
nUPLC-nESI-MS/MS	Profiling the peptide composition of ewe's milk kefir	Izquierdo-González et al. (2019)
UPLC-ESI-MRM MS/MS, HPLC, HPLC-ESI-MS	Study of the non-enzymatic PTMs of the serum protein β-lactoglobulin	Meltretter et al. (2020)
IEF-IPG + SDS-PAGE, HPLC-ESI-MS/MS	Proteomic profile of cow's milk kefir	Santini et al. (2020)
2DE: two-dimensional gel electrophoresis; MALDI-TOF MS - TOF: time-of-flight; MS: mass spectrometry; HPLC: high-performance liquid chromatography; PSD: post-source decay; QTOF: quadrupole time-of-flight	Discrimination between two different mozzarella cheeses according to the geographical region	Kandasamy et al. (2021)

However, some standardized techniques that are mandatory for this method are time-consuming. Therefore, they could be not suitable for quantification of peptides regularly. The approach for quantification of peptide was built-up by MS-based peptidomic techniques as ultra-performance liquid chromatography mass spectrophotometry (UPLC-MS/MS), which was a step forward allowing high throughput analysis (Soggiu et al., 2018). The combination of MS with chromatography besides electrophoresis is another leading technique in peptidomics/proteomics analytics. Relative quantification of proteins perhaps measured by changes in folding inside molecules among samples on determining the complete concentration is new complex.

8.3.4 The advantages of recent proteomics in milk and dairy production

The advent of shotgun proteomic technologies served as a tool that suits the discovery on identifying the bioactive peptides. Proteomics with shotgun technique tends to bottom-up proteomics approach through which digesting the complex mixture of protein onto compounds. Then, the documentation of parental proteins after bioinformatics analysis was depicted prior in grouping by enzyme, also taking place in combination with HPLC and MS (Martini, Solieri, & Tagliazucchi, 2021). However, analyzing with shotgun proteomics needs quiet improvement, as well as the difficulty on analytical properties in proteomic nutritional studies. In fact, the selectivity varied with processed food protein specificity and level of digestion conferred through multiple in vitro as well as in vivo specific that yield very complex peptide mixtures for short peptide sequences (Tilocca et al., 2020).

Bioactive peptides derived from the food completed sequences with short amino acids are inactive confidential the protein of parent origin, which can be released by endogenous proteases or digestion via gastrointestinal, processing with food storage by hydrolysis with in vitro through specific proteo-lytic enzymes (D'Alessandro & Zolla, 2012). Bioactive peptides contain 2–20 amino acid residues for each molecule. However, they may comprise more than 20 amino acids in some cases. Such peptides are typically categorized as small peptides, with less than 7 amino acid residues, being considered the most active and difficult to analyze by conventional proteomics approaches. Subsequently, medium peptides (7–25 amino acids), as well as large peptides (more than 25 amino acids) (Cunsolo, Muccilli, Saletti, & Foti, 2014), also displayed their analysis difficult. Bioactive peptides possess a wide range of biological activities, which are reported in the BIOPEP database (http://www.uwm.edu.pl/biochemia/index.php/pl/biopep). They also comprise activities, such as antimicrobial, anti-hypertensive, cholesterol-lowering, anti-inflammatory, antithrombotic, and antioxidant. In addition, functional property of bioactive peptides is potentially useful for use as components in functional foods, nutraceuticals, food-grade bio-preservatives, cosmetics, and for pharmaceuticals analysis (Almeida et al., 2015).

Overall, peptidomic studies on milk with its derivative products tend to be the most investigated matrices. The derivative product bioactive peptides are proteins, mainly casein and whey, which serve as precursors of milk. In particular, human milk covers 50% casein and 50% whey, whereas bovine milk is richer in casein (80%). Caseins are separated into α-, β-, and κ-caseins (αs_1-CN, αs_2-CN, β-CN, and κ-CN, respectively) (Bello, Simsek, Olorunnisola, Babiker, & Hammed, 2021). As the primary caseins in milk are αs_1-CN, αs_2-CN, β-CN, and κ-CN. Whey proteins are β-lactoglobulin, α-lactalbumin, immunoglobulins, glycomacropeptides, bovine serum albumin, as well as other minor proteins (D'Auria et al., 2018; Manoni, Di Lorenzo, Ottoboni, Tretola, & Pinotti, 2020).

The most studied source of bioactive peptides is human milk. However, other mammalian milk types, such as cow, sheep, goat, and donkey, are also of interest in bioactive peptide research. The main interest in animal milk resides on its resemblance toward human milk, which making it a possible substitute for infant consumption (Giacometti & Buretić-Tomljanović, 2017). Finally, given the interest in bioactive peptide research and pivotal role of milk with derivative products in this field possibly peptidomics and other modern approaches are applied on peptide analysis of milk and dairy products, particularly focused on fractionation, detection, and quantification of bioactive peptides.

8.3.5 Proteomics to investigate milk and dairy products. "Safety, authenticity, and quality"

Studies on food quality regarding nutrition were carried out to gain the importance with such techniques to analyze the complete proteome or metabolome of milk or dairy products. The food product authentication through specific marker identification from the specific geographical area was based on concept of farm to fork. Proteomics approach toward foods on post-marketing surveillance resulting from genetically modified crops (Danezis, Tsagkaris, Camin, Brusic, & Georgiou, 2016). Quality on the function of food rest on the presence of bioactive compounds that were produced throughout the manufacturing as well as exert beneficial effect on health (Tilocca et al., 2020). The plotting of peptide-based food bioactive peptides through proteomics serves as a novel tool toward peptide separation besides documentation (Balkir et al., 2021). Bioactive proteins and peptides represent large and significant class of milk products that can be isolated, purified, and characterized by several proteomic tools. Product from milk is a bioactive food component that can add value to a food and cause the prevention or treatment of diseases (Mora, Gallego, & Toldrá, 2018). Thus, the valuable attribute of bioactive peptides ranges from antioxidant, anti-microbial, and anti-hypertensive agent to modifiers and regulators, in intracellular and extracellular signaling pathways (Carrera, Cañas, & Gallardo, 2014).

Major risks related to food safety pose a key worry throughout the world concerning, not only individual yet relative economic damages, toward potential danger by consumer's health along with quality (Bello et al., 2021). Increased consumer's assurance toward reliability on supplying the food troubled by means of improper safety measures on food products. In addition, the absence of safety events in food control greatly disrupts the economy of the food industry, in addition to undermining consumer confidence. Although, food safety tends with subservient to create the established firm and also consistent systematic approaches to permit a virtuous then quick investigation toward sustenance foodstuffs throughout the food chain (Sébédio, 2017).

Analyzing with omics (proteomic) approaches epitomize an influential tool on stating this problem, owing its confirmed excellent quantitative and qualitative advantages in exploration of proteins (Thangadurai et al., 2020). Overall, this chapter illustrates the solicitations of proteomics in the former few years in milk products concentrating on dairy industry with some brief hints on other types. Thus, we aimed toward emphasizing the importance of the biotechnological knowledge as a valued tool to assess quality and safety of food products. Prominence is also posed in food processing, allergies, and possible contaminants, such as bacteria, fungi, and other pathogens.

8.4 Conclusions

Revealing of mystery on configuration and feature of milk and dairy products, there been barter with up-gradation of grown-up technologies in route for new, vigorous, and fast techniques. In this facet, the proteomics application toward milk and milk-related products shows vital part on equivalent the stride of current biotechnological developments. Proteomics stands reading of proteins which are connected by the primary part in system biology. In the present chapter, there been a summary of numerous proteomics practices with their solicitation on various fields of milk-based food products. However, the future research aspects need in simplification of sample preparation, progress the reproducibility, and validation.

The presentation of proteomic methodologies offers exceptional opportunities toward the dairy protein science field. The grouping of high-resolution parting techniques with powerful characterization still with versatile MS analysis allows the achievement of earlier unattainable information.

Meanwhile, reported studies of milk proteomics are relatively limited, the area is growing at a rapid rate, with virtually all published studies in the area seeming over the past years. In line with this upsurge in quantity, the studies completed also becoming more sophisticated. It is likely that PAGE techniques will remain dominant proteomic considers of milk for a specific period. With its extraordinary resolution, 2D-PAGE allows the separation of protein isoforms comprising the slightest differences, a key in the comprehensive analysis of individual milk samples. The application of pre-fractionation procedures, multiple separation techniques, and intelligent MS strategies are the keys to successful milk proteomics.

References

Abd El-Salam, M. H. (2014). Application of proteomics to the areas of milk production, processing and quality control—a review. *International Journal of Dairy Technology, 67*, 3832(2), 153–166. https://doi.org/10.1111/1471-0307.12116.

Abdelhamid, H. N., & Wu, H.-F. (2015). Proteomics analysis of the mode of antibacterial action of nanoparticles and their interactions with proteins. *TrAC—Trends in Analytical Chemistry, 65*, 30–46. https://doi.org/10.1016/j.trac.2014.09.010.

Agregán, R., Echegaray, N., López-Pedrouso, M., Kharabsheh, R., Franco, D., & Lorenzo, J. M. (2021). Proteomic advances in milk and dairy products. *Molecules (Basel, Switzerland), 26*(13). doi:10.3390/molecules26133832.

Agyei, D., Tsopmo, A., & Udenigwe, C. C. (2018). Bioinformatics and peptidomics approaches to the discovery and analysis of food-derived bioactive peptides. *Analytical and Bioanalytical Chemistry, 410*(15), 3463–3472. https://doi.org/10.1007/s00216-018-0974-1.

Almeida, A. M., Bassols, A., Bendixen, E., Bhide, M., Ceciliani, F., Cristobal, S., et al. (2015). Animal board invited review: Advances in proteomics for animal and food sciences. *Animal, 9*(1), 1–17. https://doi.org/10.1017/S1751731114002602.

Anagnostopoulos, A. K., Katsafadou, A. I., Pierros, V., Kontopodis, E., Fthenakis, G. C., Arsenos, G., et al. (2016). Milk of Greek sheep and goat breeds; characterization by means of proteomics. *Journal of Proteomics, 147*, 76–84. https://doi.org/10.1016/j.jprot.2016.04.008.

Arena, S., Renzone, G., D'Ambrosio, C., Salzano, A. M., & Scaloni, A. (2017). Dairy products and the Maillard reaction: A promising future for extensive food characterization by integrated proteomics studies. *Food Chemistry, 219*, 477–489. https://doi.org/10.1016/j.foodchem.2016.09.165.

Aslam, B., Basit, M., Nisar, M. A., Khurshid, M., & Rasool, M. H. (2017). Proteomics: Technologies and their applications. *Journal of Chromatographic Science, 55*(2), 182–196. https://doi.org/10.1093/chromsci/bmw167.

Baggerman, G., Vierstraete, E., De Loof, A., & Schoofs, L. (2005). Gel-based versus gel-free proteomics: A review. *Combinatorial Chemistry & High Throughput Screening, 8*(8), 669–677. https://doi.org/10.2174/138620705774962490.

Balkir, P., Kemahlioglu, K., & Yucel, U. (2021). Foodomics: A new approach in food quality and safety. *Trends in Food Science & Technology, 108*, 49–57. https://doi.org/10.1016/j.tifs.2020.11.028.

Barbé, F., Le Feunteun, S., Rémond, D., Ménard, O., Jardin, J., Henry, G., et al. (2014). Tracking the in vivo release of bioactive peptides in the gut during digestion: Mass spectrometry peptidomic characterization of effluents collected in the gut of dairy matrix fed mini-pigs. *Food Research International, 63*, 147–156. https://doi.org/10.1016/j.foodres.2014.02.015.

Bello, I., Simsek, M., Olorunnisola, S., Babiker, F., & Hammed, A. M. (2021). Proteomics for food authentication. In C. M. Galanakis (Ed.), *Food Authentication and Traceability* (pp. 247–277). USA: Elsevier. https://doi.org/10.1016/B978-0-12-821104-5.00005-2.

Bhattacharya, M., Salcedo, J., Robinson, R. C., Henrick, B. M., & Barile, D. (2019). Peptidomic and glycomic profiling of commercial dairy products: Identification, quantification and potential bioactivities. *Npj Science of Food, 3*(1)(4), 1–13. doi:10.1038/s41538-019-0037-9.

Birgit, K., Bürkle, L., Kuhn, E. J., Giavalisco, P., Lueking, A., Walter, G., et al. (2002). Large-scale plant proteomics. *Plant Molecular Biology, 48*, 133–141.

Boland, M., & Singh, H. (2019). In Boland, M., & Singh, H. (Eds.), Milk proteins (3rd ed.). Florida, USA: Academic Press.

Borgo, C., Franchin, C., Scalco, S., Bosello-Travain, V., Donella-Deana, A., Arrigoni, G., et al. (2017). Generation and quantitative proteomics analysis of CK2α/α′(−/−) cells. *Scientific Reports, 7*(1), 1–13. doi:10.1038/srep42409.

Carrera, M., Cañas, B., & Gallardo, J. M. (2014). Proteomics tools for food fingerprints: Addressing new food quality and authenticity challenges. In: Barceló, D. (Ed.), Proteomics tools for food fingerprints: Addressing new food quality and authenticity challenges. *Comprehensive Analytical Chemistry, 64*, 201–222. https://doi.org/10.1016/B978-0-444-62650-9.00008-7.

Chandramouli, K., & Qian, P.-Y. (2009). Proteomics: Challenges, techniques and possibilities to overcome biological sample complexity. *Human Genomics and Proteomics, 1*(1), 1–14. doi:10.4061/2009/239204.

Chen, Y.-C., Chen, C.-C., & Hsieh, J.-F. (2016). Propylene glycol alginate-induced coacervation of milk proteins: A proteomics approach. *Food Hydrocolloids, 53*, 233–238. https://doi.org/10.1016/j.foodhyd.2015.01.018.

Cho, C.-K. J., Shan, S. J., Winsor, E. J., & Diamandis, E. P. (2007). Proteomics analysis of human amniotic fluid. *Molecular & Cellular Proteomics, 6*(8), 1406–1415. https://doi.org/10.1074/mcp.M700090-MCP200.

Cufaro, M. C., Pieragostino, D., Lanuti, P., Rossi, C., Cicalini, I., Federici, L., et al. (2019). Extracellular vesicles and their potential use in monitoring cancer progression and therapy: The contribution of proteomics. *Journal of Oncology, 2019*, 1–19. https://doi.org/10.1155/2019/1639854.

Cunsolo, V., Muccilli, V., Saletti, R., & Foti, S. (2014). Mass spectrometry in food proteomics: A tutorial. *Journal of Mass Spectrometry, 49*(9), 768–784. https://doi.org/10.1002/jms.3374.

D'Alessandro, A., & Zolla, L. (2012). We are what we eat: Food safety and proteomics. *Journal of Proteome Research, 11*(1), 26–36. https://doi.org/10.1021/pr2008829.

Danezis, G. P., Tsagkaris, A. S., Camin, F., Brusic, V., & Georgiou, C. A. (2016). Food authentication: Techniques, trends & emerging approaches. *TrAC—Trends in Analytical Chemistry, 85*, 123–132. https://doi.org/10.1016/j.trac.2016.02.026.

D'Auria, E., Mameli, C., Piras, C., Cococcioni, L., Urbani, A., Zuccotti, G. V., et al. (2018). Precision medicine in cow's milk allergy: Proteomics perspectives from allergens to patients. *Journal of Proteomics, 188*, 173–180. https://doi.org/10.1016/j.jprot.2018.01.018.

Dickinson, E. (1998). Stability and rheological implications of electrostatic milk protein–polysaccharide interactions. *Trends in Food Science & Technology, 9*(10), 347–354. https://doi.org/10.1016/S0924-2244(98)00057-0.

Doublier, J.-L., Garnier, C., Renard, D., & Sanchez, C. (2000). Protein–polysaccharide interactions. *Current Opinion in Colloid & Interface Science, 5*(3–4), 202–214. https://doi.org/10.1016/S1359-0294(00)00054-6.

Ebner, J., Baum, F., & Pischetsrieder, M. (2016). Identification of sixteen peptides reflecting heat and/or storage induced processes by profiling of commercial milk samples. *Journal of Proteomics, 147*, 66–75. https://doi.org/10.1016/j.jprot.2016.03.021.

Felfoul, I., Jardin, J., Gaucheron, F., Attia, H., & Ayadi, M. A. (2017). Proteomic profiling of camel and cow milk proteins under heat treatment. *Food Chemistry, 216*, 161–169. https://doi.org/10.1016/j.foodchem.2016.08.007.

Fox, P. F., Uniacke-Lowe., T., McSweeney., P. L. H., & O'Mahony, J. A. (2015). Milk proteins. In P. F. Fox, T. UniackeLowe, P. L. H. McSweeney, & J. A. O'Mahony (Eds.), *Dairy chemistry and biochemistry* (2nd ed.) (pp. 145–239). New York, USA: Springer International Publishing.

Fox, Pat F. (2008). Milk: An overview. In A. Thompson, M. Boland, & H. Singh (Eds.), *Milk proteins* (pp. 1–54). USA: Academic Press. https://doi.org/10.1016/B978-0-12-374039-7.00001-5.

Gagnaire, V., Jardin, J., Jan, G., & Lortal, S. (2009). Invited review: Proteomics of milk and bacteria used in fermented dairy products: From qualitative to quantitative advances. *Journal of Dairy Science, 92*(3), 811–825. https://doi.org/10.3168/jds.2008-1476.

Giacometti, J., & Buretić-Tomljanović, A. (2017). Peptidomics as a tool for characterizing bioactive milk peptides. *Food Chemistry, 230*, 91–98. https://doi.org/10.1016/j.foodchem.2017.03.016.

Gilchrist, A., Au, C. E., Hiding, J., Bell, A. W., Fernandez-Rodriguez, J., Lesimple, S., et al. (2006). Quantitative proteomics analysis of the secretory pathway. *Cell, 127*(6), 1265–1281. https://doi.org/10.1016/j.cell.2006.10.036.

Guetouache, M., Guessas, B., & Medjekal, S. (2014). Composition and nutritional value of raw milk. *Issues in Biological Sciences and Pharmaceutical Research, 2*(10), 115–122. https://doi.org/10.15739/ibspr.005.

Guglielmetti, C., Brusadore, S., Manfredi, M., Sciuto, S., Esposito, G., Cucca, L., et al. (2019). A proteomic approach to the safeguard of a typical agri-food product: Fiore Sardo PDO. *Advances in Dairy Research, 07*(03), 1–6. doi:10.35248/2329-888x.19.7.228.

Hagenstein, M. C., Mussgnug, J. H., Lotte, K., Plessow, R., Brockhinke, A., Kruse, O., et al. (2003). Affinity-based tagging of protein families with reversible inhibitors: A concept for functional proteomics. *Angewandte Chemie International Edition, 42*(45), 5635–5638. https://doi.org/10.1002/anie.200352084.

Izquierdo-González, J. J., Amil-Ruiz, F., Zazzu, S., Sánchez-Lucas, R., Fuentes-Almagro, C. A., & Rodríguez-Ortega, M. J. (2019). Proteomic analysis of goat milk kefir: Profiling the fermentation-time dependent protein digestion and identification of potential peptides with biological activity. *Food Chemistry, 295*, 456–465. https://doi.org/10.1016/j.foodchem.2019.05.178.

Ji, X., Li, X., Ma, Y., & Li, D. (2017). Differences in proteomic profiles of milk fat globule membrane in yak and cow milk. *Food Chemistry, 221*, 1822–1827. https://doi.org/10.1016/j.foodchem.2016.10.097.

Jin, D., Liu, H., Bu, L., Ke, Q., Li, Z., Han, W., et al. (2021). Comparative analysis of whey proteins in human milk using a data-independent acquisition proteomics approach during the lactation period. *Journal of Agricultural and Food Chemistry, 69*(14), 4319–4330. https://doi.org/10.1021/acs.jafc.1c00186.

Kandasamy, S., Yoo, J., Yun, J., Kang, H. B., Seol, K. H., & Ham, J. S. (2021). Rapid discrimination and authentication of Korean farmstead mozzarella cheese through MALDI-TOF and multivariate statistical analysis. *Metabolites, 11*(6), 333–339. doi:10.3390/metabo11060333.

Kinsella, J. E., & Morr, C. V. (1984). Milk proteins: Physicochemical and functional properties. *Critical Reviews in Food Science and Nutrition, 21*(3), 197–262. https://doi.org/10.1080/10408398409527401.

Lawrance, I. C., Klopcic, B., & Wasinger, V. C. (2005). Proteomics: An overview proteomics: What does it mean? *Inflammatory Bowel Diseases, 11*(10), 927–936. http://www.nigms.nih.gov.

Le, T. T., Deeth, H. C., & Larsen, L. B. (2017). Proteomics of major bovine milk proteins: Novel insights. *International Dairy Journal, 67*, 2–15. https://doi.org/10.1016/j.idairyj.2016.11.016.

Le, T. T., Nielsen, S. D., Villumsen, N. S., Kristiansen, G. H., Nielsen, L. R., Nielsen, S. B., et al. (2016). Using proteomics to characterise storage-induced aggregates in acidic whey protein isolate drinks. *International Dairy Journal, 60*, 39–46. https://doi.org/10.1016/j.idairyj.2016.01.028.

Li, S.-S., Wang, J.-Q., Wei, H.-Y., Yang, Y.-X., Deng-Pan, B. U., Zhang, L.-Y., et al. (2012). Identification of Bovine Casein phosphorylation using titanium dioxide enrichment in combination with nano electrospray ionization tandem mass spectrometry. *Journal of Integrative Agriculture, 2012*(3), 439–445.

Li, W., Li, M., Cao, X., Han, H., Kong, F., & Yue, X. (2020). Comparative analysis of whey proteins in donkey colostrum and mature milk using quantitative proteomics. *Food Research International, 127*(1), 220–226. doi:10.1016/j.foodres.2019.108741.

Lönnerdal, B. (2003). Nutritional and physiologic significance of human milk proteins. *The American Journal of Clinical Nutrition, 77*(6), 1437S–1543S. https://doi.org/10.1093/ajcn/77.6.1537S.

Low, T. Y., van Heesch, S., van den Toorn, H., Giansanti, P., Cristobal, A., Toonen, P., et al. (2013). Quantitative and qualitative proteome characteristics extracted from in-depth integrated genomics and proteomics analysis. *Cell Reports, 5*(5), 1469–1478. https://doi.org/10.1016/j.celrep.2013.10.041.

Lu, W., Liu, J., Gao, B., Lv, X., & Yu, L., (Lucy) (2017). Technical note: Nontargeted detection of adulterated plant proteins in raw milk by UPLC-quadrupole time-of-flight mass spectrometric proteomics combined with chemometrics. *Journal of Dairy Science, 100*(9), 6980–6986. https://doi.org/10.3168/jds.2017-12574.

Manoni, M., Di Lorenzo, C., Ottoboni, M., Tretola, M., & Pinotti, L. (2020). Comparative proteomics of milk fat globule membrane (MFGM) proteome across species and lactation stages and the potentials of MFGM fractions in infant formula preparation. *Foods, 9*(9), 1251. doi:10.3390/foods9091251.

Manso, M. A., Léonil, J., Jan, G., & Gagnaire, V. (2005). Application of proteomics to the characterisation of milk and dairy products. *International Dairy Journal, 15*(6–9), 845–855. https://doi.org/10.1016/j.idairyj.2004.07.021.

Martini, S., Solieri, L., & Tagliazucchi, D. (2021). Peptidomics: New trends in food science. Current opinion in food science, *39*, 51–59. https://doi.org/10.1016/j.cofs.2020.12.016

McGrath, B. A., Kinsella, M., Huppertz, T., McSweeney, P. L. H., & Kelly, A. L. (2016). Proteomic characterisation of heat-induced hydrolysis of sodium caseinate. *International Dairy Journal, 53*, 51–59. https://doi.org/10.1016/j.idairyj.2015.09.006.

McKenzie, H. A. (1967). Milk proteins. *Advances in Protein Chemistry, 22*, 55–234. https://doi.org/10.1016/S0065-3233(08)60041-8.

McKenzie, H. (1970). Milk proteins: Chemistry and molecular biology (H. McKenzie (ed.); (1st ed.)., Vol. *1*). Elsevier.

Meltretter, J., Wüst, J., Dittrich, D., Lach, J., Ludwig, J., Eichler, J., et al. (2020). Untargeted proteomics-based profiling for the identification of novel processing-induced protein modifications in milk. *Journal of Proteome Research, 19*(2), 805–818. https://doi.org/10.1021/acs.jproteome.9b00630.

Miller, G. D., Jarvis, J. K., & McBean, L. D. (2007). *Handbook of dairy foods and nutrition* (3rd ed.). Florida, USA: CRC Press.

Minkiewicz, P., Dziuba, J., Darewicz, M., Iwaniak, A., Dziuba, M., & Nałecz, D. (2008). Food peptidomics. *Food Technology & Biotechnology, 46*(1), 1–10.

Mora, L., Gallego, M., & Toldrá, F. (2018). New approaches based on comparative proteomics for the assessment of food quality. Current Opinion in Food Science, *22*, 22–27. https://doi.org/10.1016/j.cofs.2018.01.005

Nardiello, D., Natale, A., Palermo, C., Quinto, M., & Centonze, D. (2018). Milk authenticity by ion-trap proteomics following multi-enzyme digestion. *Food Chemistry, 244*, 317–323. https://doi.org/10.1016/j.foodchem.2017.10.052.

Nedelkov, D., Kiernan, U. A., Niederkofler, E. E., Tubbs, K. A., & Nelson, R. W. (2006). Population proteomics. *Molecular & Cellular Proteomics, 5*(10), 1811–1818. https://doi.org/10.1074/mcp.R600006-MCP200.

Nguyen, H. T. H., Ong, L., Hoque, A., Kentish, S. E., Williamson, N., Ang, C. S., et al. (2017). A proteomic characterization shows differences in the milk fat globule membrane of buffalo and bovine milk. *Food Bioscience, 19*, 7–16. https://doi.org/10.1016/j.fbio.2017.05.004.

O'Donnell, R., Holland, J. W., Deeth, H. C., & Alewood, P. (2004). Milk proteomics. *International Dairy Journal, 14*(12), 1013–1023. https://doi.org/10.1016/j.idairyj.2004.04.004.

Otto, A., Becher, D., & Schmidt, F. (2014). Quantitative proteomics in the field of microbiology. *Proteomics, 14*(4–5), 547–565. https://doi.org/10.1002/pmic.201300403.

Pan, S., Brentnall, T. A., & Chen, R. (2015). Proteomics analysis of bodily fluids in pancreatic cancer. *Proteomics, 15*(15), 2705–2715. https://doi.org/10.1002/pmic.201400476.

Peng, J., Schwartz, D., Elias, J. E., Thoreen, C. C., Cheng, D., Marsischky, G., et al. (2003). A proteomics approach to understanding protein ubiquitination. *Nature Biotechnology, 21*(8), 921–926. https://doi.org/10.1038/nbt849.

Pihlanto, A. (2006). Antioxidative peptides derived from milk proteins. *International Dairy Journal, 16*(11), 1306–1314. https://doi.org/10.1016/j.idairyj.2006.06.005.

Piovesana, S., Capriotti, A. L., Cavaliere, C., La Barbera, G., Samperi, R., Zenezini Chiozzi, R., et al. (2015). Peptidome characterization and bioactivity analysis of donkey milk. *Journal of Proteomics, 119*, 21–29. https://doi.org/10.1016/j.jprot.2015.01.020.

Reinhardt, T. A., Lippolis, J. D., Nonnecke, B. J., & Sacco, R. E. (2012). Bovine milk exosome proteome. *Journal of Proteomics, 75*(5), 1486–1492. https://doi.org/10.1016/j.jprot.2011.11.017.

Roncada, P., Piras, C., Soggiu, A., Turk, R., Urbani, A., & Bonizzi, L. (2012). Farm animal milk proteomics. *Journal of Proteomics, 75*(14), 4259–4274. https://doi.org/10.1016/j.jprot.2012.05.028.

Sánchez-Rivera, L., Martínez-Maqueda, D., Cruz-Huerta, E., Miralles, B., & Recio, I. (2014). Peptidomics for discovery, bioavailability and monitoring of dairy bioactive peptides. *Food Research International, 63*, 170–181. https://doi.org/10.1016/j.foodres.2014.01.069.

Santini, G., Bonazza, F., Pucciarelli, S., Polidori, P., Ricciutelli, M., Klimanova, Y., et al. (2020). Proteomic characterization of kefir milk by two-dimensional electrophoresis followed by mass spectrometry. *Journal of Mass Spectrometry, 55*(11), e4635. https://doi.org/10.1002/jms.4635.

Savaryn, J. P., Catherman, A. D., Thomas, P. M., Abecassis, M. M., & Kelleher, N. L. (2013). The emergence of top-down proteomics in clinical research. *Genome Medicine, 5*(6), 53. https://doi.org/10.1186/gm457.

Schmitt, C., Sanchez, C., Desobry-Banon, S., & Hardy, J. (1998). Structure and technofunctional properties of protein-polysaccharide complexes: A review. *Critical Reviews in Food Science and Nutrition, 38*(8), 689–753. https://doi.org/10.1080/10408699891274354.

Schneider, T., & Riedel, K. (2010). Environmental proteomics: Analysis of structure and function of microbial communities. *Proteomics, 10*(4), 785–798. https://doi.org/10.1002/pmic.200900450.

Sébédio, J.-L. (2017). Metabolomics, nutrition, and potential biomarkers of food quality, intake, and health status. In: Granato, D. (Ed.), Metabolomics, nutrition, and potential biomarkers of food quality, intake, and health status. *Advances in Food and Nutrition Research, 82*, 83–116. https://doi.org/10.1016/bs.afnr.2017.01.001.

Soggiu, A., Roncada, P., & Piras, C. (2018). Proteomics in milk and dairy products. In A. M. de Almeida, D. EckersallIngrid, & I. Miller (Eds.), *Proteomics in domestic animals: From farm to systems biology* (pp. 169–193). New York, USA: Springer International Publishing. https://doi.org/10.1007/978-3-319-69682-9_9.

Sui, S., Zhao, J., Wang, J., Zhang, R., Guo, C., Yu, T., et al. (2014). Comparative proteomics of milk fat globule membrane proteins from transgenic cloned cattle. *Plos One, 9*(8), e105378. https://doi.org/10.1371/journal.pone.0105378.

Sun, Y., Wang, C., Sun, X., & Guo, M. (2019). Comparative proteomics of whey and milk fat globule membrane proteins of Guanzhong goat and Holstein cow mature milk. *Journal of Food Science, 84*(2), 244–253. https://doi.org/10.1111/1750-3841.14428.

Thangadurai, D., Kordrostami, M., Islam, S., Shettar, A. K., Sangeetha, J., Mafakheri, M., et al. (2020). Targeted proteomics for food authentication. In L. M. L. Nollet, & S. Ötleş (Eds.), *Proteomics for food authetication* (pp. 61–78). Florida, USA: CRC Press.

Tilocca, B., Costanzo, N., Morittu, V. M., Spina, A. A., Soggiu, A., Britti, D., et al. (2020). Milk microbiota: Characterization methods and role in cheese production. *Journal of Proteomics, 210*. doi:10.1016/j.jprot.2019.103534.

Tolstoguzov, V. B. (1997). Protein-polysaccharide interactions. In S. Damodaran, & A. Paraf (Eds.), *Food proteins and their applications* (pp. 171–199). Florida, USA: CRC Press.

Wang, W., Tai, F., & Chen, S. (2008). Optimizing protein extraction from plant tissues for enhanced proteomics analysis. *Journal of Separation Science, 31*(11), 2032–2039. https://doi.org/10.1002/jssc.200800087.

Wang, X., Zhao, X., Huang, D., Pan, X., Qi, Y., Yang, Y., et al. (2017). Proteomic analysis and cross species comparison of casein fractions from the milk of dairy animals. *Scientific Reports, 7*(1), 1–9. https://doi.org/10.1038/srep43020.

Weinbreck, F., Nieuwenhuijse, H., Robijn, G. W., & de Kruif, C. G. (2004). Complexation of whey proteins with carrageenan. *Journal of Agricultural and Food Chemistry, 52*(11), 3550–3555. https://doi.org/10.1021/jf034969t.

Wong, D. W. S., Camirand, W. M., Pavlath, A. E., Parris, N., & Friedman, M. (1996). Structures and functionalities of milk proteins∗. *Critical Reviews in Food Science and Nutrition, 36*(8), 807–844. https://doi.org/10.1080/10408399609527751.

Wright, I., & Van Eyk, J. E. (2017). A roadmap to successful clinical proteomics. *Clinical Chemistry, 63*(1), 245–247. https://doi.org/10.1373/clinchem.2016.254664.

Yang, M., Cao, X., Wu, R., Liu, B., Ye, W., Yue, X., et al. (2017a). Comparative proteomic exploration of whey proteins in human and bovine colostrum and mature milk using iTRAQ-coupled LC-MS/MS. *International Journal of Food Sciences and Nutrition, 68*(6), 671–681. https://doi.org/10.1080/09637486.2017.1279129.

Yang, Y., Zheng, N., Zhao, X., Zhang, Y., Han, R., Zhao, S., et al. (2017b). N-glycosylation proteomic characterization and cross-species comparison of milk whey proteins from dairy animals. *Proteomics, 17*(9), 2792–2800. https://doi.org/10.1002/pmic.201600434.

Yates III, J. R., Gilchrist, A., Howell, K. E., & Bergeron, J. J. M. (2005). Proteomics of organelles and large cellular structures. *Nature Reviews Molecular Cell Biology, 6*(9), 702–714. https://doi.org/10.1038/nrm1711.

Zhang, Z., Wu, S., Stenoien, D. L., & Paša-Tolić, L. (2014). High-throughput proteomics. *Annual Review of Analytical Chemistry, 7*(1), 427–454. https://doi.org/10.1146/annurev-anchem-071213-020216.

Zhao, Y., & Jensen, O. N. (2009). Modification-specific proteomics: Strategies for characterization of post-translational modifications using enrichment techniques. *Proteomics, 9*(20), 4632–4641. https://doi.org/10.1002/pmic.200900398.

SECTION III

Applications of proteomic in food challenges

CHAPTER

9

Proteomic analysis of food allergens

Francisco Javier Salgado Castro[a], *Juan José Nieto-Fontarigo*[a,b], *Francisco Javier González-Barcala*[c]

[a]Department of Biochemistry and Molecular Biology, Centro de Investigaciones Biológicas de la Universidad de Santiago (CIBUS), Universidad de Santiago de Compostela. TRIAD Research Group, Instituto de Investigación Sanitaria de Santiago de Compostela (IDIS). A Coruña, Spain

[b]Respiratory Immunopharmacology unit, Department of Experimental Medical Science, Lund University, Sweden

[c]Department of Medicine, Universidad de Santiago de Compostela. Department of Respiratory Medicine, Complejo Hospitalario Universitario de Santiago de Compostela (CHUS). Head of TRIAD Research Group, Instituto de Investigación Sanitaria de Santiago de Compostela (IDIS). A Coruña, Spain

9.1 Introduction

Adverse reactions to food can be classified as toxic (presence of toxins or poisons) and non-toxic. We have two types within the non-toxic reactions: (1) food allergies, which are mediated by immune mechanisms; and (2) food intolerances, which involve non-immune mechanisms (e.g., lactose intolerance) (Yu et al., 2016; Lopes and Sicherer, 2020; Sicherer and Sampson, 2018). The prevalence of food allergies in developed countries is 6%–8% in children and 2%–3% in adults (Gupta et al., 2011; Michelet et al., 2021); this is due to certain food allergies, such as cow's milk allergy, which are overcome in children without the need for any therapeutic intervention (Kim et al., 2011; Leonard et al., 2015; Savage et al., 2016). The prevalence of food allergies displayed an increase over the last

Food Proteomics
DOI: https://doi.org/10.1016/B978-0-323-90889-4.00003-8

10–20 years (Michelet et al., 2021) probably related to either the industrialization or the economic expansion of developing countries (Koplin et al., 2012; Leung et al., 2018). Therefore, food allergies are a growing health problem that generate high costs for healthcare systems.

The most common treatment for food allergies is to avoid the food that causes the problem. However, the wide distribution of certain allergens and the poor labeling of many processed products make this strategy difficult to carry out. According to the United States (US) and the European Union (EU regulation 1169/2011) regulations, up to 8 and 14 allergens, respectively, must be clearly indicated on food labels (Marzano et al., 2020). However, this requires the food industry to be able to detect minute amounts of allergens in their products. The identification of allergens using proteomic techniques can help the generation of rapid allergen detection tests in food industry and the design of new therapeutic strategies.

9.2 Immunological mechanism of food allergies

To understand food allergies, we must address some definitions and the study of the immune response in the gut. Thus, according to the conditions dictionary of the American Academy of Allergy Asthma & Immunology the term "atopy" defines "the genetic tendency to develop allergic diseases such as allergic rhinitis, asthma and atopic dermatitis (eczema)" (https://www.aaaai.org/conditions-and-treatments/conditions-dictionary/). Atopy is typically associated with heightened immune responses to common allergens, especially inhaled and food antigens (Vaillant et al., 2021). Among these atopic manifestations, atopic dermatitis, and inflammation of the upper (allergic rhinitis) and lower airways (allergic asthma) have been described, being part of what has been oversimplified as "the atopic march" (Yang et al., 2020). Therefore, one of the first issues to highlight is that food allergies usually coexist with other types of atopic diseases (Yang et al., 2020).

A hypersensitivity reaction is an excessive response against environmental antigens (e.g., food compounds) that leads to inflammation and tissue damage (Dispenza, 2019). Hypersensitivity reactions arise because of the total or partial loss of oral tolerance to food antigens, which depends, among others, on the presence of an intact mucosal barrier in the gastrointestinal tract. The mucosa surface is formed by a layer of epithelial cells (enterocytes) covered by glycoproteins (mucus) that hinder the contact with commensal/pathogenic microbes and food molecules (Boyaka and Fujihashi, 2019). Interspersed between the enterocytes there are intraepithelial lymphocytes (IELs) with diverse origins (αβ T cells, γδ T cells, group 1 innate lymphoid cells (ILC1s), or ILC1-like cells) and important in antigen presentation (McDonald et al., 2018). However, sometimes this intestinal epithelial barrier loses its integrity due to birth defects (e.g., filaggrin mutations), the presence of acids (e.g., gastric reflux), or environmental exposures to viruses, alcohol, or drugs (e.g., non-steroidal anti-inflammatory drugs/NSAIDs). This disruption of the epithelial barrier leads to increased permeation of food compounds, a major cause of allergic sensitization to food (Krempski et al., 2020; Renz et al., 2018).

Allergen sensitization can also occur with an intact epithelial barrier along the gut. This is because food allergens or antigens interact with the mucosal-associated lymphatic tissue, also called gut-associated lymphoid tissue in the digestive tract, leading to different immune responses in the lamina propria (LP), Peyer's patches (PPs), and the intestinal epithelium. LP lies just underneath the epithelium and contains diffuse groups of lymphoid cells (e.g., B lymphocytes, T cells, ILCs, NK, LTi). High endothelial venules (HEVs) in LP work as entry points for memory or effector T/B lymphocytes, including IgA-secreting plasma cells. IgA is found in mucous secretions as a dimeric molecule called secretory IgA (sIgA). sIgA is important for inhibiting the adherence of commensal bacteria to the epithelium and neutralizing viruses. Moreover, sIgA exhibits anti-inflammatory activity, inhibiting complement activation or phagocyte activities, or reducing IgE-dependent anaphylactic activities and mucosa inflammation/damage (Boyaka and Fujihashi, 2019).

On the other hand, PPs are dome-shaped structures covered by a thinner mucus layer, which facilitates the interaction with antigens and microorganisms (Boyaka and Fujihashi, 2019). Within PPs, microfold cells (M cells) form a specialized structure that favors the food antigen interaction with antigen-presenting cells (APCs; i.e., macrophages and immature dendritic cells (DCs)) (Boyaka and Fujihashi, 2019). Immature DCs can also capture food antigens by extending their dendrites to the intestine lumen through the paracellular space between enterocytes (Bird et al., 2019). In the dome region of PPs, there are also scattered plasma cells secreting IgM or IgA and several follicles with germinal centers (Boyaka and Fujihashi, 2019). These follicles contain follicular dendritic cells, which are important for the isotype switching from IgM to IgA. In this way, germinal centers in secondary follicles are active sites of generation of IgA^+ memory and IgA-secreting plasma cells from naïve B lymphocytes (Boyaka and Fujihashi, 2019). In contrast, the inter-follicular regions of PPs contain mature DCs and T lymphocytes, which are predominantly $\alpha\beta$ T helper cells (TH cells). HEVs facilitate the entry of naïve T/B lymphocytes to PPs. Once inside, these cells become activated lymphocytes upon encountering the specific antigen and turn into memory or effector cells (e.g., TH1, TH2, TH17, follicular T helper cells/TfH, TH22). There are also IgG/IgM/IgA-producing plasma cells, but IgE-secretion is an issue of debate. IgE is a very important class of immunoglobulin in many food allergies, but whether these PPs are the site of IgE class-switching during sensitization to food remains to be determined. However, it is known that B cells go through IgE class switching from upstream isotypes (IgM, IgD, and especially IgG/IgA1) in secondary lymphoid tissues (Hoh et al., 2020; Michelet et al., 2021).

Turning again to the TH lymphocytes, TH1 effector lymphocytes mainly produce IFNγ or TNFα and activate cell-mediated immune mechanisms where macrophages, NK cells, and cytotoxic T lymphocytes are important; they also cooperate with the generation of plasma cells that produce IgG1 and IgG3, with opsonizing and complement-fixing capacities. In contrast, TH2 effector lymphocytes produce cytokines such as IL-4, IL-5, IL-6, IL-9, IL-10, IL-13, which are characteristic of allergic responses. TH2 cells also collaborate with B cells to produce IgE, IgA, and IgG4. On the other hand, TH17 cells secrete IL-17A, IL-17F, or IL-22, all of them involved in inflammatory processes and responses to commensal bacteria and fungi, where neutrophils play an important role. TfH lymphocytes produce IL-21 and are associated with the formation of germinal centers and the T cell-dependent class switching and affinity-maturation of antibody responses. Finally, $FoxP3^+$ regulatory T cells control inflammatory responses and produce key cytokines such as IL-10 or TGFβ that stimulate, for example, IgA production (Boyaka and Fujihashi, 2019). These cells are key in responses to dietary peptides, as the higher or lower activity of regulatory T lymphocytes (as well as other existing suppressor cells) determines to what extent food-derived peptides will generate immune or tolerogenic responses.

Apart from T-cell dependent class-switching in B cells, the production of specific antibodies in response to food allergens can be induced by T-cell independent mechanisms (i.e., outside the germinal centers). For example, the immune response to a carbohydrate moiety on a protein uses to be T-cell independent. Although it was thought that T-cell independent antigens could only generate plasma cells of short duration (days) that secrete low-affinity IgM, it is currently known that affinity maturation and class-switching can also occur in a T cell-independent manner to produce long-life plasma cells (Allman et al., 2019). Thus, there can be IgE antibodies recognizing T-cell independent antigens such as the IgE specific for the galactose-alpha-1,3-galactose (alpha-gal) epitope in red meat allergy (Steinke et al., 2015).

Gell and Coombs classified hypersensitivities according to the immune mechanisms involved, describing four main types (I-IV) (Dispenza, 2019). Cells and molecules from both the innate and adaptive branches participate, and two types of hypersitivity reactions seem to be the most important in food allergies: Type I (Immunoglobulin E/IgE; Immediate hypersensitivity) and type IV (delayed hypersensitivity) hypersensitivity reactions (De Angelis et al., 2010; Dispenza, 2019; Valenta et al., 2015). The prevalence of IgE-mediated food allergy is higher compared to non-IgE-mediated hypersensitivity reactions, but in both types there must be first a stage of sensitization to the food allergen, where the mechanisms of oral tolerance induction fail, followed by the second stage of antigen provocation.

9.2.1 Immediate food allergies

As we have just commented, food allergies are mostly characterized by type I (immediate) hypersensitivity reactions where there are two phases: (1) sensitization phase (primary response), which involves the first encounter with the allergen, takes several days and no symptoms or signs of a food allergy are present; (2) provocation stage (secondary response), which occurs upon repeated exposures to the same allergen and gives rise to the typical symptoms of an allergic reaction (Lopes and Sicherer, 2020; Sicherer and Sampson, 2018; Valenta et al., 2015; Yu et al., 2016).

Based on the sensitization route, food allergens can be classified into two types: Class I food allergens, which cause sensitization via gastrointestinal tract; and class II food allergens, which used alternative sensitization routes (e.g., the respiratory tract) (Valenta et al., 2015). During the *sensitization phase* with class I food allergens, naïve B lymphocytes from the bone marrow reach the lymph nodes that drain the gastrointestinal tract, where they become activated. Food antigens arrive at the lymph nodes in two ways: (1) by the above-mentioned DCs, which migrate from gut to the lymph nodes; (2) in a soluble form, being subsequently captured, processed, and presented to TH cells by resident DCs. In the lymph nodes, naïve B lymphocytes mature into long-lived memory B cells and IgE-secreting plasma cells in the presence of secondary signals (CD40 / CD40L) and cytokines (IL-4, IL-13) from Tfh lymphocytes. On the other hand, sensitization with class II food allergens (aeroallergens) occurs in the respiratory epithelium (Valenta et al., 2015). An example of this is the pollen-fruit allergy syndrome (PFAS), which will be extensively explained later (Carlson and Coop, 2019; Yagami and Ebisawa, 2019). Alternatively, sensitization can occur through the skin, leading for example to peanuts or red meat allergy (tick bite) (Apostolovic et al., 2020; Tordesillas et al., 2017).

IgE produced by plasma cells is bound to FcεRI receptors (high-affinity IgE receptor) in the surface of mast cells (tissues) and basophils (blood), which generated sensitized cells. During the *provocation stage*, the cross-linking of IgE by allergenic polyvalent proteins from the gut causes a rapid release (minutes) (mast cell/basophils degranulation) of pre-formed inflammatory/vasoactive mediators, chemokines, cytokines, and enzymes; e.g., histamine, tryptase, chymase, carboxypeptidase, heparin, and platelet activating factor (PAF) (Renz et al., 2018). Since prostaglandins and leukotrienes are generated from arachidonic acid, they are generated somewhat later, but still within minutes, while the release of newly synthesized cytokines requires hours. This process results in vasodilation, increased vascular permeability, glandular hypersecretion, and smooth muscle contraction (Michelet et al., 2021; Renz et al., 2018).

The signs of food allergies are varied. In some cases, such as oral allergy syndrome, they are very localized and limited to the lips, mouth, and pharynx. In others, they manifest in the skin and subcutaneous tissues (urticaria, angioedema), the digestive tract (e.g., vomiting, diarrhoea), the ventilatory system (e.g., bronchoconstriction), or more systemically (anaphylaxis) (Cardona et al., 2020). These signs usually appear quickly (minutes) after ingestion of the food containing the allergen (hence the adjective "immediate"), and only on rare occasions this reaction begin after several hours (Bird et al., 2019).

Tissue infiltration of inflammatory cells such as eosinophils occurs at a late stage (2–4 h) and usually plays a secondary role in food allergies. The number of eosinophils increases greatly in allergic processes. However, these cells participate in antigen presentation to T cells rather than in IgE-mediated degranulation processes. In contrast, eosinophils degranulation can occur upon high-affinity interaction of FcαRI (CD89) with sIgA, although, as mentioned above, this type of antibody is usually anti-inflammatory. Similarly, eosinophils are also capable of capturing IgG with high (FcγRI/CD64) and low (FcγRII/CD32) affinity (Kovalszki and Weller, 2019). However, natural resolution of food allergies has been also associated with increasing food-specific IgG levels. Furthermore, the sequential administration of cumulative doses of food allergens (oral immunotherapy) is accompanied by the induction of allergens-specific IgG antibodies in both human subjects and murine models (Kanagaratham et al., 2020).

Last, but not least, there are so many activating receptors in mast cells, basophils, and eosinophils that it is possible to generate type I hypersensitivity reactions to food allergens where degranulation is not mediated by IgE (i.e., there is no sensitization); these are called "pseudo-allergic reactions" (Kumar et al., 2021). Thus, mast cells may be activated through receptors such as Mas-related G-protein coupled receptor member X2 (MRGPRX2) by food additives (Kumar et al., 2021). Furthermore, we should be aware of the existence of epigenetically mediated and memory-like responses (trained immunity) in cells from the innate immune system (e.g., innate lymphocytes, DCs, and monocytes) (Eljaszewicz et al., 2021).

9.2.2 Delayed hypersensitivity reactions to food allergens

Cell-mediated type IV hypersensitivity plays an important role in delayed food allergies (Valenta et al., 2015). However, type IV hypersensitivity reactions are so closely linked to type I mechanisms that it is impossible to separate each other. Moreover, this kind of hypersensitivity reactions are heterogenous. The most important ones in food allergy are perhaps type IVb hypersensitivity reactions, which are mediated by TH2 lymphocytes and type 2 cytokines (e.g., IL-4, IL-5, IL-13) and where innate (mast cells, basophils, ILC2, eosinophils) and adaptive cells (B lymphocytes producing IgE and IgG4) play a role (Bird et al., 2019; Castells and Bonamichi-Santos, 2019; Dispenza, 2019). Additional hypersensitivity reactions are the type IVa, where TH1 lymphocytes and the cytokines IL-12, IFNγ, and TNFα are crucial. In this type of reactions, IFNγ participates in the activation of monocytes and cytotoxic $CD8^+$ T cells; B lymphocytes participate as well through the secretion of complement-fixing antibodies that lead to tissue damage (Castells and Bonamichi-Santos, 2019; Dispenza, 2019). On the other hand, type IVc reactions are mainly related to cytotoxic mechanisms produced by $CD8^+$ T lymphocytes, but other innate leukocytes such as monocytes, eosinophils, and neutrophils might also participate (Castells and Bonamichi-Santos, 2019; Dispenza, 2019). Finally, type IVd reactions are orchestrated by TH17 cells and characterized by neutrophilic inflammation and chemokines/cytokines such as CXCL8 (IL-8), and GM-CSF (Castells and Bonamichi-Santos, 2019; Dispenza, 2019).

The role of eosinophil in this group of diseases seems important but remains obscure. As anticipated above, eosinophils express reduced amounts of the high-affinity receptor FcεRI, so IgE does not usually induce their degranulation. However, these cells have high levels of both IgG (FcγRI; CD64) and sIgA (FcαRI; CD89) receptors (Aoki et al., 2021; Klion et al., 2020), and these antibodies classes appear to participate in the capture of antigenic peptides and their presentation to $CD4^+$ T cells via MHC class II (Aoki et al., 2021; Klion et al., 2020). Moreover, it is worth mentioning that eosinophils respond to multiple stimuli, such as cytokines (e.g., IL-5, IL-3, GM-CSF, IL-1α, IL-2, IL-4, IL-13, IL-17, IL-23, IL-27, IL-33, TSLP, IFNγ, TNFα), chemokines (e.g., eotaxins, MIP-1α, MCP-3, RANTES, SDF1α), damage-associated molecular patterns, pathogen-associated molecular patterns, lipids (e.g., PAF, LTB4), complement factors (C1q, C3b, C4b), anaphylatoxins (C3a, C5a), or cleavage of protease-activated receptors (Aoki et al., 2021; Klion et al., 2020). Some examples of delayed hypersensitivity to food allergens where eosinophils play an important (but not clear) role are eosinophilic esophagitis (EoE), eosinophilic gastroenteritis, food protein-induced enterocolitis, eosinophilic proctocolitis, and the well-known coeliac disease (Bird et al., 2019).

EoE is a chronic disease whose prevalence (around 35 cases/100,000 inhabitants) has experienced a dramatic increase in the last decade. It affects predominantly to young males (Fernandez-Becker et al., 2020), and does not show any association with genetic variants in the MHC I or II (Kottyan et al., 2019). Under normal conditions, there are no eosinophils in the oesophagus. However, EoE is characterized by an intense (but diffuse) eosinophilic infiltration in this part of the digestive tract. This phenomenon, together with the presence of increased numbers of mast cells, is accompanied by inflammation (oedema, white exudates) and subepithelial (LP) fibrosis. Oesophagus remodeling, in turn, generates rings, rings, longitudinal furrows, strictures, and a narrow calibre (stenosis), leading to a difficult swallowing of bulky foods (dysphagia); additional signs are vomiting, eating disorders, and abdominal pain (Bird et al., 2019; Bird, 2018; Lim et al., 2021). EoE is associated with other atopic conditions (asthma, atopic dermatitis, rhinitis, eczema, food allergy) in 50%–81% of cases (Lim et al., 2021), being predominantly a non-IgE mediated disorder where antigen-specific TH2 cells and their cytokines/chemokines (IL-4, IL-5, IL-13, eotaxin-3, TSLP) recruit eosinophils to the oesophagus (Fernandez-Becker et al., 2020). Products of both animal (milk, egg) and vegetable (wheat, soy) origin have been most related to this pathology, although other factors have been pointed out (Bird et al., 2019). However, no proteomic studies have been conducted on food allergens that could trigger this disease.

Eosinophilic gastroenteritis (EGE) is a rare (1–9 cases/100,000 inhabitants) inflammatory disorder that begins with vomiting, abdominal pain, diarrhea, nausea, dyspepsia, weight loss, and growth retardation (Orphanet, ORPHA: 2070). Eosinophilia is frequently limited to the distal oesophagus, stomach, and duodenum, and can affect several layers of the gastrointestinal wall. This disease often starts in middle age individuals (30–50 years old) with overweight/obesity and tendency to suffer from gastric reflux (Fernandez-Becker et al., 2020). Specifically, this acid reflux disrupts the epithelial barrier and causes the erosive esophagitis, Barrett's oesophagus, and peptic structures (Fernandez-Becker et al., 2020). In addition, the pathophysiological mechanism seems to be mediated by a hypersensitivity reaction in which IgE may participate (20%–80% of cases show elevated IgE) and where there is eosinophilic infiltration and concomitant tissue damage (Orphanet, ORPHA: 2070). The sensitization to different types of food allergens is common in this disease, although dietary elimination of the most frequent allergens (milk, egg, soy, wheat, peanut, fish, shellfish, etc.) is usually less effective than for EoE (Bird et al., 2019). Once again, there are also no proteomic studies on food allergens that can trigger EGE.

Food protein-induced enterocolitis syndrome (FPIES), by its part, is a non-IgE-mediated and delayed food allergic disease characterized by the appearance of several signs and symptoms after 2–4 h of eating. The most common culprits are cow's milk or soy, and less frequently grains (rice, oats), fruits, or vegetables; there can be single or multiple triggers. This disease courses with repeated vomiting (sometimes with diarrhea) and severe dehydration, hypotension, lethargy and, in severe cases, hypovolemic shock and intestinal perforation. It usually occurs in infancy, when a baby is introduced to solid food or formula, and disappears after 3 years of age, except in a few cases (Bird et al., 2019). Lastly, other non-IgE-mediated (cellular) disorders are food protein-induced proctocolitis, food protein-induced enteropathy (FPE) (Marzano et al., 2020), or food allergies induced by gluten, such as coeliac disease, gluten hypersensitivity, wheat allergy, and wheat-dependent exercise-induced anaphylaxis (WDEIA) (Renz et al., 2018). Once again, given the secondary role of IgE in these types of diseases, there are no proteomic studies on the allergens that trigger these pathologies, except for celiac disease, which we will describe later in this chapter.

9.3 Food allergens

9.3.1 Allergenicity and cross-reactivity

The European Food Safety Authority (EFSA) glossary defines *allergenicity* as the "ability to trigger an abnormal immune response that causes an allergic reaction in a person." Within the large number of proteins that make up the human diet, only a limited set of them are potential inducers of food allergies. It is unknown what converts some dietary proteins into allergens, as well as what makes an allergenic protein a major or a secondary trigger. However, it seems that many allergenic proteins share certain general features, such as their abundance in the offending food, a relatively small molecular mass (<70 kDa), the presence of glycosylated residues, their solubility in water, and their resistance to gastrointestinal proteolysis or heat (Bird et al., 2019). For example, lipid transfer proteins (LTPs) from Rosaceae fruits, peanuts, tree nuts, or cereals, and conglutins like the major peanut allergens Ara h 2 and Ara h 6, resist the high cooking temperatures, the extreme acid pH in the stomach, and the hydrolytic actions of gastric, pancreatic, and intestinal peptidases (Tordesillas et al., 2017). However, there are food allergens where there is no direct correlation between digestion-resistance and allergenicity (Marzano et al., 2020).

According to the United States (US) Food and Drug Administration, the foods that trigger 90% of allergies comprise a group named as "The Big 8," which includes eggs, fish, milk, peanuts, soybeans, tree nuts, and wheat (https://www.fda.gov/food/food-labeling-nutrition/food-allergies). However, their impact worldwide is not homogeneous. Thus, the prevalence of fish and shellfish allergies is higher in Asian countries, while the prevalence of allergies to milk, eggs, and nuts occurs more frequently in the United States, Australia, and Western Europe. Some of these allergies tend to persist over time, such as peanuts, tree nuts, fish, and shellfish (e.g. crustaceans), while others, such as milk allergies, tend to disappear in adulthood.

The severity and duration of food allergy is also influenced by the epitope type. Thus, conformational epitopes (e.g., allergy to cow's milk or eggs) appear to be related to a more transient disease. In addition, structural changes during food processing could affect proteins allergenicity, which explains why children allergic to these foods tolerate them after cooking (i.e., the conformational epitopes are destroyed). In contrast, epitope-recognition based on the primary sequence of cow's milk or egg white proteins is associated with persistence and severity of the food allergy (Tordesillas et al., 2017). Additionally, isoforms and post-translational modifications (PTMs) could affect protein allergenicity as well. An example is the above-mentioned allergy to mammalian meat (Danijela Apostolovic et al., 2020). This allergens heterogeneity can be observed at the AllerBase (http://bioinfo.unipune.ac.in/AllerBase/Home.html), a database of experimentally validated allergens from animals, plants, fungi, bacteria, and viruses (Kadam et al., 2017).

Cross-reactivity, on the other hand, is defined by the EFSAglossary as "A situation where an allergic reaction to one substance also leads to an allergic reaction to another substance." This usually takes place when the similarity of the primary sequence of allergens is greater than 50%–70% (Dodig and Čepelak, 2018). It is common to find this type of cross-reactivities between tree nuts, such as cashew and pistachio or walnut and pecan. However, this phenomenon is not exclusive to proteins of plant origin, as can also be found with proteins from different animal sources. An example are tropomyosins in crustaceans, so that the presence of allergy to one species generates a 75% probability of developing cross-reactivity to a second one.

9.3.2 Food-allergy diagnosis

Diagnosis of food allergies relies on the clinical history (anamnesis), the use of specific allergen extracts in skin prick tests (SPT), or detection of total (tIgE) and specific (sIgE) IgE by immunoassays in serum samples (Peters et al., 2021; Yu et al., 2016). SPT and sIgE are very sensitive, but generate a high rate of false positives, especially with the cut-off points commonly used (wheal ≥ 3 mm and ≥ 0.35 kU/L, respectively). Hence, a negative SPT/sIgE result could indicate that these people are probably food tolerant, but also that they may present antibodies and be sensitized to food allergens without suffering any symptom (Peters et al., 2021). Moreover, SPT and sIgE are not suited for diagnosing disorders such as FPIES, which are not mediated by IgE, or EoE, where IgE appears to have a limited role (Yu et al., 2016). Therefore, the double-blind placebo-controlled oral food challenge (OFC) remains as the gold standard and second-line approach for the diagnosis of food allergies. However, this diagnostic method is time-consuming, costly, requires medical guidance, and bears the risk of inducing severe anaphylactic reactions; all this together makes that the use of OFC is currently under debate (D'Auria et al., 2018; Peters et al., 2021).

Other second-line tests are the so-called cellular assays, such as the "basophil activation" or "mast cell activation" tests. In basophil activation test (BAT), fresh whole blood of a patient is exposed to different food allergens or control molecules; afterwards, different activation/degranulation markers (e.g., CD63 or histamine release) are measured in basophils (SSC^{low}, $CD203c^{+}$, $CD123^{+}$) by flow cytometry (Commins et al., 2014; Santos et al., 2021). In mast cell activation test (MAT), primary mast cells are firstly obtained and sensitized with previously stored serum/plasma samples from patients. Then, these cells are activated with food allergens and activation/degranulation is measured by flow cytometry (e.g., CD63) (Bahri et al., 2018). Both BAT and MAT display a good performance, with high sensitivity and specificity values, but are time-consuming (Peters et al., 2021) and therefore impractical.

A major advance in food allergy diagnosis has been allergen component-resolved diagnostics (CRD). During the determination of sIgE by CRD, crude allergen extracts containing both allergenic and non-allergenic compounds are substituted for purified native or recombinant allergenic molecules (Peters et al., 2021; Yu et al., 2016). Detection of sIgE antibodies can be achieved with single-plex systems such as ImmunoCAP (Thermo Fisher Scientific), or multiplex systems such as the ImmunoCAP ISAC (Thermo Fisher Scientific). They currently offer greater diagnostic accuracy compared with determination of sIgE with extract-based analyses (Peters et al., 2021), but it is certain that improvements in the design of peptide and protein microarrays will boost molecular diagnosis of food allergies in the future.

In the absence of an effective treatment and since routine immunotherapy has not been implemented, allergen avoidance and emergency medication are considered the best practice. However, this is very difficult to attain considering the broad distribution of allergens or the defective safety labeling of many processed products, which do not indicate the presence of trace amounts of allergenic substances. DNA-based techniques such as PCR can be used to detect DNA encoding food allergens, but this methodology can present problems like cross-reactivity and inter-assay variability in the case of processed foods; besides, it does not directly detect allergenic proteins and does not guarantee their presence (Carrera et al., 2012; Marzano et al., 2020; Stella et al., 2020).

Identification and quantification of allergens in fresh and processed foods also can be carried out by commercially available enzyme-linked immunosorbent assay (ELISA). ELISA is a sensitive technique that does not require a high level of expertise (Croote and Quake, 2016). Nonetheless, this technique presents several drawbacks: (1) epitope degradation or masking/unmasking due to protein denaturation/aggregation; (2) altered solubility of food allergens; (3) reaction with matrix components or non-enzymatic glycation (Maillard reaction) of allergens during food processing; (4) antibody cross-reactivities with matrix compounds and matrix effects; (5) differences of analytical performance amongst kits (low reproducibility); (6) lack of multiplex capability; (7) and high cost of this technology. All together can lead to inter-assay variability, false-negatives/false-positives, or widespread application problems (Croote and Quake, 2016; Lopez-Pedrouso et al., 2020; Stella et al., 2020). Some solutions have been provided, such as the use of ELISA kits based on antibodies targeted against denatured proteins (Lopez-Pedrouso et al., 2020). In any case, all these techniques require the a priori identification of allergenic molecules in food samples, a task where proteomic approaches are gaining importance.

9.3.3 Proteomics in food-allergy: from discovery to targeted methods

Proteomic techniques have been used to identify and characterize the major allergenic components in fresh and processed foods. *Bottom-up* approaches, where food protein/s are separated by two-dimensional gel electrophoresis (2-DE), capillary electrophoresis, or liquid chromatography (LC), and converted into peptides (e.g., with trypsin) prior to mass spectrometry (MS) analysis, have been extensively employed (Ahsan et al., 2016; Carrera et al., 2020; Dhondalay et al., 2018; Lopez-Pedrouso et al., 2020; Marzano et al., 2020). Indeed, such application of proteomic techniques to food analysis led to coin two terms: *foodomics*, "a discipline that studies the food and nutrition domains through the application of advanced omics technologies to improve consumers' well-being, health and knowledge" (D'Auria et al., 2018); and the more specific term *allergomics*, a field that uses efficient fractionation and sample simplification technologies, high-throughput proteomic analysis, and bio-informatics tools to identify and perform structural characterization of the whole repertoire of allergens in food (Lopez-Pedrouso et al., 2020; Shahali et al., 2019; Wang et al., 2020). One way of performing qualitative/semiquantitative and sensitive/specific detection of allergens in food samples is *immunoproteomics*, were 2-DE is combined with immunoblotting with IgE-containing sera from patients. However, this tool requires specialized equipment, trained staff, special handling of the antibody, and subsequent protein identification of IgE-reactive allergens by proteolytic cleavage and matrix-assisted laser desorption ionization-time of flight MS (MALDI-TOF/TOF) or LC coupled to tandem MS (LC–MS/MS) analysis (Carrera et al., 2018; De Angelis et al., 2010). Besides, alternative techniques such as ELISA or targeted proteomics (see below) are required for results confirmation. Strikingly, there are no works that use immunoprecipitation or affinity chromatography assays with IgE from allergic patient's sera to identify reactive allergens (for example, within the proteome of bovine milk or white from the egg). This is probably due to the null and low affinity, respectively, of protein G and protein A to capture human IgE. However, it is worth noting that anti-IgE antibodies such as omalizumab could be used in this type of assays instead. In addition, the recent use of the pan allergen phage libraries (Allerscan) for the detection of key wheat allergic proteins deserves attention (Mohan et al., 2018; Monaco et al., 2021).

Alternatively to immunoproteomics, *discovery approaches/shotgun proteomics* can also be used to identify new food allergens that generate responses mediated by B (IgE, IgG) or T cells. Thus, food proteins are digested either through unique proteases (e.g., trypsin) or more elaborate enzymatic systems that attempt to emulate in vivo conditions and generate proteolysis-resistant peptides ("digestome"). Then, LC coupled to high-resolution MS/MS (e.g., Q-TOF, Q-Orbitrap), combined with either data-dependent (DDA) or data-independent (DIA; e.g., sequential window acquisition of all theoretical mass spectra or SWATH) acquisition modes, is used for the qualitative/quantitative description of the food proteome. In these methods, MS/MS scans of product ions are acquired, and specialized algorithms, search engines (e.g., SEQUEST, X!Tandem, Mascot, Andromeda), and databases (e.g., NCBI, UniProtKB) are used to compute peptide sequences and select proteotypic peptides, which can be subsequently used in targeted proteomic strategies (Marzano et al., 2020). However, full characterization of a food digestome is not as simple as it might seem at first, due to the low quality of the annotation of some genomes (e.g., in plants), the complexity of the digestomes, or preanalytical/analytical issues (e.g., varying sample preparation and fractionation/enrichment methods, low solubility or ionization of peptides, suppression ion effects due to overrepresented peptides, presence of PTMs or disulfide bonds, or the use of search engines that tend to under-represent small peptides) (De Cicco et al., 2019).

Shotgun proteomics analysis of food "digestomes" ("digestomics") allows the accurate determination of the molecular weight of protease-resistant peptides (MS1) as well as the identification of both the primary sequence and other structural details (MS/MS) (Apostolovic, Stanic-Vucinic, et al., 2016; Di Stasio et al., 2017; Marzano et al., 2020). An important objective of these assays is the identification of "proteotypic peptides," i.e., peptides that uniquely associate to a protein, like for example a food allergen. These proteotypic peptides should meet the following requirements: (1) a length between 7 and 25 amino acids; (2) a reproducible digestion pattern (i.e., there must not be missing cleavage sites); (3) a reproducible fragmentation pattern and a high ion-current response to achieve good detection sensitivity; (4) a C-terminal arginine instead of a

C-terminal lysine to prevent the formation of Maillard's reaction products; (5) no methionine and/or cysteine residues (Ahsan et al., 2016; Marzano et al., 2020; Stella et al., 2020). Besides, detection of those proteotypic peptides must not be influenced by variations in a food matrix, sample preparation protocols, or MS instrumentation (Croote and Quake, 2016), and the absence of orthologous (i.e., in different species) or paralogous (i.e., in the same organism) sequences must be verified by search engines (e.g., the Basic Local Alignment Search Tool/BLAST algorithm) combined with publicly available databases such as NCBI (https://www.ncbi.nlm.nih.gov/protein/) or/and UniProtKB (https://www.uniprot.org/).

Still, all these steps do not guarantee the uniqueness of the pre-selected proteotypic peptides (Marzano et al., 2020) neither their biological meaning in terms of antigenicity/allergenicity. Identification of T-cell and B-cell immune epitopes in food allergens can be achieved in different ways. The first step is searching databases such as the Immune Epitope Database (http://www.iedb.org/), that stores information about T-cell and B-cell immune epitopes in food allergens. If such information is not available, the second possibility is to predict those B- or T-cell-dependent epitopes using Bioinformatic tools. In the case of type I (IgE-mediated) or type II (IgG/IgM directed against cell surface antigens) hypersensitivity reactions, lineal B-cell epitopes in a primary sequence of an allergen can be identified with machine learning-based Bioinformatic tools like BepiPred 2.0 (http://www.cbs.dtu.dk/services/BepiPred) or ABCPred (http://crdd.osdd.net/raghava/abcpred) (Sanchez-Trincado et al., 2017), while prediction of discontinuous/conformational B-cell epitopes based on the allergenic protein's 3D structure (either in PDB format or pre-modelled primary sequences) can be performed with ElliPro (http://tools.iedb.org/ellipro/help/), PEPITO (http://pepito.proteomics.ics.uci.edu/), or SEPPA (http://lifecenter.sgst.cn/seppa/) (Sanchez-Trincado et al., 2017). In the case of delayed-hypersensitivity responses to food allergens (e.g., coeliac disease), data-driven methods can be used to predict the most likely peptides from an allergenic protein (and sometimes the binding affinity) that fit with the groove of MHC class I or class II alleles prior to presentation to $CD8^+$ or $CD4^+$ T cells, respectively (Sanchez-Trincado et al., 2017). The predictions made by this type of Bioinformatic tools can be confirmed by assays such as ELISPOT (Enzyme-Linked ImmunoSpot Assays), which can detect the presence of specific IgE-secreting B lymphocytes or IL-4/5-producing TH2 lymphocytes in response to specific allergenic peptides.

Targeted proteomics is a hypothesis-driven strategy focused on the analysis of proteotypic peptides that represent a selected subset of food allergens. Therefore, despite their advantages and a notable expansion in recent years (especially in the field of plant allergens), hypothesis-driven analyses require prior knowledge of food allergens and their proteotypic peptides. Targeted methodologies have been extensively used to detect potentially allergenic proteins in processed food, since many of them have been listed in freely accessible databases, like the above-mentioned AllerBase; (http://bioinfo.unipune.ac.in/AllerBase/Home.html), AllergenOnline (http://www.allergenonline.org/), Structural Database of Allergenic Proteins (SDAP, http://fermi.utmb.edu/), Allergome (http://www.allergome.org/), or Allergen Peptide Browser (http://www.AllergenPeptideBrowser.org/) (Ahsan et al., 2016; Carrera et al., 2018). However, the information contained in these databases may be redundant, not fully up to date, or refers to proteins that may represent only a "potential risk" of allergenic cross-reactivity; i.e., they are not recognized as reliable allergenic proteins, but they contain either amino acid sequences or domains similar to well-known allergenic proteins (Ahsan et al., 2016); besides, proteotypic peptides have not been described for the whole set of food allergens (Croote and Quake, 2016). Once chosen the best-performing proteotypic peptides in terms of resistance to matrix and processing effects, sensitivity, and signal linearity to identify/quantify particular food allergens, targeted proteomics methods can be used for their specific detection. Different acquisition modes and instruments can be employed, like selected/multiple reaction monitoring (SRM/MRM) on triple quadrupoles (QQQ) or hybrid quadrupole-ion trap (Q-Trap) mass spectrometers; additionally, detection and quantitation of allergens in foods can be achieved by parallel reaction monitoring (PRM) in hybrid quadrupole-Orbitrap or ToF mass analyzers (Ahsan et al., 2016; Carrera et al., 2018; Croote and Quake, 2016; Marzano et al., 2020). When proteins are analyzed by SRM/MRM on a QQQ mass spectrometer the first quadrupole (Q1) is used as a mass filter to select a given precursor. Then, such precursor is fragmented in the second quadrupole 2 (q2), meanwhile the third quadrupole (Q3) uses narrow m/z windows to detect a reduced number of product ions (>3 ion transitions). In PRM all fragment ions are measured after fragmentation of a selected precursor in q2 (Ahsan et al., 2016; Carrera et al., 2018; Croote and Quake, 2016; Marzano et al., 2020). In this workflow, +2/+3 charged precursor ions and 3–5 transitions are used to detect allergens both in raw and manufactured food products. These technologies provide low limits of detection, reproducibility, multiplexing capacity, and the capability of absolute quantification (AQUA) by sample spiking of different amounts of stable heavy isotope-labelled synthetic peptides and the use of calibration curves (Ahsan et al., 2016). However, a major drawback of MS is the time-consuming sample preparation (Stella et al., 2020). In addition, the use of alternative methodologies (e.g., ELISA) and validation/verification assays (e.g., selectivity, accuracy, precision, LOQ, LOD, etc.) according to current regulations to ensure data quality and good laboratory practices is still required (Stella et al., 2020).

9.3.4 Proteomic studies and allergens in plants

9.3.4.1 *Gluten-related disorders*

Gluten is a mix of hundreds of proteins with storage functions, most of them from two families: prolamins and glutelins. Prolamins are monomeric proteins that play an important role in the extensibility and cohesiveness of flour. Wheat (*Triticum aestivum L.*) has the highest content of prolamins, where they receive the name of gliadins. Prolamins are also the main storage proteins in rye (secalin), barley (hordein), corn (zein), and sorghum (kafirin), while lower amounts are present in oat (avenin) and rice (orzein). On the other hand, glutelins (or glutenins in wheat) are polymeric proteins with inter chain disulfide bonds that confer elasticity to wheat flour (Scherf et al., 2016). According to their molecular weight and sequence homology, gluten proteins are divided into low (LMW), medium (MMW), and high molecular weight (HMW) groups (Table 9.1) (Lexhaller et al., 2019; Scherf et al., 2016; Scherf, 2019). LMW group consist of both prolamins and glutelins, whereas MMW and HMW groups are made up of monomeric prolamins and glutelins, respectively.

In wheat, prolamins are subdivided into four types (α-, γ-, ω5-, and ω1,2-gliadins), while glutelins are classified in low (LMW-GS) and high (HMW-GS) molecular weight subtypes. In rye, prolamins are framed within the MMW (ω-) and LMW (γ-75k-, γ-40k-) secalins, whereas there are only HMW-glutelins (HMW-secalins). Finally, prolamins in barley present a medium (C-hordeins) and a low (γ-hordeins) molecular weight, while glutelins can be found in both the HMW (D-hordeins) and LMW (B-hordeins) categories (Table 9.1) (Martínez-Esteso et al., 2016; Scherf et al., 2016; Wieser et al., 1998).

TABLE 9.1 Types of proteins in gluten.

Group	Wheat	Barley	Rye
HMW (>70 kDa)			
Glutelins	HMW-GS	D-hordeins	HMW-secalins
MMW (40–70 kDa)			
Prolamins	ω1,2-gliadins	–	–
	ω5-gliadins	C-hordeins	ω-secalins
LMW (<50 kDa)			
Glutelins	LMW-GS	B-hordeins	–
Prolamins	α-gliadins	–	γ-75k-secalins
	γ-gliadins	γ-hordeins	γ-40k-secalins

GS, glutelins; HMW, high molecular weight; LMW, low molecular weight; MMW, medium molecular weight.

Apart from being very important for quality of different food products, gluten proteins are known for causing the so-called gluten-related disorders (GRDs). GDRs encompasses several diseases, including gluten hypersensitivity (or non-celiac gluten sensitivity /NCGS), wheat allergy (WA), and WDEIA (Tovoli, 2015). However, the most studied and prevalent GRD is the coeliac disease (CD). Prolamins (gliadins in wheat) are the main immunogenic proteins in gluten-causing CD. Gliadins are proteins with high content of proline and glutamine residues. These two amino acid residues confer resistance to the cleavage by gastric and pancreatic endopeptidases, including pepsin, intestinal brush border peptidases (i.e., dipeptidyl peptidase IV, dipeptidyl carboxypeptidase I), or pancreatic proteases (Cebolla et al., 2018). In healthy intestinal cells, gliadin fragments are transported through endocytosis to be completely degraded in lysosomes (Ciccocioppo et al., 2005; Matysiak-Budnik et al., 2003). However, protease-resistant peptides from gliadins in coeliac patients lead to an increased expression of Zonulin (haptoglobin 2 precursor) in epithelial cells, causing and enhanced permeability of tight junctions and a concomitant enhanced paracellular transport of gliadine peptides toward the LP.

Gliadin peptides in LP induce both innate (toxic peptides) and adaptive (immunogenic peptides) immune responses. A database with the different toxic and immunogenic peptides causing CD has been created by the Food Allergy Research and Resource Program of the University of Nebraska–Lincoln, USA (http://www.allergenonline.org/celiachome.shtml). Toxic peptides from α-gliadine (i.e., α1-30, α3-24, α25-55, α31-43, α31-49, α31-55, α44-55, α51-70, α56-75, and α206-217) can activate the production of IL-15, which in turn led to the increased expression of MICA and MICB (stress markers) in epithelial cells and their C-type lectin-like receptor NK-G2D on IELs (NK-G2D$^+$ lymphocytes). Ligation of NK-G2D triggers the activation of IELs and the subsequent destruction of intestinal epithelium, causing enhanced permeability. On the other hand, 28 disease-relevant HLA-DQ-restricted epitopes in gliadin and glutenin have been described (Sollid et al., 2020), as well as hundreds of immunogenic peptides containing those epitopes (Juhász et al., 2015). In the LP, the enzyme Transglutaminase 2 (TG2) deaminates gliadin immunogenic peptides, which in turn are phagocyted by APCs, processed into smaller peptides, and presented to CD4$^+$ T cells by HLA-DQ2 and HLA-DQ8 molecules. This will result in a generation of a cellular immune response characterized by the activation of naïve TH lymphocytes and their differentiation into TH1 cells, which will produce in turn high amounts of IFN-γ and TNF-α. In the end, this will result in the activation of several matrix metalloproteases (MMPs) and the subsequent epithelial remodeling. Finally, plasma cells will be also activated, resulting in the release of TG2-specific IgA, which forms complexes with gliadin peptides that are retrotranscytosed through the transferrin receptor (CD71). This, together with the release of TNF-α and IFN-γ from gluten-activated TH1 cells, also contributes to the increased permeability of the intestine in patients with CD (Balakireva and Zamyatnin, 2016; Papista et al., 2012).

GRDs affect about 5% of world population and their prevalence is continuously increasing (Bascuñán et al., 2018). Nowadays, the gold standard therapy (and almost the only treatment available) for CD is a "gluten-free" diet. In this regard, it is of particular importance to develop techniques to identify and quantify all these gluten proteins from commercially available products to ensure gluten-free food. The official method for gluten protein detection recommended by the Codex Alimentarius is the ELISA R5 Mendez method (CXS 118-1979) (Shatalova et al., 2020). The main problem of ELISAs is that they are exclusively focused on gliadins (wheat prolamins), so they do not take into account species-specific changes in gluten content. Moreover, they presuppose the same ratio gliadin/glutenin under different conditions, including different extraction procedures or different cultivars. Therefore, novel high-sensitive and reliable methodologies are necessary to address these issues. In this context, several MS-based methods have recently been developed (Alves et al., 2019).

In order to perform the identification of the different peptide sequences by using proteomics, the existence of sequence databases is necessary. In 2017, Bromilow et al. created the first curated wheat gluten protein sequence database (GluPro v1.0), an open-source database to support the application of proteomic technologies for gluten identification and quantification (Bromilow et al., 2017). This database was updated to include avenin sequences from bread wheat (GluPro v1.1 and GluPro v 1.2), and sequences from pasta wheat (*Triticum turgidum ssp* durum; GluPro V2.0 and V2.1). Multiple efforts have been made to create reference material of gluten proteins from other cereals (Lexhaller et al., 2019). Finally, curated sequence databases for barley (GluPro v3.0), rye (GluPro v4.0), and oat (GluPro v5.0) have been created and combined with the previous ones to contain a broader database of gluten-protein sequences (n = 1041) in GluPro v 6.0 and 6.1 (Daly et al., 2020a; 2020b).

In the last years, several proteomic studies have been developed for the identification and quantification of gluten peptides in several food matrices by MS (Table 9.2), being LC-ESI-MS/MS the preferred methodology (Allred et al., 2020; Alves et al., 2019; Cao, Baumert, & Downs, 2020; Daly et al., 2020a; Geisslitz, Longin, Koehler, & Scherf, 2020; Li, Byrne, et al., 2018; Liao et al., 2017; Manfredi et al., 2015; Martínez-Esteso et al., 2016; Ogilvie et al., 2020a; Sealey-Voyksner et al., 2010; Tanner et al., 2013). Thus, several proteomic studies have been aimed to identify grain-specific peptide markers of gluten (Bönick et al., 2017; Colgrave, Byrne, Blundell, Heidelberger, & Lane, 2016a; Colgrave, Byrne, Blundell, & Howitt, 2016b; Colgrave et al., 2015; Fiedler et al., 2014; Lexhaller et al., 2019;

TABLE 9.2 Proteomic approaches used for the study of gluten allergens.

Food	Methodologies	Aim	References[a]
Different types of flour, pasta, and other processed food (powdered ice tea, cheerios, orzo, hot sauce, goldfish crackers, vinegar, tooth-paste, body lotion, beer, vodka, gin, rum, wine)	-MRM by HPLC-ESI-QqQ-MS/MS -nanoUPLC-ESI-MRM-MS -HPLC-IT-MS/MS -UHPLC-MS/MS in DDA mode	Identification and Quantification of celiac disease immunogenic and toxic peptides in food	Martinez-Esteso et al. (2016), Daly et al. (2020a), Sealey-Voyksner et al. (2010), Manfredi et al. (2015), Ogilvie et al. (2020a), Tanner et al. (2013), Cao, Baumert, & Downs, 2020, Allred et al. (2020), Geisslitz, Longin, Koehler, & Scherf, 2020, Liao et al. (2017), Li et al. (2018)
Wheat	-1-DE and 2-DE immunoblotting followed by nanoLC-MS/MS -nanoUPLC-QTOF-MS/MS -On-line 2D nanoLC–MS/MS -UPLC-MRM-MS/MS -nanoUPLC-MSE -nanoUPLC-UDMSE -MALDI-TOF-MS	To compare the allergenicity between different wheat varieties/cultivars	van Den Broeck et al. (2015), Larre et al. (2011), Uvackova et al. (2013), Rogniaux et al. (2015), Schalk et al. (2017), Prandi et al. (2017), Malalgoda et al. (2018), Jang et al. (2020), Lakhneko et al. (2020), Alves et al. (2018)
Genetic modified wheat	-1-DE immunoblotting followed by nanoLC-QTOF-MS/MS	Comparison of the number of allergenic peptides in 5 GM wheats	Lupi et al. (2014)
Wheat, Barley, Rye, Oats, Amaranth, Chia, Quinoa, Sorghum, Tef, Buckwheat, Soy, Millet, Maize.	- RP-HPLC -nanoHPLC-ESI-pSRM; MS/MS -nanoUPLC-ESI-MS/MS -MRM by HPLC-ESI-MS/MS - SWATH-MS	To characterize grain-specific peptide markers for the gluten detection	Lexhaller et al. (2019), Colgrave et al. (2016b), Fiedler et al. (2014), Seki et al. (2021), Colgrave et al. (2015), Schalk et al. (2018), Colgrave et al. (2016a)

(Continued)

TABLE 9.2 Cont'd

Food	Methodologies	Aim	References[a]
Gluten-free food	-ELISA -LC-QTOF-MS/MS -HPLC-IT-MS/MS -MRM by HPLC-ESI-QqQ-MS/MS	To detect gluten contamination in gluten-free food To characterize the efficiency of gluten detoxification strategies	Sealey-Voyksner et al. (2010), Manfredi et al. (2015), Pilolli et al. (2020), Allred et al. (2014)
Wheat; spelt; barley; maize; oat; rye	-UPLC-ESI-MRM-MS/MS	To check the authenticity of wheat, spelt and rye in bread products	Bonick et al. (2017)
Durum wheat, bread	-ELISA -UHPLC-ESI-MS/MS -PRM by UPLC-ESI-MS/MS -uHPLC-MS/MS in DDA mode -HPLC/ESI-MS/MS -uHPLC-LTQ-OrbiTrap	Identification of peptides generated by in vitro gastrointestinal digestion protocols of wheat Identified of celiac disease immunogenic peptides during bread digestion by discovery proteomics	Ogilvie et al. (2020a), Ogilvie et al. (2020b), Boukid et al. (2019), Prandi et al. (2014), Ogilvie et al. (2021)
Barley	-MRM by nanoLC-TripleTOF-MS/MS	To study α-amylase trypsin inhibitors in barley implicated in non-celiac gluten sensitivity	Bose et al. (2019)
Couscous malt bread	-1D-SDS-PAGE-iTRAQ-MALDI-TOF-MS/MS -PRM by UPLC-ESI-MS/MS	To study the changes of cereal allergenic proteins after food processing. To study the effect of food processing (e.g., temperature, baking time) in celiac peptide digestibility	Ogilvie et al. (2021), Flodrova et al. (2015)

[a]Reviewed in Alves et al. until 2019 (Alves et al., 2019). New references to date have been included. 1-DE, one-dimensional gel electrophoresis; 2-DE, two-dimensional gel electrophoresis; DDA, Data dependent adquisition; ESI, electrospray ionization; HPLC, high performance liquid chromatography; IT, ionic trap; QqQ, triple quadrupole; LC, liquid chromatography; MALDI-TOF, matrix-assisted laser desorption ionization-time of flight; MRM, multiple reaction monitoring; MS, mass spectrometry; MS/MS, tandem mass spectrometry; PRM, parallel reaction monitoring; pSMR, pseudo-selected reaction monitoring; SWATH-MS, sequential window acquisition of all theoretical mass spectra; UDMSE, ultra definition mass spectrometry; UHPLC, ultra high performance liquid chromatography; UPLC, ultra-performance liquid chromatography.

Martínez-Esteso et al., 2016; Schalk et al., 2018; Seki et al., 2021), of great importance to generate reference databases, but also for inferring the presence of gluten proteins and check the authenticity, miss-labeling, or adulteration of cereal-based products (Table 9.2) (Bönick et al., 2017). In addition, the group of Akiyama and coworkers developed in 2021 a HPLC-ESI-MRM-MS/MS method for the simultaneous detection of multiple cereals containing gluten in processed foods, including wheat, barley, spelt, rye, oats, and buckwheat (Table 9.2) (Seki et al., 2021).

The use of MRM or PRM approaches allows the quantification of immunogenic peptides even at trace concentrations (Table 9.2) (van Den Broeck et al., 2015). MRM protocols for the identification and quantification of gluten marker peptides in wheat, barley, rye, and oats have shown comparable results to other methodologies such as R5 ELISA and gel-permeation high-performance liquid chromatography with fluorescence detection (GP-HPLC-FLD) (Table 9.2). MRM-LC-MS/MS strategy was also implemented to assess α-amylase trypsin inhibitors in barley potentially implicated in NCGS (Bose et al., 2019). Although MRM is considered the gold-standard methodology for allergenic peptide quantification, other methodologies such as SWATH-MS showed comparable results (Colgrave, Byrne, Blundell, Heidelberger, et al., 2016a).

The use of in vitro digestion protocols, such as INFOGEST or *in vitro* simulated human gastroduodenal digestion, have been implemented in some procedures before the application of LC-MS/MS (Boukid et al., 2019; Ogilvie et al., 2020b; Pilolli et al., 2020; Prandi et al., 2014). Of note, Ogilvie et al. (2020a) have developed a targeted PRM protocol for the quantification of six immunogenic gluten peptides produced from food matrices using simulated digestion (INFOGEST). By using this technology, it was demonstrated an effect of both temperature and baking time in the release of these six immunogenic gluten peptides (Ogilvie et al., 2021). Furthermore, a different protocol using *in vitro* simulated human gastroduodenal digestion followed by LC-MS/MS has been used for the evaluation of gluten detoxification strategies in prototype gluten-free breads (Pilolli et al., 2020).

Other multiple proteomic studies have been recently performed with different aims (Table 9.2); for example, to compare the allergenicity between different wheat cultivars (Alves et al., 2018; Jang et al., 2020; Lakhneko et al., 2020; Larré et al., 2011; Malalgoda et al., 2018; Prandi et al., 2017; Rogniaux et al., 2015; Schalk et al., 2017; Uvackova et al., 2013; van Den Broeck et al., 2015), to study the effect of genetic modification of wheat in the number of allergenic peptides (Lupi et al., 2014), to detect gluten contamination in gluten-free food and to check gluten detoxification strategies (Allred et al., 2014; Manfredi et al., 2015; Pilolli et al., 2020; Sealey-Voyksner et al., 2010), or to study the changes of cereal allergenic proteins after food processing (Flodrová et al., 2015; Ogilvie et al., 2021) (Table 9.2).

9.3.4.2 *Pollen-fruit allergy syndrome*

Fruits and vegetables are other allergenic foods widely studied. The majority of allergic reactions result from "cross-allergenicity" with pollen allergens, causing a disease called "oral-allergy syndrome (OAS)" or PFAS. This is an IgE-mediated or type-I hypersensitivity reaction to fruit and vegetable allergens that occurs after sensitization to an inhalant allergen. Therefore, it is driven by pollen proteins that generate allergen-specific IgE antibodies which cross-react with proteins with similar structures in raw fruits and vegetables (Carlson and Coop, 2019; Yagami and Ebisawa, 2019). PFAS is mainly characterized by oral-pharyngeal pruritus or burning sensation, while gastrointestinal symptoms are not common due to the digestibility of allergenic proteins (class II food allergens). Indeed, allergen activity of raw fruits/vegetables is greatly reduced after heating and processing. Anaphylactic reactions in PFAS are also a rare event (1.7% of patients) (Carlson and Coop, 2019). The prevalence of PFAS ranges from 13% to 58% in adults, although it is dependent on the geographical region and the main aeroallergen in these regions. For example, 50%–93% of birch allergic patients are affected by PFAS in areas where birch pollen is endemic (i.e., Europe) (Skypala et al., 2013). Importantly, allergy to specific aeroallergens is highly associated with allergy to specific fruits or vegetables (https://www.aaaai.org/conditions-and-treatments/library/allergy-library/outdoor-food-allergies-relate) (Table 9.3).

A list of common clinically relevant allergens causing PFAS is presented in Table 9.4. The major protein families responsible for PFAS are pathogenesis-related (PR) proteins, profilins, and prolamins (Jenkins et al., 2005; Skypala et al., 2013). Among them, PR-10 family (i.e., Bet v 1-like allergens) includes the most important allergenic proteins causing PFAS. Birch pollen allergen Bet v 1 was shown to produce cross-reactive reactions with many fruits, vegetables, and nuts allergens; it is responsible for the well-known *birch-fruit-vegetable-syndrome* (Popescu, 2015) (Table 9.4). Of particular interest is the *birch-apple syndrome* (i.e., Bet v 1 and Mal d 1 cross-reactivity), being the most common PFAS. Another syndrome with particular importance is the *celery-mugwort-spice syndrome,* which could be associated with complicated cases of curry spice allergy (Egger et al., 2006; Yagami et al., 2009), involving this time the profilin family protein Art v 4 from mugwort pollen, Api g 4 from celery, Api g 5 from fennel, and other multiple allergens in different spices (Egger et al., 2006). An alternative example is the *Platanus pollen-fruit allergy syndrome,* which involves LTPs from Platanus tree pollen (Pla a 3) and banana (Mus a 3), hazelnut (Cor a 8), celery (Api g 2), peach (Pru a 3), or peanut (Ara h 9) (Popescu, 2015).

So far, several proteomic studies have been performed for the identification and quantification of different allergens responsible for PFAS (Table 9.5).

TABLE 9.3 Pollens and cross-reacting foods causing pollen-fruit allergy syndrome (PFAS).

Pollen	Food
Birch pollen	Fruits: Apples, apricots, cherries, almonds, peaches, pears, plums, kiwi, strawberry
	Vegetables: avocados, carrots, celery, parsley
	Legumes: peanuts, soybean
	Nuts: almond, hazelnut
Olive pollen	Fruits: peach, pear, melon, kiwi
Grass pollen	Fruits: peaches, watermelon, orange, tomato
	Vegetables: White potato
Ragweed pollen	Fruits: Banana, cantaloupe, honeydew, watermelon
	Vegetables: Cucumber, white potato, zucchini
Mugwort pollen	Vegetables: Bell pepper, broccoli, cabbage, cauliflower, garlic, onion, parsley
	Spices: Aniseed, caraway, coriander, fennel, black pepper

TABLE 9.4 Important allergen families involved in pollen-fruit allergy syndrome (PFAS).

Allergen families	Primary sensitization	Cross-reactive allergens
Pathogenesis-related 10 (PR-10) family proteins	Bet v 1 (Birch), Que a 1 (Oak), Que i 1 (Holm oak)	Mal d 1 (apple), Pyr c 1 (pear), Pru ar 1 (apricot), Pru c 1 (plum), Pru p 1 (peach), Pru av 1 (cherry), Cor a 1 (hazelnut), Dau c 1 (carrot), Api g 1 (celery), or Gly m 4 (soy bean)
Profilins	Bet v 2 (Birch), Ole e 2 (olive), Art v 4 (mugwort), Amb a 8 (ragweed)	Act d 9 (kiwi), Ana c 1 (pineapple), Mal d 4 (apple), Pru p 4 (peach), banana (Mus xp 1), Cuc m 2 (melon), Api g 4 (celery), Dau c 4 (carrot), or Ara h 5 (peanut)
LTPs (lipid transfer proteins)	Ole e 7 (olive), Pla a 3 (Platanus pollen), Art v 3 (mugwort), Amb a 6 (ragweed)	Pru p 3 (peach), Mus a 3 (banana), Api g 2 (celery), Ara h 9 (peanut), Act d 10 (kiwi), Cor a 8 (hazelnut), Cuc m LTP (melon), or Sin a 3 (mustard)
Cysteine proteases	Fic c Ficin (Ficus)	Act d 1 (kiwi), Ana c 2 (pineapple), Car p 1 (papaya)

TABLE 9.5 Proteomic approaches for identification and quantification of fruit allergens.

Food	Allergen	Family	Proteomic approaches	References
Kiwi	Act d 1 Act d 5 Act d 11	Actinidin Kiwellin Ripening-related protein	MRM by LC-ESI-QTOF-MS/MS	Shih et al. (2019)
Apple	Mal d 1	PR-10	CPLL followed by SDS-PAGE-nanoLC-MS/MS 2-DE	Lerma-Garcia et al. (2019), Herndl et al. (2007)
	Mal d 2	TLP		
	Mal d 3	LTP		
	Mal d 4	Profilin		
Strawberry	Fra a 1	PR-10	1-DE immunobloting and nanoLC-MS/MS	Franz-Oberdorf et al. (2017)
Banana	Mus a 7	Catalase	CPLL followed by 2-DE immunoblotting and LC-MS/MS	Nikolić et al. (2018)
Mango	Man i 1	G3PDH	CPLL followed by 2-DE immunoblotting and nano LC-MS/MS	Cardona et al. (2018)
	Man i 2	G3PDH		
	Mus a 1	Profilin		
	Mus a 2	Chitinase		
	Mus a 5	β-1,3-glucanase		
Peach	Pru p 1	PR-10	2-DE immunoblotting and LC-MS/MS	Reuter et al. (2005), Pasini et al. (2012)

1-DE, one-dimensional gel electrophoresis; 2-DE, two-dimensional gel electrophoresis; CPLLs, combinatorial peptide ligand libraries; LC, liquid chromatography; LTP, lipid transfer proteins; MRM, multiple reaction monitoring; MS, mass spectrometry; MS/MS, tandem mass spectrometry; PR-10, pathogenesis-related 10; TLP, Thaumatin-like protein.

Thus, different allergens with high homology to Bet v 1, including Que a 1 (oak), Que i 1 (Holm oak), Car b 1 (hornbeam), Cor a 1 (hazel), or Aln g 1 (alder), have been identified by using ELISA and MS approaches (Hauser et al., 2011; Pedrosa et al., 2020; Wallner et al., 2009). In particular, the novel allergen Que i 1 was associated with PFAS symptoms related to banana, melon, apple, watermelon, walnut, pear, kiwi, lentils, and peach (Pedrosa et al., 2020), and it might be responsible for PFAS in birch pollen-free areas (Pedrosa et al., 2020). Proteomics has been also used to identify isoforms and variants of fruits allergens, such as the Bet v 1-homologous protein Pru p 1 in peach Pas (Pasini et al., 2012), Pru av 1 isoforms in cherries (Reuter et al., 2005), or Mal d 1 isoforms in apple (Helsper et al., 2002, Herndl et al., 2007).

Identification of allergenic proteins in foods requires to be aware that even trace amounts of fruit allergens are able to elicit allergic reactions. Therefore, the use of combinatorial peptide ligand libraries (CPLLs) for the enrichment of such low-abundance proteins is a common step before MS approaches (Lerma-García et al., 2019). Examples of this include the identification of new allergens in banana (Mus a 7) and mango (Cardona et al., 2018; Nikolić et al., 2018), and the main allergens in apple (Mal d 1, Mal d 2, Mal d 3, and Mal d 4) (Lerma-García et al., 2019). Other approaches such as MRM have been also used to quantify specific allergens in fruits, such as Act d 1, Act d 5, and Act d 11 in kiwi fruit (Shih et al., 2019).

Proteomics approaches were not only used for identification of allergens implicated in PFAS, but for the selection of hypoallergenic fruit varieties. For instance, Franz-Oberdorf et al. used in 2017 1D-immunoblot followed by LC-MS/MS to measure the Fra a 1 allergen (Bet v 1 homologue) and identificate hypoallergenic varieties of strawberries (Franz-Oberdorf et al., 2017). In addition, quantitative proteomics is suited for assessing the stability of different allergens in fruit food. For example, Act d 1 and Act d 5 kiwi allergens were decreased in kiwi jam and hot-air-dried kiwi compared to other kiwi-containing foods (i.e., pasteurized purée, juice, lyophilized kiwi, or high-pressure-processed purée) (Shih et al., 2019).

9.3.4.3 *Legume allergy*

Legumes are one of the major sources of protein in Mediterranean areas and also part of Africa and Asia. Legumes are seeds of plants from the Fabaceae family, and they are a heterogeneous group of foods with high allergenic potential. Peanuts and soybean account for the most significant food allergies in Europe and the United States, following by lupine, lentils, chickpeas, peas, or green beans. The high resistance to gastrointestinal enzymes in combination to their heat endurance could explain the association of legume allergens (class I food allergens) with severe anaphylactic reactions.

9.3.4.3.1 Peanut allergy

Peanut allergy is one of the most common food allergies in the world, affecting up to 4.5% of children (Shaker and Greenhawt, 2019). This allergy persists in 70%–80% of patients entering adulthood, being a major concern in developed countries and causing more severe health problems than other food allergies; for example, this type I hypersensitivity reaction (IgE-mediated allergy) is responsible for the 39% of anaphylaxis admissions in North America countries (i.e., Canada, United States, and Mexico) (Lieberman et al., 2021). Peanut's allergy can start locally after skin contact with peanut allergens or systemically after ingestion. Moreover, cross-reactions can also occur with other legumes and tree nuts.

To date, 18 allergens have been described that cause peanut allergy. These allergens belong to several families (WHO/IUIS Allergen Nomenclature Sub-Committee); http://www.allergen.org: cupins (Ara h 1, 3, 4), prolamins (2S-albumins; Ara h 2, 6, 7), oleosins (Ara h 10, 11, 13, 14, 15), LTPs (Ara h 9, 16, 17), defensins (Ara h 12, 13), cyclophilin (Ara h 18), profilins (Ara h 5), and PR-10 proteins (Ara h 8). Ara h 1, Ara h 2, Ara h 3, and Ara h 6 represent the major allergens, responsible for more than 50% of allergic reactions (Mueller et al., 2014; Ozias-Akins and Breiteneder, 2019; Bignardi et al., 2010; Chassaigne et al., 2007). Interestingly, 2-DE followed by immunoblotting and nLC-MS/MS has shown that 2S albumins Ara h 2 and Ara h 6 are highly resistant to pepsin digestion, whereas cupins (Ara h 1 and Ara h 3) are partially digested. Despite this difference, all of them maintain a high allergenic potential after simulated gastric digestion (Prodic et al., 2018).

New therapies, including the allergen immunotherapy, are currently available. However, avoiding the consumption of peanut-containing foods remains the most common solution for peanut allergy. Although ELISAs are widely used for peanut allergen detection (Costa et al., 2014; Pandey et al., 2019), the usage of MS approaches is currently increasing. Several proteomic approaches have been employed in order to detect trace amounts of peanuts proteins in food (Ogura et al., 2019; Bignardi et al., 2010) (Table 9.6). For example, Monaci et al. developed a method for peanut allergen detection in a mixture of nuts with a high sensitivity (4 mg/g of total protein) by using UHPLC-ESI-MS/MS (Monaci et al., 2015). In addition, by using a DIA method coupled with ion mobility (IM)-MS/MS, Johnson et al. have characterized the peanut allergen composition of raw and roasted peanut flours, being Ara h 1, Ara h 2, Ara h 3, Ara h 6, and Ara h 7 the most abundant allergens. Moreover, Ara h 1 and Ara h 3 levels decreased after roasting, but they still remained as the most abundant allergens. On the other hand, Ara h 8, Ara h 10, and Ara h 11 were moderately presented in raw peanuts and experienced a high reduction after roasting (Johnson et al., 2016). Importantly, Hebling et al. highlighted the necessity of untargeted LC-MS/MS approaches to select proper analytes for targeted approaches as thermal processing of peanuts generates allergen modifications (Hebling et al., 2013).

Proteomics was also used for the characterization of peanut proteome at different developmental stages (Li et al., 2020). Of note, Li et al. have concluded that allergen abundance in peanuts is associated with the maturity of seeds, being greater at later developmental stages (R5-R7). Hence, peanut seeds at early development (R1-R4) could be used to produce foods containing peanut protein for allergen-sensitive populations (Li et al., 2020).

TABLE 9.6 Proteomic approaches used for the study of legume allergens.

Food	Allergen	Family	Proteomic approaches	References
Peanut	Ara h 1	Cupin	-MRM by HPLC-MS/MS	Prodic et al. (2018), Johnson (2016), Li et al. (2020), Heick et al. (2011), Chassaigne et al. (2007), Bignardi et al. (2010)
	Ara h 2	2S-albumin	-2-DE followed by MALDI-TOF/TOF-MS	
	Ara h 3/4	Cupin	-DIA-IM-MS/MS	
	Ara h 6	2S-albumin	-nano-ESI Q-TOF MS/MS	
	Ara h 7	2S-albumin	-LC-ESI-LIT-MS	
	Ara h 8	PR-10	-2DE immunoblot followed by nLC-MS/MS	
	Ara h 9	LTP		
	Ara h 10	Oleosin		
	Ara h 11	Oleosin		
	Ara h 13	Defensin		
Soy	Gly m 6	Glycinin	-MRM by HPLC-MS/MS	Chen et al. (2019), Chen et al. (2021), McClain et al. (2018), Heick et al. (2011)
	Jug r1	2S-albumin	-PRM by UHPLC-MS/MS	
	Gly m 5	β-conglycinin	-SRM by nanoLC-MS/MS	
	Gly m Bd 28k	Vincilin		
	Gly m Bd 30k	Cysteine protease		
	Kunitz	Trypsin inhibitor		
	Trypsin Inhibitor 1	Trypsin inhibitor		
	Kunitz	Trypsin inhibitor		
	Trypsin Inhibitor 3	Trypsin inhibitor		
Lupin	Lup an 1	β-Conglutin (7S Vicilin)	-2-DE followed by LC-MS/MS	Islam et al. (2012)
	Lup an α-Conglutin	α-Conglutin (11S Globulin)		
	Lup an δ-Conglutin	δ-Conglutin (2S Albumin)		
	Lup an γ-Conglutin	γ-Conglutin (7S globulin)		
Lentil	Len c 1	Vincillin	-MRM by LC–MS/MS	Shaheen et al. (2019)
	Len c 2	LTP		

2-DE, two-dimensional gel electrophoresis; DIA, data independent acquisition; ESI, electrospray ionization; HPLC, high performance liquid chromatography; IM, ion mobility; LC, liquid chromatography; MALDI-TOF, matrix-assisted laser desorption ionization-time of flight; MRM, multiple reaction monitoring; MS, mass spectrometry; MS/MS, tandem mass spectrometry; PRM, parallel reaction monitoring; SMR, selected reaction monitoring; UHPLC, ultra high performance liquid chromatography.

9.3.4.3.2 Soybean allergy

After peanuts and wheat, soybean is the third plant responsible for the most significant food allergies in the United States and Europe (Sicherer and Sampson, 2010). Soybean is widely used in the world due to its high protein content, especially with the rise of the vegan diet. Soybean allergy (type-I hypersensibility reaction) has a prevalence of 0.4% (Sicherer and Sampson, 2010; Verma et al., 2013), but does not usually cause lifethreatening events.

To date, only eight soybean allergens have been described. They are listed in the database http://www.allergen.org (WHO/IUIS Allergen Nomenclature Sub-Committee): Gly m 1 (LTP), Gly m 2 (Defensin), Gly m 3 (Profilin), Gly m 4 (PR-10), Gly m 5 (β-conglycinin; 7S globulin), Gly m 6 (Glycinin; legumin; 11S globulin), Gly m 7 (Seed biotinylated protein), and Gly m 8 (2S albumin). Interestingly, soy allergenicity protein can be reduced by several thermal processing methods, including pressure cooking and boiling (Cabanillas et al., 2018), and also through non-thermal processes, such as high hydrostatic pressure (Li, Jia, et al., 2018).

The wide variety of soy-derived ingredients in combination with the high number of processing methods to produce soy-including foods makes allergen detection a real challenge. Hence, several proteomic studies were undertaken to identify soybean peptide markers stable after food processing, which could then potentially be used for further targeted analyses (Cucu et al., 2012) (Table 9.6). Indeed, targeted proteomic approaches (PRM) have been recently developed to quantify total soy protein in different types of processed food/matrices (Chen et al., 2019; Chen et al., 2021), showing a better performance than ELISA kits (Chen et al., 2021). Other targeted strategies, including SRM, have been also used to assess changes in allergen content between different soy cultivars (McClain et al., 2018), so that those with the lowest allergen content can be selected.

9.3.4.3.3 Lupin allergy

Lupin and lupin flour are valuable food products because of their potential health benefits, including their anticholesterolemic/antihypertensive properties as well as a high protein content (Islam et al., 2012; Siger et al., 2012). The use of lupin has, therefore, significantly increased during the last decade, in parallel to the incidence of the allergy to this legume. Moreover, between 28% and 35% of peanut-allergic patients also present lupin allergy, which indicates a high cross-reactivity between these two legumes (Moneret-Vautrin et al., 1999; Peeters et al., 2009). There are four main lupin species used as human food, including white lupin (*Lupinus albus*), very common in Mediterranean regions, pearl lupin (*Lupinus mutabilis*), commonly used in America, the narrow-leaved blue lupin or Australian sweet lupin (*Lupinus angustifolius*), with increasing use worldwide, and the European yellow lupin (*Lupinus luteus*).

Several allergens have been described so far in the different lupine species (Table 9.7; http://www.allergome.org /) (Jappe et al., 2021), but only one allergen from white lupine, Lup a 5 (Profilin), and two from narrow-leaved blue lupin, Lup an 1 (Conglutin beta; 7S seed storage globulin; vicilin) and Lup an 3 (nsLTP), are listed in the database from the WHO/IUIS Allergen Nomenclature Sub-Committee (http://www.allergen.org). Moreover, it has been demonstrated that these allergens are thermostable (Guillamón et al., 2008).

Similar to what happens with other allergens, different ELISAs have been developed for the detection of common lupin allergens. However, proteomic approaches have been evolved as well (Table 9.6). For example, by using 2-DE immunoblot followed by MALDI-MS/MS, Goggin D et al. have identified Lup an 1 as a major allergen in *L. angustifolius* (Goggin et al., 2008). A method for lupin seed flour allergens identification by using MALDI-TOF MS/MS was also described in 2021 (Calvano et al., 2021). In addition, Ma et al. have used MALDI-TOF MS and 2-DE approaches to study seed lupin proteins between different narrowed-leaved lupin cultivars in Australia (Islam et al., 2011, 2012). Recently, Czubinsky et al. have analyzed (2-DE followed by MS/MS) the proteome of *L. angustifolius* and *L. luteus* seeds after in vitro digestion protocols to mimic gastrointestinal processing of lupin proteins; alternatively, they employed hydrolysis with trypsin as well. In both cases, this group demonstrated that lupin major allergens, β-conglutins, are resistant to

TABLE 9.7 Description of allergens and protein families causing lupin allergy.

Specie	Allergen	Family
White Lupin (*Lupinus albus*)	Lup a 1	β-Conglutin (7S Vicilin)
	Lup a 4	PR-10 protein (Bet v 1-like)
	Lup a 5	Profilin
	Lup a α-Conglutin	α-Conglutin (11S Globulin)
	Lup a δ-Conglutin	δ-Conglutin (2S Albumin)
	Lup a γ-Conglutin	γ-Conglutin (7S globulin)
Narrow-leaved blue lupin (*Lupinus angustifolius*)	Lup an 1	β-Conglutin (7S Vicilin)
	Lup an 3	nsLTP
	Lup an α-Conglutin	α-Conglutin (11S Globulin)
	Lup an δ-Conglutin	δ-Conglutin (2S Albumin)
	Lup an γ-Conglutin	γ-Conglutin (7S globulin)
European Yellow Lupin (*Lupinus luteus*)	Lup l 4	PR-10 protein (Bet v 1-like)

hydrolysis (Czubinski et al., 2017). Finally, LC-ESI-MS/MS approaches were also used for the identification of conglutins in different food matices, including pasta and biscuits (Mattarozzi et al., 2012).

9.3.4.3.4 Lentil allergy

Lentils are another important source of proteins in Mediterranean countries and in some Asian regions. However, the number of studies on lentil allergens is limited compared to other legume species such as soybean or peanuts. So far, only three different allergens have been described, including the vicilin Len c 1, the seed biotinylated protein Len c 2, and the LTP Len c 3 (http://www.allergome.org/). To our knowledge, only a few proteomic studies have been carried out to study the proteome and allergen content of lentil seeds (Ialicicco et al., 2012; Scippa et al., 2010; Shaheen et al., 2019) (Table 9.6). However, Shaheen et al. have demonstrated the potential of MS approaches, particularly MRM based LC-MS/MS, to characterize the allergen content of different lentils cultivars (Shaheen et al., 2019). Moreover, they have also identified novel ideal peptides that could potentially be used for further targeted quantitative analyses (Shaheen et al., 2019).

9.3.4.4 *Tree nuts allergy*

Tree nuts allergy (type-I hypersensibility reactions; IgE-mediated allergy) is one of the most commonly reported food allergies in children and adults. Indeed, it was reported that up to 4.9% of the world's population suffers from this type of allergy (Geiselhart et al., 2018). The main tree nuts responsible for this type of allergic reactions are hazelnuts (*Corylus avellana*), followed by walnuts (*Juglans regia*), pecans (*Carya illinoinensis*), almonds (*Prunus dulcis*), cashews (*Anacardium occidentale*), pistachios (*Pistacia vera*), Brazil nuts (*Bertholletia excelsa*), coconuts (*Cocos nucifera*), and macadamia nuts (*Macadamia integrifolia*).

Tree nuts are rich on seed storage proteins, many of them highly allergenic. The majority of allergens causing tree nuts allergy belong to the prolamin superfamily (2S albumins and nsLTPs), vicilins, legumins, oleosins, PR-10 proteins (Bet v 1-like proteins), and profilins (Renz et al., 2018). In the case of hazelnut, only eight food allergens have been listed in WHO/IUIS database (http://www.allergen.org), including Cor a 2, Cor a 8, Cor a 9, Cor a 11, Cor a 12, Cor a 13, Cor a 14, and Cor a 15 (Table 9.8), although a thaumatin-like protein (Cor a TLP) has been reported to trigger severe allergic reactions (Palacín et al., 2012). Between them, Cor a 8, Cor a 9, Cor a 11, and Cor a 14 are thermostable and trigger severe allergic reactions (Geiselhart et al., 2018).

Similar to other allergens, several proteomic approaches have been used for the identification and quantification of these allergens (Table 9.8). Moreover, LC-MS/MS was used to characterize the allergenic potential of different hazelnut cultivars, showing minor differences between them (Korte et al., 2017a; Ribeiro et al., 2020). Proteomic analyses have been carried out not only to identify the different hazelnut allergens, but also to characterize them in terms of digestibility and allergenic potential after gastro-intestinal digestion (Korte et al., 2017b; Di Stasio et al., 2020). In this regard, Korte et al. have shown that Cor a 8 and Cor a 14 are highly resistant to GI digestion, the oleosins Cor a 12 and Cor a 13 display a two-step degradation, whereas Cor a 9 and Cor a 11 are intensively degraded after in vitro GI digestion. Moreover, they have also studied the immunogenic and sensitization capability of the different peptides identified by MS after GI digestion, being the highest immunogenic allergen Cor a 9 (although there were only available epitope data for Cor a 9, Cor a 11, and Cor a 14) (Korte et al., 2017b). Finally, by using MALDI-TOF/MS, it has been shown that Cor a 9 and Cor a 11 are insolubilized by high hydrostatic pressure (Prieto et al., 2014), highlighting the potential effect of different cooking techniques in the allergenic potential of the diverse food products.

Walnut is the second most common tree nut causing allergic reactions, with special prevalence in the United States (Sicherer et al., 2010). Only three allergens have been described in black walnut (*Juglans nigra*), including Jug n 1 (2S albumin), Jug n 2 (Vicilin) and Jug n 4 (Legumin), although the most common walnut cultivated in the world is the English walnut (*Juglans regia*) (http://www.allergen.org). The International Union of Immunological Societies included to date 8 different allergens belonging to English walnut (http://www.allergen.org; Table 9.8). All these different allergens have been demonstrated to have IgE-binding epitopes (Geiselhart et al., 2018). Jug r 1 is the major allergen in walnut (>80% of walnut allergic patients contain sIgE against Jug r 1), and linear epitopes are present in certain digestion-resistant peptides (e.g., AA16-30 and AA125-139) (Guo et al., 2020). Only Jug r 2 was shown to elicited $CD4^+$ T-cell responses (Archila et al., 2015). Different targeted proteomic approaches were used for identification and quantification of walnut allergens, including MRM and MRM^3 based LC-MS/MS (Table 9.8). LC-MS/MS strategies were also used to study the effect of thermal processing on the digestibility and allergenicity of walnut allergens (Downs et al., 2016). Similarly to other allergens, thermal processing of walnuts results in alterations of their digestibility, leading to changes in detection levels by using LC-MS/MS (Downs et al., 2016). For example, it has been described that the measurement of Jur r 2 (7S globulin) and Jur r 4 (11S globulin) by label-free LC-MS/MS is significantly increased after roasting (Downs et al., 2016). Pecan (*Carya illinoinensis*) is another tree nut from same family as walnut, the Juglandaceae family. Three different pecan allergens have been included in the WHO/IUIS Allergen database (http://www.allergen.org; Table 9.8), two of them with a high

TABLE 9.8 Proteomic approaches used for the study of tree nuts allergens.

Tree nut	Allergen[a]	Allergen family	Proteomic approaches	References
Hazelnut	Cor a 2	Profilin	MRM by HPLC-MS/MS	Costa et al. (2014), Calvano et al. (2021), Heick et al. (2011), Korte et al. (2017b), Korte and Brockmeyer, 2016, Calvano, De Ceglie, & Zambonin, 2014
	Cor a 8	nsLTP	SRM by UHPLC-MS/MS	
	Cor a 9	11S-albumin (Legumin)	MALDI-TOF-TOF LC-ESI-MS/MS	
	Cor a 11	7S-globulin (vicilin)	MALDI-TOF-TOF	
	Cor a 12	17 kDa oleosin	MRM^3-based LC-MS/MS	
	Cor a 13	14–16 kDa oleosin	1DE followed by LC-MS/MS	
	Cor a 14	2S albumin		
	Cor a 15	Oleosin		
Walnut	Jug r 1	2S-albumin	MRM by HPLC-MS/MS	Heick et al. (2011), Korte and Brockmeyer, 2016
	Jug r 2	Vicilin	MRM^3-based LC-MS/MS	
	Jug r 3	nsLTP		
	Jug r 4	Legumin		
	Jug r 5	PR-10		
	Jug r 6	Vicilin-like cupin		
	Jug r 7	Profilin		
	Jug r 8	Ns-LTP-2		
Pecan	Car i 1	2S-albumin	MRM by LC-MS/MS	New, Schreiber, Stahl-Zeng, & Liu, 2018, Sealey-Voyksner et al. (2016)
	Car i 2	Vicilin		
	Car i 4	Legumin		

Almond	Pru du 3	nsLTP-1	MRM by HPLC-MS/MS	Heick et al. (2011), Inman et al. (2018), Korte & Brockmeyer, 2016, Korte, Lepski, & Brockmeyer, 2016
	Pru du 4	Profilin	SRM by LC-MS/MS	
	Pru du 5	60s acidic ribosomal prot. P2	MRM^3-based LC-MS/MS	
	Pru du 6	Amandin	LC-HRMS/MS	
	Pru du 8	Antimicrobial seed storage protein		
	Pru du 10	Mandelonitrile lyase 2		
Cashew	Ana o 1	Vicilin	LC-MS/MS and ELISA	New, Schreiber, Stahl-Zeng, & Liu, 2018, Sealey-Voyksner et al. (2016), Mattison et al. (2016)
	Ana o 2	Legumin	MRM by LC-MS/MS	
	Ana o 3	2S-albumin	MRM by LC-MS/MS	
Pistachio	Pis v 1	2S-albumin	MRM by LC-MS/MS	New, Schreiber, Stahl-Zeng, & Liu, 2018, Sealey-Voyksner et al. (2016)
	Pis v 2	11S-globulin	MRM by LC-MS/MS	
	Pis v 3	Vicilin		
	Pis v 4	manganese superoxide dismutase		
	Pis v 5	11S-globulin		

[a]In bold are highlighted the allergens detected by the different proteomic approaches. 1-DE, one-dimensional gel electrophoresis; ESI, electrospray ionization; HRMS, High resolution mass spectrometry; HPLC, high performance liquid chromatography; LC, liquid chromatography; MALDI-TOF, matrix-assisted laser desorption ionization-time of flight; MRM, multiple reaction monitoring; MS, mass spectrometry; MS/MS, tandem mass spectrometry; SMR, selected reaction monitoring; UHPLC, ultra high-performance liquid chromatography; UPLC, ultra performance liquid chromatography.

similarity to walnut allergens: Car i 1 (88% identity with Jug r 1) and Car i 4 (95% identity with Jug r 4); obviously this explains the high association between pecan and walnut allergies (Andorf et al., 2017). Car i 2 (vicilin) is, however, the allergen mostly used for proteomic detection by MRM-based LC-MS/MS analyses (Sealey-Voyksner et al., 2016; Sun New et al., 2018).

Almond (*Prunus dulcis*) was also been shown to generate allergic reactions. Six are the allergens collected in the WHO/IUIS Allergen Nomenclature Sub-Committee database (http://www.allergen.org): Pru du 3 (nsLTP-1), Pru du 4 (profilin), Pru du 5 (60s acidic ribosomal prot. P2), Pru du 6 (Amandin, 11S globulin legumin-like protein), Pru du 8 (Antimicrobial seed storage protein), Pru du 10 (Mandelonitrile lyase 2). Inman et al. in 2015 have developed an LC-MS method for the detection of marker peptides (necessary for subsequent target analyses) of almond in different spices, including cumin and paprika (Inman et al., 2018). Moreover, different targeted proteomic approaches were used for almond allergens detection in several food matrices (Heick et al., 2011; Inman et al., 2018; Korte et al., 2016; Korte and Brockmeyer, 2016 (Table 9.8).

Another tree nut allergy with increasing prevalence is cashew (*Anacardium occidentale*) allergy (Johnson et al., 2014). Cashew allergens, Ana o 1 (vicilin), Ana o 2 (legumin), and Ana o 3 (2S albumin), can cause severe allergic reactions in minutes (van der Valk et al., 2016). Several proteomic studies have been carried out to study cashew allergens (Table 9.8). For example, Mattison et al. have use LC-MS/MS approaches to evaluate the effect of heat processing in cashew allergens, showing a clear alteration of Ana o 1 and Ana o 2, but no changes in Ana o 3 (Mattison et al., 2016). Another tree nut from Anarcadiaceae family is Pistachio (Pistacia vera). There are five different allergenic proteins from pistachio recognized by WHO/IUIS committee (http://www.allergen.org), including Pis v 1 (2S albumin), Pis v 2 (legumin), Pis v 3 (vicilin), Pis v 4 (iron/manganese superoxide dismutase), and Pis v 5 (legumin) (Table 9.8). LC-MS/MS identification of IgE-reactive allergens evidences the high cross-reactivity between cashew and pistachio allergens (Andorf et al., 2017).

Other less prevalent tree nuts causing allergic reactions are Macadamia nut (*Macadamia integrifolia*) and Brazilian nut (*Bertholletia excelsa*) (Rost et al., 2020). Two allergens from Macadamia nut, Mac i 1 (vicilin) and Mac i 2 (legumin), and two from Brazilian nut, Ber e 1 (2S-albumin) and Ber e 2 (11S-globulin), are included in WHO/IUIS database (http://www.allergen.org). However, proteomic studies in these nut species are clearly under-represented in the literature compared to the other tree nuts previously commented (Rost et al., 2020; Sealey-Voyksner et al., 2016).

Nowadays, scientific community is putting great efforts in developing proteomic approaches that allow the simultaneous detection and quantification of multiple allergens. This is the case of the group of Korte et al., which used high-resolution MS for the identification of peptide markers from almond, hazelnut, cashew, peanut, pistachio, and walnut (Korte et al., 2016). MALDI-TOF MS was also used for the identification of allergens in hazelnut and lupin seeds (Damiana Calvano et al., 2014) (Calvano et al., 2021). Regarding quantitative proteomics, MRM-based LC-MS/MS strategy is still the most used methodology for multiple allergen detection (Planque et al., 2017). New et al. optimized an MRM-based LC-MS/MS procedure for the screening and quantification of almond, Brazil nut, cashew, hazelnut, pecan, pine nut, pistachio, walnut, peanut, and soy allergens (Sun New et al., 2018). In addition, the group of Voyksner et al. used a similar approach for detecting almond, cashew, peanut, pecan, hazelnut, pine nut, Brazil nut, macadamia nut, pistachio nut, walnut, chestnut, and coconut allergenic proteins (Sealey-Voyksner et al., 2016). For their part, Korte et al. developed in 2016 a highly sensitive proteomic method for the targeted detection and quantification of trace amounts of multiple nut allergens by using a multiple reaction monitoring cubed (MRM^3)-based LC-MS/MS approach (Korte and Brockmeyer, 2016). This methodology showed a sensitivity 30-fold higher than the regular MRM method (limit of detection of 1 μg/g in fortified matrix samples). Another advantage was its higher inter-sample reproducibility, allowing a precise quantification of allergens from almond, cashew, hazelnut, peanut, pistachio, and walnut (Korte and Brockmeyer, 2016).

9.3.5 Proteomic studies and allergens in animals

9.3.5.1 Cow's milk allergy

Cow's milk allergy is one of the most common food allergies during childhood, achieving a prevalence slightly lesser than 3% in children under 2 years old (D´Auria et al., 2019). This food allergy tends to disappear in most patients by the third year of life (D'Auria et al., 2018). Moreover, there is an overrepresentation of women in those patients in whom this allergy still persists into adulthood (Wal, 2002). This hypersensitivity affects the skin as well as the respiratory and gastrointestinal systems, and arises due to failure in oral tolerance mechanisms. Only 60%–75% of patients develop an IgE-mediated immune response against one (25% cases) or several cow's milk proteins (75% cases) (Wal, 2002). The remaining 25%–40% show negative results in skin prick tests, absence of lactose intolerance, and diverse immunopathogenic mechanisms (type II, III, and IV hypersensitivity) with delayed symptoms (1h to several days) (Hochwallner et al., 2010; 2011). A correct diagnosis of this food allergy is crucial to avoid severe reactions in hypersensitized patients, or unnecessary dietary exclusion of milk in healthy children.

Bovine milk contains about 3.3% protein. Acidification of milk at pH 4.6 or ultracentrifugation generates two major fractions in milk proteins:

(1) the *caseins* (coagulum; curd), accounting for 80% of total bovine milk proteins and including α-S1-casein (Bos d 9, 32% of casein proteins), α-S2-casein (Bos d 10, 10%), β-casein (Bos d 11, 28%), and κ-casein (Bos d 12, 10%); (2) the *whey proteins* (lactoserum), representing 20% of total bovine milk proteins and mainly consisting of globular proteins synthesized in the mammary gland, such as α-lactalbumin (Bos d 4) and β-lactoglobulin (Bos d 5), or coming from the blood, like bovine serum albumin (BSA, Bos d 6), immunoglobulins (Bos d 7), or lactoferrin (Bos d lactoferrin) (D'Auria et al., 2018; Wal, 2002). Protein composition of human skimmed milk differs remarkably from the corresponding in other mammals, and one major reason is the reduced content in caseins and β-lactoglobulin (D'Auria et al., 2005; Wal, 2002; Ansari et al., 2011).

Target proteins of sIgE in cow's milk allergy have been identified by different proteomic methods. They include the highly allergenic protein β-lactoglobulin, which remains intact after digestion due to its relative resistance to acid hydrolysis/proteases and its β-barrel structure (Wal, 2002), but also caseins, which have been considered more susceptible to proteases during food digestion and therefore less immunogenic (Wal, 2002). Genetic polymorphisms and posttranslational modifications (e.g., glycosylation, phosphorylation) of milk proteins can also potentiate this immunogenicity (e.g., caseins) (Wal, 2002). Natale et al. used 2-DE analysis of skimmed cow's milk followed by immunoblotting with sIgE-containing sera from allergic children and peptide mass fingerprinting (PMF) (Natale et al., 2004) (Table 9.9) to show multiple sensitization (2-7 allergens) and several milk glycoproteins as the major allergens: Bos d 12/ κ-casein (50% of prevalence of sensitization), Bos d 9/α-S1-casein (55%), and especially Bos d 10/α-S2-casein (90%) within the casein fraction; Bos d 5/β-lactoglobulin (45%), Bos d 6/BSA (45%), Bos d lactoferrin (50%), and especially Bos d 7/ immunoglobulin (95%), within the whey proteins (Natale et al., 2004). A good correlation has been observed between SPT, oral challenge, and immunoproteomics results (Natale et al., 2004), as well as between sensitivities to caseins, α-lactalbumin, and β-lactoglobulin (Wal, 2002).

As we have just seen, any protein in the milk has the potential to become a food allergen, even those in low amounts (e.g., BSA, lactoferrin, and IgG heavy chain) but with a HMW (i.e., higher likelihood of harboring more antigenic peptides) (D'Auria et al., 2005; Wal, 2002); in clear contrast, predominant proteins such as α-lactalbumin (containing many B-cell conformational epitopes) or β-casein appear to be less IgE-reactive (Natale et al., 2004). In this sense, the refinement of sample fractionation and enrichment techniques prior to proteomic analysis (e.g., CPLLs) has facilitated the detection of those low abundant whey proteins that react with IgE, such as bovine immunoglobulins (D'Amato et al., 2009).

Raw milk is very perishable and must be subjected to heat treatment to prolong shelf-life and preserve microbiological safety. Usually, raw

TABLE 9.9 Main proteomics studies aimed to the detection and quantification of new food allergens in animal foods: a non-exhaustive list.

Food	Type hypersensitivity (HS) reaction	Allergens (% de sensitized patients)	Proteotypic peptides used in SRM/MRM/PRM/SMIM assays or specific peptides detected	Proteomic Technology	Reference PMID (year)
Cow´s milk (skimmed)	Type I (immediate HS; IgE)	**Proteins highly affected by heat-dependent lactosylation: Caseins.** αS1-casein (Bos d 9) β-casein (Bos d 11) **Proteins highly affected by heat-dependent lactosylation: Whey proteins.** α-lactalbumin (Bos d 4) β-lactoglobulin (Bos d 5)	n.d.	Colorimetric, ELISA, 1-DE/immunoblots, MS, and protein sequence analyses (Edman)	Scaloni et al. (2002)
Cow´s milk (skimmed)	Type I (immediate HS; IgE)	**Caseins** αs1-casein (Bos d 9) (55%) αs2-casein (Bos d 10) (90%) β-casein (Bos d 11) (15%) κ-casein (Bos d 12) (50%) **Whey proteins** α-lactalbumin (Bos d 4) (0%) β-lactoglobulin (Bos d 5) (45%) BSA (Bos d 6) (45%) Immunoglobulin (Bos d 7) (95%) Lactoferrin (Bos d lactoferrin) (50%)	n.d.	2-DE + immunoblotting with patient sera + MALDI-TOF (PMF)	Natale et al. (2004)

(Continued)

TABLE 9.9 Cont'd

Food	Type hypersensitivity (HS) reaction	Allergens (% de sensitized patients)	Proteotypic peptides used in SRM/MRM/PRM/SMIM assays or specific peptides detected	Proteomic Technology	Reference PMID (year)
Cow´s milk (skimmed)	Type I (immediate HS; IgE)	**Caseins** αs1-casein (Bos d 9) (n.d.) β-casein (Bos d 11) (n.d.)	YLGYLEQLLR, FFVAPFPEVFGK GPFPIIV, VLPVPQK	LC-MS/MS (QTrap); SRM	Ansari et al. (2011)
		Whey proteins α-lactalbumin (Bos d 4) (n.d.) β-lactoglobulin (Bos d 5) (n.d.)	VGINYWLAHK IPAVFK, **LIVTQTMK** (cattle-specific)		
Cow´s milk (detection of caseins in white wines)	Type I (immediate HS; IgE)	**Caseins** αs1-casein (Bos d 9) (n.d.) β-casein (Bos d 11) (n.d.)	n.d.	LC/HRMS (Orbitrap)	Monaci et al. (2013)
Cow´s milk (whey proteins in swine meat or high protein processed food product)	Type I (immediate HS; IgE)	**Whey proteins** β-lactoglobulin (Bos d 5) (n.d.)	IDALNENK, TPEVDDEALEK	LC-HRMS/MS (qOrbitrap, untargeted/DDA & PRM)	Stella et al. (2020)
Cow´s milk (skimmed)	Type I (immediate HS; IgE)	**Caseins** αS1-casein (Bos d 9) (n.d.)	FFVAPFPEVFGK, YLGYLEQLLR	LC-MS/MS (QQQ; MRM)	Monaci et al. (2020)
		Whey proteins β-lactoglobulin (Bos d 5) (n.d.)	TPEVDDEALEK, VLVLDTDYK		

Food	Type hypersensitivity (HS) reaction	Allergens (% de sensitized patients)	Proteotypic peptides used in SRM/MRM/PRM/SMIM assays or specific peptides detected	Proteomic Technology	Reference PMID (year)
Red meat	Type IV (delayed HS; IgE)	**IgE-reactive proteins** Laminin γ-1 (n.d.) Collagen α-1 (VI) chain (n.d.)	n.d.	2-DE + immunoblotting with patient sera or anti-α-Gal antibody + MALDI-TOF/TOF (PMF/PFF)	Takahashi et al. (2014)
Red meat	Type IV (delayed HS; IgE)	**IgE-reactive proteins (18 proteins)** Adenylate kinase isoenzyme 1, albumin, alpha-enolase, aspartate aminotransferase, beta-enolase, carbonic anhydrase 3, creatine kinase M-type, fructose-biphosphate aldolase A, fructose-biphosphate aldolase C, glycerol-3-phosphate dehydrogenase, heat shock 70 kDa protein 2, hemoglobin subunit beta, lactate dehydrogenase A, myoglobin, phophoglycerate mutase 2, phosphatidylethanolamine-binding protein 1, phosphoglycerate kinase 1, triosephosphate isomerase (n.d.) **IgE-reactive proteins that contain α-Gal (3 proteins)** Triosephosphate isomerase, carbonic anhydrase 3, lactate dehydrogenase A (n.d.)	n.d.	2-DE + immunoblotting with patient sera or	Apostolovic et al. (2014)

(Continued)

TABLE 9.9 Cont'd

Food	Type hypersensitivity (HS) reaction	Allergens (% de sensitized patients)	Proteotypic peptides used in SRM/MRM/PRM/SMIM assays or specific peptides detected	Proteomic Technology	Reference PMID (year)
		IgE-reactive thermostable proteins that contain α-Gal (4 proteins)			
		Creatine kinase M-type, aspartate aminotransferase, α-enolase, β-enolase (n.d.)			
Egg	Type I (immediate HS; IgE)	**Egg white**	n.d.	2-DE + immunoblotting with patient sera + LC-MS/MS (IT or MALDI-TOF/TOF; DDA); RT-PCR	Suzuki et al. (2010)
		Ovotransferrin (OTr, Gal d 3) (74%)			
		Lysozyme (Lsz, Gal d 4) (50%)			
		Ch21 protein (n.d.)			
		Lipocalin-type prostaglandin D synthase (L-PGDS) (16%)			
		Cystatin (53%)			
Egg	Type I (immediate HS; IgE)	**Egg white**	n.d.	CPLLs + 1-DE/2-DE + immunoblotting with patient sera + LC-MS (MALDI-TOF/TOF; PMF)	Martos et al. (2013)
		Ovomucoid (OM; Gal d 1) (n.d.)			
		Ovoalbumin (OVA; Gal d 2) (n.d.)			
		Lysozyme (LYS; Gal d 4) (n.d.)			
		Ovoinhibitor (OvoI) (n.d.)			
		Clusterin (n.d.)			

Food	Type hypersensitivity (HS) reaction	Allergens (% de sensitized patients)	Proteotypic peptides used in SRM/MRM/PRM/SMIM assays or specific peptides detected	Proteomic Technology	Reference PMID (year)
Egg (detection of egg white proteins in white wines)	Type I (immediate HS; IgE)	**Egg white** Ovoalbumin (OVA; Gal d 2) (n.d.) Lysozyme (LYS; Gal d 4) (n.d.)	n.d.	LC/HRMS (qOrbitrap)	Monaci et al. (2013)
Egg (detection of egg white proteins in commercial ciders)	Type I (immediate HS; IgE)	**Egg white** Lysozyme (Lsz, Gal d 4) (n.d.)	n.d.	1-DE + LC-MS/MS (qOrbitrap, untargeted/DDA) 1-DE + immunoblotting	Mainente et al. (2017)
Egg	Type I (immediate HS; IgE)	**Egg yolk** Chicken serum albumin (α-livetin; Gal d 5) (n.d.) Vitellogenin I (YGP42; Gal d 6) (n.d.) Vitellogenin II (YGP40) (39.6%) Apolipoprotein B (n.d.)	n.d.	Dot-blotting; 1-DE + immunoblotting with patient sera + LC-MS/MS	Sogawa et al. (2018)
Egg (white)	Type I (immediate HS; IgE)	**Egg white** Ovomucoid/Gal d 1 (n.d.) Ovoalbumin/Gal d 2 (n.d.) Ovotransferrin/Gal d 3 (n.d.) Lysozyme/Gal d 4 (n.d.)	n.d.	Immunoblots (Gal d1-4) and peptide microarrays (Gal d 1 and 3) to detect sIgE/sIgG4 in allergic and tolerant individuals to egg proteins	Ehlers et al. (2020)

(Continued)

TABLE 9.9 Cont'd

Food	Type hypersensitivity (HS) reaction	Allergens (% de sensitized patients)	Proteotypic peptides used in SRM/MRM/PRM/SMIM assays or specific peptides detected	Proteomic Technology	Reference PMID (year)
Egg (Egg white proteins in in swine meat or high protein processed food product)	Type I (immediate HS; IgE)	**Egg white** Ovoalbumin (Gal d 2) (n.d.)	GGLEPINFQTAADQAR, HIATNAVLFFGR	LC-HRMS/MS (qOrbitrap, untargeted/DDA & PRM)	Stella et al. (2020)
Egg proteins (in cookies)	Type I (immediate HS; IgE)	**Egg white** Ovalbumin (Gal d 2) (n.d.) **Egg yolk** Vitellogenin II (YGP40) (n.d.)	ISQAVHAAHAEINEAGR, GGLEPINFQTAADQAR NIPFAEYPTYK, NIGELGVEK	LC-MS/MS (QQQ; MRM)	Monaci et al. (2020)
Fish (cod, tuna, salmon, perch, carp, and eel muscle extracts)	Type I (immediate HS; IgE)	Parvalbumin (12 kDa cod Gad c 1) (100%). High cross-reactivity with parvalbumin from tuna, salmon, perch, carp, and eel) Other minor allergens (30%).	n.d.	1-DE + immunoblotting with sera from patients or a nonallergic individual	Bugajska-Schretter et al. (1998)
Fish (carp, catfish, cod, tilapia, and yellowfin tuna muscle extracts).		Parvalbumins. Detection in carp, catfish, cod, and tilapia, but not yellowfin tuna	n.d.	2-DE + immunoblotting with anti-frog parvalbumin mouse IgG mAb	Chen et al. (2006)

Food	Type hypersensitivity (HS) reaction	Allergens (% de sensitized patients)	Proteotypic peptides used in SRM/MRM/PRM/SMIM assays or specific peptides detected	Proteomic Technology	Reference PMID (year)
Fish (sarcoplasmic proteins from species/subspecies of the *Merlucciidae* family)	Type I (immediate HS; IgE)	Parvalbumin (β-PRVB): Complete (n= 25; 100% coverage) and partial (n=16; 53.7-90.7% coverage) sequences of new parvalbumins from 13 different species/subspecies of the *Merlucciidae* family	n.d.	Bottom-up (2-DE; LC-ESI-IT-MS/MS DDA; FTICR-MS; SMIM)	Carrera et al. (2010)
Fish (*Merlucciidae* family)	Type I (immediate HS; IgE)	Parvalbumin. Use of high intensity focused ultrasound (HIFU) to accelerate in-solution digestions with trypsin and a targeted proteomic methodology in a linear ion trap (LC-MS/MS SMIM) to detect 11 parvalbumin peptides and identify different commercial species from the *Merlucciidae* family in any seafood products in less than 2 h	F-MER604 (*Merlucciidae* family; LFLQVFSAGAR); G-MER517 (*Merluccius* genus; VFGIIDQDK); G-MAC524 (*Macruronus* genus; VFAIIDQDK); S-MER967 (American hakes; AGDSDGDGAIGVDEFAVLVK); S-MER794 (*Merluccius bilinearis*; N-Acyl-AFSGILADADIAAALK); S-MER612 (*Merluccius productus*; IGVDEFTAMLK); S-MER736 (*Merluccius australis australis, M. australis australis polylepis*; AEGTFTHGEFFTK); S-MER721 (*Merluccius australis australis*; AEGTFTHGVFFTK); S-MER590 (*Merluccius capensis or M. senegalensis*; IGVDEFAAMVK); S-MER973 (*Merluccius paradoxus*; AGDSDGDGAIGVDEWAA LVK); S-MER987 (*Merluccius poli*; AGDSDGDGAIGVDEWAV LVK)	Targeted proteomics (HIFU-LC-MS/MS SMIM)	Carrera et al. (2011)

(Continued)

TABLE 9.9 Cont'd

Food	Type hypersensitivity (HS) reaction	Allergens (% de sensitized patients)	Proteotypic peptides used in SRM/MRM/PRM/SMIM assays or specific peptides detected	Proteomic Technology	Reference PMID (year)
Fish	Type I (immediate HS; IgE)	Parvalbumin. Use of 19 biomarker peptides (11 exclusive to different fish species and 5 shared with organisms such as frogs), HIFU, and a targeted proteomic methodology in a linear ion trap (LC-MS/MS SMIM) to rapidly identify fish β-parvalbumins in any foodstuff product in less than 2 h. The peptides encompassed residues 46-77, which correspond to the Ca^{2+} binding domain (EF-hand), important in the immunoreactivity of these proteins.	Fish-specific parvalbumin peptides: B3 (SDFVEEDELK), B5 (LFLQTFSAGAR), B7 (VIDQDASGFIEVEELK), B9 (FFAIIDQDHSGFIEEEELK), B10 (LFLQNFAAGAR); B11 (LFLQNFCPK), B12 (AFAIIDQDNSGFIEEDELK), B13 (LFLQTFGAGAR), B14 (EGFIEEDELK), B15 (AFAIIDQDNSGFIEEEELK), B16 (AFAIIDQDISGFIEEEELK), B17 (AFHLLDADNSGFI EEEELK)	Targeted proteomics (HIFU-LC-MS/MS SMIM)	Carrera et al. (2012)
Fish (longtail tuna/*Thunnus tonggol* muscle extracts)		Parvalbumin (10-13 kDa; thermoresistant) (100%) Creatine kinase (42 kDa; thermolabile) (55.2%) Enolase (51 kDa; thermolabile) (52.2%)	n.d.	1-DE/2-DE + immunoblotting with sera from fish allergic patients and from non-allergic subjects+ PMF (MALDI-TOF) + PFF (qTOF)	Rosmilah et al. (2013)

Food	Type hypersensitivity (HS) reaction	Allergens (% de sensitized patients)	Proteotypic peptides used in SRM/MRM/PRM/SMIM assays or specific peptides detected	Proteomic Technology	Reference PMID (year)
Fish (muscle extracts from *Lates niloticus* and *Gadus morhua*)		**Nile perch (*Lates niloticus*)** Enolase 3 (beta, muscle) (Lat n 2) (66.6%); Creatine kinase 1 (Lat n 4.2) (25%); Fructose-bisphosphate Aldolase A (Lat n 5) (8%); Fast myotomal muscle actin (Lat n 3) (8%); Creatine kinase muscle isoform 2 (Lat n 4.1) (8%); Apolipoprotein AI (Lat n 6) (8%); Adenylate kinase 1-2 (Lat n 7) (8%); Phosphoglucomutase 1 (Lat n 1) (8%) **Cod (*Gadus morhua*)** Alpha-enolase (Gad m 2.1) (100%); Enolase 3-1 (Gad m 2.2) (100%); Tropomyosin (Gad m 3) (8%); Myosin light chain 1 (Gad m 5) (8%); Nucleoside diphosphate kinase B (Gad m 6) (8%)	n.d.	2-DE + immunoblotting with sera from fish allergic patients and from a non-atopic subject+ LC-MS/MS (Q Orbitrap)	Tomm et al. (2013)

(Continued)

TABLE 9.9 Cont'd

Food	Type hypersensitivity (HS) reaction	Allergens (% de sensitized patients)	Proteotypic peptides used in SRM/MRM/PRM/SMIM assays or specific peptides detected	Proteomic Technology	Reference PMID (year)
Fish (muscle; cod, salmon, and tuna)	Type I (immediate HS; IgE)	Cod parvalbumin (thermostable) (72.6%) Cod fructose-bisphosphate aldolase A (thermolabile) (50%) Cod β-enolase (thermolabile) (62.9%)	n.d.	1-DE + immunoblotting with sera from fish allergic patients or non-atopic subjects + N-terminal Edman sequencing + PMF (MALDI-TOF); ELISA	Kuehn et al. (2013)
Fish (muscle extracts from gilthead sea bream/*Sparus aurata*)	Type I (immediate HS; IgE)	Parvalbumin: Higher content in farmed fish compared to wild specimens.	n.d.	LC-MS/MS (shotgun proteomics; label-free or dimethyl labeling)	Piovesana et al. (2016)
Fish	Type I (immediate HS; IgE)	Parvalbumins beta (β-PRVBs)	B-cell epitopes LKLFLQV, ACAHLCK, FAVLVKQ, and LFLQNFV	LC-MS/MS (Shotgun proteomics; LTQ-Orbitrap) + bioinformatic analysis + ELISA	Carrera et al. (2019)

Food	Type hypersensitivity (HS) reaction	Allergens (% de sensitized patients)	Proteotypic peptides used in SRM/MRM/PRM/SMIM assays or specific peptides detected	Proteomic Technology	Reference PMID (year)
Fish (farmed gilthead seabream)	Type I (immediate HS; IgE)	Parvalbumin: Reduced abundance of parvalbumin in muscle proteome and lower IgE-reactivity in farmed fish on EDTA-supplemented diets	n.d.	1-DE + immunoblotting with sera from fish allergic patients. 2-DE (DIGE). Protein identification by MALDI-TOF/TOF (PMF/PFF)	De Magalhaes et al. (2020)
Fish (Atlantic salmon and *Pangasius*/ catfish)	Type I (immediate HS; IgE)	**IgE-reactive proteins in atlantic Salmon (*S. salar*)** Heat-stable allergens: Parvalbumin (Sal s 1; 49%), tropomyosin (Sal s 4; 13%), and collagen α (Sal s 6) Heat-labile allergens: Enolase (Sal s 2; 34%), aldolase (Sal s 3; 26%), creatine kinase (Sal s 7; 14%), triose phosphate isomerase (Sal s 8; 34%) **IgE- reactive proteins in catfish (*P. hypophthalmus*)** Heat-stable allergens: Parvalbumin (Pan h 1; 52-58%), tropomyosin (Pan h 4; 38%)	n.d.	1-DE + immunoblotting with sera from fish allergic patients or non-atopic/atopic but tolerant controls. Protein identification by MS	Ruethers et al. (2021)

(Continued)

TABLE 9.9 Cont'd

Food	Type hypersensitivity (HS) reaction	Allergens (% de sensitized patients)	Proteotypic peptides used in SRM/MRM/PRM/SMIM assays or specific peptides detected	Proteomic Technology	Reference PMID (year)
		Heat-labile allergens: enolase (Pan h 2; 21%), aldolase (Pan h 3; 21%), creatine kinase (Pan h 7; 10%), triose phosphate isomerase (Pan h 8; 19%), pyruvate kinase (Pan h 9; 6%), L-lactate dehydrogenase (Pan h 10; 13%), and glucose-6-phosphate isomerase (Pan h 11; 8%)			
Shellfish (Shrimp; sarcoplasmic extracts from black tiger shrimp; *Penaeus monodon*)	Type I (immediate HS; IgE)	Tropomyosin (72%) Arginine kinase (27%)	n.d.	2-DE + immunoblotting with sera from shrimp allergic patients. Protein identification by MALDI-TOF (PMF) and postsource decay (PSD) analyses.	Yu et al. (2003)
Shellfish (Shrimp species; sarcoplasmic extracts)	Type I (immediate HS; IgE)	Arginine kinase from: *Penaeus monodon* (giant tiger prawn) *Litopenaeus vannamei* (pacific white shrimp); *Fenneropenaeus indicus* (indian white prawn); *Farfantepenaeus notialis* (southern pink shrimp) *Pleoticus muelleri* (argentine red shrimp) *Pandalus borealis* (northern shrimp)	n.d.	Bottom-up proteomics (2-DE; MALDI-TOF MS, PMF; ESI- MS/MS, IT, *de novo*)	Ortea et al. (2009a)

Food	Type hypersensitivity (HS) reaction	Allergens (% de sensitized patients)	Proteotypic peptides used in SRM/MRM/PRM/SMIM assays or specific peptides detected	Proteomic Technology	Reference PMID (year)
Shellfish (Shrimp species)	Type I (immediate HS; IgE)	Arginine kinase from: *Penaeus monodon* (giant tiger prawn) *Litopenaeus vannamei* (pacific white shrimp); *Fenneropenaeus indicus* (indian white prawn); *Fenneropenaeus merguiensis* (banana prawn); *Farfantepenaeus notialis* (southern pink shrimp) *Pleoticus muelleri* (argentine red shrimp) *Pandalus borealis* (northern shrimp)	Superfamily/family-specific peptides: FLQAANACR (superfamily Penaeoidea) TFLVWVNEEDHLR (family Penaeidae) LVWVCNEEDHLR (family Solenoceridae) SFLVWVNEEDQLR (family Pandalidae) Species-specific peptides: AVFDQLKEK (*P. monodon*) ALFDQLKDKK (*F. merguiensis*) SFLVWVNEEDQLR (*P. borealis*)	Bottom-up proteomics (2-DE, MALDI-TOF MS/PMF; ESI-MS/MS, IT)	Ortea et al. (2009b)
Shellfish (Shrimp)	Type I (immediate HS; IgE)	Arginine kinase from: *Penaeus monodon* (giant tiger prawn) *Litopenaeus vannamei* (pacific white shrimp); *Fenneropenaeus indicus* (indian white prawn); *Fenneropenaeus merguiensis* (banana prawn); *Farfantepenaeus notialis* (southern pink shrimp) *Pleoticus muelleri* (argentine red shrimp) *Pandalus borealis* (northern shrimp)	Peptides used in SMIM assays: AVFDQLKEK (*P. monodon*) ALFDQLKDKK (*F. merguiensis*) SFLVWVNEEDQLR (*P. borealis*) LTNAVNEIEKR (*P. muelleri*) VSSTLSSLEGELK (*P. monodon, L. vannamei, F. notialis*) TFLVWVNEEDHLR (*P. monodon, L. vannamei, F. merguiensis, F. indicus, F. notialis*) LEEVAGKYNLQVR (*P. monodon, L. vannamei, F. merguiensis, F. indicus*)	Targeted proteomics (LC-MS/MS SMIM)	Ortea et al. (2011)

(Continued)

TABLE 9.9 Cont'd

Food	Type hypersensitivity (HS) reaction	Allergens (% de sensitized patients)	Proteotypic peptides used in SRM/MRM/PRM/SMIM assays or specific peptides detected	Proteomic Technology	Reference PMID (year)
Shellfish (crustacean's proteins in swine meat or high protein processed food product)	Type I (immediate HS; IgE)	Tropomyosin (n.d.)	IVELEEELR, LAEASQAADESER	LC-HRMS/MS (qOrbitrap, untargeted/DDA & PRM)	Stella et al. (2020)

1-DE, one-dimensional gel electrophoresis; 2-DE, two-dimensional gel electrophoresis; CPLL, combinatorial hexapeptide ligand library; DDA, data dependent analysis; DIA, data-independent analysis; FTICR, Fourier-transform ion cyclotron resonance; HIFU, high-intensity focused ultrasound; HR, high resolution; HS, hypersensitivity; ICP, inductively coupled plasma; IM, ion-mobility; IT, ionic trap; LC, liquid chromatography; MALDI-TOF, matrix-assisted laser desorption ionization-time of flight; MRM, multiple reaction monitoring; MS, mass spectrometry; MS/MS, tandem mass spectrometry; PMF, peptide mass fingerprinting; PFF, peptide fragment fingerprinting; PRM, parallel reaction monitoring; SIM, selected ion monitoring; SMIM, selected tandem mass spectrometry ion monitoring; SRM, selected reaction monitoring.

milk undergoes a process called "sanitization," consisting of a treatment for 10–20 s at 62–65°C, after which temperature is rapidly lowered (up to 4°C). After that, milk can be subjected to different pasteurization protocols (e.g., 72°C, 15 s) to obtain fresh pasteurized milk with an expiry period of several days at 4°C. Milk can also be treated more intensively with heat, which preserves its integrity for a longer period (e.g., 6 months at 20°C); an example is ultra-high temperature (UHT) processed milk, which is heated above 135°C for 2–5 s. However, these treatments can alter milk proteins (e.g., causing protein denaturation/aggregation and Maillard products). Indeed, a trend toward a diminished reactivity against whey proteins has been found, whereas the opposite occurs with BSA and caseins (Wal, 2002); among others, this switch in the allergenicity pattern could be related to processing methods of milk/dairy products (Natale et al., 2004). For instance, UHT milk is more popular in southern Europe (except for Greece) compared to northern countries, and Scaloni et al. have reported decreased amounts of milk caseins upon intense heat treatments (UHT, 142–145°C; bottle sterilized, 115–120°C) and increased lactosylation of unstructured α-S1 (Bos d 9) and β caseins (Bos d 11), which could reduce their digestibility and makes them more allergenic; in contrast, less abundant proteins, such as α-lactalbumin (Bos d 4) and β-lactoglobulin (Bos d 5), were less affected (Scaloni et al., 2002) (Table 9.9).

In 2013, Monaci et al. combined ultrafiltration, tryptic digestion, and LC/high-resolution mass spectrometry (LC/HRMS) to detect traces of αs1-casein (Bos d 9) and β-casein (Bos d 11) in white wines (Monaci et al., 2013). In 2020 the same group confirmed by LC-MS/MS (MRM) the absence of cross-contamination with caseins (αS1-casein/Bos d 9) and whey proteins (β-lactoglobulin/Bos d 5) in commercial cookies (Monaci et al., 2020) (Table 9.9). For their part, Stella et al. developed a targeted proteomic method based on LC-MS/MS (PRM) for the accurate detection of specific food allergens (amongst them, peptides from milk proteins) in swine meat or processed food products with an animal origin (e.g., ham, fried fish and fish sticks, frankfurter, mortadella salami, crab claws) and above a threshold value (e.g., 5 μg/g milk allergens) (Stella et al., 2020). However, even though the possible presence of milk allergens was listed in 11 out of 18 products tested, PRM results (in agreement with ELISA data) only revealed their presence (β-lactoglobulin peptides) in 1 out of 18 (5.6%) processed foods (Stella et al., 2020) (Table 9.9). Moreover, the interaction between milk proteins and food matrix components during the manufacture (e.g., baking) of food products could reduce their allergenicity (Bavaro et al., 2019).

9.3.5.2 Red meat allergy

This relatively new food allergy has been extensively investigated by Apostolovic et al. (Apostolovic et al., 2020; Apostolovic et al., 2014, 2017,

2018, 2015; Apostolovic, Tran, et al., 2016; Tjernberg, Hamsten, Apostolovic, & van Hage, 2017). It is mediated by IgE antibodies specific for an immunogenic oligosaccharide characteristic of non-primate mammals: Galactose-α-1,3-galactose (α-Gal). There is also small percentage of patients displaying sIgE to other allergens, such as grass and tree pollen, PR-10 proteins, and animal dander due to cross-reactivity (Kiewiet et al., 2021; Platts-Mills et al., 2020). Compared to healthy donors, red meat-allergic patients showed higher levels of IgE, IgG1, and IgG4 specific to α-Gal, which is characteristic of a type 2 immune response (Apostolovic et al., 2018). Besides, B-blood group donors have significantly reduced antibody responses to α-Gal, due to similarities with the B-antigen (Apostolovic et al., 2018). Sensitization to α-Gal antigen appears to be caused by a tick bite (Apostolovic et al., 2020; Bird et al., 2019; Platts-Mills et al., 2020), because α-Gal–containing proteins are present in saliva from different species from South America, Japan, and Europe: *Amblyomma sculptum, Haemaphysalis longicornis, Ixodes ricinus,* and *Hyalomma marginatum* (Apostolovic et al., 2020). Strikingly, this food allergy courses with an IgE-mediated delayed response, since clinical signs/symptoms (e.g., urticaria, angioedema, gastrointestinal symptoms) appear 3-6 h after beef, pork, lamb, or venison intake (Bird et al., 2019; Platts-Mills et al., 2020). It is characterized by severe allergic reactions and almost half of the patients reports anaphylaxis (Kiewiet et al., 2021; Platts-Mills et al., 2020). The pathophysiology of this disease and the reason for its delayed onset and such severity are not clear, but the kinetics of allergen processing or the presence of immunopathological mechanisms away from the usual immediate-type food allergy seems to play an important role (Levin et al., 2019; Platts-Mills et al., 2020).

The α-Gal epitope is often present in N-linked glycoproteins with basic pI values (pI ≥7) (D. Apostolovic et al., 2014), and several immunoproteomics works have been undertaken to bring out these allergenic proteins. Thus, Takahashi et al. demonstrated the presence of α-Gal epitopes on two beef proteins: laminin γ-1 and collagen α-1(VI) chain (Takahashi et al., 2014). Apostolovic et al. carried out an immunoproteomic study with beef samples subjected to different thermal treatments (raw, boiled, fried, and undercooked beef). They detected 18 IgE-binding proteins, among them myoglobin and BSA, two well-known beef allergens. Seven of those proteins (triosephosphate isomerase, carbonic anhydrase 3, lactate dehydrogenase A, creatine kinase M-type, aspartate aminotransferase, α-enolase, and β-enolase) were α-Gal-containing cytoplasmic enzymes with important roles in metabolism. In addition, four proteins (creatine kinase M-type, aspartate aminotransferase, α-enolase, and β-enolase) preserved their α-Gal reactivity even after thermal treatment (Apostolovic et al., 2014) (Table 9.9). Other α-Gal-containing proteins were type I bovine collagen, thyroglobulin, fibrinogen, IgG, fibronectin, and integrin (β1 subunit).

Finally, red meat-allergic patients also displayed α-Gal-independent IgE-binding (Apostolovic et al., 2014).

9.3.5.3 *Egg allergy*

Hen´s eggs are part of the diet of many cultures. Egg proteins are present in many manufactured food products, which means that their exclusion from the diet is problematic. Allergy to chicken´s eggs is frequently observed in children, but it is also present as a persistent or late-onset disease in 0.02-0.6% of adult Europeans (Ehlers et al., 2020). The white is by far the main source of allergens, and the major allergenic protein is the thermo-stable ovomucoid (OM, Gal d 1), followed by heat-labile proteins such as ovoalbumin (OA, Gal d 2), ovotransferrin (OTr, Gal d 3), and lysozyme (Lsz, Gal d 4) (Holen and Elsayed, 1996). Yolk also contains sIgE-reactive proteins, and the best-known examples are heat-sensitive chicken serum albumin (α-livetin; Gal d 5), vitellogenin I (YGP42; Gal d 6; heat-stable), vitellogenin II (YGP40), and apolipoprotein B (Sogawa et al., 2018). Several proteomic works have been performed on eggs allergens (Table 9.9). Thus, Suzuki et al. detected in 2010 various unknown IgE-reactive proteins in egg white: Ch21, lipocalin-type prostaglandin D synthase, and cystatin (Suzuki et al., 2010) (Table 9.9). Three years later, Martos et al. used CPLLs-treated samples of egg-white proteins and 1-DE/2-DE combined with immunoblotting with serum from allergic patients and LC-MS to confirm ovomucoid (OM; Gal d 1), ovoalbumin (OVA; Gal d 2), and lysozyme (LYS; Gal d 4) as the major allergens, pinpointing ovoinhibitor and clusterin as additional IgE-targets (Martos et al., 2013). Moreover, this group used simulated gastrointestinal digestions to show that ovoalbumin and lysozyme (but not ovoinhibitor or clusterin) remained unaltered upon enzymatic digestion (Martos et al., 2013) (Table 9.9). More recently, Sogawa et al. found in 2018 a new allergen in egg yolk, YGP40, which is a vitellogenin II fragment recognized by 39.6% of sera´s patients, as well as α-livetin/Gal d 5 and vitellogenin I/YGP42/Gal d 6 (Sogawa et al., 2018). In 2020, Ehlers et al. determined, by means of immunoblots, the levels of sIgE and sIgG4 (a tolerogenic antibody; (Trampert et al., 2018)) against native ovomucoid/Gal d 1, ovoalbumin/Gal d 2, ovotransferrin/Gal d 3, and lysozyme/Gal d 4 in both allergic and sensitised (but tolerant) adults (Ehlers et al., 2020); unexpectedly, only a trend toward a higher sIgE/sIgG4 ratio for native ovomucoid/Gal d 1 in allergic adults was reported, probably due to a sample size issue (Ehlers et al., 2020). Additionally, these authors combined immunoblot and 15mer peptides microarrays to reveal multiple linear epitopes in ovomucoid/Gal d 1 and mainly conformational epitopes in ovotransferrin/Gal d 3 (Ehlers et al., 2020). Indeed, the presence of sIgE specific for linear epitopes in ovomucoid/Gal d 1 (especially the aa 30–41, aa 39-50, and aa 84-95 epitopes) helped to differentiate allergic from tolerant individuals, while aa 45-56 (previously related to egg allergy persistence in children) was just an immunodominant epitope (Ehlers et al., 2020).

The presence of egg proteins has also been detected in beverages and manufactured foods. In 2013, Monaci et al. combined ultrafiltration, tryptic digestion, and LC/HRMS (DIA) to detect traces of egg-white powders (ovoalbumin/Gal d 2 and lysozyme/Gal d 4) in white wines (Monaci et al., 2013) (Table 9.9). Similarly, Mainente et al. found traces of hen egg-white lysozyme without label declaration in 12 out of 18 commercial Italian ciders samples by means of LC-MS/MS (DDA) (Mainente et al., 2017); strikingly, immunoblots, and ELISAs only detected the presence of lysozyme in 2 out of 18 cider samples, which highlights the high analytical sensitivity of MS (Mainente et al., 2017) (Table 9.9). In contrast, by employing trypsin digestion, solid-phase extraction (C18-SPE), and LC-MS/MS (MRM) in a triple quadrupole mass analyzer, Monaci et al. confirmed in 2020 the inexistence of cross-contamination with white (ovoalbumin) or yolk (vitellogenin II) derived peptides in bakery products (commercial cookies) declared as prepared without adding egg and milk proteins (Monaci et al., 2020) (Table 9.9). The previously mentioned analytical method developed by Stella et al. was able to detect egg allergens (ovoalbumin peptides) above a 5 μg/g threshold level in 1 out of 18 (5.6%) processed foods (in agreement with ELISA data), when the labels themselves listed their presence in 8 out of 18 food products (44.4%) (Stella et al., 2020) (Table 9.9).

9.3.5.4 Fish and shellfish

Fish and shellfish are other major sources of food allergens (Ruethers et al., 2018). The application of proteomic technologies in the detection and quantification of seafood allergens was described in depth by Carrera et al. (2020). Regarding fish allergies, they are IgE-mediated/immediate hypersensitivity reactions leading to diarrhea, angioedema, and anaphylaxis in sensitized individuals (mainly children and young adults). β-sarcoplasmic parvalbumins are the major cross-reactive allergen in saltwater and freshwater fish. These glycoproteins present in muscles from different fish species are small (around 12 kDa), acidic (pI 3.0-5.0), heat-stable, and denaturation/proteolytic-resistant. Moreover, fish β-parvalbumins are prone to form amyloid fibers, which confers them resistance to proteolysis, and contain three EF-hand motifs, two of them with Ca^{2+}-binding capability (Carrera et al., 2019). Indeed, IgE-binding to parvalbumins relies on cross-reactive and conformational epitopes that are highly conserved and are Ca^{2+}-dependent (Bugajska-Schretter et al., 1998) (Table 9.9). This explains why farmed gilthead seabream on ethylenediaminetetraacetic acid (EDTA)-supplemented diets (a chelating agent that binds to Ca^{2+}) show a reduced abundance of parvalbumin in muscle proteome as well as a lower IgE reactivity (De Magalhães et al., 2020). Cross-reactive parvalbumins have been detected in multiple fish species (e.g., cod, tuna, salmon, perch, carp, eel, etc.) (Bugajska-Schretter et al., 1998) (Table 9.9) and amphibians (Hilger et al., 2002; 2004), but not in higher vertebrates. Thus, based on this cross-reactivity between frog and fish parvalbumins, an anti-frog parvalbumin mouse IgG mAb has been used in 2-DE immunoblots to detect fish parvalbumins in carp, catfish, cod, and tilapia (Chen et al., 2006).

In addition, there are patients who are not sensitized to parvalbumins but to alternative cross-reactive allergens present in both saltwater (e.g., longtail tuna, cod) and freshwater (e.g., pangasius, Nile tilapia, Nile perch) fish. Among these additional allergens we have thermolabile enolase (also an allergen in some molds and plants) and creatine kinase, even though enolase-sensitization can be found in atopic control individuals (Ebo et al., 2010; Kuehn et al., 2013; Rosmilah et al., 2013; Tomm et al., 2013) (Table 9.9). Moreover, not only the type of fish species but also whether they come from a fish farm seems to influence the allergenicity of their proteins. Thus, Piovesana et al. demonstrated in 2015, by means of shotgun proteomics (LC-MS/MS) combined with either label-free or dimethyl labeling, that the sarcoplasmic proteome from farmed gilthead sea bream (*Sparus aurata*) displayed a higher content in parvalbumin compared to wild specimens (Piovesana et al., 2016).

To detect specific allergenic proteins is mandatory to know their primary sequence. However, the genome of many fish species is not yet sequenced, which has prompted the use of alternative strategies. Thus, in 2010 Carrera et al. designed a proteomic strategy to know the primary sequence of the major fish allergen (parvalbumin) in different species. Their strategy was based on the extraction of sarcoplasmic proteins and their separation by 2-DE, in-gel protein digestions of excised parvalbumins spots, LC-ESI-IT-MS/MS (DDA) combined with PFF (SEQUEST)/de novo peptide sequencing (PEAKS, DeNovoX)/manual interpretation, and the use of Fourier-transform ion-cyclotron resonance (FTICR) to determine the exact Mr of both intact parvalbumins and proteolytic fragments (Carrera et al., 2010). In addition, "sequence gaps" were filled in by a targeted proteomics called selected MS/MS ion monitoring (SMIM). These authors obtained both complete ($n = 25$) and partial ($n = 16$) sequences of new parvalbumins from 13 different species/subspecies of the Merlucciidae family, a relevant information for subsequent immunological studies (Carrera et al., 2010) (Table 9.9). In 2011, the same group used a short heat treatment step to deplete thermolabile proteins, a high intensity focused ultrasound (HIFU) process to accelerate trypsin digestion, and a targeted proteomic methodology (LC-MS/MS SMIM) in a linear ion trap (LIT) for the fast (< 2 h) detection of 11 parvalbumin peptides from different commercial Merlucciidae species in 10 commercial fish products (e.g., battered precooked sticks frozen) (Carrera et al., 2011) (Table 9.9). A year later, these researchers used 19 peptide biomarkers (11 exclusive to different fish species with commercial interest and 5 shared with organisms such as frogs) and the same optimized targeted proteomic protocol to rapidly detect possible cross-contaminations with fish β-parvalbumins in commercial sea-food products (or any other foods) (Carrera et al., 2012); moreover, the peptide biomarkers encompassed residues 46–77, which correspond to the Ca^{2+}-binding domain (EF-hand), as above commented an important immunoreactive epitope of these proteins (Carrera et al., 2012) (Table 9.9).

Sometimes, even after extensive freezing or cooking, the ingestion of fish by sensitized people triggers a type I allergic reaction that is dependent on proteins from fish-borne parasites like the nematode *Anisakis* simplex, whose 3rd stage larvae possesses numerous allergenic proteins. Application of proteomic approaches to study this food allergy can be observed in different works (Carrera et al., 2016; Fæste et al., 2016; Kochanowski et al., 2020; Robertson et al., 2020). Furthermore, this was also extensively reviewed in different papers (Carrera et al., 2020; D'Amelio, Lombardo, Pizzarelli, Bellini, & Cavallero, 2020).

To conclude the comments regarding studies on fish allergies, Ruethers et al. used 1-DE, immunoblotting with patients´ sera or in-house (parvalbumin, tropomyosin) or commercial (aldolase, collagen) antibodies, and MS to detect allergens in raw and heated muscle tissue extracts from salmon (*Salmo salar*) and catfish (*Pangasianodon hypophthalmus*) (Ruethers et al., 2021). Thus, patients´sera contained IgE antibodies specific to thermostable parvalbumins (Sal s 1, Pan h 1), tropomyosins (Sal s 4, Pan h 4), and collagen α (Sal s 6), but also to heat-labile enolase (Sal s 2, Pan h 2), aldolase (Sal s 3, Pan h 3), creatine kinase (Sal s 7, Pan h 7), triose phosphate isomerase (Sal s 8, Pan h 8), pyruvate kinase (Pan h 9), L-lactate dehydrogenase (Pan h 10), and glucose-6-phosphate isomerase (Pan h 11) (Ruethers et al., 2021) (Table 9.9).

Shellfish (molluscs, crustaceans) are also an important source of food allergens (Carrera et al., 2020). In 2003, Yu et al. detected by means of immunoproteomic and MALDI-TOF MS analysis different allergens in muscle extracts from *Penaeus monodon shrimps* (Yu et al., 2003). They confirmed tropomyosin as the major allergen, but also identified for the first time arginine kinase (Pen m 2) (Yu et al., 2003). Arginine kinase fulfils a very similar role to the fish allergen creatinine kinase (see above) and displays interspecific variability and cross-reactivity (Yu et al., 2003) (Table 9.9). Six years later, Ortea et al. analysed tryptic digests from sarcoplasmic proteins belonging to six commercially relevant shrimp species (Ortea et al., 2009a; Ortea et al., 2009b). They used 2-DE, MALDI-TOF MS, and sequencing (ESI-IT MS/MS) to circumvent the problem of the absence of sequenced genomes in the Order Decapoda. This group detected an interspecific variation in the pI of a 40 kDa spot, which resulted to be arginine kinase by MS (PMF) and MS/MS. In addition, these authors demonstrated the usefulness of PMF during shrimp species identification in commercial samples (Ortea et al., 2009a) (Table 9.9). An additional work from the same group in 2009 was aimed to the identification of seven commercial and related shrimp species through superfamily-, family-, and species-specific arginine kinase peptides (Ortea et al., 2009b) (Table 9.9). Some of these arginine kinase peptides were subsequently used in 2011 by Ortea et al. for the discrimination of the same seven shrimp species, although in this case, they combined HIFU-assisted tryptic digestion and targeted proteomics (LC-MS/MS, SMIM) for a rapid identification (< 90 min) (Ortea et al., 2011) (Table 9.9). Finally, even though the labels listed the potential presence of crustacean allergens (tropomyosin) in 5 out of 18 (27.8%) processed food products, the previously commented targeted-method (LC-MS/MS, PRM) developed by Stella et al. only detected a significant level (>10 μg/g) in 1 out of 18 (5.6%) cases, in agreement with ELISA data (Stella et al., 2020). This finding, once again, underlines the issue of "it may contain" warnings in food labeling, which restrict food options to allergy sufferers. More worrying, their results also revealed the presence of undeclared allergens (soy, but not milk, egg, or crustacean proteins) in processed foods (Stella et al., 2020) (Table 9.9).

9.4 Conclusions

In the European Union, Regulation (EU) 1169/2011 stipulates that any food ingredient (from a list of 14 constituents including egg, milk, gluten-containing cereals, peanuts, tree nuts, crustacean, soy, fish, etc.) with the potential of causing allergic responses must be carefully listed on the food labels. Current European legislation, in contrast, does not face the unintentional cross-contamination with allergens along the food chain. Nevertheless, the presence of unintended traces of food allergens should be indicated at least with a precautionary allergen labeling, since commercialization of food products with a potential health risk is forbidden (European Union Regulation (EU) 178/2002) (Stella et al., 2020). The disproportionate use of precautionary allergen labeling (PAL) in food industry is driving the implementation of reliable technologies for determining minute amounts of hidden allergens in simple or more complex food matrices. Targeted MS methods display a high potential to detect food allergens and several examples of this type of efforts can be found in the works of Planque et al. (2019, 2017), Sun New et al. (2018), Stella et al. (2020), the Voluntary Incidental Trace Allergen Labeling (VITAL) system (used by Australia, New Zealand and also some countries at the European

Union) (Linda Monaci et al., 2020), or the ThRAll project (funded by the European Food Safety Agency/EFSA) (Mills et al., 2019).

However, it should be emphasized that the detection of these contaminating and proteolysis-resistant proteotypic peptides at the end of a food manufacturing process does not necessarily imply the capacity of triggering an allergic response. Thus, this feature also depends on the total amount of cross-contaminating allergen consumed, which involves knowing the minimum eliciting dose (Monaci et al., 2020). Additionally, the presence of proteotypic peptides in sufficient amounts in a food product does not guarantee an allergenic response, since it is equally important to consider the immunogenic capacity of proteotypic (and undetected non-proteotypic) peptides derived from "stowaway" allergenic proteins, the lineal/conformational or T-cell/B-cell nature of epitopes, or whether they are recognized by pathogenic sIgE or protective sIgG4 in sensitized subjects, amongst others. An example of this kind of incoming studies is the work of Carrera et al., where they used heated sarcoplasmic extracts from 15 different fish species to purify the major fish allergen, β-parvalbumin (Carrera et al., 2019). Parvalbumins were then analyzed by a combination of shotgun proteomics and bioinformatic analysis to identify 243 non-redundant parvalbumin peptides. Finally, 35 of them were selected as putative linear B-cell epitopes by means of the Kolaskar and Tongaonkar method (IEDB; http://tools.immuneepitope.org/bcell/), but only 12 (as well as control peptides) were tested by ELISA with sera from healthy and fish allergic individuals. Four linear B-cell epitopes (LKLFLQV, ACAHLCK, FAVLVKQ, and LFLQNFV) were finally selected based on their high reactivity and the potential of becoming peptide vaccine candidates (Carrera et al., 2019). To go further, the possibility of knowing the B (or T) cell epitopes in a molecule will complement the vision focused on the simple identification of specific allergen peptides in food with a deeper look oriented to the type of patient´s response as well (e.g., Based Epitope Assay; BBEA) (Suprun et al., 2020, 2019). More complex studies are therefore needed to accurately quantify the risk faced by potential consumers of food products sold by supermarkets and grocery shops.

References

Ahsan, N., Rao, R. S. P., Gruppuso, P. A., Ramratnam, B., & Salomon, A. R. (2016). Targeted proteomics: current status and future perspectives for quantification of food allergens. *Journal of Proteomics, 143*, 15–23. https://doi.org/10.1016/j.jprot.2016.04.018.

Allman, D., Wilmore, J. R., & Gaudette, B. T. (2019). The continuing story of T-cell independent antibodies. *Immunological Reviews, 288*(1), 128–135. https://doi.org/10.1111/imr.12754.

Allred, L. K., Nye-Wood, M. G., & Colgrave, M. L. (2020). Analysis of gluten in dried yeast and yeast-containing products. Foods, 1790. doi:10.3390/foods9121790.

Allred, L. K., Voyksner, J. A. S., & Voyksner, R. D. (2014). Evaluation of qualitative and quantitative immunoassays to detect barley contamination in gluten-free beer with confirmation using LC/MS/MS. *Journal of AOAC International, 97*(6), 1615–1625. https://doi.org/10.5740/jaoacint.14-058.

Alves, T. O., D'Almeida, C. T. S., Scherf, K. A., & Ferreira, M. S. L (2019). Modern Approaches in the Identification and Quantification of Immunogenic Peptides in Cereals by LC-MS/MS. *Frontiers in Plant Science, 10*. https://doi.org/10.3389/fpls.2019.01470.

Alves, T. O., D'Almeida, C. T. S., Victorio, V. C. M., Souza, G. H. M. F., Cameron, L. C., & Ferreira, M. S. L (2018). Immunogenic and allergenic profile of wheat flours from different technological qualities revealed by ion mobility mass spectrometry. *Journal of Food Composition and Analysis, 73*, 67–75. https://doi.org/10.1016/j.jfca.2018.07.012.

Andorf, S., Borres, M. P., Block, W., Tupa, D., Bollyky, J. B., Sampath, V., et al. (2017). Association of clinical reactivity with sensitization to allergen components in multifood-allergic children. *Journal of Allergy and Clinical Immunology: In Practice, 5*(5), 1325–1334. e4. https://doi.org/10.1016/j.jaip.2017.01.016.

Ansari, P., Stoppacher, N., Rudolf, J., Schuhmacher, R., & Baumgartner, S. (2011). Selection of possible marker peptides for the detection of major ruminant milk proteins in food by liquid chromatography-tandem mass spectrometry. *Analytical and Bioanalytical Chemistry, 399*(3), 1105–1115. https://doi.org/10.1007/s00216-010-4422-0.

Aoki, A., Hirahara, K., Kiuchi, M., & Nakayama, T. (2021). Eosinophils: Cells known for over 140 years with broad and new functions. *Allergology International, 70*(1), 3–8. https://doi.org/10.1016/j.alit.2020.09.002.

Apostolovic, D., Krstic, M., Mihailovic, J., Starkhammar, M., Cirkovic Velickovic, T., Hamsten, C., et al. (2017). Peptidomics of an in vitro digested α-Gal carrying protein revealed IgE-reactive peptides. *Scientific Reports, 7*(1). https://doi.org/10.1038/s41598-017-05355-4.

Apostolovic, D., Rodrigues, R., Thomas, P., Starkhammar, M., Hamsten, C., & van Hage, M. (2018). Immunoprofile of α-Gal- and B-antigen-specific responses differentiates red meat-allergic patients from healthy individuals. *Allergy, 73*(7), 1525–1531. https://doi.org/10.1111/all.13400.

Apostolovic, D., Stanic-Vucinic, D., De Jongh, H. H. J., De Jong, G. A. H., Mihailovic, J., Radosavljevic, J., et al. (2016). Conformational stability of digestion-resistant peptides of peanut conglutins reveals the molecular basis of their allergenicity. *Scientific Reports, 6*, 29249. https://doi.org/10.1038/srep29249.

Apostolovic, D., Tran, T. A. T., Hamsten, C., Starkhammar, M., Cirkovic Velickovic, T., & van Hage, M. (2014). Immunoproteomics of processed beef proteins reveal novel galactose-α-1,3-galactose-containing allergens. *Allergy, 69*(10), 1308–1315. https://doi.org/10.1111/all.12462.

Apostolovic, D., Tran, T. A. T., Sánchez-Vidaurre, S., Cirkovic Velickovic, T., Starkhammar, M., Hamsten, C., et al. (2015). Red meat allergic patients have a selective IgE response to the α-Gal glycan. *Allergy, 70*(11), 1497–1500. https://doi.org/10.1111/all.12695.

Apostolovic, D., Tran, T. A. T., Starkhammar, M., Sánchez-Vidaurre, S., Hamsten, C., & van Hage, M. (2016). The red meat allergy syndrome in Sweden. *Allergo Journal International, 25*(2), 49–54. https://doi.org/10.1007/s40629-016-0098-0.

Apostolovic, D., Mihailovic, J., Commins, S. P., Wijnveld, M., Kazimirova, M., Starkhammar, M., et al. (2020). Allergenomics of the tick Ixodes ricinus reveals important α-Gal–carrying IgE-binding proteins in red meat allergy. *Allergy, 75*(1), 217–220. https://doi.org/10.1111/all.13978.

Archila, L. D., Jeong, D., Pascal, M., Bartra, J., Juan, M., Robinson, D., et al. (2015). Jug r 2-reactive CD4+ T cells have a dominant immune role in walnut allergy. *Journal of Allergy and Clinical Immunology, 136*(4), 983–992.e7. https://doi.org/10.1016/j.jaci.2015.01.029.

Bahri, R., Custovic, A., Korosec, P., Tsoumani, M., Barron, M., Wu, J., et al. (2018). Mast cell activation test in the diagnosis of allergic disease and anaphylaxis. *Journal of Allergy and Clinical Immunology, 142*(2), 485–496.e16. https://doi.org/10.1016/j.jaci.2018.01.043.

Balakireva, A. V., & Zamyatnin, A. A. (2016). Properties of gluten intolerance: Gluten structure, evolution, pathogenicity and detoxification capabilities. *Nutrients, 8*(10). https://doi.org/10.3390/nu8100644.

Bascuñán, K. A., Roncoroni, L., Branchi, F., Doneda, L., Scricciolo, A., Ferretti, F., et al. (2018). The 5 Ws of a gluten challenge for gluten-related disorders. *Nutrition Reviews, 76*(2), 79–87. https://doi.org/10.1093/nutrit/nux068.

Bavaro, S. L., De Angelis, E., Barni, S., Pilolli, R., Mori, F., Novembre, E. M., et al. (2019). Modulation of milk allergenicity by baking milk in foods: a proteomic investigation. *Nutrients, 11*(7), 1536. https://doi.org/10.3390/nu11071536.

Bignardi, C., Elviri, L., Penna, A., Careri, M., & Mangia, A. (2010). Particle-packed column versus silica-based monolithic column for liquid chromatography-electrospray-linear ion trap-tandem mass spectrometry multiallergen trace analysis in foods. *Journal of Chromatography A, 1217*(48), 7579–7585. https://doi.org/10.1016/j.chroma.2010.10.037.

Bird, J. A. (2018). Food allergy. *Immunology and Allergy Clinics of North America, 38*(1), xv–xvi. https://doi.org/10.1016/j.iac.2017.10.001.

Bird, J. A., Jones, S., & Burks, W. (2019). Food Allergy. In R. R. Rich, T. A. Fleisher, W. T. Shearer, H. W. Schroeder, A. J. Frew, & C. M. Weyand (Eds.), *Clinical Immunology* (5th ed., p. 625–631.e1). Amsterdam, Netherlands: Elsevier.

Bönick, J., Huschek, G., & Rawel, H. M. (2017). Determination of wheat, rye and spelt authenticity in bread by targeted peptide biomarkers. *Journal of Food Composition and Analysis, 58*, 82–91. https://doi.org/10.1016/j.jfca.2017.01.019.

Bose, U., Byrne, K., Howitt, C. A., & Colgrave, M. L. (2019). Targeted proteomics to monitor the extraction efficiency and levels of barley α-amylase trypsin inhibitors that are implicated in non-coeliac gluten sensitivity. *Journal of Chromatography A, 1600*, 55–64. https://doi.org/10.1016/j.chroma.2019.04.043.

Boukid, F., Prandi, B., Faccini, A., & Sforza, S. (2019). A complete mass spectrometry (MS)-based peptidomic description of gluten peptides generated during in vitro gastrointestinal digestion of durum wheat: implication for celiac disease. *Journal of the American Society for Mass Spectrometry, 30*(8), 1481–1490. https://doi.org/10.1007/s13361-019-02212-8.

Boyaka, P. N., & Fujihashi, K. (2019). Host defenses at mucosal surfaces. In R. R. Rich, T. A. Fleisher, W. T. Shearer, H. W. Schroeder, A. J. Frew, & C. M. Weyand (Eds.), *Clinical Immunology* (5th ed.) (pp. 285–298.e1). Amsterdam, Netherlands: Elsevier BV. https://doi.org/10.1016/b978-0-7020-6896-6.00020-x.

Bromilow, S., Gethings, L. A., Buckley, M., Bromley, M., Shewry, P. R., Langridge, J. I., et al. (2017). A curated gluten protein sequence database to support development of proteomics methods for determination of gluten in gluten-free foods. *Journal of Proteomics, 163*, 67–75. https://doi.org/10.1016/j.jprot.2017.03.026.

Bugajska-Schretter, A., Elfman, L., Fuchs, T., Kapiotis, S., Rumpold, H., Valenta, R., et al. (1998). Parvalbumin, a cross-reactive fish allergen, contains IgE-binding epitopes sensitive to periodate treatment and Ca2+ depletion. *Journal of Allergy and Clinical Immunology, 101*(1), 67–74. https://doi.org/10.1016/S0091-6749(98)70195-2.

Cabanillas, B., Cuadrado, C., Rodriguez, J., Dieguez, M. C., Crespo, J. F., & Novak, N. (2018). Boiling and pressure cooking impact on IgE reactivity of soybean allergens. *International Archives of Allergy and Immunology, 175*(1–2), 36–43. https://doi.org/10.1159/000485894.

Calvano, C. D., Bianco, M., Losito, I., & Cataldi, T. R. I. (2021). Proteomic analysis of food allergens by MALDI TOF/TOF mass spectrometry. *Methods in Molecular Biology, 2178*, 357–376. https://doi.org/10.1007/978-1-0716-0775-6_24.

Calvano, C. D., De Ceglie, C., & Zambonin, C. G. (2014). Proteomic analysis of complex protein samples by MALDI-TOF mass spectrometry. *Methods in Molecular Biology*, 1129, 365–380. https://doi.org/10.1007/978-1-62703-977-2_27.

Cao, W., Baumert, J. L., & Downs, M. L. (2020). Compositional and immunogenic evaluation of fractionated wheat beers using mass spectrometry. Food Chemistry, 333, 127379. https://doi.org/10.1016/j.foodchem.2020.127379.

Cardona, E. E. G., Heathcote, K., Teran, L. M., Righetti, P. G., Boschetti, E., & D'Amato, A (2018). Novel low-abundance allergens from mango via combinatorial peptide libraries treatment: a proteomics study. *Food Chemistry, 269,* 652–660. https://doi.org/10.1016/j.foodchem.2018.06.113.

Cardona, V., Ansotegui, I. J., Ebisawa, M., El-Gamal, Y., Fernandez Rivas, M., Fineman, S., et al. (2020). World allergy organization anaphylaxis guidance 2020. *World Allergy Organization Journal, 13*(10), Article 100472. https://doi.org/10.1016/j.waojou.2020.100472.

Carlson, G., & Coop, C. (2019). Pollen food allergy syndrome (PFAS): a review of current available literature. *Annals of Allergy, Asthma and Immunology, 123*(4), 359–365. https://doi.org/10.1016/j.anai.2019.07.022.

Carrera, M., Cañas, B., & Gallardo, J. M. (2018). Advanced proteomics and systems biology applied to study food allergy. Current Opinion in Food Science, 22, 9–16. https://doi.org/10.1016/j.cofs.2017.12.001

Carrera, M., Cañas, B., & Gallardo, J. M. (2012). Rapid direct detection of the major fish allergen, parvalbumin, by selected MS/MS ion monitoring mass spectrometry. *Journal of Proteomics, 75*(11), 3211–3220. https://doi.org/10.1016/j.jprot.2012.03.030.

Carrera, M., Cañas, B., López-Ferrer, D., Piñeiro, C., Vázquez, J., & Gallardo, J. M. (2011). Fast monitoring of species-specific peptide biomarkers using high-intensity-focused-ultrasound-assisted tryptic digestion and selected MS/MS ion monitoring. *Analytical Chemistry, 83*(14), 5688–5695. https://doi.org/10.1021/ac200890w.

Carrera, M., Cañas, B., Vázquez, J., & Gallardo, J. M. (2010). Extensive de Novo sequencing of new parvalbumin isoforms using a novel combination of bottom-up proteomics, accurate molecular mass measurement by FTICR-MS, and selected MS/MS ion monitoring. *Journal of Proteome Research, 9*(9), 4393–4406. https://doi.org/10.1021/pr100163e.

Carrera, M., Gallardo, J. M., Pascual, S., González, Á. F., & Medina, I. (2016). Protein biomarker discovery and fast monitoring for the identification and detection of Anisakids by parallel reaction monitoring (PRM) mass spectrometry. *Journal of Proteomics, 142,* 130–137. https://doi.org/10.1016/j.jprot.2016.05.012.

Carrera, M., González-Fernández, Á., Magadán, S., Mateos, J., Pedrós, L., Medina, I., et al. (2019). Molecular characterization of B-cell epitopes for the major fish allergen, parvalbumin, by shotgun proteomics, protein-based bioinformatics and IgE-reactive approaches. *Journal of Proteomics, 200,* 123–133. https://doi.org/10.1016/j.jprot.2019.04.005.

Carrera, M., Pazos, M., & Gasset, M. (2020). Proteomics-based methodologies for the detection and quantification of seafood allergens. *Foods, 9*(8). doi:10.3390/foods9081134.

Castells, M., & Bonamichi-Santos, R. (2019). *Drug hypersensitivity*. In R. R. Rich, T. A. Fleisher, W. T. Shearer, H. W. Schroeder, A. J. Frew, & C. M. Weyand (Eds.), *Clinical Immunology* (5th ed., pp. 649–667.e1). Amsterdam, Netherlands: Elsevier.

Cebolla, Á., Moreno, M.d. L., Coto, L., & Sousa, C. (2018). Gluten immunogenic peptides as standard for the evaluation of potential harmful prolamin content in food and human specimen. *Nutrients, 10*(12). https://doi.org/10.3390/nu10121927.

Chassaigne, H., Nørgaard, J. V., & Van Hengel, A. J. (2007). Proteomics-based approach to detect and identify major allergens in processed peanuts by capillary LC-Q-TOF (MS/MS). *Journal of Agricultural and Food Chemistry, 55*(11), 4461–4473. https://doi.org/10.1021/jf063630e.

Chen, L., Hefle, S. L., Taylor, S. L., Swoboda, I., & Goodman, R. E. (2006). Detecting fish parvalbumin with commercial mouse monoclonal anti-frog parvalbumin IgG. *Journal of Agricultural and Food Chemistry, 54*(15), 5577–5582. https://doi.org/10.1021/jf060291g.

Chen, S., Yang, C. T., & Downs, M. L. (2019). Detection of six commercially processed soy ingredients in an incurred food matrix using parallel reaction monitoring. *Journal of Proteome Research, 18*(3), 995–1005. https://doi.org/10.1021/acs.jproteome.8b00689.

Chen, S., Yang, C., & Downs, M (2021). Targeted mass spectrometry quantification of total soy protein residues from commercially processed ingredients for food allergen management. *Journal of Proteomics, 239*, Article 104194. https://doi.org/10.1016/j.jprot.2021.104194.

Ciccocioppo, R., Di Sabatino, A., & Corazza, G. R. (2005). The immune recognition of gluten in coeliac disease. *Clinical and Experimental Immunology, 140*(3), 408–416. https://doi.org/10.1111/j.1365-2249.2005.02783.x.

Colgrave, M. L., Byrne, K., Blundell, M., Heidelberger, S., Lane, C. S., et al. (2016a). Comparing multiple reaction monitoring and sequential window acquisition of all theoretical mass spectra for the relative quantification of barley gluten in selectively bred barley lines. *Analytical Chemistry, 88*(18), 9127–9135. https://doi.org/10.1021/acs.analchem.6b02108.

Colgrave, M. L., Byrne, K., Blundell, M., & Howitt, C. A. (2016b). Identification of barley-specific peptide markers that persist in processed foods and are capable of detecting barley contamination by LC-MS/MS. *Journal of Proteomics, 147*, 169–176. https://doi.org/10.1016/j.jprot.2016.03.045.

Colgrave, M. L., Goswami, H., Byrne, K., Blundell, M., Howitt, C. A., & Tanner, G. J. (2015). Proteomic profiling of 16 cereal grains and the application of targeted proteomics to detect wheat contamination. *Journal of Proteome Research, 14*(6), 2659–2668. https://doi.org/10.1021/acs.jproteome.5b00187.

Commins, S. P., James, H. R., Stevens, W., Pochan, S. L., Land, M. H., King, C., et al. (2014). Delayed clinical and ex vivo response to mammalian meat in patients with IgE to galactose-alpha-1,3-galactose. *Journal of Allergy and Clinical Immunology, 134*(1), 108–e11. https://doi.org/10.1016/j.jaci.2014.01.024.

Costa, J., Ansari, P., Mafra, I., Oliveira, M. B. P. P., & Baumgartner, S. (2014). Assessing hazelnut allergens by protein- and DNA-based approaches: LC-MS/MS, ELISA and real-time PCR. *Analytical and Bioanalytical Chemistry, 406*(11), 2581–2590. https://doi.org/10.1007/s00216-014-7679-x.

Croote, D., & Quake, S. R. (2016). Food allergen detection by mass spectrometry: The role of systems biology. *Npj Systems Biology and Applications, 2*. https://doi.org/10.1038/npjsba.2016.22.

Cucu, T., De Meulenaer, B., & Devreese, B. (2012). MALDI based identification of soybean protein markers—psossible analytical targets for allergen detection in processed foods. *Peptides, 33*(2), 187–196. https://doi.org/10.1016/j.peptides.2011.12.007.

Czubinski, J., Montowska, M., Pospiech, E., & Lampart-Szczapa, E. (2017). Proteomic analysis of Lupinus angustifolius (var. Zeus and Bojar) and Lupinus luteus (var. Lord and Parys) seed proteins and their hydrolysates. *Journal of the Science of Food and Agriculture, 97*(15), 5423–5430. https://doi.org/10.1002/jsfa.8436.

D'Auria, E., Salvatore, S., Pozzi, E., Mantegazza, C., Sartorio, M. U. A., Pensabene, L., et al. (2019). Cow's Milk Allergy: immunomodulation by dietary intervention. *Nutrients, 11*(6), 1399. https://doi.org/10.3390/nu11061399.

Daly, M., Bromilow, S. N., Nitride, C., Shewry, P. R., Gethings, L. A., & Mills, E. N. C. (2020a). Mapping coeliac toxic motifs in the prolamin seed storage proteins of barley, rye, and oats using a curated sequence database. *Frontiers in Nutrition, 7*, 87. https://doi.org/10.3389/fnut.2020.00087.

Daly, M., Bromilow, S., Nitride, Shewry, P., Gethings, L., & Mills, E (2020b). The GluPro suite of curated cereal seed storage prolamin protein sequence datasets. *figshare*.

D'Amato, A., Bachi, A., Fasoli, E., Boschetti, E., Peltre, G., Sénéchal, H., et al. (2009). In-depth exploration of cow's whey proteome via combinatorial peptide ligand libraries. *Journal of Proteome Research, 8*(8), 3925–3936. https://doi.org/10.1021/pr900221x.

Damiana Calvano, C., De Ceglie, C., & Zambonin, C. G. (2014). Proteomic analysis of complex protein samples by MALDI–TOF mass spectrometry. *Methods in Molecular Biology, 1129*, 365–380. https://doi.org/10.1007/978-1-62703-977-2_27.

D'Amelio, S., Lombardo, F., Pizzarelli, A., Bellini, I., & Cavallero, S. (2020). Advances in Omic studies drive discoveries in the biology of Anisakid Nematodes. *Genes (Basel), 11*(7), 801. https://doi.org/10.3390/genes11070801.

D'Auria, E., Agostoni, C., Giovannini, M., Riva, E., Zetterström, R., Fortin, R., et al. (2005). Proteomic evaluation of milk from different mammalian species as a substitute for breast milk. *Acta Paediatrica, International Journal of Paediatrics, 94*(12), 1708–1713. https://doi.org/10.1080/08035250500434793.

D'Auria, E., Mameli, C., Piras, C., Cococcioni, L., Urbani, A., Zuccotti, G. V., et al. (2018). Precision medicine in cow's milk allergy: proteomics perspectives from allergens to patients. *Journal of Proteomics, 188*, 173–180. https://doi.org/10.1016/j.jprot.2018.01.018.

De Angelis, M., Di Cagno, R., Minervini, F., Rizzello, C. G., & Gobbetti, M. (2010). Two-dimensional electrophoresis and IgE-mediated food allergy. *Electrophoresis, 31*(13), 2126–2136. https://doi.org/10.1002/elps.201000101.

De Cicco, M., Mamone, G., Di Stasio, L., Ferranti, P., Addeo, F., & Picariello, G. (2019). Hidden \digestome\: current analytical approaches provide incomplete peptide inventories of food digests. *Journal of Agricultural and Food Chemistry, 67*(27), 7775–7782. https://doi.org/10.1021/acs.jafc.9b02342.

De Magalhães, C. R., Schrama, D., Fonseca, F., Kuehn, A., Morisset, M., Ferreira, S. R., et al. (2020). Effect of EDTA enriched diets on farmed fish allergenicity and muscle quality; a proteomics approach. *Food Chemistry, 305*. https://doi.org/10.1016/j.foodchem.2019.125508.

Dhondalay, G. K., Rael, E., Acharya, S., Zhang, W., Sampath, V., Galli, S. J., et al. (2018). Food allergy and omics. *Journal of Allergy and Clinical Immunology, 141*(1), 20–29. https://doi.org/10.1016/j.jaci.2017.11.007.

Di Stasio, L., d'Acierno, A., Picariello, G., Ferranti, P., Nitride, C., & Mamone, G. (2020). In vitro gastroduodenal and jejunal brush border membrane digestion of raw and roasted tree nuts. *Food Research International, 136*, 109597. http://doi.org/10.1016/j.foodres.2020.109597.

Di Stasio, L., Picariello, G., Mongiello, M., Nocerino, R., Berni Canani, R., Bavaro, S., et al. (2017). Peanut digestome: identification of digestion resistant IgE binding peptides. *Food and Chemical Toxicology, 107*, 88–98. https://doi.org/10.1016/j.fct.2017.06.029.

Dispenza, M. C. (2019). Classification of hypersensitivity reactions. *Allergy and Asthma Proceedings, 40*(6), 470–473. https://doi.org/10.2500/aap.2019.40.4274.

Dodig, S., & Čepelak, I. (2018). The potential of component-resolved diagnosis in laboratory diagnostics of allergy. *Biochemia Medica, 28*(2). https://doi.org/10.11613/BM.2018.020501.

Downs, M. L., Baumert, J. L., Taylor, S. L., & Mills, E. N. C. (2016). Mass spectrometric analysis of allergens in roasted walnuts. *Journal of Proteomics, 142*, 62–69. https://doi.org/10.1016/j.jprot.2016.04.045.

Ebo, D. G., Kuehn, A., Bridts, C. H., Hilger, C., Hentges, F., & Stevens, W. J. (2010). Monosensitivity to pangasius and tilapia caused by allergens other than parvalbumin. *Journal of Investigational Allergology and Clinical Immunology, 20*(1), 84–88. http://www.jiaci.org/issues/vol20issue1/13.pdf.

Egger, M., Mutschlechner, S., Wopfner, N., Gadermaier, G., Briza, P., & Ferreira, F. (2006). Pollen-food syndromes associated with weed pollinosis: an update from the molecular point of view. *Allergy, 61*(4), 461–476. https://doi.org/10.1111/j.1398-9995.2006.00994.x.

Ehlers, A. M., Otten, H. G., Wierzba, E., Flügge, U., Le, T., Knulst, A. C., et al. (2020). Detection of specific IgE against linear epitopes from Gal d 1 has additional value in diagnosing hen's egg allergy in adults. *Clinical & Experimental Allergy, 50*(12), 1415–1423. https://doi.org/10.1111/cea.13730.

Eljaszewicz, A., Ruchti, F., Radzikowska, U., Globinska, A., Boonpiyathad, T., Gschwend, A., et al. (2021). Trained immunity and tolerance in innate lymphoid cells, monocytes, and dendritic cells during allergen-specific immunotherapy. *Journal of Allergy and Clinical Immunology, 147*(5), 1865–1877. https://doi.org/10.1016/j.jaci.2020.08.042.

Fæste, C. K., Moen, A., Schniedewind, B., Haug Anonsen, J., Klawitter, J., & Christians, U. (2016). Development of liquid chromatography-tandem mass spectrometry methods for the quantitation of Anisakis simplex proteins in fish. *Journal of Chromatography A, 1432*, 58–72. https://doi.org/10.1016/j.chroma.2016.01.002.

Fernandez-Becker, N. Q., Raja, S., Scarpignato, C., Lynch, K. L., Ahuja, N. K., & Horsley-Silva, J. L. (2020). Eosinophilic esophagitis: updates on key unanswered questions. *Annals of the New York Academy of Sciences, 1481*(1), 30–42. https://doi.org/10.1111/nyas.14421.

Fiedler, K. L., McGrath, S. C., Callahan, J. H., & Ross, M. M. (2014). Characterization of grain-specific peptide markers for the detection of gluten by mass spectrometry. *Journal of Agricultural and Food Chemistry, 62*(25), 5835–5844. https://doi.org/10.1021/jf500997j.

Flodrová, D., Benkovská, D., & Laštovičková, M. (2015). Study of quantitative changes of cereal allergenic proteins after food processing. *Journal of the Science of Food and Agriculture, 95*(5), 983–990. https://doi.org/10.1002/jsfa.6773.

Franz-Oberdorf, K., Eberlein, B., Edelmann, K., Bleicher, P., Kurze, E., Helm, D., et al. (2017). White-fruited strawberry genotypes are not per se hypoallergenic. *Food Research International, 100*, 748–756. https://doi.org/10.1016/j.foodres.2017.07.076.

Geiselhart, S., Hoffmann-Sommergruber, K., & Bublin, M. (2018). Tree nut allergens. *Molecular Immunology, 100*, 71–81. https://doi.org/10.1016/j.molimm.2018.03.011.

Geisslitz, S., Longin, CFH., Koehler, P., & Scherf, KA. (2020). Comparative quantitative LC-MS/MS analysis of 13 amylase/trypsin inhibitors in ancient and modern Triticum species. *Scientific Reports, 10*(1), 14570. https://doi.org/10.1038/s41598-020-71413-z.

Goggin, D. E., Mir, G., Smith, W. B., Stuckey, M., & Smith, P. M. C. (2008). Proteomic analysis of lupin seed proteins to identify conglutin Beta as an allergen, Lup an 1. *Journal of Agricultural and Food Chemistry, 56*(15), 6370–6377. https://doi.org/10.1021/jf800840u.

Guillamón, E., Burbano, C., Cuadrado, C., Muzquiz, M., Pedrosa, M. M., Sánchez, M., et al. (2008). Effect of an instantaneous controlled pressure drop on in vitro allergenicity to lupins (Lupinus albus var Multolupa). *International Archives of Allergy and Immunology, 145*(1), 9–14. https://doi.org/10.1159/000107461.

Guo, X., Jiang, S., Li, X., Yang, S., Cheng, L., Qiu, J., et al. (2020). Sequence analysis of digestion-resistant peptides may be an efficient strategy for studying the linear epitopes of Jug r 1, the major walnut allergen. *Food Chemistry, 322*. https://doi.org/10.1016/j.foodchem.2020.126711.

Gupta, R. S., Springston, E. E., Warrier, M. R., Smith, B., Kumar, R., Pongracic, J., et al. (2011). The prevalence, severity, and distribution of childhood food allergy in the United States. *Pediatrics, 128*(1), e9–e17. https://doi.org/10.1542/peds.2011-0204.

Hauser, M., Asam, C., Himly, M., Palazzo, P., Voltolini, S., Montanari, C., et al. (2011). Bet v 1-like pollen allergens of multiple Fagales species can sensitize atopic individuals. *Clinical & Experimental Allergy, 41*(12), 1804–1814. https://doi.org/10.1111/j.1365-2222.2011.03866.x.

Hebling, C. M., McFarland, M. A., Callahan, J. H., & Ross, M. M. (2013). Global proteomic screening of protein allergens and advanced glycation endproducts in thermally processed peanuts. *Journal of Agricultural and Food Chemistry, 61*(24), 5638–5648. https://doi.org/10.1021/jf303554t.

Heick, J., Fischer, M., & Pöpping, B. (2011). First screening method for the simultaneous detection of seven allergens by liquid chromatography mass spectrometry. *Journal of Chromatography A, 1218*(7), 938–943. https://doi.org/10.1016/j.chroma.2010.12.067.

Helsper, J. P. F. G., Gilissen, L. J. W. J., Van Ree, R., America, A. H. P., Cordewener, J. H. G., & Bosch, D. (2002). Quadrupole time-of-flight mass spectrometry: a method to study the actual expression of allergen isoforms identified by PCR cloning. *Journal of Allergy and Clinical Immunology, 110*(1), 131–138. https://doi.org/10.1067/mai.2002.125599.

Herndl, A., Marzban, G., Kolarich, D., Hahn, R., Boscia, D., Hemmer, W., Maghuly, F., et al. (2007). Mapping of Malus domestica allergens by 2-D electrophoresis and IgE-reactivity. *Electrophoresis, 28*(3), 437–448. https://doi.org/10.1002/elps.200600342.

Hilger, C., Grigioni, F., Thill, L., Mertens, L., & Hentges, F. (2002). Severe IgE-mediated anaphylaxis following consumption of fried frog legs: definition of alpha-parvalbumin as the allergen in cause. *Allergy, 57*(11), 1053–1058. https://doi.org/10.1034/j.1398-9995.2002.23677.x.

Hilger, C., Thill, L., Grigioni, F., Lehners, C., Falagiani, P., Ferrara, A., et al. (2004). IgE antibodies of fish allergic patients cross-react with frog parvalbumin. *Allergy, 59*(6), 653–660. https://doi.org/10.1111/j.1398-9995.2004.00436.x.

Hochwallner, H., Schulmeister, U., Swoboda, I., Balic, N., Geller, B., Nystrand, M., et al. (2010). Microarray and allergenic activity assessment of milk allergens. *Clinical & Experimental Allergy, 40*(12), 1809–1818. https://doi.org/10.1111/j.1365-2222.2010.03602.x.

Hochwallner, H., Schulmeister, U., Swoboda, I., Twaroch, T. E., Vogelsang, H., Kazemi-Shirazi, L., et al. (2011). Patients suffering from non-IgE-mediated cow's milk protein intolerance cannot be diagnosed based on IgG subclass or IgA responses to milk allergens. *Allergy, 66*(9), 1201–1207. https://doi.org/10.1111/j.1398-9995.2011.02635.x.

Hoh, R. A., Joshi, S. A., Lee, J. Y., Martin, B. A., Varma, S., Kwok, S., et al. (2020). Origins and clonal convergence of gastrointestinal IgE+ B cells in human peanut allergy. *Science Immunology, 5*(45). https://doi.org/10.1126/sciimmunol.aay4209.

Holen, E., & Elsayed, S. (1996). Specific T cell lines for ovalbumin, ovomucoid, lysozyme and two OA synthetic epitopes, generated from egg allergic patients' PBMC. *Clinical and Experimental Allergy, 26*(9), 1080–1088. https://doi.org/10.1111/j.1365-2222.1996.tb00647.x.

Ialicicco, M., Viscosi, V., Arena, S., Scaloni, A., Trupiano, D., Rocco, M., et al. (2012). Lens culinaris Medik. seed proteome: analysis to identify landrace markers. *Plant Science, 197*, 1–9. https://doi.org/10.1016/j.plantsci.2012.08.010.

Inman, S. E., Groves, K., McCullough, B., Quaglia, M., & Hopley, C. (2018). Development of a LC-MS method for the discrimination between trace level Prunus contaminants of spices. *Food Chemistry, 245*, 289–296. https://doi.org/10.1016/j.foodchem.2017.10.101.

Islam, S., Ma, W., Ma, J., Buirchell, B. J., Appels, R., & Yan, G. (2011). Diversity of seed protein among the Australian narrow-leafed lupin (Lupinus angustifolius L.) cultivars. *Crop and Pasture Science, 62*(9), 765–775. https://doi.org/10.1071/CP11046.

Islam, S., Yan, G., Appels, R., & Ma, W. (2012). Comparative proteome analysis of seed storage and allergenic proteins among four narrow-leafed lupin cultivars. *Food Chemistry, 135*(3), 1230–1238. https://doi.org/10.1016/j.foodchem.2012.05.081.

Jang, Y. R., Cho, K., Kim, S. W., Altenbach, S. B., Lim, S. H., Sim, J. R., et al. (2020). Development of an optimized MALDI-TOF-MS method for high-throughput identification of high-molecular-weight glutenin subunits in wheat. *Molecules (Basel, Switzerland), 25*(18). https://doi.org/10.3390/molecules25184347.

Jappe, U., Karstedt, A., Warneke, D., Hellmig, S., Böttger, M., Riffelmann, F. W., et al. (2021). Identification and purification of novel low-molecular-weight lupine allergens as components for personalized diagnostics. *Nutrients, 13*(2), 1–21. https://doi.org/10.3390/nu13020409.

Jenkins, J. A., Griffiths-Jones, S., Shewry, P. R., Breiteneder, H., & Mills, E. N. C. (2005). Structural relatedness of plant food allergens with specific reference to cross-reactive allergens: An in silico analysis. *Journal of Allergy and Clinical Immunology, 115*(1), 163–170. https://doi.org/10.1016/j.jaci.2004.10.026.

Johnson, J., Malinovschi, A., Alving, K., Lidholm, J., Borres, M. P., & Nordvall, L. (2014). Ten-year review reveals changing trends and severity of allergic reactions to nuts and other foods. *Acta Paediatrica, International Journal of Paediatrics, 103*(8), 862–867. https://doi.org/10.1111/apa.12687.

Johnson, P. E., Sayers, R. L., Gethings, L. A., Balasundaram, A., Marsh, J. T., Langridge, J. I., et al. (2016). Quantitative proteomic profiling of peanut allergens in food ingredients used for oral food challenges. *Analytical Chemistry, 88*(11), 5689–5695. https://doi.org/10.1021/acs.analchem.5b04466.

Juhász, A., Haraszi, R., & Maulis, C. (2015). ProPepper: a curated database for identification and analysis of peptide and immune-responsive epitope composition of cereal grain protein families. *Database, 2015*(1). https://doi.org/10.1093/database/bav100.

Kadam, K., Karbhal, R., Jayaraman, V. K., Sawant, S., & Kulkarni-Kale, U. (2017). AllerBase: a comprehensive allergen knowledgebase. *Database, 2017*. https://doi.org/10.1093/database/bax066.

Kanagaratham, C., El Ansari, Y. S., Lewis, O. L., & Oettgen, H. C. (2020). IgE and IgG antibodies as regulators of mast cell and basophil functions in food allergy. *Frontiers in Immunology, 11*. https://doi.org/10.3389/fimmu.2020.603050.

Kiewiet, M. B. G., Grundström, J., Apostolovic, D., Andersson, M., Borres, M. P., Hamsten, C., et al. (2021). Elucidating the α-Gal syndrome at the molecular allergen level. *Allergy: European Journal of Allergy and Clinical Immunology, 76*(5), 1576–1578. https://doi.org/10.1111/all.14660.

Kim, J. S., Nowak-Węgrzyn, A., Sicherer, S. H., Noone, S., Moshier, E. L., & Sampson, H. A. (2011). Dietary baked milk accelerates the resolution of cow's milk allergy in children. *Journal of Allergy and Clinical Immunology, 128*(1), 125–131. e2. https://doi.org/10.1016/j.jaci.2011.04.036.

Klion, A. D., Ackerman, S. J., & Bochner, B. S. (2020). Contributions of eosinophils to human health and disease. *Annual Review of Pathology: Mechanisms of Disease, 15*, 179–209. https://doi.org/10.1146/annurev-pathmechdis-012419-032756.

Kochanowski, M., Różycki, M., Dąbrowska, J., Bełcik, A., Karamon, J., Sroka, J., et al. (2020). Proteomic and bioinformatic investigations of heat-treated Anisakis simplex third-stage larvae. *Biomolecules, 10*(7), 1–36. https://doi.org/10.3390/biom10071066.

Koplin, J. J., Dharmage, S. C., Ponsonby, A. L., Tang, M. L. K., Lowe, A. J., Gurrin, L. C., et al. (2012). Environmental and demographic risk factors for egg allergy in a population-based study of infants. *Allergy: European Journal of Allergy and Clinical Immunology, 67*(11), 1415–1422. https://doi.org/10.1111/all.12015.

Korte, R., & Brockmeyer, J. (2016). MRM3-based LC-MS multi-method for the detection and quantification of nut allergens. *Analytical and Bioanalytical Chemistry, 408*(27), 7845–7855. https://doi.org/10.1007/s00216-016-9888-y.

Korte, R., Happe, J., Brümmer, I., & Brockmeyer, J. (2017a). Structural characterization of the allergenic 2S Albumin Cor a 14: comparing proteoform patterns across hazelnut cultivars. *Journal of Proteome Research, 16*(2), 988–998. https://doi.org/10.1021/acs.jproteome.6b00924.

Korte, R., Lepski, S., & Brockmeyer, J. (2016). Comprehensive peptide marker identification for the detection of multiple nut allergens using a non-targeted LC-HRMS multi-method. *Analytical and Bioanalytical Chemistry, 408*(12), 3059–3069. https://doi.org/10.1007/s00216-016-9384-4.

Korte, R., Bräcker, J., & Brockmeyer, J. (2017b). Gastrointestinal digestion of hazelnut allergens on molecular level: Elucidation of degradation kinetics and resistant immunoactive peptides using mass spectrometry. *Molecular Nutrition & Food Research, 61*(10), Article 1700130. https://doi.org/10.1002/mnfr.201700130.

Kottyan, L. C., Maddox, A., Braxton, J. R., Stucke, E. M., Mukkada, V., Putnam, P. E., et al. (2019). Genetic variants at the 16p13 locus confer risk for eosinophilic esophagitis. *Genes and Immunity, 20*(4), 281–292. https://doi.org/10.1038/s41435-018-0034-z.

Kovalszki, A., & Weller, P. F. (2019). Eosinophils and eosinophilia. In R. R. Rich, T. A. Fleisher, W. T. Shearer, H. W. Schroeder, A. J. Frew, & C. M. Weyand (Eds.), *Clinical immunology* (5th ed., pp. 349–361.e1). Amsterdam, Netherlands: Elsevier.

Krempski, J. W., Dant, C., & Nadeau, K. C. (2020). The origins of allergy from a systems approach. *Annals of Allergy, Asthma and Immunology, 125*(5), 507–516. https://doi.org/10.1016/j.anai.2020.07.013.

Kuehn, A., Hilger, C., Lehners-Weber, C., Codreanu-Morel, F., Morisset, M., Metz-Favre, C., et al. (2013). Identification of enolases and aldolases as important fish allergens in cod, salmon and tuna: component resolved diagnosis using parvalbumin and the new allergens. *Clinical & Experimental Allergy, 43*(7), 811–822. https://doi.org/10.1111/cea.12117.

Kumar, M., Duraisamy, K., & Chow, B. K. C. (2021). Unlocking the Non-IgE-mediated Pseudo-Allergic reaction puzzle with mas-related G-protein coupled receptor member X2 (MRGPRX2). *Cells, 10*(5). https://doi.org/10.3390/cells10051033.

Lakhneko, O., Danchenko, M., Morgun, B., Kováč, A., Majerová, P., & Škultéty, L. (2020). Comprehensive comparison of clinically relevant grain proteins in modern and traditional bread wheat cultivars. *International Journal of Molecular Sciences, 21*(10). https://doi.org/10.3390/ijms21103445.

Larré, C., Lupi, R., Gombaud, G., Brossard, C., Branlard, G., Moneret-Vautrin, D. A., et al. (2011). Assessment of allergenicity of diploid and hexaploid wheat genotypes: Identification of allergens in the albumin/globulin fraction. *Journal of Proteomics, 74*(8), 1279–1289. https://doi.org/10.1016/j.jprot.2011.03.014.

Leonard, S. A., Caubet, J. C., Kim, J. S., Groetch, M., & Nowak-Wegrzyn, A. (2015). Baked milk- and egg-containing diet in the management of milk and egg allergy. *Journal of Allergy and Clinical Immunology: In Practice, 3*(1), 13–23. https://doi.org/10.1016/j.jaip.2014.10.001.

Lerma-García, M. J., Nicoletti, M., Simó-Alfonso, E. F., Righetti, P. G., & Fasoli, E. (2019). Proteomic fingerprinting of apple fruit, juice, and cider via combinatorial peptide ligand libraries and MS analysis. *Electrophoresis, 40*(2), 266–271. https://doi.org/10.1002/elps.201800320.

Leung, A. S. Y., Wong, G. W. K., & Tang, M. L. K. (2018). Food allergy in the developing world. *Journal of Allergy and Clinical Immunology, 141*(1), 76–78.e1. https://doi.org/10.1016/j.jaci.2017.11.008.

Levin, M., Apostolovic, D., Biedermann, T., Commins, S. P., Iweala, O. I., Platts-Mills, T. A. E., et al. (2019). Galactose α-1,3-galactose phenotypes: Lessons from various patient populations. *Annals of Allergy, Asthma and Immunology, 122*(6), 598–602. https://doi.org/10.1016/j.anai.2019.03.021.

Lexhaller, B., Colgrave, M. L., & Scherf, K. A. (2019). Characterization and relative quantitation of wheat, rye, and barley gluten protein types by liquid chromatography–tandem mass spectrometry. *Frontiers in Plant Science, 10*. https://doi.org/10.3389/fpls.2019.01530.

Li, H., Byrne, K., Galiamov, R., Mendoza-Porras, O., Bose, U., Howitt, C. A., et al. (2018). Using LC-MS to examine the fermented food products vinegar and soy sauce for the presence of gluten. *Food Chemistry, 254*, 302–308. https://doi.org/10.1016/j.foodchem.2018.02.023.

Li, H., Jia, Y., Peng, W., Zhu, K., Zhou, H., & Guo, X. (2018). High hydrostatic pressure reducing allergenicity of soy protein isolate for infant formula evaluated by ELISA and proteomics via Chinese soy-allergic children's sera. *Food Chemistry, 269*, 311–317. https://doi.org/10.1016/j.foodchem.2018.07.001.

Li, H., Liang, X., Zhou, B., Chen, X., Hong, Y., Zhou, R., et al. (2020). A proteomic analysis of peanut seed at different stages of underground development to understand the changes of seed proteins. *Plos One, 15*(12), Article e0243132. https://doi.org/10.1371/journal.pone.0243132.

Liao, Y. S., Kuo, J. H., Chen, B. L., Tsuei, H. W., Lin, C. Y., Lin, H. Y., et al. (2017). Development and validation of the detection method for wheat and barley glutens using mass spectrometry in processed foods. *Food Analytical Methods, 10*(8), 2839–2847. https://doi.org/10.1007/s12161-017-0827-0.

Lieberman, J. A., Gupta, R. S., Knibb, R. C., Haselkorn, T., Tilles, S., Mack, D. P., et al. (2021). The global burden of illness of peanut allergy: a comprehensive literature review. *Allergy: European Journal of Allergy and Clinical Immunology, 76*(5), 1367–1384. https://doi.org/10.1111/all.14666.

Lim, A. H., Wong, S., & Nguyen, N. Q. (2021). Eosinophilic Esophagitis and IgG4: is there a relationship? In *Digestive Diseases and Sciences*. doi:10.1007/s10620-020-06788-0.

Lopes, J. P., & Sicherer, S. (2020). Food allergy: epidemiology, pathogenesis, diagnosis, prevention, and treatment. *Current Opinion in Immunology, 66*, 57–64. https://doi.org/10.1016/j.coi.2020.03.014.

Lopez-Pedrouso, M., Lorenzo, J., Gagaoua, M., & Franco, D. (2020). Current trends in proteomic advances for food allergen analysis. *Biology (Basel), 9*(9), 247.

Lupi, R., Masci, S., Rogniaux, H., Tranquet, O., Brossard, C., Lafiandra, D., et al. (2014). Assessment of the allergenicity of soluble fractions from GM and commercial genotypes of wheats. *Journal of Cereal Science, 60*(1), 179–186. https://doi.org/10.1016/j.jcs.2014.02.009.

Mainente, F., Simonato, B., Pasini, G., Franchin, C., Arrigoni, G., & Rizzi, C. (2017). Hen egg white lysozyme is a hidden allergen in Italian commercial ciders. *Food Additives and Contaminants - Part A Chemistry, Analysis, Control, Exposure and Risk Assessment, 34*(2), 145–151. https://doi.org/10.1080/19440049.2016.1265673.

Malalgoda, M., Meinhardt, S. W., & Simsek, S. (2018). Detection and quantitation of immunogenic epitopes related to celiac disease in historical and modern hard red spring wheat cultivars. *Food Chemistry, 264*, 101–107. https://doi.org/10.1016/j.foodchem.2018.04.131.

Manfredi, A., Mattarozzi, M., Giannetto, M., & Careri, M. (2015). Multiplex liquid chromatography-tandem mass spectrometry for the detection of wheat, oat, barley and rye prolamins towards the assessment of gluten-free product safety. *Analytica Chimica Acta, 895*, 62–70. https://doi.org/10.1016/j.aca.2015.09.008.

Martínez-Esteso, M. J., Nørgaard, J., Brohée, M., Haraszi, R., Maquet, A., & O'Connor, G (2016). Defining the wheat gluten peptide fingerprint via a discovery and targeted proteomics approach. *Journal of Proteomics, 147*, 156–168. https://doi.org/10.1016/j.jprot.2016.03.015.

Martos, G., López-Fandiño, R., & Molina, E. (2013). Immunoreactivity of hen egg allergens: influence on in vitro gastrointestinal digestion of the presence of other egg white proteins and of egg yolk. *Food Chemistry, 136*(2), 775–781. https://doi.org/10.1016/j.foodchem.2012.07.106.

Marzano, V., Tilocca, B., Fiocchi, A. G., Vernocchi, P., Levi Mortera, S., Urbani, A., et al. (2020). Perusal of food allergens analysis by mass spectrometry-based proteomics. *Journal of Proteomics, 215*, Article 103636. https://doi.org/10.1016/j.jprot.2020.103636.

Mattarozzi, M., Bignardi, C., Elviri, L., & Careri, M. (2012). Rapid shotgun proteomic liquid chromatography-electrospray ionization-tandem mass spectrometry-based method for the lupin (Lupinus albus L.) multi-allergen determination in foods. *Journal of Agricultural and Food Chemistry, 60*(23), 5841–5846. https://doi.org/10.1021/jf302105r.

Mattison, C. P., Bren-Mattison, Y., Vant-Hull, B., Vargas, A. M., Wasserman, R. L., & Grimm, C. C. (2016). Heat-induced alterations in cashew allergen solubility and IgE binding. *Toxicology Reports, 3*, 244–251. https://doi.org/10.1016/j.toxrep.2015.12.009.

Matysiak-Budnik, T., Candalh, C., Dugave, C., Namane, A., Cellier, C., Cerf-Bensussan, N., et al. (2003). Alterations of the intestinal transport and processing of gliadin peptides in celiac disease. *Gastroenterology, 125*(3), 696–707. https://doi.org/10.1016/S0016-5085(03)01049-7.

McClain, S., Stevenson, S. E., Brownie, C., Herouet-Guicheney, C., Herman, R. A., Ladics, G. S., et al. (2018). Variation in seed allergen content from three varieties of soybean cultivated in nine different locations in iowa, illinois, and indiana. *Frontiers in Plant Science, 9*. https://doi.org/10.3389/fpls.2018.01025.

McDonald, B. D., Jabri, B., & Bendelac, A. (2018). Diverse developmental pathways of intestinal intraepithelial lymphocytes. *Nature Reviews Immunology*, 514–525. https://doi.org/10.1038/s41577-018-0013-7.

Michelet, M., Balbino, B., Guilleminault, L., & Reber, L. L. (2021). IgE in the pathophysiology and therapy of food allergy. *European Journal of Immunology, 51*(3), 531–543. https://doi.org/10.1002/eji.202048833.

Mills, E. N. C., Adel-Patient, K., Bernard, H., De Loose, M., Gillard, N., Huet, A. C., et al. (2019). Detection and quantification of allergens in foods and minimum eliciting doses in food-allergic individuals (ThRAll). *Journal of AOAC International, 102*(5), 1346–1353. https://doi.org/10.1093/jaoac/102.5.1346.

Mohan, D., Wansley, D. L., Sie, B. M., Noon, M. S., Baer, A. N., Laserson, U., et al. (2018). PhIP-Seq characterization of serum antibodies using oligonucleotide-encoded peptidomes. *Nature Protocols, 13*(9), 1958–1978. https://doi.org/10.1038/s41596-018-0025-6.

Monaci, L., Losito, I., De Angelis, E., Pilolli, R., & Visconti, A. (2013). Multi-allergen quantification of fining-related egg and milk proteins in white wines by high-resolution mass spectrometry. *Rapid Communications in Mass Spectrometry, 27*(17), 2009–2018. https://doi.org/10.1002/rcm.6662.

Monaci, L., De Angelis, E., Bavaro, S. L., & Pilolli, R. (2015). High-resolution OrbitrapTM-based mass spectrometry for rapid detection of peanuts in nuts. *Food Additives & Contaminants: Part A, 32*(10), 1607–1616. https://doi.org/10.1080/19440049.2015.1070235.

Monaci, L., De Angelis, E., Guagnano, R., Ganci, A. P., Garaguso, I., Fiocchi, A., et al. (2020). Validation of a MS based proteomics method for milk and egg quantification in cookies at the lowest VITAL levels: an alternative to the use of precautionary labeling. *Foods, 9*(10), 1489. https://doi.org/10.3390/foods9101489.

Monaco, D. R., Sie, B. M., Nirschl, T. R., Knight, A. C., Sampson, H. A., Nowak-Wegrzyn, A., et al. (2021). Profiling serum antibodies with a pan allergen phage library identifies key wheat allergy epitopes. *Nature Communications, 12*(1). https://doi.org/10.1038/s41467-020-20622-1.

Moneret-Vautrin, D. A., Guérin, L., Kanny, G., Flabbee, J., Frémont, S., & Morisset, M. (1999). Cross-allergenicity of peanut and lupine: the risk of lupine allergy in patients allergic to peanuts. *Journal of Allergy and Clinical Immunology, 104*(4 I), 883–888. https://doi.org/10.1016/S0091-6749(99)70303-9.

Mueller, G. A., Maleki, S. J., & Pedersen, L. C. (2014). The molecular basis of peanut allergy topical collection on allergens. *Current Allergy and Asthma Reports, 14*(5). https://doi.org/10.1007/s11882-014-0429-5.

Natale, M., Bisson, C., Monti, G., Peltran, A., Garoffo, L. P., Valentini, S., et al. (2004). Cow's milk allergens identification by two-dimensional immunoblotting and mass spectrometry. *Molecular Nutrition and Food Research, 48*(5), 363–369. https://doi.org/10.1002/mnfr.200400011.

New, L. S., Schreiber, A., Stahl-Zeng, J., & Liu, H.-F. (2018). Simultaneous Analysis of Multiple Allergens in Food Products by LC-MS/MS. *Journal of AOAC International, 101*(1), 132–145. https://doi.org/10.5740/jaoacint.17-0403.

Nikolić, J., Nešić, A., Kull, S., Schocker, F., Jappe, U., & Gavrović-Jankulović, M. (2018). Employment of proteomic and immunological based methods for the identification of catalase as novel allergen from banana. *Journal of Proteomics, 175*, 87–94. https://doi.org/10.1016/j.jprot.2018.01.007.

Ogilvie, O., Larsen, N., Sutton, K., Domigan, L., Gerrard, J., Demarais, N., et al. (2020a). A targeted mass spectrometry method for the accurate label-free quantification of immunogenic gluten peptides produced during simulated digestion of food matrices. *MethodsX, 7*, 101076. https://doi.org/10.1016/j.mex.2020.101076.

Ogilvie, O., Roberts, S., Sutton, K., Domigan, L., Larsen, N., Gerrard, J., et al. (2020b). Proteomic modelling of gluten digestion from a physiologically relevant food system: a focus on the digestion of immunogenic peptides from wheat implicated in celiac disease. *Food Chemistry, 333*, Article 127466. https://doi.org/10.1016/j.foodchem.2020.127466.

Ogilvie, O., Roberts, S., Sutton, K., Gerrard, J., Larsen, N., & Domigan, L. (2021). The effect of baking time and temperature on gluten protein structure and celiac peptide digestibility. *Food Research International, 140*, Article 109988. https://doi.org/10.1016/j.foodres.2020.109988.

Ogura, T., Clifford, R., & Oppermann, U. (2019). Simultaneous detection of 13 allergens in thermally processed food using targeted LC-MS/MS Approach. *Journal of AOAC International, 102*(5), 1316–1329. https://doi.org/10.5740/jaoacint.19-0060.

Ortea, I., Cañas, B., Calo-Mata, P., Barros-Velázquez, J., & Gallardo, J. M. (2009a). Arginine kinase peptide mass fingerprinting as a proteomic approach for species identification and taxonomic analysis of commercially relevant shrimp species. *Journal of Agricultural and Food Chemistry, 57*(13), 5665–5672. https://doi.org/10.1021/jf900520h.

Ortea, I., Cañas, B., & Gallardo, J. M. (2009b). Mass spectrometry characterization of species-specific peptides from arginine kinase for the identification of commercially relevant shrimp species. *Journal of Proteome Research, 8*(11), 5356–5362. https://doi.org/10.1021/pr900663d.

Ortea, I., Cañas, B., & Gallardo, J. M. (2011). Selected tandem mass spectrometry ion monitoring for the fast identification of seafood species. *Journal of Chromatography A, 1218*(28), 4445–4451. https://doi.org/10.1016/j.chroma.2011.05.032.

Ozias-Akins, P., & Breiteneder, H. (2019). The functional biology of peanut allergens and possible links to their allergenicity. *Allergy, 74*(5), 888–898. https://doi.org/10.1111/all.13719.

Palacín, A., Rivas, L. A., Gómez-Casado, C., Aguirre, J., Tordesillas, L., Bartra, J., et al. (2012). The involvement of thaumatin-like proteins in plant food cross-reactivity: a multicenter study using a specific protein microarray. *Plos One, 7*(9). https://doi.org/10.1371/journal.pone.0044088.

Pandey, A. K., Sudini, H. K., Upadhyaya, H. D., Varshney, R. K., & Pandey, M. K. (2019). Hypoallergen peanut lines identified through large-scale phenotyping of global diversity panel: Providing hope toward addressing one of the major global food safety concerns. *Frontiers in Genetics, 10*. https:/doi.org/10.3389/fgene.2019.01177.

Papista, C., Gerakopoulos, V., Kourelis, A., Sounidaki, M., Kontana, A., Berthelot, L., et al. (2012). Gluten induces coeliac-like disease in sensitised mice involving IgA, CD71 and transglutaminase 2 interactions that are prevented by probiotics. *Laboratory Investigation, 92*(4), 625–635. https://doi.org/10.1038/labinvest.2012.13.

Pasini, G., Curioni, A., Vegro, M., Pagani, M., Masi, A., Schievano, E., et al. (2012). Extraction and mass spectrometry identification of a major peach allergen Pru p 1. *Journal of the Science of Food and Agriculture, 92*(3), 570–576. https://doi.org/10.1002/jsfa.4609.

Pedrosa, M., Guerrero-Sanchez, V. M., Canales-Bueno, N., Loli-Ausejo, D., Castillejo, M.Á., Quirce, S., et al. (2020). Quercus ilex pollen allergen, Que i 1, responsible for pollen food allergy syndrome caused by fruits in Spanish allergic patients. *Clinical & Experimental Allergy, 50*(7), 815–823. https://doi.org/10.1111/cea.13679.

Peeters, K. A. B. M., Koppelman, S. J., Penninks, A. H., Lebens, A., Bruijnzeel-Koomen, C. A. F. M., Hefle, S. L., et al. (2009). Clinical relevance of sensitization to lupine in peanut-sensitized adults. *Allergy: European Journal of Allergy and Clinical Immunology, 64*(4), 549–555. https://doi.org/10.1111/j.1398-9995.2008.01818.x.

Peters, R. L., Krawiec, M., Koplin, J. J., & Santos, A. F. (2021). Update on food allergy. *Pediatric Allergy and Immunology, 32*(4), 647–657. https://doi.org/10.1111/pai.13443.

Pilolli, R., De Angelis, M., Lamonaca, A., De Angelis, E., Rizzello, C. G., Siragusa, S., et al. (2020). Prototype gluten-free breads from processed durum wheat: Use of monovarietal flours and implications for gluten detoxification strategies. *Nutrients, 12*(12), 1–19. https://doi.org/10.3390/nu12123824.

Piovesana, S., Capriotti, A. L., Caruso, G., Cavaliere, C., La Barbera, G., Zenezini Chiozzi, R., et al. (2016). Labeling and label free shotgun proteomics approaches to characterize muscle tissue from farmed and wild gilthead sea bream (Sparus aurata). *Journal of Chromatography A, 1428*, 193–201. https://doi.org/10.1016/j.chroma.2015.07.049.

Planque, M., Arnould, T., Delahaut, P., Renard, P., Dieu, M., & Gillard, N. (2019). Development of a strategy for the quantification of food allergens in several food products by mass spectrometry in a routine laboratory. *Food Chemistry, 274,* 35–45. https://doi.org/10.1016/j.foodchem.2018.08.095.

Planque, M., Arnould, T., Dieu, M., Delahaut, P., Renard, P., & Gillard, N. (2017). Liquid chromatography coupled to tandem mass spectrometry for detecting ten allergens in complex and incurred foodstuffs. *Journal of Chromatography A, 1530,* 138–151. https://doi.org/10.1016/j.chroma.2017.11.039.

Platts-Mills, T. A. E., Commins, S. P., Biedermann, T., van Hage, M., Levin, M., Beck, L. A., et al. (2020). On the cause and consequences of IgE to galactose-α-1,3-galactose: A report from the National Institute of Allergy and Infectious Diseases Workshop on Understanding IgE-Mediated Mammalian Meat Allergy. *Journal of Allergy and Clinical Immunology, 145*(4), 1061–1071. https://doi.org/10.1016/j.jaci.2020.01.047.

Popescu, F.-D. (2015). Cross-reactivity between aeroallergens and food allergens. *World Journal of Methodology, 5(2),* 31–50. https://doi.org/10.5662/wjm.v5.i2.31.

Prandi, B., Faccini, A., Tedeschi, T., Cammerata, A., Sgrulletta, D., D'Egidio, M. G., et al. (2014). Qualitative and quantitative determination of peptides related to celiac disease in mixtures derived from different methods of simulated gastrointestinal digestion of wheat products. *Analytical and Bioanalytical Chemistry, 406*(19), 4765–4775. https://doi.org/10.1007/s00216-014-7858-9.

Prandi, B., Tedeschi, T., Folloni, S., Galaverna, G., & Sforza, S. (2017). Peptides from gluten digestion: A comparison between old and modern wheat varieties. *Food Research International, 91,* 92–102. https://doi.org/10.1016/j.foodres.2016.11.034.

Prieto, N., Burbano, C., Iniesto, E., Rodríguez, J., Cabanillas, B., Crespo, J. F., et al. (2014). A novel proteomic analysis of the modifications induced by high hydrostatic pressure on hazelnut water-soluble proteins. *Foods, 3*(2), 279–289. https://doi.org/10.3390/foods3020279.

Prodic, I., Stanic-Vucinic, D., Apostolovic, D., Mihailovic, J., Radibratovic, M., Radosavljevic, J., et al. (2018). Influence of peanut matrix on stability of allergens in gastric-simulated digesta: 2S albumins are main contributors to the IgE reactivity of short digestion-resistant peptides. *Clinical & Experimental Allergy, 48*(6), 731–740. https://doi.org/10.1111/cea.13113.

Renz, H., Allen, K. J., Sicherer, S. H., Sampson, H. A., Lack, G., Beyer, K., et al. (2018). Food allergy. *Nature Reviews Disease Primers, 4,* 17098. https:/doi.org/10.1038/nrdp.2017.98.

Reuter, A., Fortunato, D., Garoffo, L. P., Napolitano, L., Scheurer, S., Giuffrida, M. G., et al. (2005). Novel isoforms of Pru av 1 with diverging immunoglobulin E binding properties identified by a synergistic combination of molecular biology and proteomics. *Proteomics, 5*(1), 282–289. https://doi.org/10.1002/pmic.200400874.

Ribeiro, M., Costa, J., Mafra, I., Cabo, S., Silva, A. P., Gonçalves, B., et al. (2020). Natural variation of hazelnut allergenicity: Is there any potential for selecting hypoallergenic varieties? *Nutrients, 12*(7), 1–20. https://doi.org/10.3390/nu12072100.

Robertson, L., Arcos, S. C., Ciordia, S., Carballeda-Sanguiao, N., Mena, M., Sanchez-Alonso, I., et al. (2020). Immunoreactive proteins in the esophageal gland cells of anisakis simplex Sensu Stricto detected by MALDI-TOF/TOF Analysis. *Genes (Basel), 11*(6), 683. https://doi.org/10.3390/genes11060683.

Rogniaux, H., Pavlovic, M., Lupi, R., Lollier, V., Joint, M., Mameri, H., et al. (2015). Allergen relative abundance in several wheat varieties as revealed via a targeted quantitative approach using MS. *Proteomics, 15*(10), 1736–1745. https://doi.org/10.1002/pmic.201400416.

Rosmilah, M., Shahnaz, M., Meinir, J., Masita, A., Noormalin, A., & Jamaluddin, M. (2013). Identification of parvalbumin and two new thermolabile major allergens of thunnus tonggol using a proteomics approach. *International Archives of Allergy and Immunology, 162*(4), 299–309. https://doi.org/10.1159/000354544.

Rost, J., Muralidharan, S., & Lee, N. A. (2020). A label-free shotgun proteomics analysis of macadamia nut. *Food Research International, 129.* https://doi.org/10.1016/j.foodres.2019.108838.

Ruethers, T., Taki, A. C., Johnston, E. B., Nugraha, R., Le, T. T. K., Kalic, T., et al. (2018). Seafood allergy: a comprehensive review of fish and shellfish allergens. *Molecular Immunology, 100*, 28–57. https://doi.org/10.1016/j.molimm.2018.04.008.

Ruethers, T., Taki, A. C., Karnaneedi, S., Nie, S., Kalic, T., Dai, D., et al. (2021). Expanding the allergen repertoire of salmon and catfish. *Allergy, 76*(5), 1443–1453. https://doi.org/10.1111/all.14574.

Sanchez-Trincado, J. L., Gomez-Perosanz, M., & Reche, P. A. (2017). Fundamentals and methods for T- and B-cell epitope prediction. *Journal of Immunology Research, 2017*. https://doi.org/10.1155/2017/2680160.

Santos, A. F., Alpan, O., & Hoffmann, H. (2021). Basophil activation test: mechanisms and considerations for use in clinical trials and clinical practice. *Allergy, 76*(8), 2420–2432. https://doi.org/10.1111/all.14747.

Savage, J., Sicherer, S., & Wood, R. (2016). The Natural History of Food Allergy. *Journal of Allergy and Clinical Immunology: In Practice, 4*(2), 196–203. https://doi.org/10.1016/j.jaip.2015.11.024.

Scaloni, A., Perillo, V., Franco, P., Fedele, E., Froio, R., Ferrara, L., et al. (2002). Characterization of heat-induced lactosylation products in caseins by immunoenzymatic and mass spectrometric methodologies. *Biochimica et Biophysica Acta - Protein Structure and Molecular Enzymology, 1598*(1–2), 30–39. https://doi.org/10.1016/S0167-4838(02)00290-X.

Schalk, K., Koehler, P., & Scherf, K. A. (2018). Quantitation of specific barley, rye, and oat marker peptides by targeted liquid chromatography-mass spectrometry to determine gluten concentrations. *Journal of Agricultural and Food Chemistry, 66*(13), 3581–3592. https://doi.org/10.1021/acs.jafc.7b05286.

Schalk, K., Lang, C., Wieser, H., Koehler, P., & Scherf, K. A. (2017). Quantitation of the immunodominant 33-mer peptide from α-gliadin in wheat flours by liquid chromatography tandem mass spectrometry. *Scientific Reports, 7*, 45092. https://doi.org/10.1038/srep45092.

Scherf, K. A. (2019). Immunoreactive cereal proteins in wheat allergy, non-celiac gluten/wheat sensitivity (NCGS) and celiac disease. Current Opinion in Food Science, 25, 35–41. https://doi.org/10.1016/j.cofs.2019.02.003

Scherf, K. A., Koehler, P., & Wieser, H. (2016). Gluten and wheat sensitivities—an overview. *Journal of Cereal Science, 67*, 2–11. https://doi.org/10.1016/j.jcs.2015.07.008.

Scippa, G. S., Rocco, M., Ialicicco, M., Trupiano, D., Viscosi, V., Di Michele, M., et al. (2010). The proteome of lentil (Lens culinaris Medik.) seeds: discriminating between landraces. *Electrophoresis, 31*(3), 497–506. https://doi.org/10.1002/elps.200900459.

Sealey-Voyksner, J. A., Khosla, C., Voyksner, R. D., & Jorgenson, J. W. (2010). Novel aspects of quantitation of immunogenic wheat gluten peptides by liquid chromatography-mass spectrometry/mass spectrometry. *Journal of Chromatography A, 1217*(25), 4167–4183. https://doi.org/10.1016/j.chroma.2010.01.067.

Sealey-Voyksner, J., Zweigenbaum, J., & Voyksner, R. (2016). Discovery of highly conserved unique peanut and tree nut peptides by LC-MS/MS for multi-allergen detection. *Food Chemistry, 194*, 201–211. https://doi.org/10.1016/j.foodchem.2015.07.043.

Seki, Y., Nakamura, K., Arimoto, C., Kikuchi, H., Yamakawa, H., Nagai, H., et al. (2021). Development of a simple and reliable high-performance liquid chromatography–tandem mass spectrometry approach to simultaneously detect grains specified in food allergen labeling regulation on processed food commodities. *Journal of Chromatography A, 1639*, Article 461877. https://doi.org/10.1016/j.chroma.2021.461877.

Shahali, Y., Sénéchal, H., & Poncet, P. (2019). The use of combinatorial hexapeptide ligand library (CPLL) in allergomics. *Methods in Molecular Biology, 1871*, 393–403. https://doi.org/10.1007/978-1-4939-8814-3_21.

Shaheen, N., Halima, O., Akhter, K. T., Nuzhat, N., Rao, R. S. P., Wilson, R. S., et al. (2019). Proteomic characterization of low molecular weight allergens and putative allergen proteins in lentil (Lens culinaris) cultivars of Bangladesh. *Food Chemistry, 297*. https://doi.org/10.1016/j.foodchem.2019.06.003.

Shaker, M., & Greenhawt, M. (2019). *Peanut allergy: burden of illness*. Allergy and Asthma Proceedings 40 (pp. 290–294). East Providence, United States: OceanSide Publications Inc. https://doi.org/10.2500/aap.2019.40.4240.

Shatalova, A., Shatalov, I., & Lebedin, Y. (2020). X6: A novel antibody for potential use in gluten quantification. *Molecules (Basel, Switzerland), 25*(14). https://doi.org/10.3390/molecules25143107.

Shih, Y. C., Hsiao, J. T., & Sheu, F. (2019). Feasibility of utilizing stable-isotope dimethyl labeling in liquid chromatography–tandem mass spectrometry-based determination for food allergens—case of kiwifruit. *Molecules (Basel, Switzerland), 24*(10). https://doi.org/10.3390/molecules24101920.

Sicherer, S. H., Muñoz-Furlong, A., Godbold, J. H., & Sampson, H. A. (2010). US prevalence of self-reported peanut, tree nut, and sesame allergy: 11-year follow-up. *Journal of Allergy and Clinical Immunology, 125*(6), 1322–1326. https://doi.org/10.1016/j.jaci.2010.03.029.

Sicherer, S. H., & Sampson, H. A. (2010). Food allergy. *Journal of Allergy and Clinical Immunology, 125*(2), S116–S125. https://doi.org/10.1016/j.jaci.2009.08.028.

Sicherer, S. H., & Sampson, H. A. (2018). Food allergy: a review and update on epidemiology, pathogenesis, diagnosis, prevention, and management. *Journal of Allergy and Clinical Immunology, 141*(1), 41–58. https://doi.org/10.1016/j.jaci.2017.11.003.

Siger, A., Czubinski, J., Kachlicki, P., Dwiecki, K., Lampart-Szczapa, E., & Nogala-Kalucka, M. (2012). Antioxidant activity and phenolic content in three lupin species. *Journal of Food Composition and Analysis, 25*(2), 190–197. https://doi.org/10.1016/j.jfca.2011.10.002.

Skypala, I. J., Bull, S., Deegan, K., Gruffydd-Jones, K., Holmes, S., Small, I., et al. (2013). The prevalence of PFS and prevalence and characteristics of reported food allergy; a survey of UK adults aged 18-75 incorporating a validated PFS diagnostic questionnaire. *Clinical & Experimental Allergy, 43*(8), 928–940. https://doi.org/10.1111/cea.12104.

Sogawa, K., Takahashi, Y., Shibata, Y., Satoh, M., Kodera, Y., Nomura, F., et al. (2018). Search for a novel allergen in Hen's egg allergy using an IgE immunoblotting assay. *International Archives of Allergy and Immunology, 176*(3–4), 189–197. https://doi.org/10.1159/000488144.

Sollid, L. M., Tye-Din, J. A., Qiao, S. W., Anderson, R. P., Gianfrani, C., & Koning, F. (2020). Update 2020: nomenclature and listing of celiac disease–relevant gluten epitopes recognized by CD4+ T cells. *Immunogenetics, 72*(1–2), 85–88. https://doi.org/10.1007/s00251-019-01141-w.

Steinke, J. W., Platts-Mills, T. A. E., & Commins, S. P. (2015). The alpha-gal story: lessons learned from connecting the dots. *Journal of Allergy and Clinical Immunology, 135*(3), 589–596. https://doi.org/10.1016/j.jaci.2014.12.1947.

Stella, R., Sette, G., Moressa, A., Gallina, A., Aloisi, A. M., Angeletti, R., et al. (2020). LC-HRMS/MS for the simultaneous determination of four allergens in fish and swine food products. Food Chemistry, 331. doi:10.1016/j.foodchem.2020.127276.

Sun New, L., Schreiber, A., Stahl-Zeng, J., & Liu, H. F. (2018). Simultaneous analysis of multiple allergens in food products by LC-MS/MS. *Journal of AOAC International, 101*(1), 132–145. https://doi.org/10.5740/jaoacint.17-0403.

Suprun, M., Getts, R., Grishina, G., Tsuang, A., Suárez-Fariñas, M., & Sampson, H. A. (2020). Ovomucoid epitope-specific repertoire of IgE, IgG 4, IgG 1, IgA 1, and IgD antibodies in egg-allergic children. *Allergy, 75*(10), 2633–2643. https://doi.org/10.1111/all.14357.

Suprun, M., Getts, R., Raghunathan, R., Grishina, G., Witmer, M., Gimenez, G., et al. (2019). Novel bead-based epitope assay is a sensitive and reliable tool for profiling epitope-specific antibody repertoire in food allergy. *Scientific Reports, 9*(1), 18425. https://doi.org/10.1038/s41598-019-54868-7.

Suzuki, M., Fujii, H., Fujigaki, H., Shinoda, S., Takahashi, K., Saito, K., et al. (2010). Lipocalin-type prostaglandin D synthase and egg white cystatin react with IgE antibodies from children with egg allergy. *Allergology International, 59*(2), 175–183. https://doi.org/10.2332/allergolint.09-OA-0121.

Takahashi, H., Chinuki, Y., Tanaka, A., & Morita, E. (2014). Laminin γ-1 and collagen α-1 (VI) chain are galactose-α-1,3-galactose-bound allergens in beef. *Allergy: European Journal of Allergy and Clinical Immunology, 69*(2), 199–207. https://doi.org/10.1111/all.12302.
Tanner, G. J., Colgrave, M. L., Blundell, M. J., Goswami, H. P., & Howitt, C. A. (2013). Measuring hordein (Gluten) in Beer—A Comparison of ELISA and mass spectrometry. *Plos One, 8*(2), e56452. https://doi.org/10.1371/journal.pone.0056452.
Tjernberg, I., Hamsten, C., Apostolovic, D., & van Hage, M. (2017). IgE reactivity to alpha-Gal in relation to Lyme borreliosis. *Plos One, 12*(9), e0185723. https://doi.org/10.1371/journal.pone.0185723.
Tomm, J. M., van Do, T., Jende, C., Simon, J. C., Treudler, R., von Bergen, M., et al. (2013). Identification of new potential allergens from nile perch (Lates niloticus) and Cod (Gadus morhua). *Journal of Investigational Allergology and Clinical Immunology, 23*(3), 159–167. http://www.jiaci.org/issues/vol23issue3/4.pdf.
Tordesillas, L., Berin, M. C., & Sampson, H. A. (2017). Immunology of food allergy. *Immunity, 47*(1), 32–50. https://doi.org/10.1016/j.immuni.2017.07.004.
Tovoli, F. (2015). Clinical and diagnostic aspects of gluten related disorders. *World Journal of Clinical Cases, 3*(3), 275–284. https://doi.org/10.12998/wjcc.v3.i3.275.
Trampert, D. C., Hubers, L. M., van de Graaf, S. F. J., & Beuers, U. (2018). On the role of IgG4 in inflammatory conditions: lessons for IgG4-related disease. *Biochimica et Biophysica Acta - Molecular Basis of Disease, 1864*(4), 1401–1409. https://doi.org/10.1016/j.bbadis.2017.07.038.
Uvackova, L., Skultety, L., Bekesova, S., McClain, S., & Hajduch, M. (2013). MSE based multiplex protein analysis quantified important allergenic proteins and detected relevant peptides carrying known epitopes in wheat grain extracts. *Journal of Proteome Research, 12*(11), 4862–4869. https://doi.org/10.1021/pr400336f.
Vaillant, J., Modi, P., & Jan, A. (2021). *Atopy. StatPearls [Internet]*. Treasure Island, FL: StatPearls Publishing.
Valenta, R., Hochwallner, H., Linhart, B., & Pahr, S. (2015). Food allergies: the basics. *Gastroenterology, 148*(6), 1120–1131.e4. https://doi.org/10.1053/j.gastro.2015.02.006.
van Den Broeck, H. C., Cordewener, J. H. G., Nessen, M. A., America, A. H. P., & van der Meer, I. M. (2015). Label free targeted detection and quantification of celiac disease immunogenic epitopes by mass spectrometry. *Journal of Chromatography A, 1391*(1), 60–71. https://doi.org/10.1016/j.chroma.2015.02.070.
van der Valk, J. P. M., Gerth van Wijk, R., Baumert, J. L., Nordlee, J. A., Vlieg-Boerstra, B. J., de Groot, H., et al. (2016). Threshold dose distribution and eliciting dose of cashew nut allergy. *Annals of Allergy, Asthma and Immunology, 117*(6), 712–714. https://doi.org/10.1016/j.anai.2016.09.007.
Verma, A. K., Kumar, S., Das, M., & Dwivedi, P. D. (2013). A comprehensive review of legume allergy. *Clinical Reviews in Allergy and Immunology, 45*(1), 30–46. https://doi.org/10.1007/s12016-012-8310-6.
Wal, J. M. (2002). Cow's milk proteins/allergens. *Annals of Allergy, Asthma and Immunology, 89*(6), 3–10. https://doi.org/10.1016/S1081-1206(10)62115-1.
Wallner, M., Erler, A., Hauser, M., Klinglmayr, E., Gadermaier, G., Vogel, L., et al. (2009). Immunologic characterization of isoforms of Car b 1 and Que a 1, the major hornbeam and oak pollen allergens. *Allergy, 64*(3), 452–460. https://doi.org/10.1111/j.1398-9995.2008.01788.x.
Wang, C., Wang, Y., Liu, G., & Fu, L. (2020). Food allergomics based on high-throughput and bioinformatics technologies. *Food Research International, 130*, Article 108942. https://doi.org/10.1016/j.foodres.2019.108942.
Wieser, H., Antes, S., & Seilmeier, W. (1998). Quantitative determination of gluten protein types in wheat flour by reversed-phase high-performance liquid chromatography. *Cereal Chemistry, 75*(5), 644–650. https://doi.org/10.1094/CCHEM.1998.75.5.644.
Yagami, A., & Ebisawa, M. (2019). New findings, pathophysiology, and antigen analysis in pollen-food allergy syndrome. *Current Opinion in Allergy and Clinical Immunology, 19*(3), 218–223. https://doi.org/10.1097/ACI.0000000000000533.

Yagami, A., Nakazawa, Y., Suzuki, K., & Matsunaga, K. (2009). Curry spice allergy associated with pollen-food allergy syndrome and latex fruit-syndrome. *Journal of Dermatology, 36*(1), 45–49. https://doi.org/10.1111/j.1346-8138.2008.00584.x.

Yang, L., Fu, J., & Zhou, Y. (2020). Research progress in atopic march. *Frontiers in Immunology, 11*, 1907. https://doi.org/10.3389/fimmu.2020.01907.

Yu, C. J., Lin, Y. F., Chiang, B. L., & Chow, L. P. (2003). Proteomics and immunological analysis of a novel shrimp allergen, Pen m 2. *Journal of Immunology, 170*(1), 445–453. https://doi.org/10.4049/jimmunol.170.1.445.

Yu, W., Freeland, D. M. H., & Nadeau, K. C. (2016). Food allergy: Immune mechanisms, diagnosis and immunotherapy. *Nature Reviews Immunology, 16*(12), 751–765. https://doi.org/10.1038/nri.2016.111.

CHAPTER

10

Proteomic approaches for authentication of foods of animal origin

Rituparna Banerjee[a], Naveena Basappa Maheswarappa[a], Kiran Mohan[b], Subhasish Biswas[c]

[a]ICAR-National Research Centre on Meat, Chengicherla, Hyderabad, Telangana, India

[b]Department of Livestock Products Technology, Veterinary College, KVAFSU, Bidar, Karnataka, India

[c]Department of Livestock Products Technology, West Bengal University of Animal and Fishery Sciences, Kolkata, India

10.1 Introduction

Food authentication is a process that involves procedures for verifying that the food is in compliance with its label statements and it conforms to the provision of applicable laws and regulations. Consumers have the right to know what is in the food they are consuming so that they can make informed choices about their diet. Producers also have the obligation to provide a clear and reliable information and not knowingly deceive the consumers over the source or content of the food. Authenticity has always been a major concern for consumers, food producers, regulators, scientific community, importers, and exporters, since ancient times (Drivelos & Georgiou, 2012; Posudin, Peiris, & Kays, 2015). Authenticity is important to gain confidence in what we are eating, to pay the correct price of the product, to safeguard health, and to prevent the unlawful activities (Walker, 2017). Assuring the authenticity of food products is a primary concern for preventing economic fraud, which affects both consumers and the processors. Additionally, it is critical for health and safety

Food Proteomics
DOI: https://doi.org/10.1016/B978-0-323-90889-4.00001-4

reasons, since a nondeclared food ingredient may pose a potential public health risk to a specific group of consumers (Spink & Moyer, 2011).

Legal authorities must control that standard requirements are effectively carried out by food producers or traders by means of robust and reliable methodologies in order to assure that fraudulent or inadvertent mislabeling does not happen, which is often done to obtain additional economic gain (Spink & Moyer, 2011). Scientific expertise and technologies in testing the authenticity of different food products are continuously evolving in order to protect consumer interests and public health and to combat the growing problems of food fraud and adulteration. The use of rapid, effective, and reliable analytical methods represents a valuable and irreplaceable tool for the authorities to establish control over food products in market circulation. In the last few decades, several authentication technologies for foods of animal origin have been established or optimized, namely, polymerase chain reaction (PCR) based on deoxyribonucleic acids (DNAs), m-RNA based transcriptomic technologies, immunological technologies based on antigen-antibody reactions, proteomic technologies based on protein identification, spectroscopic technologies based on specific metabolites, and so on. These technologies have their own characteristics and have some advantages as well as disadvantages regarding authentication of food.

Proteomics, the large-scale analysis of proteins in a particular biological system at a specific time (Pandey & Mann, 2000), has been projected as a rapid and sensitive technique for food authenticity assessment (Piñeiro, Barros-Velázquez, Vázquez, Figueras, & Gallardo, 2003). Proteomics approaches can help to overcome the drawbacks of the conventional methods, improving accuracy and sensitivity (Gallardo, Ortea, & Carrera, 2013). Proteins can act as suitable markers for many food quality characteristics from farm to fork. Proteomics-based tools can be applied for the discovery of new protein markers, such as species-specific peptides and then sensitive analytical assays targeting these markers can be developed and validated for qualitative and quantitative assessment of food adulteration and deceptive practices (Gallardo et al., 2013). One of the major drawbacks of DNA-based methods, that is the degradation of DNA in highly processed samples, can be overcome by proteomics-based methods since peptides are quite stable against processing (Buckley, Melton, & Montgomery, 2013). Hence, heat-stable proteins can be selected as targets and nonenzymatic post-translational modifications due to food processing can be monitored. These facts encouraged researchers to develop standard proteomics-based protocols to assess food authenticity.

The current chapter will discuss different proteomic tools and their application for detection of food fraud and authentication in animal sourced foods including determination of species/breed, geographical origin, adulteration or substitution of original ingredients with additives and determination of processing techniques. The chapter will mainly discuss meat and poultry products, but also covers milk, fish, and seafoods authentication by using different proteomic techniques.

10.2 Proteomic approaches

Although many strategies have been implemented to verify the authenticity of foods of animal origin such as protected designation of origin (PDO), protected geographical indication (PGI), certificate of specific characteristics, and so on (Abbas et al., 2018), rapid, effective, accurate, and reliable detection technologies are the fundamental prerequisites to achieve this goal. The most commonly employed proteomic methods for the evaluation of authenticity issues of meat, milk, fish, and seafoods are depicted in Fig. 10.1.

10.2.1 Electrophoresis-based approach

The electrophoretic methods are based on the separation of proteins in the electric field following their extraction from the source. Initially, starch gels were utilized, and later the starch was replaced by polyacrylamide for protein separation. Presently, the electrophoretic separation is conducted on polyacrylamide gels (PAGE) containing a denaturing agent sodium dodecyl sulfate (SDS-PAGE) or by isoelectric focusing (IEF) on the agar or polyacrylamide gel (PAGIF). The IEF method consists of the

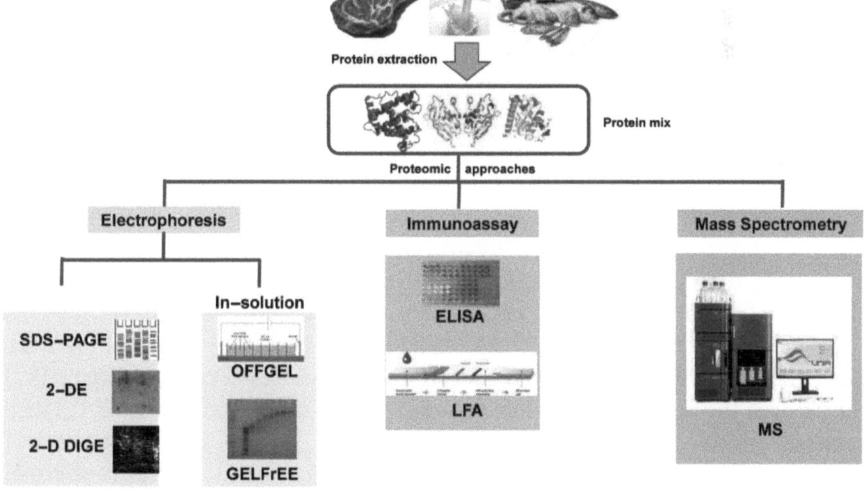

FIG. 10.1 Proteomic approaches for authentication of animal origin foods.

separation of proteins on immobilized pH gradient (IPG) gel which works on the principle of pH gradient resulting from the addition of ampholyte. Individual proteins move toward the pH value corresponding to their isoelectric point (pI).

10.2.1.1 Sodium dodecyl sulfate-polyacrylamide gel electrophoresis (SDS-PAGE)

The single most widely used protein separation method according to their size, or molecular weight (MW) is sodium dodecyl sulfate-polyacrylamide gel electrophoresis (SDS-PAGE), often referred to as one-dimensional electrophoresis (1-DE) (Laemmli, 1970). Anionic surfactant SDS present in sample buffer binds and denatures individual protein molecules and imparts an equal negative charge density for all proteins (Aslam, Basit, Nisar, Khurshid, & Rasool, 2017). When an electric field is applied, proteins migrate on polyacrylamide gel; the rate/velocity of which depends on the ratio between the protein's charge and mass. Complex mixture of proteins can be resolved into multiples of bands on a gel. However, the resolution of the gels is very low; multiple proteins with similar MW may appear in a single protein band. Interference of SDS in mass spectrometric (MS) analysis may be a limiting factor but its ability in ensuring size-based separation and protein solubilization outweigh the disadvantages. The SDS-PAGE can resolve the complex protein mix into few protein bands and hence it fails to fractionate hundreds of proteins on a single gel for further characterization in proteomic studies (Table 10.1).

10.2.1.2 2-dimensional gel electrophoresis (2-DE)

It is possible to separate thousands of proteins simultaneously on a single gel using two-dimensional electrophoresis. During the first dimension, proteins are separated with the assistance of iso-electric focusing depending on their isoelectric point (pI) while the second dimension uses SDS-PAGE to separate proteins depending on their size or MW. The combination of the above-mentioned techniques results in the development of 2-D maps with uniform distribution of protein spots visualized across the gel. Upto 10,000 proteins can be simultaneously separated by this technique and the spots that contain less than 1 ng proteins can be identified and determined qualitatively and quantitatively as well (Lopez, 2007) Moreover, intact proteins, changes in protein expression levels, isoforms, and post-translational modifications can also be analyzed.

Potential of the 2-DE technique was first explained by O'Farrell (1975). Thanks to the introduction of ready-made immobilized pH gradient (IPG) strips and the availability of computer software providing tools for the segmentation, alignment, and matching of 2-DE images greatly facilitated the separation, spot detection, and quantitative analysis with great credibility and precision. In-spite of certain limitations associated with the

TABLE 10.1 Summary of selected proteomic studies for identification of origin of animal-based foods.

		Proteomic techniques	Significant findings	References
	Chicken in meat mixes	OFFGEL isoelectric focusing, LC-MS/MS	Detect 0.5% contaminating chicken in pork meat with high confidence	Montowska et al. (2015)
Species identification	Raw and cooked meat and meat mixes from six meat animal species	2DE, MALDI-TOF MS	Myosin light chain (MLC) isoforms as markers for meat species identification	Montowska and Pospiech (2012)
	Detection and identification of rabbit meat-specific peptide-marker	LC-QTOF-MS/MS	49 heat-stable rabbit meat specific peptides	Stachniuk, Sumara, Montowska, and Fornal (2021)
	Differentiation of cheese from cow, sheep, and goat milk and their mixes	HPLC-chip/MS/MS microfluidic system	kappa-casein as a potential marker for detecting origin of milk from cheese	Franck et al. (2010)
	Characterization of species-specific peptides from different seafood	Tandem mass spectrometry (MS/MS) ion monitoring (SMIM)	Differential classification of seven commercial, closely related, species of Decapoda shrimps	Ortea, Cañas, and Gallardo (2011)
	Identification of commercial prawn and shrimp species	Native isoelectric focusing, Tandem mass spectrometry (MS/MS)	Sarcoplasmic calcium-binding proteins (SCPs) were identified by tandem mass spectrometry (MS/MS) as the major species-specific proteins	Ortea et al. (2010)
	Egg yolks of three poultry species (duck, quail, and chicken)	2-DE, MALDI-TOF-MS/MS	The unique protein expression pattern of each species was obtained by analyzing the amino acid sequences of the proteins	Liu et al. (2018)

(Continued)

TABLE 10.1 Cont'd

		Proteomic techniques	Significant findings	References
	Identification and comparative proteomic study of quail and duck egg	2-DE, MALDI-TOF-MS/MS	Interspecies diversities may be associated with the egg white protein functions	Hu et al. (2016)
Geographical origin/ production method	Quantitative proteomic distances between three Spanish bovine breeds	2-DE, MALDI-TOF and MALDI TOF/TOF MS	Candidate biomarkers for molecular breeding programs and authentication of the three Spanish beef breeds	Rodríguez-Vázquez et al. (2020)
	Lamb from different regions of South Africa	Isotope ratio mass spectrometry	Differentiated using ^{15}N and ^{13}C as regions differ in aridity as well as amounts of C3 fodder plants vs savanna-type C4 grasses	Erasmus, Muller, Butler, and Hoffman (2018)
	Organic and conventional Irish beef	Isotope ratio mass spectrometry	Seasonality in animal feeding practices modulated by tissue turnover rates may	Bahar et al. (2008)
	Korean Native Chickens and Commercial Broilers	2-DE, MALDI TOF MS	Understanding of the molecular mechanisms responsible for breed-specific difference in postmortem degradation of protein and meat flavor	Liu et al. (2012)
	Post translational modifications of milk proteins in geographically diverse goat breeds	2-DE, Q-Exactive LC-MS/MS	Investigated PTM in casein and non-casein goat milk proteins and their interactions	Rout and Verma (2021)
	Brazilian artisanal "Coalho" cheese	SDS-PAGE, MS	Revealed peptides that may represent molecular markers of "Coalho" cheese	Silva et al. (2016)
	Differentiation of specialty eggs, such as omega-3-enriched eggs, organic eggs, and free-range eggs	2-DE, LC-MS/MS	A quiescence-specific protein precursor in egg white was identified	Wang et al. (2012)

application of the 2-DE method, this versatile technique is most commonly used in proteome analysis (Table 10.2).

10.2.1.3 2D-differential gel electrophoresis (2D-DIGE)

The advantage of 2D-DIGE over two-dimensional-based technologies is that it enables the comparison of two or more protein samples simultaneously on a single gel (Gade, Thiermann, Markowsky, & Rabus, 2003; Karp, Feret, Rubtsov, & Lilley, 2008). This method is performed by covalently tagging two protein samples (control and treatment) with two distinct N-hydroxysuccinamide derivatives (Cy3, Cy5), which have identical masses and electrophoretic mobility but different fluorescent emission spectra. The labeled extracts are pooled and then separated by 2-DE. An imager scans the gel at the two Cy-dye emission wavelengths, and the image analysis software can identify spots of different intensity by superimposing the images. However, in larger studies while comparing multiple conditions, a novel approach of using a third dye (Cy2) was suggested by Alban et al. (2003) to solve the issues with gel-to-gel variation. The third dye was used to label a third sample, which acts as an internal standard, comprising of equal amounts of each sample to be compared. This standard is then mixed with two of the protein samples that have to be compared and subjected to 2-DE, which enables the spot matching and relative quantification much easier and more accurate.

However, 2DE gel can only visualize 30%–50% of the entire proteome depending on the type of tissue. Low dynamic range of proteins is one challenging problem as highly abundant proteins may obscure the detection of low-abundance ones (Chevalier, 2010). Limited loading capacity of 2-DE is another drawback and depending on the sample, as little as 1.0 mg of protein on a single gel can result in band smearing and degraded resolution (Zuo & Speicher, 2000). Moreover, excision, digestion, and subsequent analysis of multiple gel spots make it a labor-intensive and time-consuming task.

An alternative to circumvent these problems is offered by solution-based separations, which are much easier to automate. Simplification and automation are important prerequisites while characterizing complex biological samples to get fast and affordable results. A wide range of in-solution protein or peptide separation strategies, if used collectively can potentially provide full proteome coverage.

10.2.1.4 OFFGEL electrophoresis

The OFFGEL electrophoresis allows fractionation of proteins or peptides according to their p*I* in a two-phase system followed by their direct recovery in solution. The upper liquid phase is divided in compartments and the lower phase is a conventional rehydrated IPG strip. The sample is diluted in the focusing buffer followed by loading into 12 or 24 well

TABLE 10.2 Summary of selected proteomic studies applied for detection of adulteration/substitution of animal-based foods.

	Proteomic techniques	Significant findings	References
Buffalo, sheep, and goat meat mix	2DE and OFFGEL electrophoresis, MALDI TOF MS	Authenticating meat mixes spiked at minimum 1.0% and 0.1% levels in both raw and cooked meat	Naveena et al. (2018)
Raw and cooked ground sheep meat and buffalo meat up to 0.5% substitution	OFFGEL electrophoresis, Quantitative label-free mass spectrometry	Efficient in detecting a minimum of 0.5% and 1.0% when buffalo meat was spiked with sheep meat in raw and cooked meat mixes	Naveena et al. (2017)
Fresh and cooked chicken in meat mixes	OFFGEL isoelectric focusing, MALDI-TOF MS and LC-ESI-MS/MS	Detection up to 0.5% w/v contaminating chicken in pork meat based on chicken specific tryptic peptide arising from MLC-3	Sentandreu et al. (2010)
Commercially available highly processed meat products containing chicken, duck, goose meat, pork, beef, and turkey	LC–ESI–QTOF MS/MS and LC–ESI–QQQ MS/MS	Species specific peptides of duck, goose, chicken	Fornal and Montowska (2019)
Bovine and porcine meat and bone meal (MBM)	MALDI-TOF MS and HR-QTOF MS	Detection of species specific osteocalcium derived maker peptides with high sensitivity	Balizs et al. (2011)
Porcine (meat and collagen) residues	Lateral flow assay	Rapid identification of porcine residues derived from raw meat, cooked meat, and gelatin down to 0.01%, 1.0%, and 2.5% contamination, respectively	Masiri et al. (2016)
Chicken in meat mixtures	Lateral flow assay	Rapid detection of 0.063–0.125% (w/w) chicken in raw beef, pork, lamb, and rabbit meat mixtures	Hendrickson et al. (2021)

	Proteomic techniques	Significant findings	References
Cow milk in goat milk	lateral flow colloidal gold immunoassay	Rapid detection up to 0.07% of cow milk adulterated with goat milk	Liu et al. (2019)
Cow milk with buffalo milk	Carbon nanoparticle based lateral flow assay	Detect the presence of buffalo's milk in cow's milk at less than a 5% adulteration level	Sharma et al. (2021)
Mixing sheep and goat milk with cow milk	MALDI-TOF	Intact phospholipids as markers of illegal bovine milk adulteration of high-quality milk	Calvano, De Ceglie, Aresta, Facchini, and Zambonin (2013a)
Detection of cow milk in goat milk	Capillary zone electrophoresis	Different migration times for the αs1-casein (αs1-CN) fraction	Lee et al. (2001)
Detection of sheep and goat milk adulteration	MALDI-TOF	Detect cow and goat milk peptide markers up to 5% level of adulteration	Calvano et al. (2012)
Detecting sheep milk in goat and cow milk cheese	LC/ESI-MS/MS	Accurate and suitable for detecting up to 2% of sheep's milk in cow/goat milk cheese	Guarino et al. (2010)
Powdered milk in fresh cow milk	2 DE, MALDI-TOF-MS	Liquid milk samples adulterated with powdered milk at different percentages, and diagnostic peptides were detected down to 1% of adulteration level	Calvano, Monopoli, Loizzo, Faccia, and Zambonin (2013b)
Skim milk powder with plant proteins	HPLC MS	Numerous peptides originating from the major seed proteins of soy and pea could be identified by MS/MS analysis	Luykx et al. (2007)

(Continued)

TABLE 10.2 Cont'd

	Proteomic techniques	Significant findings	References
Discrimination of bovine milk from non-dairy milk by lipids fingerprinting	MALDI-TOF MS	Novel and simple method without organic solvent extractions for the detection of bovine and non-dairy milks based on lipids fingerprint	England, Tang, Kostrzewa, Shahrezaei, and Larrouy-Maumus (2020)
Detection of plant protein adulterated in fluid milk	2 DE, MALDI-TOF MS	β-conglycinin and glycinin were detected in milk adulterated with soy protein, while legumin, vicilin, and convicilin indicated the addition of pea protein, and β-amylase and serpin marked wheat protein	Yang et al. (2018)
Bovine milk powder with vegetable oils or fats	MALDI-QTOF MS	MALDI-QTOF MS as an alternative to classical methods of milk analysis	Gracia et al. (2012)
Detection of possible milk adulteration with plant proteins	UPLC-QTOF-MS	Nontargeted detection of adulterated plant proteins in raw milk	Lu et al. (2017)

TABLE 10.3 Application of proteomic techniques for miscellaneous authenticity issues of animal-based foods.

	Proteomic techniques	Significant findings	References
Unravelling variation in tender and less tender muscle	2-DE, MALDI TOF-TOF MS	Revealed 123 differentially abundant proteins	Kiran et al. (2015)
Breast and thigh muscles of poultry	2-DE, MALDI TOF MS	16 protein spots were differentially expressed in the breast and thigh meat	Liu et al. (2012)
Variation in traditional halal slaughter without stunning versus slaughter with electrical stunning	2D-DIGE MALDI-TOF/TOF MS	Identification of important proteins/peptides that correlates with meat texture and pre-slaughter stress	Kiran, Naveena, Smrutirekha, Baswa Reddy, and Banerjee (2019)
Chicken eggshell brownness	iTRAQ and SCX SDS-PAGE, LC-ESI-MS/MS	Demonstrated the feasibility and sensitivity of iTRAQ-based quantitative proteomics analysis	Li et al. (2016)
Differences between fertilized and unfertilized chicken egg white proteins	2-DE, MALDI-TOF-MS/MS	Six protein spots represented higher intensity in fertilized eggs through 2-DE analysis	Qiu, Liu, Ma, Zhao, and Li (2013)

frames. As there is no fluidic connection between the wells, the proteins are focused through IPG strip and migrate from well to well under the applied electric field until it is positioned at the right well with pH value corresponding to its p*I* and permeates into the solution-filled compartment above the gel strip. High resolution, high sample loading, well-defined pH range, and the flexibility of choosing the desired pH gradient are some of the noted advantages of using IPG technology in OFFGEL electrophoresis (Manadas, Mendes, English, & Dunn, 2010). An additional benefit of OFFGEL electrophoresis is that multiple samples can be fractionated in the same run (Horth, Miller, Preckel, & Wenz, 2006). The time required for the fractionation varies between 24 and 32 h depending on sample composition.

10.2.1.5 Gel-eluted liquid fraction entrapment electrophoresis (GELFrEE)

GELFrEE is a size-based protein separation platform similar to SDS-PAGE to isolate distinct MW fractions with a high degree of resolution and high sample recovery. The sole difference with SDS-PAGE is that at the end of the run, proteins can be recovered in the solution phase for downstream analysis. The GELFrEE 8100 fractionation system consists of a bench top instrument including cathode chamber, sample loading chamber, specially designed single-use eight channels, fraction collection chamber, and anode chamber. The channels consist of precision-cast gel column which measures 2.5 cm length (including stacking and resolving gel). Upon application of an electric current, charged protein molecules start migrating in the stacking gel and separates according to their MW in the resolving gel. Smaller proteins move faster and elute into the fractionation collection chamber, while the larger proteins end up in the later fractions. The three different proprietary Cartridge Kits (5%, 8%, and 12%) from Expendeon Corporation employ large, medium, and short resolving gels to afford partitioning protein complex across the mass range of 3.5–500 kDa, 3.5–150 kDa, and 3.5–50 kDa, respectively. Moreover, particular fractions may be targeted, relating the time of collection to a particular mass range of the sample (Maheswarappa et al., 2021; Banerjee, Naveena, Dasoju, & Ande, 2021).

10.2.2 Immunoassay-based approach

Immunoassay based on antigen-antibody reaction are widely been exploited for food authentication. Among different immunoassay formats, the selection of methods depends upon the target molecule and application. The most commonly used methodologies used for food authentication are enzyme-linked immunosorbent assay (ELISA) and lateral flow assay (LFA).

ELISA is the most widely applied immunoassay method for the detection of food adulteration. An enzyme conjugate is prepared by using a known antigen or antibody adsorbed on the surface of the solid phase, and the enzyme conjugate specifically bind to samples containing antibodies or antigens. Addition of an enzyme substrates produces a visible signal or color, which is measured with a spectrophotometer. The development of color is positively correlated with the amount of the antibody or antigen in the samples. The commonly used ELISA methods for authenticity detection are direct ELISA, sandwich ELISA, and indirect competitive ELISA. High specificity, sensitivity, easy sample preparation, low cost, and less time consumption are some of the advantages of ELISA. In addition, it does not require any complex equipment and is also easily feasible for onsite monitoring (Mandli, El Fatimi, Seddaoui, & Amine, 2018; Thienes, Masiri, Benoit, Barrios-Lopez, & Samuel, 2019). However, this technique shows some limitations due to false positives caused by cross-reactivity in closely related species and proteolysis caused by severe heat treatment. Researchers are therefore dedicated to develop more sensitive, point-of-care (PoC) protein-based methods for food authentication.

LFA or lateral flow strip (LFS) or lateral flow dipstick is a very common technique that has been used to develop PoC devices for application in various fields including food safety and food authenticity (Lim et al., 2017; Matabaro, Ishimwe, Uwimbabazi, & Lee, 2017; Sajid, Kawde, & Daud, 2015). The simple, rapid, user-friendly, and cost-effective characteristics have made them one of the most widespread PoC sensors (Bahadır & Sezgintürk, 2016; Parolo & Merkoçi, 2013). The LFAs consist of a sample pad, mainly from glass fiber, nitrocellulose membrane containing the test and control zones, and conjugate and absorbent pads from cellulose. These components are fitted, laminated, or embedded on a sheet of polyvinyl chloride backing card or a housing chamber. Sample or analyte encounters the sample pad, migrates through the membrane (Ahmed, Bui, & Abbas, 2016) by the action of capillary force, and reaches the conjugate pad. The conjugate pad releases the labeled detection bioreceptor (e.g., an antibody, often conjugated to a nanoparticle for detection) and allows for first analyte-bioreceptor interaction. Analyte-bioreceptor complex continues to migrate and is captured by biorecognition reagents at the test line (T-line) and control line (C-line). The accumulation of detection tags at T- and C-lines produces visible color that can be detected by naked eyes. The visible signal at C-line represents the correctness of the test. The absorbent pad maintains the capillary flow rate and stops back flow of the sample.

10.2.3 Mass spectrometry based approach

Mass spectrometry technologies based on protein and peptide analysis have rapidly evolved and have been increasingly applied for food authentication. Since the amino acid sequence of peptides is more stable than DNA during food processing, they have an incomparable advantage in food adulteration identification, especially for highly processed food products and closely related species (Prandi et al., 2017). Mass spectrometry allows the determination of the exact molecular mass of charged molecules. A mass spectrometer is composed of an ionization source that converts analyte molecule into gas-phase ions, a mass analyzer for the separation of the ionized analyte according to their mass-to-charge ratio, and a detector for the precise quantification of each species coming out the analyzer. The ionization source is preferentially an electrospray ionization (ESI) or a matrix-assisted laser desorption ionization (MALDI) source and the analyzer is commonly a quadrupole (Q), a time of flight (TOF) or an orbitrap analyzer. Mass spectrometers are either one-stage instruments (MS) if they are composed of only one analyzer, or multi-stage instruments (MS^n) if they combine several analyzers in tandem (for instance Q3) or in hybrid (Q-TOF) configurations. The MS^n allow two consecutive stage of molecular mass analysis and by the way are particularly adapted for protein sequencing. For that purpose, the protein is first hydrolyzed by selective proteolytic enzymes (trypsin, chymotrypsin) into a set of peptides, which are then separated based on their mass-to-charge ratio by the mass analyzer. Each peptide is subsequently fragmented into fragment ions that are characteristic of the precursor ion and are separated and detected with a second mass analyzer. Advanced progress in the analysis of protein at the sequence/peptide level using mass spectrometry technologies became possible thanks to the simultaneous advancement in liquid chromatography (LC) and capillary electrophoresis (CE) for in-line separation. These new technological advances in mass spectrometry methodology enable the identification of peptides with femtomole ($I0^{-15}$ mol) to attomole (10^{-18} mol) sensitivity in complex biological samples (Mallick & Kuster, 2010).

10.3 Authentication of animal origin foods

The following section describes different identification/detection methods for authentication of animal-sourced foods. Various issues related to authentication of foods and detailed proteomics workflow is outlined in Fig. 10.2.

10.3.1 Identification of species/breed

A list of selected proteomic studies for identification of origin of animal-based foods is depicted in Table 10.1. A number of different electrophoretic methods have been employed for species/breed identification of foods of animal origin. Hofmann (1985) and Parisi and Aguiari (1985) are among the first researchers who successfully utilized IEF and SDS-PAGE for species identification of aqueous extracts of mammalian muscles. The SDS-PAGE method was also employed by Savage, Richardson, Jolley, Hargin, and Steward (1995) to discriminate frozen mechanically recovered meat (MRM) and hand deboned meat and 5%–10% addition of MRM to red meat and 25% addition of MRM to poultry meat. A multivariate analysis of sarcoplasmic protein profile using IEF allowed species identification in minced beef, pork, and poultry mix with 10% accuracy (Scarpeid, Kraal, & Hildrum, 1998). Even, electrophoretic profile of muscle proteins was exploited to differentiate heated samples and closely related three shrimp species (An, Marshall, Otwell, & Wei, 1989) SDS-PAGE and IEF succeeded in differentiating shrimp from crab (Civera & Parisi, 1991) and from fish, lobster, and cephalopod meat (Rehbein, 1995). However, no single extraction method and electrophoretic system were found to be effective for all applications: for differentiating raw samples, SDS–PAGE and IEF of the sarcoplasmic proteins worked better, while SDS performed better for the cooked one.

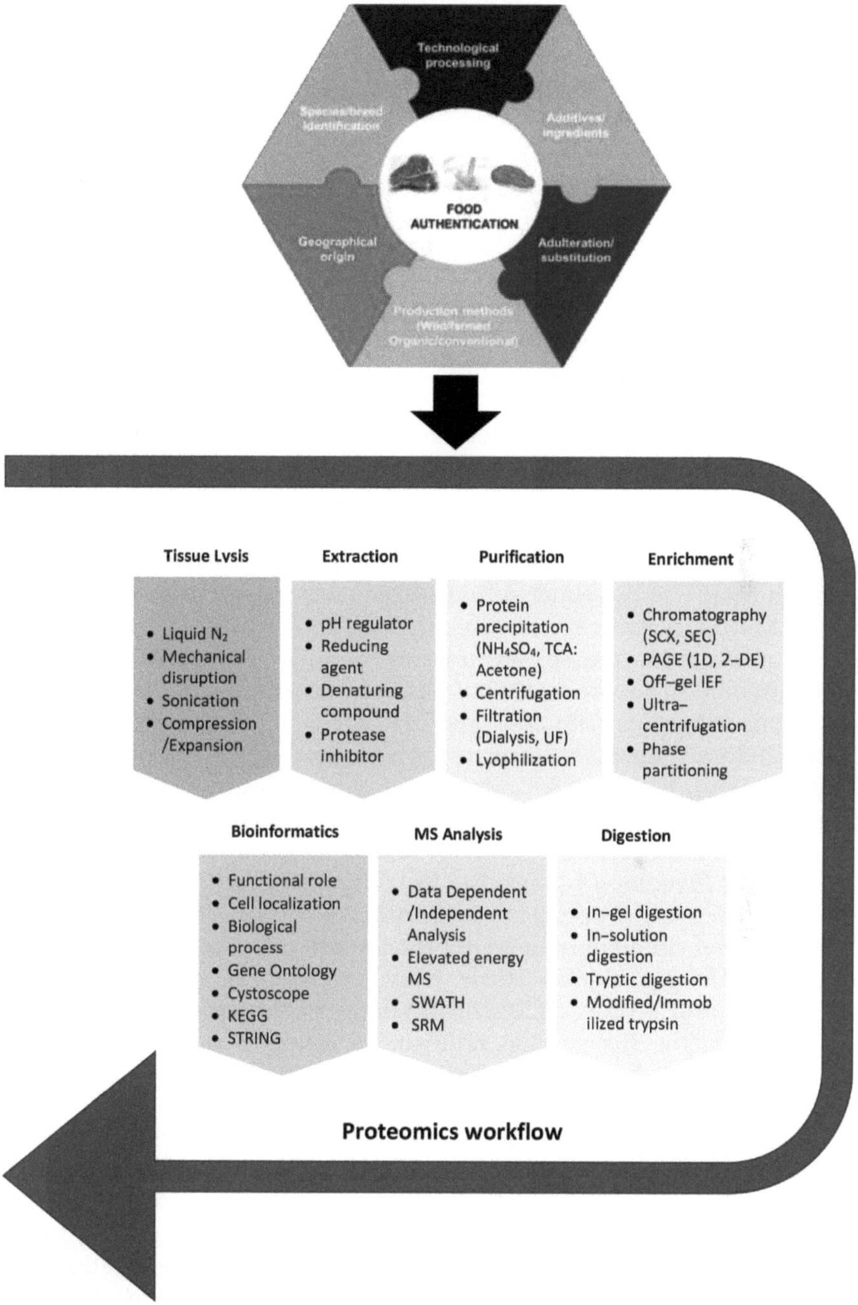

FIG. 10.2 Schematic representation of proteomic work flows used in authentication of animal-based foods.

Parvalbumins, a group of heat-stable sarcoplasmic calcium-binding proteins, found in fish and crustacean muscle were used as markers proteins for the identification of fourteen shrimp and prawn species using native IEF analysis of the water-soluble muscular proteins (Ortea, Cañas, Gallardo, Calo-Mata, & Calo-Mata, 2010). Differences in the amino acid sequences of the parvalbumins were employed in the differentiation of gadoid fishes (Piñeiro, Barros-Velázquez, Sotelo, Pérez-Martín, & Gallardo, 1998), flatfishes (Piñeiro, Barros-Velázquez, Sotelo, & Gallardo, 1999), and hake species (Piñeiro, Vázquez, Marina, Barros-Velázquez, & Gallardo, 2001) to assess for intraspecies variability, as well as geographical location. Nucleoside diphosphate kinase A (NDK A) was also identified as a marker protein for the identification of hake species (Carrera, Cañas, Piñeiro, Vázquez, & Gallardo, 2007; Piñeiro et al., 2001).

Authenticity was investigated much more comprehensively using 2-DE and advancements in mass spectrometry. Four freshwater fish species which are sold as fillets under the generic label of "perch" were distinguished by Berrini, Tepedino, Borromeo, and Secchi (2006) using IEF and 2-DE approach. The species-specific pattern of these four species was revealed by IEF of the water-soluble proteins extracted from fish muscle. López, Marina, Vázquez, and Álvarez (2002) analyzed 2-DE protein profiles of two marine mussel species and identified 37 protein spots that were differentially expressed, 15 of which by means of MS analysis. Analysis of parvalbumins with 2-DE and MALDI-TOF MS allowed identification of 10 closely related fish species from the Merlucciidae family (Carrera, Cañas, Piñeiro, Vázquez, & Gallardo, 2006). Some other proteins are also proposed as potential markers for discriminating tuna species by 2-DE; triose phosphate isomerase, pyruvate kinase, troponin T, and beta-enolase for *Thunnus thynnus, Thunnus alalunga, Thunnus albacares,* and *Thunnus. obesus*, respectively (Pepe et al., 2010, 2012).

Interspecies differences in 2-DE protein patterns of raw and processed meat of cattle, pig, chicken, turkey, duck, and goose were reported by Montowska and Pospiech (2012, 2013) Some of the proteins which were found to be stable during meat aging and resistant to thermal processing, and even identified in highly processed products, were proposed as suitable markers. A set of 25 heat-stable beef, pork, horse, and poultry peptide markers derived from salt-soluble and water-soluble proteins in various processed meat products were identified by Montowska, Alexander, Tucker, and Barrett (2015) using rapid ambient liquid extraction surface analysis mass spectrometry (LESA-MS). Eighteen unique heat-stable peptide biomarkers, which could be used for species identification and adulteration studies, were identified by LC-MS in cooked pork, chicken, duck, beef, and sheep (Wang et al., 2018). LC–MS analyses using either a high-resolution Q-TOF mass analyzer or low-resolution a Q3 mass analyzer was able to successfully identify and monitor unique peptide markers of three poultry species (duck, goose, and chicken) along with pork and beef in highly processed mixed meat samples (Fornal & Montowska, 2019).

Casein and whey protein, the major milk protein fractions serve as biomarkers for authentication of milk and milk products. Variations in their molecular structure may help in differentiating the milk species. Separation and identification of the most abundant milk proteins using different types of electrophoresis is the most common practice in the authentication of dairy products. Among different fractions, αS1-casein was found to have the highest mobility, while performing capillary zone electrophoresis in bovine milk (Lee, Chen, & Lin, 2001). In Bierce-type cheese, the ß-casein of bovine and the para-κ-caseins of caprine milk proved to be efficient markers in the detection of these species' milk (Molina, de Frutos, & Ramos, 2000). Nissen, Bendixen, Ingvartsen, and Rontved (2012) performed different fractionation techniques (acidification, filtration, and centrifugation) followed by LC–MS/MS and identified 635 bovine whey proteins. Similarly, Molle et al. (2009) fractionated bovine milk proteins in combination with ESI-MS/MS and MALDI-MS/MS. Using LC–MS/MS, Zhang et al. (2015) reported change in abundance of acute-phase protein abundance in colostrum and mature milk.

The protein expression pattern of duck, quail, and chicken egg yolks was identified and compared using 2-DE and MALDI-TOF-MS/MS to explore unique biomarkers for authentication (Liu, Qiu, Gao, & Ma, 2018). A total of 32, 35, and 32 protein spots representing 12 different proteins were identified successfully in the egg yolk of duck, quail, and chicken, respectively. Several common proteins, such as vitellogenin (VTG)-related proteins, serum albumin and ovalbumin, found in all three species have different positional distributions on 2D gels. However, specific proteins such as hemopexin and riboflavin-binding protein were only identified in duck egg yolk and β-2-glycoprotein 1 and VTG-3 were only observed in that of chicken. Apoptosis antagonizing transcription factor found in quail and duck were absent from that of chicken. Research findings indicated that chicken and quail have closer evolutionary relationship which may be due to the fact that both chicken and quail are Galliformes, while duck is Anseriformes. The variations in protein expression may be related to the diverse physiological needs, such as cell defence for adapting to distinct living environments.

10.3.2 Identification of geographical origin/production method

Specific culinary practices or organoleptic qualities associated with regional products are compelling the consumers to show an increased interest in the provenance of the foods they consume. Concern about animal welfare and "environment friendly" practices adopted by the local producers also decreasing their confidence in the quality and safety of products produced outside their local region. Animal-derived foods have a distinct value that is linked to the geographical origin or production system from which they are derived, e.g. "PDO," "PGI," "traditional speciality guaranteed," "organic," "free range," or "grass-fed" (Camin et al., 2017; Daley, Abbott, Doyle, Nader, & Larson, 2010). Consumers are even ready to pay a premium for these products (Henchion, McCarthy, Resconi, & Troy, 2014).

The four most common elements in living matter are hydrogen, carbon, nitrogen, and oxygen. Each of these elements has two stable isotopic forms ($^{2}H/^{1}H$, $^{13}C/^{12}C$, $^{15}N/^{14}N$, and $^{18}O/^{16}O$) whose ratio can be of value in authenticity studies (Primrose, Woolfe, & Rollinson, 2010). Stable isotope ratios change with the climatic conditions, geographical origin, soil pedology, and geology of the locations of food ingredients origin (Camin et al., 2017). The H and O isotopic data in food organic matter are related to the same data of water from source region having geographical variability. N and C isotopic data are linked to the climate and agricultural practices (Kelly, Heaton, & Hoogewerff, 2005), whereas geology, distance from the sea, and certain anthropogenic practices affect the S isotopic data (Krouse & Grinenko, 1991).

H and O isotopic data in food organic matter are linked to the H and O isotope data of water from the source region which have geographical variability, N and C isotopes are related to the climate and the agricultural practices, whereas S isotopes are affected by geology, volcanism, distance from the sea, and certain anthropogenic effects (Krouse & Grinenko, 1991). Additionally, isotopic fingerprinting can be combined with other indicators (such as elemental analysis, nuclear magnetic resonance, and gas chromatography) to improve the identification of various food sources.

The characteristic of Iberian pork with PDO is that its diet is based on acorns (González-Martin, González-Pérez, Mendez, Marqués-Macias, & Poveda, 1999), while pigs of different breeds that receive different diets (acorns or feed) are differentiated based on the (González-Martín et al., 2001) ^{13}C and ^{34}S of adipose and liver tissues. Stable isotope ratios were first reported by Boner and Forstel (2004) to determine if a product was organically or conventionally farmed. A $^{13}C/^{12}C$ value above 20% was the limit for European organic farming, as organic cattle fodder consists mainly of C_3 plants while in conventional farming, use of C_4 plants is more usual. Stable ratio mass spectrometry of C, N and S isotopes was used by Schmidt et al. (2005) to differentiate between organic and conventional Irish beef. The differences in their isotopic composition are due in part to the difference in the feed intake (Schmidt et al., 2005). Higher content of ^{15}N in conventional beef compared with organic beef (Bahar et al., 2008) might be a due to mineral fertilizers applied to the soil where conventionally grown animals are fed (Watzka, Buchgraber, & Wanek, 2006). Stable isotope ratios was also used by Heaton, Kelly, Hoogewerff, and Woolfe (2008), to distinguish Brazilian from the UK and Irish beef. In order to comply with EEC Regulation 1906/90, the diet of "corn-fed" chickens must contain at least 50% maize. Rhodes et al. (2010) identified ^{13}C of protein to be a reliable marker of the dietary status of the chickens analyzing stable isotope ratio and distinguished between chicken from birds fed varying levels of maize. Zhao et al. (2013) classified beef coming from different Chinese regions and succeeded to correctly determine the origin of 60%–80% of samples by measuring C and N stable isotope composition by isotope ratio mass spectrometry. Another study aiming to detect the origin of beef was conducted by Liu, Guo, Wei, Shi, and Sun (2013), where researchers used cattle tail hair to measure stable isotope ratios.

In the dairy industry, the monitoring of the geographical origin of a product is a primary concern especially for PDO cheese (Camembert in France, Mozzarella in Italy, and Emmental in Switzerland) products and their different varieties that are manufactured in many countries but are marketed under the same trademark. Extensive research has been carried out for analytical traceability of the geographical origins of these cheese products. Characterization of major milk proteins and determination of post-translational modifications of abundant protein fractions provide a sound basis of cheese making, identifying the origin and quality evaluation of dairy products (Matéos et al., 2009; Roncada et al., 2002; Sørensen, Møller, Vinther, Petersen, & Rasmussen, 2003). The stable isotope ratios of C and N of casein proteins were able to distinguish between Pecorino cheeses produced from sheep milk in Sardinia from those produced in Sicily and Apulia (Manca et al., 2001), and between buffalo mozzarella originating from two regions of Southern Italy (Brescia, Monfreda, Buccolieri, & Carrino, 2005). Stable isotope ratio determinations of the light elements (C, N, O, S) and the heavy element, strontium can enable the reliable detection of regional provenance of butter (Rossmann et al., 2000). Some peptides are highly specific for the cheese type and milk origin. Pappa et al. (2008) distinguished the milk source used for cheese manufacturing by identifying the peptides specific to source of milk. Organic milk was differentiated from conventional milk in Korea (Chung, Park, Yoon, Yang, & Kim, 2014) and Germany (Kaffarnik, Schröder, Lehnert, Baars, & Vetter, 2014; Molkentin, 2009) based on ^{13}C of protein in combination with other isotopes (^{15}N) or other analytical parameters.

Martínez, Slizyte, and Dauksas (2007) differentiated wild from farmed cod using 2-DE and found some specific spots in farmed cod samples with different electrophoretic ability. These spots were not found in wild samples; however, the spots were not sequenced and the proteins were not identified by the authors. Use of isotopic analysis to determine the authenticity of seafood, was less reported especially when compared to the research on meat or dairy products. Carbon and nitrogen stable isotope ratios were used to distinguish farmed from wild Atlantic salmon (Dempson & Power, 2004). The researchers reported that aquaculture salmon was more significantly enriched in ^{13}C but depleted in ^{15}N compared with wild fish. Similarly, few researchers later differentiated between different wild and farmed sea and freshwater fish species (turbot, sea bass, cod, cachara fish, gilthead sea bream) using ^{13}C and ^{15}N with or without other analyses (fatty acid profile and chemical composition).

10.3.3 Identification of adulteration/substitution

Nutritional and functional properties of cheap and easily available soy proteins are mostly exploited for fraudulent substitution of animal protein, which can be detected by ELISA methods (Koppelman, Lakemond, Vlooswijk, & Hefle, 2004). Macedo-Silva, Shimokomaki, Vaz, Yamamoto, and Tenuta-Filho (2001) used ELISA to detect the presence of soy proteins in hamburgers prepared from beef, chicken, and pork. A triplex immunoassay containing microsphere-based flow cytometric immunoassay with MultiAnalyte Profiling for simultaneous detection of pea, wheat, and soy protein has been reported in milk (Haasnoot & du Pre, 2007), and it could be relevant to meat and meat products as well. The adulteration/substitution of meat has also been explored by the use of MS techniques. Using 2DE coupled with LC-MS/MS, Leitner, Castro-Rubio, Marina, and Lindner (2006) detected the soybean proteins added to commercial heat-processed meat products (from pork, chicken, turkey, and beef) due to the presence of unique peptides from a glycinin subunit in all products. Fraudulent substitution of meat with collagen and offal is another concern for the meat industry. The unique amino acid, 4-hydroxyproline (hyp) present in collagen, can be used to calculate the content of collagen in meat and meat products (Anon, 2008). More advanced methods, namely, spectroscopic techniques or chromatographic methods such as LC-MS/MS (Colgrave, Allingham, & Jones, 2008) can also be employed. A proteomic strategy using MALDI-TOF MS was developed by Balizs et al. (2011) to detect the presence of meat and bone meal in foodstuffs together with its animal origin. Osteocalcin was used as the target protein and the molecular mass of intact bovine and porcine osteocalcin were determined to discriminate between these two species.

Sentandreu, Fraser, Halket, Patel, and Bramley (2010) were able to detect the presence of 0.5% w/v contaminating chicken in a mixture with pork meat by enriching the target muscle proteins with OFFGEL IEF followed by analysis of species-specific peptide biomarkers by MALDI–TOF MS and LC–ESI–MS/MS. Similarly, Sentandreu et al. (2010)

differentiated raw and cooked chicken and turkey meats using species-specific peptides obtained from myosin light chain 3 (MLC-3). An adulteration/contamination of horse or pork in a beef matrix was detected at the level of 0.55% using high-performance liquid chromatography–mass spectrometry (HPLC–MS) analysis of tryptic digests of protein extracts (Von Bargen, Brockmeyer, & Humpf, 2014). Naveena et al. (2017) succeeded in identification of MLC-1/MLC-2, a peptide marker for authenticating contaminating buffalo meat in sheep meat mix as low as 0.5% in both raw and cooked meat mixtures involving OFFGEL fractionation and tandem mass spectrometry. In another study, authentication of water buffalo, sheep, and goat meat (raw and cooked) was compared by both in-gel and OFFGEL-based proteomic approach (Naveena et al. 2018). The researchers reported that 2DE and MS-based in-gel method can detect up to 1.0% substitution, whereas OFFGEL electrophoresis and MALDI-PMF can detect even up to 0.1% substitution of sheep and goat meat in buffalo meat.

Immunosensors for detection of meat adulteration have been reported by few researchers (Kuswandi, Gani, & Ahmad, 2017; Lim & Ahmed, 2016; Mandli et al., 2018; Masiri et al., 2016). An electrochemical competitive immunosensor was developed based on antipig IgG polyclonal antibody to identify as low as 0.01% pork adulteration within 20 min. Compared with competitive ELISA methods, which can detect 0.1% pork adulteration within 45 min, the detection limit and detection time were greatly improved by using immunosensor (Mandli et al., 2018). LFA was able to detect 0.01%, 0.1%, and 1% pork from raw meat, beef meatballs, and cooked meat, respectively (Kuswandi et al., 2017; Masiri et al., 2016). Yayla, Karakaya, Akcael, and Yucel (2016) produced monoclonal antibodies (mAbs) against swine IgG to detect porcine materials in food. The mAbs showed no cross-reactivity with beef, sheep, human, horse, mouse, and rabbit by indirect and sandwich ELISA, and these purified antibodies were conjugated with colloidal gold, which was used for the preparation of LFA strips.

Milk-related food fraud with deliberate addition of vegetable fats and oils, melamine urea, and anhydrous milk products is very common (Abernethy & Higgs, 2013; Garcia et al., 2012). In addition, adulteration of high-value goat and buffalo milk with low-priced milk has also been an area of important concern (Calvano et al., 2013a). Thus, to check these unethical practices, and safeguard the interests of consumers, different methodologies such as HPLC (Rodriguez, Ortiz, Sarabia, & Gredilla, 2010), immunoassays, nuclear magnetic resonance (Lamanna, Braca, Di Paolo, & Imparato, 2011) and CE have been used. In recent years, application of MS-based proteomics was found to be a more robust and competent technique in detection of milk adulterations. 2-DE gels coupled with MALDI-TOF have enabled to detect cow milk adulteration in mixtures of buffalo, yak, camel milk mixtures by observing the distribution patterns of α-lactalbumin and β-lactoglobulin, and αs1-casein (Yang et al., 2014). Garcia et al. (2012) used MALDI-TOF MS to identify adulteration of milk with vegetable fats and oils by studying the intact triacylglycerols profile. Adulteration of cheese with less expensive milk is another major concern in the dairy industry. The presence of cow milk in ewe and/or goat milk cheese was verified based on the detection of cow γ2- and γ3-caseins bands on an IEF gel (Kim, Atalla, Mallard, Robert, & Karrow, 2011). MALDI-TOF protein profiling was used to identify whey proteins as biomarkers for investigating the addition of bovine milk to ewe and buffalo milk and in the adulteration of mozzarella cheese (Cozzolino et al., 2001, 2002). Similarly, adulteration of donkey and goat milk with cow, ewe, and buffalo milk was also reported by Di Girolamo et al., 2014. β-lactoglobulin protein was used for the detection and quantification of bovine milk in goat (Chen, Chang, Chung, Lee, & Ling, 2004) caprine or ovine milk (Müller et al., 2008) by means of HPLC-ESI-MS or CE-MS. Detection of 5% level of cow and goat milk adulteration was reported by Calvano, De Ceglie, Monopoli, and Zambonin (2012) by employing in-solution digestion of whole milk samples followed by MALDI-TOF analysis of the tryptic digests. Non-targeted UPLC Q-TOF mass spectrometry approach was able to differentiate authentic milks from their counterparts adulterated with nonmilk proteins (soybean and pea proteins) as low as 1% (Lu, Liu, Gao, Lv, & Yu, 2017). Addition of reconstituted milk to fresh milk is not permitted in some situations, such as during production of PDO milk products. Although the powder derivative, would have the same protein composition as liquid milk, high-temperature processing during milk powder preparation may lead to protein modification. Lactosylated diagnostic peptides for powdered milk were identified using 2-DE and MALDI MS and by detecting these marker peptides, as little as 1% adulteration of powdered milk in liquid milk was reported (Calvano et al. 2013). The use of LFAs have been widely reported for detection of milk adulteration. Martín-Hernández et al. (2009) used monoclonal antibodies against bovine k-casein in a sandwich immunoassay format to develop a rapid dipstick test for the qualitative or semiquantitative detection of adulterated bovine rennet whey in liquid milk or milk powder. Competitive LFS using commercially available polyclonal antibodies conjugated to gold nanoparticles detected the presence of soymilk in milk at 1.75% level within 5 min (Gautam, Sharma, Lata, Rajput, & Mann, 2017). Lateral flow immunoassay technique was used to detect the presence of cow milk in goat, sheep or buffalo milk (Galan-Malo, Mendiara, Razquin, & Mata, 2018; Liu et al., 2019). An on-site carbon nano-particle based competitive lateral flow immunoassay has been developed by Sharma et al. (2021) for the rapid detection of adulteration of cow milk with buffalo milk. Proteomic studies explored for detection of adulteration of animal-based foods are summarized in Table 10.2.

10.3.4 Identification of additives/ingredients

In the meat industry, a vast number of additives namely colorant, aromas, preservatives, or some reducing chemicals can be used to increase fresh meat appearance and shelf-life. Smoke aroma can fraudulently be used as a substitute of natural smoking of meat. Preservatives can make the meat appear fresh much longer than normal. Enzymes that can be used as blood-based binding agents are also used as additives. These are added to meat cuts or minced meat to form portions of desired mass and shape. During the binding process, thrombin cleaves fibrinogen to fibrinopeptides A and B, which was detected in concentrations down to 5% by LC-MS/MS (Grundy et al., 2007; Grundy et al., 2008).

Addition of water in meat is a real problem (Elliot, 2007) as fraudulent addition amounts to selling water in the price of meat. Determination of water/protein ratio serve as a clear indication of water addition. However, the addition of protein and phosphate increases the water-binding leading to water/protein ratio close to normal. In this case, detection of foreign protein or phosphate might offer evidence of fraudulent practice (Ballin, 2010).

10.3.5 Identification of processing treatment

Fresh meat and fish products are still preferred by consumers who are willing to pay more for fresh than frozen–thawed products. Meat or fish on thawing loses exudate and characteristic flavor and nutritional value (water-soluble vitamins, minerals) are partially lost, during thawing. The texture of meat is also altered by the formation of ice crystals, which damages the muscle structure. Subsequently, water activity on the surface is increased supporting the growth of microorganisms and a shorter shelf-life. Therefore, meat or fish that had been frozen prior to marketing must be labeled correctly (i.e., previously frozen or "defrosted"); otherwise, it is considered as fraudulent practice (Bozzetta et al., 2012; Ghidini, Varrà, & Zanardi, 2019). To detect and avoid such fraud, different analytical approaches have been investigated. Enzymatic methods commonly used for distinguishing fresh and thawed meat are β-hydroxyacyl-CoA-dehydrogenase (HADH) method (Gottesmann & Hamm, 1983), which measures the conversion rate of NADH to NAD+ by monitoring the decrease in absorption at 340 nm by spectrophotometer and the API-ZYM system, which is based on a semi-quantitative method for simultaneous measurement of 19 different enzymes (Ellerbroek, Lichtenberg, & Weise, 1995; Toldra, Torrero, & Flores, 1991). Toldra et al. (1991) reported that enzymatic activity of esterase-lipase, β-glucuronidase, and α-glucosidase differed significantly between fresh and thawed pork. To discriminate between fresh and freeze-thawed pork, Kim, Liesse, Kemp, and Balan (2015) identified a total of 450 protein spots from meat exudates out of which 22 proteins, mainly myofibrillar protein, myoglobin, were selected using 2DE-MS-based proteomics approach.

During post-harvest processing food proteins are subjected changes that influence the nutritional value, shelf-life, as well as sensory properties (color, flavor, texture/tenderness, juiciness). Thermal processing is one of the most commonly employed method to improve the nutritional and organoleptic properties and improve the food safety. Oxidation modifications such as carbonylation, thiol oxidation, and Maillard glycation are the protein modifications most frequently reported in foods that are subjected to thermal processing (Pischetsrieder & Baeuerlein, 2009). Protein–carbonyl compounds formed during carbonylation are markers of protein oxidation, and several carbonylated proteins and protein oxidation sites in milk (Meyer et al., 2011), meat (Promeyrat et al., 2011), and fish (Pazos et al., 2011) have been detected using 2-DE and MS/MS. At high temperature (140–165°C) processing, Maillard reaction products are formed, which make it possible to differentiate between long-term low temperature and short-term high temperature cooking. Boiled, fried and grilled meat can be distinguished by development of Maillard products (Skog, Johansson, & Jagerstad, 1998) during heating. Acrylamide, which is formed above 120°C (Mottram, Wedzicha, & Dodson, 2002), can be used as an indicator of products exposed to high temperature using HPLC (Paleologos & Kontominas, 2007) or LC-MS/MS (Granby & Fagt, 2004). Montowska and Spychaj (2018) developed a label-free quantification method utilizing high-resolution MS to authentify cooked and smoked sausages.

During milk processing, phosphorylation, glycosylation, deamidation or other modifications produce changes in molecular mass and/or net charge of proteins; thus, 2-DE is a good technique for separating protein isoforms formed by these changes (Manso, Léonil, Jan, & Gagnaire, 2005). Holland, Deeth, and Alewood (2004) characterized glycosylation and phosphorylation sites in bovine κ-caseins using 2-DE coupled with MALDI-TOF-MS and nano ESI-MS/MS. 2-DE was used by Larsen, Wedholm-Pallas, Lindmark-Månsson, and Andrén (2010) for visualization of disulfide-linked complexes in sweet whey and rennet casein fractions made from low-pasteurized milk and also to detect the changes in milk heat treated at 90°C upto 3 min (Chevalier, Hirtz, Sommerer, & Kelly, 2009). Lactosylation sites on milk proteins in processed milk products were characterized with MALDI-TOF-MS and nLC-ESI-LIT-MS/MS (Arena et al., 2010). MALDI-TOF MS can detect the addition of UHT milk to raw or pasteurized milk with a limit of detection of 10% (Pinto et al., 2012). An integrated MALDI-TOF-MS platform for the combined peptidomic and proteomic profiling of bovine, water buffalo, ovine, and goat milk was done by Sassi, Arena, and Scaloni (2015) to identify markers of thermal treatments. A scheduled multiple reaction monitoring method was adopted by Meltretter, Wust, and Pischetsrieder (2013), 2014) to semi-quantify site-specific post-translationally modified β-lactoglobulin in heated milk and dairy products to detect marker candidates for milk processing. Proteomic techniques utilized to address several other authenticity issues in animal-origin foods is depicted in Table 10.3.

10.4 Conclusion

Food products of animal origin are more often vulnerable to food fraud incidents. Authentication of foods is of importance to gain consumers' trust in a specific brand and to prevent economic losses in an already sensitive sector. A wide variety of proteomic techniques are being developed to safeguard the consumers and screen for adulteration in the food chain. However, the major challenges regarding the application of proteomics in food authentication are extraction of protein from complex matrices. As pre-analytical factors largely affect the outcome of proteomic studies, processing, sample extraction and storage should be described in detail. Efforts should be directed to standardize the extraction and enrichment steps for identification of low-abundant proteins in order to increase the sensitivity as well as reproducibility of those techniques. Sample collection should be done in a correct way to avoid intraspecies variability, season changes within a year, or developmental status (age) of animal for studies involving geographical origin of food products. It is anticipated that optimization of high-throughput MS methods with high-resolution power will further increase the specificity and sensitivity of proteomic assays in future. The development of new bioinformatic tools and the application of existing ones to proteomic studies will definitely play a key role in the food authentication studies. Future research should also be directed toward development of ideal ultralow-cost, affordable PoC devices for on-site detection of food adulteration. Another important future research aspect is the multiple use of a single device to save the time, cost and resources. Stability of such multi-use device along with the enhancement of stability of reagents and probes against the backdrop of temperature and humidity variation should also be studied comprehensively. Finally, integration of genomics, transcriptomics, and metabolomics with proteomics will become part of the routine pipeline work in the very near future and will definitely add to the quality and reliability of proteomics results.

References

Abbas, O., Zadravec, M., Baeten, V., Mikuš, T., Lešić, T., Vulić, A., et al. (2018). Analytical methods used for the authentication of food of animal origin. *Food Chemistry, 246*, 6–17.

Abernethy, G., & Higgs, K. (2013). Rapid detection of economic adulterants in fresh milk by liquid chromatography–tandem mass spectrometry. *Journal of Chromatography A, 1288*, 10–20.

Ahmed, S., Bui, M. P. N., & Abbas, A. (2016). Paper-based chemical and biological sensors: Engineering aspects. *Biosensors & Bioelectronics, 77*, 249–263.

Alban, A., David, S. O., Bjorkesten, L., Andersson, C., Sloge, E., Lewis, S., et al. (2003). A novel experimental design for comparative two-dimensional gel analysis: Two-dimensional difference gel electrophoresis incorporating a pooled internal standard. *Proteomics, 3*, 36–44.

An, H., Marshall, M. R., Otwell, W. S., & Wei, C. I. (1989). Species identification of raw and boiled shrimp by a urea gel isoelectric focusing technique. *Journal of Food Science, 54*(2), 233–236.

Anon (2008). Commission Regulation (EC) No 903/2008 of 17 September 2008 on special conditions for granting export refunds on certain pigmeat products. *L249. Official Journal of the European Union*, 3–8.

Arena, S., Renzone, G., Novi, G., Paffetti, A., Bemardini, G., Santucci, A., et al. (2010). Modern proteomic methodologies for the characterization of lactosylation protein targets in milk. *Proteomics, 10*, 3414–3434.

Aslam, B., Basit, M., Nisar, M. A., Khurshid, M., & Rasool, M. H. (2017). Proteomics: Technologies and their applications. *Journal of Chromatographic Science, 55*, 182–196.

Bahadır, E. B., & Sezgintürk, M. K. (2016). Lateral flow assays: Principles, designs and labels. *Trends Analyt. Chem., 82*, 286–306.

Bahar, B., Schmidt, O., Moloney, A. P., Scrimgeour, C. M., Begley, I. S., & Monahan, F. J. (2008). Seasonal variation in the C, N and S stable isotope composition of retail organic and conventional Irish beef. *Food Chemistry, 106*(3), 1299–1305.

Balizs, G., Weise, B., Rozycki, C., Opialla, T., Sawada, S., Zagon, J., et al. (2011). Determination of osteocalcin in meat and bone meal of bovine and porcine origin using matrix assisted laser desorption ionization/time-of-flight mass spectrometry and high resolution hybrid mass spectrometry. *Analytica Chimica Acta, 693*, 89–99.

Ballin, N. Z. (2010). Authentication of meat and meat products. *Meat Science, 86*(3), 577–587.

Banerjee, R., Naveena, B. M., Dasoju, S., & Ande, S. (2021). OFFGEL and GELFrEE fractionation: Novel liquid-phase protein recovery strategies in proteomics studies. *TrAC Trends in Analytical Chemistry, 140*, Article 116282.

Berrini, A., Tepedino, V., Borromeo, V., & Secchi, C. (2006). Identification of freshwater fish commercially labelled "perch" by isoelectric focusing and two-dimensional electrophoresis. *Food Chemistry, 96*, 163–168.

Boner, M., & Förstel, H. (2004). Stable isotope variation as a tool to trace the authenticity of beef. *Anal. Bioanal. Chem., 378*(2), 301–310.

Bozzetta, E., Pezzolato, M., Cencetti, E., Varello, K., Abramo, F., Mutinelli, F., et al. (2012). Histology as a valid and reliable tool to differentiate fresh from frozen-thawed fish. *Journal of Food Protection, 75*, 1536–1541.

Brescia, M. A., Monfreda, M., Buccolieri, A., & Carrino, C. (2005). Characterisation of the geographical origin of buffalo milk and mozzarella cheese by jeans of analytical and spectroscopic determinations. *Food Chemistry, 89*, 139–147.

Buckley, M., Melton, N. D., & Montgomery, J. (2013). Proteomics analysis of ancient food vessel stitching reveals >4000-year-old milk protein. *Rapid Communications in Mass Spectrometry, 27*(4), 531–538.

Calvano, C. D., De Ceglie, C., Aresta, A., Facchini, L. A., & Zambonin, C. G. (2013a). MALDI-TOF mass spectrometric determination of intact phospholipids as markers of illegal bovine milk adulteration of high-quality milk. *Anal Bioanal Chem, 405*(5), 1641–1649.

Calvano, C. D., De Ceglie, C., Monopoli, A., & Zambonin, C. G. (2012). Detection of sheep and goat milk adulterations by direct MALDI-TOF MS analysis of milk tryptic digests. *Journal of Mass Spectrometry, 47*(9), 1141–1149.

Calvano, C. D., Monopoli, A., Loizzo, P., Faccia, M., & Zambonin, C. (2013b). Proteomic approach based on MALDI-TOF MS to detect powdered milk in fresh cow's milk. *Journal of Agricultural and Food Chemistry, 61*(8), 1609–1617.

Camin, F., Boner, M., Bontempo, L., Fauhl-Hassek, C., Kelly, S. D., Riedl, J., et al. (2017). Stable isotope techniques for verifying the declared geographical origin of food in legal cases. *Trends in Food Science & Technology, 61*, 176–187.

Carrera, M., Cañas, B., Piñeiro, C., Vázquez, J., & Gallardo, J. M. (2006). Identification of hake and grenadier species by proteomic analysis of the parvalbumin fraction. *Proteomics, 6*, 5278–5287.

Carrera, M., Cañas, B., Piñeiro, C., Vázquez, J., & Gallardo, J. M. (2007). De novo mass spectrometry sequencing and characterization of species-specific peptides from nucleoside diphosphate kinase B for the classification of commercial fish species belonging to the family Merlucciidae. *Journal of Proteome Research, 6*(8), 3070–3080.

Chen, R. K., Chang, L. W., Chung, Y. Y., Lee, M. H., & Ling, Y. C. (2004). Quantification of cow milk adulteration in goat milk using high-performance liquid chromatography with electrospray ionization mass spectrometry. *Rapid Communications in Mass Spectrometry, 18*(10), 1167–1171.

Chevalier, F. (2010). Highlights on the capacities of "Gel-based" proteomics. *Proteome Science, 28*(8), 23.

Chevalier, F., Hirtz, C., Sommerer, N., & Kelly, A. L. (2009). Use of reducing/nonreducing two-dimensional electrophoresis for the study of disulfide-mediated interactions between proteins in raw and heated bovine milk. *Journal of Agricultural and Food Chemistry, 57*, 5948–5955.

Chung, I. M., Park, I., Yoon, J. Y., Yang, Y. S., & Kim, S. H. (2014). Determination of organic milk authenticity using carbon and nitrogen natural isotopes. *Food Chemistry, 160*, 214–218.

Civera, T., & Parisi, E. (1991). The use of sodium dodecyl sulphate polyacrylamide gel electrophoresis for the identification of raw and cooked shrimps and crabs. *Italian Journal of Food Science, 3*, 149–157.

Colgrave, M. L., Allingham, P. G., & Jones, A. (2008). Hydroxyproline quantification for the estimation of collagen in tissue using multiple reaction monitoring mass spectrometry. *J. Chromatogr. A., 1212*(1–2), 150–153.

Cozzolino, R., Passalacqua, S., Salemi, S., & Garozzo, D. (2002). Identification of adulteration in water buffalo mozzarella and in ewe cheese by using whey proteins as biomarkers by matrix-assisted laser desorption/ionization mass spectrometry. *Journal of Mass Spectrometry, 37*(9), 985–991.

Cozzolino, R., Passalacqua, S., Salemi, S., Malvagna, P., Spina, E., & Garozzo, D. (2001). Identification of adulteration in milk by matrix-assisted laser desorption/ionization time-of-flight mass spectrometry. *Journal of Mass Spectrometry, 36*(9), 1031–1037.

Daley, C. A., Abbott, A., Doyle, P. S., Nader, G. A., & Larson, S. (2010). A review of fatty acid profiles and antioxidant content in grass-fed and grain-fed beef. *Nutrition journal, 9*(1), 10.

Dempson, J. B., & Power, M. (2004). Use of stable isotopes to distinguish farmed from wild Atlantic salmon, *Salmo salar*. *Ecology of freshwater fish, 13*, 176–184.

Di Girolamo, F., Masotti, A., Salvatori, G., Scapaticci, M., Muraca, M., & Putignani, L. (2014). A sensitive and effective proteomic approach to identify she-donkey's and goat's milk adulterations by MALDI-TOF MS fingerprinting. *International Journal of Molecular Sciences, 15*(8), 13697–13719.

Drivelos, S. A., & Georgiou, C. A. (2012). Multi-element and multi-isotope-ratio analysis to determine the geographical origin of foods in the European Union. *TrAC - Trends in Analytical Chemistry, 40*, 38–51.

Ellerbroek, L. I., Lichtenberg, G., & Weise, E. (1995). Differentiation between fresh and thawed meat by an enzyme profile test. *Meat Science, 40*(2), 203–209.

Elliot, V. (2007). Chicken fillets 'secretly pumped up with water' to increase weight. www.timesonline.co.uk/tol/life_and_style/food_and_drink/article1980529.ece (Accessed 7 June 2021) pp. 25–26.

England, P., Tang, W., Kostrzewa, M., Shahrezaei, V., & Larrouy-Maumus, G. (2020). Discrimination of bovine milk from non-dairy milk by lipids fingerprinting using routine matrix-assisted laser desorption ionization mass spectrometry. *Science Reports, 10*(1), 5160.

Erasmus, S. W., Muller, M., Butler, M., & Hoffman, L. C. (2018). The truth is in the isotopes: Authenticating regionally unique South African lamb. *Food Chemistry, 239*, 926–934.

Fornal, E., & Montowska, M. (2019). Species-specific peptide-based liquid chromatography–mass spectrometry monitoring of three poultry species in processed meat products. *Food Chemistry, 283*, 489–498.

Franck, M., Krízek, T., Coufal, P., & Stulík, K. (2010). Differentiation among various kinds of cheese by identification of casein using HPLC-chip/MS/MS. *Journal of Separation Science, 33*(16), 2515–2519.

Gade, D., Thiermann, J., Markowsky, D., & Rabus, R. (2003). Evaluation of two dimensional difference gel electrophoresis for protein profiling. *Journal of Microbiology and Biotechnology, 5*, 240–251.

Galan-Malo, P., Mendiara, I., Razquin, P., & Mata, L. (2018). Validation of a rapid lateral flow method for the detection of cows' milk in water buffalo, sheep or goat milk. *Food Addit. Contaminants: Part A, 35*(4), 609–614.

Gallardo, J. M., Ortea, I., & Carrera, M. (2013). Proteomics and its applications for food authentication and food-technology research. *Trends in Analytical Chemistry, 52*, 135–141.

Garcia, J. S., Sanvido, G. B., Saraiva, S. A., Zacca, J. J., Cosso, R. G., & Eberlin, M. N. (2012). Bovine milk powder adulteration with vegetable oils or fats revealed by MALDI–QTOF MS. *Food Chemistry, 131*, 722–726.

Gautam, P. B., Sharma, R., Lata, K., Rajput, Y. S., & Mann, B. (2017). Construction of a lateral flow strip for detection of soymilk in milk. *Journal of Food Science and Technology, 54*(13), 4213–4219.

Ghidini, S., Varrà, M. O., & Zanardi, E. (2019). Approaching authenticity issues in fish and seafood products by qualitative spectroscopy and approaching authenticity issues in fish and seafood products by qualitative spectroscopy and chemometrics products by qualitative spectroscopy and chemomet. *Molecules., 24*, 1812.

González-Martin, I., González-Pérez, C., Mendez, J. H., Marqués-Macias, E., & Poveda, F. S. (1999). Use of isotope analysis to characterize meat from Iberian-breed swine. *Meat Science, 52*(4), 437–441.

González-Martín, I., González Pérez, C., Hernández Méndez, J., Sánchez González, C. (2001). Differentiation of dietary regimene of Iberian swine by means of isotopic analysis of carbon and sulphur in hepatic tissue. *Meat Science, 58*(1), 25–30.

Gottesmann, P., & Hamm, R. (1983). New biochemical methods of differentiating between fresh meat and thawed frozen meat. *Fleischwirtsch, 63*(2), 219–221.

Gracia, J. C., Savido, G. B., Saraiva, Zacca, J. J., Cosso, R. G., & Eberlin, M. (2012). Bovine milk powder adulteration with vegetable oils or fats revealed by MALDI-QTOF MS. *Food Chemistry, 131*(2), 722–726.

Granby, K., & Fagt, S. (2004). Analysis of acrylamide in coffee and dietary exposure to acrylamide from coffee. *Analytica Chimica Acta, 520*(1–2), 177–182.

Grundy, H. H., Reece, P., Sykes, M. D., Clough, J. A., Audsley, N., & Stones, R. (2007). Screening method for the addition of bovine blood-based binding agents to food using liquid chromatography triple quadrupole mass spectrometry. *Rapid Communications in Mass Spectrometry, 21*(18), 2919–2925.

Grundy, H. H., Reece, P., Sykes, M. D., Clough, J. A., Audsley, N., & Stones, R. (2008). Method to screen for the addition of porcine blood-based binding products to foods using liquid chromatography/triple quadrupole mass spectrometry. *Rapid Communications in Mass Spectrometry, 22*(12), 2006–2008.

Haasnoot, W., & du Pre, J. G. (2007). Luminex-based triplex immunoassay for the simultaneous detection of soy, pea, and soluble wheat proteins in milk powder. *Journal of Agricultural and Food Chemistry, 55*(10), 3771–3777.

Heaton, K., Kelly, S. D., Hoogewerff, J., & Woolfe, M. (2008). Verifying the geographical origin of beef: The application of multi-element isotope and trace element analysis. *Food Chemistry, 107*(1), 506–515.

Henchion, M., McCarthy, M., Resconi, V. C., & Troy, D. (2014). Meat consumption: Trends and quality matters. *Meat Science, 98*(3), 561–568.

Hendrickson, O. D., Zvereva, E. A., Vostrikova, N. L., Chernukha, I. M., Dzantiev, B. B., & Zherdev, A. V. (2021). Lateral flow immunoassay for sensitive detection of undeclared chicken meat in meat products. *Food Chemistry, 15*(344), Article 128598.

Hofmann, K. (1985). *Principal problems in the identification of meat species of slaughter animals using electrophoretic methods.* In *Biochemical identification of meat species* (pp. 9–31). London and New York, NY: Elsevier Applied Science Publishers. Patterson R.L.S. Ed.

Holland, J. W., Deeth, H. C., & Alewood, P. F. (2004). Proteomic analysis of kappa-casein micro-heterogeneity. *Proteomics, 4*(3), 743–752.

Horth, P., Miller, C. A., Preckel, T., & Wenz, C. (2006). Efficient fractionation and improved protein identification by peptide OFFGEL electrophoresis. *Mol Cell Proteomics, 5*, 1968–1974.

Hu, S., Qiu, N., Liu, Y., Zhao, H., Gao, D., Song, R., et al. (2016). Identification and comparative proteomic study of quail and duck egg white protein using 2-dimensional gel electrophoresis and matrix-assisted laser desorption/ionization time-of-flight tandem mass spectrometry analysis. *Poultry Science, 95*(5), 1137–1144.

Kaffarnik, S., Schröder, M., Lehnert, K., Baars, T., & Vetter, W. (2014). $\delta^{13}C$ values and phytanic acid diastereomer ratios: Combined evaluation of two markers suggested for authentication of organic milk and dairy products. *European Food Research and Technology, 238*(5), 819–827.

Karp, N. A., Feret, R., Rubtsov, D. V., & Lilley, K. S. (2008). Comparison of DIGE and post-stained gel electrophoresis with both traditional and Same Spots analysis for quantitative proteomics. *Proteomics, 8*, 948–960.

Kelly, S., Heaton, K., & Hoogewerff, J. (2005). Tracing the geographical origin of food: The application of multi-element and multi-isotope analysis. *Trends in Food Science & Technology, 16*(12), 555–567.

Kim, Y. H. B., Liesse, C., Kemp, R., & Balan, P. (2015). Evaluation of combined effects of ageing period and freezing rate on quality attributes of beef loins. *Meat Science, 110*, 40–45.

Kim, Y., Atalla, H., Mallard, B., Robert, C., & Karrow, N. (2011). Changes in Holstein cow milk and serum proteins during intramammary infection with three different strains of Staphylococcus aureus. *BMC Vet Res, 7*, 51.

Kiran, M., Naveena, B. M., Reddy, S., Shashikumar, K., Ravinder Reddy, M., Kulkarni, V., et al. (2015). Muscle-specific variation in buffalo (*Bubalus bubalis*) meat texture: Biochemical, ultrastructural and proteome characterization. *Journal of Texture Studies, 46*(4), 254–261.

Kiran, M., Naveena, B. M., Smrutirekha, M., Baswa Reddy, P., Banerjee, R., et al. (2019). Traditional halal slaughter without stunning versus slaughter with electrical stunning of sheep *(Ovis aries)*. *Meat Science, 148*, 127–136.

Koppelman, S. J., Lakemond, C. M., Vlooswijk, R., & Hefle, S. L. (2004). Detection of soy proteins in processed foods: Literature overview and new experimental work. *Journal of Aoac International, 87*(6), 1398–1407.

Krouse, H. R., & Grinenko, V. A. (1991). *Stable isotopes: Natural and anthropogenic Sulphur in the environment (SCOPE 43)*. Chichester; New York; Brisbane; Toronto; Singapore: Wiley.

Kuswandi, B., Gani, A. A., & Ahmad, M. (2017). Immunostrip test for detection of pork adulteration in cooked meatballs. *Food bioscience, 19*, 1–6.

Laemmli, U. K. (1970). Cleavage of structural proteins during the assembly of the head of bacteriophage T4. *Nature, 227*, 680–685. https://doi.org/10.1038/227680a0.

Lamanna, R., Braca, A., Di Paolo, E., & Imparato, G. (2011). Identification of milk mixtures by ^{1}H NMR profiling. *Magnetic Resonance in Chemistry, 49*(Suppl. 1), S22–S26.

Larsen, L. B., Wedholm-Pallas, A., Lindmark-Månsson, H., & Andrén, A. (2010). Different proteomic profiles of sweet whey and rennet casein obtained after preparation from raw versus heat treated skimmed milk. *Dairy Science and Technology, 90*, 641–656.

Lee, S. C., Chen, M. C., & Lin, C. W. (2001). Detection of cows' milk in goats' milk by capillary zone electrophoresis. *Australian Journal of Dairy Technology, 56*, 24–27.

Leitner, A., Castro-Rubio, F., Marina, M. L., & Lindner, W. (2006). Identification of marker proteins for the adulteration of meat products with soybean proteins by multidimensional liquid chromatography–tandem mass spectrometry. *Journal of Proteome Research, 5*(9), 2424–2430.

Li, G., Sun, C., Wu, G., Shi, F., Liu, A., & Yang, N. (2016). iTRAQ-based quantitative proteomics identifies potential regulatory proteins involved in Chicken Eggshell brownness. *Plos One, 11*(12), Article e0168750.

Lim, D. K., Long, N. P., Mo, C., Dong, Z., Cui, L., Kim, G., et al. (2017). Combination of mass spectrometry-based targeted lipidomics and supervised machine learning algorithms in detecting adulterated admixtures of white rice. *Food Research International, 100*, 814–821.

Lim, S. A., & Ahmed, M. U. (2016). A label free electrochemical immunosensor for sensitive detection of porcine serum albumin as a marker for pork adulteration in raw meat. *Food Chemistry, 206*, 197–203.

Liu, B., Si, J., Zhao, F., Wang, Q., Wang, Y., Li, J., et al. (2019). Rapid detection of cow milk adulteration/contamination in goat milk by a lateral flow colloidal gold immunoassay strip. *Journal of Dairy Research, 86*(1), 94–97.

Liu, X., Guo, B., Wei, Y., Shi, J., & Sun, S. (2013). Stable isotope analysis of cattle tail hair: A potential tool for verifying the geographical origin of beef. *Food Chemistry, 140*(1–2), 135–140.

Liu, X., Jayasena, D. D., Jung, Y., Heo, K., Lee, J. H., & Jo, C. (2012). Differential proteome analysis of breast and thigh muscles between Korean native chickens and commercial broilers. *Anim Biosci, 25*(6), 895–902.

Liu, Y., Qiu, N., Gao, D., & Ma, M. (2018). Comparative proteomic analysis of chicken, duck, and quail egg yolks. *International Journal of Food Properties, 21*, 1311–1321.

López, J. L., Marina, A., Vázquez, J., & Álvarez, G. (2002). A proteomic approach to the study of the marine mussels Mytilus edulis and M. galloprovincialis. *Marine Biology, 141*(2), 217–223.

Lopez, J. L. (2007). Two-dimensional electrophoresis in proteome expression analysis. *Journal of Chromatography. B, Analytical Technologies in the Biomedical and Life Sciences, 849*, 190–202.

Lu, W., Liu, J., Gao, B., Lv, X., & Yu, L. (2017). Technical note: Nontargeted detection ofadulterated plant proteins in raw milk by UPLC-quadrupole time-of-flight mass spectrometric proteomics combined with chemometrics. *Journal of Dairy Science, 100*(9), 6980–6986.

Luykx, D. M., Cordewener, J. H., Ferranti, P., Frankhuizen, R., Bremer, M. G., Hooijerink, H., et al. (2007). Identification of plant proteins in adulterated skimmed milk powder by high-performance liquid chromatography—mass spectrometry. *Journal of Chromatography A, 1164*(1-2), 189–197.

Macedo-Silva, A., Shimokomaki, M., Vaz, A. J., Yamamoto, Y. Y., & Tenuta-Filho, A. (2001). Textured soy protein quantification in commercial hamburger. *Journal of Food Composition and Analysis, 14*(5), 469–478.

Maheswarappa, N. B., Banerjee, R., Muthupalani, M., Barbuddhe, S., Dasoju, S., & Varakumar, P. (2021). Optimization of novel GELFrEE fractionation for molecular weight–based in-solution protein separation from buffalo meat, pork, and chicken. *Food Analysis Methods, 14*(1), 88–97.

Mallick, P., & Kuster, B. (2010). Proteomics: A pragmatic perspective. *Nature Biotechnology, 28*, 695–709.

Manadas, B., Mendes, V. M., English, J., & Dunn, M. J. (2010). Peptide fractionation in proteomic approaches. *Expert Rev. Proteomics., 7*, 655–663.

Manca, G., Camin, F., Coloru, G. C., Del Caro, A., Depentori, D., Franco, M. A., et al. (2001). Characterization of the geographical origin of Pecorino Sardo cheese by casein stable isotope ($^{13}C/C^{12}$ and $^{15}N/^{14}N$) ratios and free amino acid ratios. *Journal of Agricultural and Food Chemistry, 49*, 1404–1409.

Mandli, J., El Fatimi, I., Seddaoui, N., & Amine, A. (2018). Enzyme immunoassay (ELISA/immunosensor) for a sensitive detection of pork adulteration in meat. *Food Chemistry, 255*, 380–389.

Manso, M. A., Léonil, J., Jan, G., & Gagnaire, V. (2005). Application of proteomics for characterization of milk and dairy products. *International Dairy Journal, 15*(6–9), 845–855.

Martínez, I., Slizyte, R., & Dauksas, E. (2007). High resolution two-dimensional electrophoresis as a tool to differentiate wild from farmed cod (*Gadus morhua*) and to assess the protein composition of klipfish. *Food Chemistry, 102*, 504–510.

Martín-Hernández, C., Muñoz, M., Daury, C., Weymuth, H., Kemmers-Voncken, A. E. M., Corbatón, V., et al. (2009). Immunochromatographic lateral-flow test strip for the rapid detection of added bovine rennet whey in milk and milk powder. *International Dairy Journal, 19*, 205–208.

Masiri, J., Benoit, L., Barrios-Lopez, B., Thienes, C., Meshgi, M., Agapov, A., et al. (2016). Development and validation of a rapid test system for detection of pork meat and collagen residues. *Meat Science, 121*, 397–402.

Matabaro, E., Ishimwe, N., Uwimbabazi, E., & Lee, B. H. (2017). Current immunoassay methods for the rapid detection of aflatoxin in milk and dairy products. *Comprehensive Reviews in Food Science and Food Safety, 16*(5), 808–820.

Matéos, A., Girardet, J. M., Mollé, D., Dary, A., Miclo, L., & Gaillard, J. L. (2009). Two-dimensional cartography of equine β-casein variants achieved by isolation of phosphorylation isoforms and control of the deamidation phenomenon. *Journal of Dairy Science, 92*(6), 2389–2399.

Meltretter, J., Wust, J., & Pischetsrieder, M. (2013). Comprehensive analysis of nonenzymatic post-translational beta-lactoglobulin modifications in processed milk by ultrahigh-performance liquid chromatography-tandem mass spectrometry. *Journal of Agricultural and Food Chemistry, 61*, 6971–6981.

Meltretter, J., Wust, J., & Pischetsrieder, M. (2014). Modified peptides as indicators for thermal and nonthermal reactions in processed milk. *Journal of Agricultural and Food Chemistry, 62*, 10903–10915.

Meyer, B., Al-Diab, D., Vollmer, G., & Pischetsrieder, M. (2011). Mapping the glycoxidation product Nε-carboxymethyllysine in the milk proteome. *Proteomics, 11*, 420–428.

Molina, E., de Frutos, M., & Ramos, M. (2000). Capillary electrophoresis characterization of the casein fraction of cheeses made from cows', ewes' and goats' milks. *Journal of Dairy Research, 67*(2), 209–216.

Molkentin, J. (2009). Authentication of Organic Milk Using $\delta^{13}C$ and the α-Linolenic Acid Content of Milk Fat. *Journal of Agricultural and Food Chemistry, 57*(3), 785–790.

Molle, D., Jardin, J., Piot, M., Pasco, M., Leonil, J., & Gagnaire, V. (2009). Comparison of electrospray and matrix-assisted laser desorption ionization on the same hybrid quadrupole time-of-flight tandem mass spectrometer: Application to bidimensional liquid chromatography of proteins from bovine milk fraction. *Journal of Chromatography A, 1216*, 2424–2432.

Montowska, M., Alexander, M. R., Tucker, G. A., & Barrett, D. A. (2015). Authentication of processed meat products by peptidomic analysis using rapid ambient mass spectrometry. *Food Chemistry, 15*(187), 297–304.

Montowska, M., & Pospiech, E. (2012). Is authentication of regional and traditional food made of meat possible. *Critical Reviews in Food Science and Nutrition, 52*, 475–487.

Montowska, M., & Pospiech, E. (2013). Species-specific expression of various proteins in meat tissue: Proteomic analysis of raw and cooked meat and meat products made from beef, pork and selected poultry species. *Food Chemistry, 136*(3-4), 1461–1469.

Montowska, M., & Spychaj, A. (2018). Quantification of species-specific meat proteins in cooked and smoked sausages using infusion mass spectrometry. *Journal of Food Science and Technology, 55*(12), 4984–4993.

Montowskaa, M., & Pospiech, E. (2013). Species-specific expression of various proteins in meat tissue: Proteomic analysis of raw and cooked meat and meat products made from beef, pork and selected poultry species. *Food Chemistry, 136*, 1461–1469.

Mottram, D. S., Wedzicha, B. L., & Dodson, A. T. (2002). Food chemistry: Acrylamide is formed in the Maillard reaction. *Nature, 419*(6906), 448–449.

Müller, L., Barták, P., Bednář, P., Fryšová, I., Ševčík, J., & Lemr, K. (2008). Capillary electrophoresis-mass spectrometry—A fast and reliable tool for the monitoring of milk adulteration. *Electrophoresis, 29*(10), 2088–2093.

Naveena, B. M., Jagadeesh, D. S., Jagadeesh Babu, A., Madhava Rao, T., Kamuni, V., Vaithiyanathan, S., et al. (2017). OFFGEL electrophoresis and tandem mass spectrometry approach compared with DNA-based PCR method for authentication of meat species from raw and cooked ground meat mixtures containing cattle meat, water buffalo meat and sheep meat. *Food Chemistry, 233*, 311–320.

Naveena, B. M., Jagadeesh, D. S., Kamuni, V., Muthukumar, M., Kulkarni, V. V., Kiran, M., et al. (2018). In-gel and OFFGEL-based proteomic approach for authentication of meat species from minced meat and meat products. *Journal of the Science of Food and Agriculture, 98*(3), 1188–1196.

Nissen, A., Bendixen, E., Ingvartsen, K. L., & Rontved, C. M. (2012). In-depth analysis of low abundant proteins in bovine colostrum using different fractionation techniques. *Proteomics, 12*, 2866–2878.

O'Farrell, P. H. (1975). High resolution two-dimensional electrophoresis of proteins. *Journal of Biological Chemistry, 250*, 4007–4021.

Ortea, I., Cañas, B., & Gallardo, J. M. (2011). Selected tandem mass spectrometry ion monitoring for the fast identification of seafood species. *Journal of chromatography A, 1218*, 4445–4451.

Ortea, I., Cañas, B., Gallardo, J. M., Calo-Mata, P., & Calo-Mata, J. (2010). Identification of commercial prawn and shrimp species of food interest by native isoelectric focusing. *Food Chemistry, 121*, 569–574.

Paleologos, E. K., & Kontominas, M. G. (2007). Effect of processing and storage conditions on the generation of acrylamide in precooked breaded chicken products. *Journal of Food Protection, 70*(2), 466–470.

Pandey, A., & Mann, M. (2000). Proteomics to study genes and genomes. *Nature, 405*, 837–846.

Pappa, E. C., Robertsona, J. A., Rigbya, N. M., Mellona, F., Kandarakisb, I., & Mills, E. N. C. (2008). Application of proteomic techniques to protein and peptide profiling of Teléme cheese made from different types of milk. *International Dairy Journal, 18*, 605–614.

Parisi, E., & Aguiari, D. (1985). Methods of differentiating meats of different species of animals. In R. L. S. Patterson (Ed.), *Biochemical identification of meat species* (pp. 40–49). London and New York: Elsevier Applied Science Publishers.

Parolo, C., & Merkoçi, A. (2013). Paper-based nanobiosensors for diagnostics. *Chemical Society reviews, 42*, 450–457.

Pazos, M., da Rocha, A. P., Roepstorff, P., & Rogowska-Wrzesinska, A. (2011). Fish proteins as targets of ferrous-catalyzed oxidation: identification of protein carbonyls by fluorescent labeling on two-dimensional gels and MALDI-TOF/ TOF mass spectrometry. *Journal Agriculture Food Chemistry, 59*, 7962–7977.

Pepe, T., Ceruso, M., Carpentieri, A., Ventrone, I., Amoresano, A., & Anastasio, A. (2010). Proteomics analysis for the identification of three species of Thunnus. *Veterinary Research Communications, 34*, S153–S155.

Pepe, T., Ceruso, M., Carpentieri, A., Ventrone, I., Amoresano, A., Anastasio, A., et al. (2012). Differentiation of four tuna species by two-dimensional electrophoresis and mass spectrometric analysis. In J. L. Heazlewood, & C. J. Petzold (Eds.), *Proteomic Applications in Biology*. London, UK: InTech.

Piñeiro, C., Barros-Velázquez, J., Sotelo, C. G., & Gallardo, J. M. (1999). The use of two-dimensional electrophoresis in the characterization of the water-soluble protein fraction of commercial flat fish species. *Zeitschrift Für Leb. Und-Forsch. A, 208*, 342–348.

Piñeiro, C., Barros-Velázquez, J., Sotelo, C. G., Pérez-Martín, R. I., & Gallardo, J. M. (1998). Two-dimensional electrophoretic study of the water-soluble protein fraction in white muscle of gadoid fish species. *Journal of Agricultural and Food Chemistry, 46*(10), 3991–3997.

Piñeiro, C., Barros-Velázquez, J., Vázquez, J., Figueras, A., & Gallardo, J. M. (2003). Proteomics as a tool for the investigation of seafood and other marine products. *Journal of Proteome Research, 2*, 127–135.

Piñeiro, C., Vázquez, J., Marina, A., Barros-Velázquez, J., & Gallardo, J. M. (2001). Characterization and partial sequencing of species-specific sarcoplasmic polypeptides from commercial hake species by mass spectrometry following two-dimensional electrophoresis. *Electrophoresis, 22*(8), 1545–1552.

Pinto, G., Caira, S., Cuollo, M., Fierro, O., Nicolai, M. A., Chianese, L., et al. (2012). Lactosylated casein phosphopeptides as specific indicators of heated milks. *Analytical and Bioanalytical Chemistry, 402*, 1961–1972.

Pischetsrieder, M., & Baeuerlein, R. (2009). Proteome research in food science. *Chemical Society reviews, 38*(9), 2600–2608.

Posudin, Y. I., Peiris, K. H. S., & Kays, S. J. (2015). Non-destructive detection of food adulteration to guarantee human health and safety. *Ukrainian Food J, 4*(2), 207–260. 344, 353.

Prandi, B., Lambertini, F., Faccini, A., Suman, M., Leporati, A., Tedeschi, T., et al. (2017). Mass spectrometry quantification of beef and pork meat in highly processed food: Application on Bolognese sauce. *Food Control, 74*, 61–69.

Primrose, S., Woolfe, M., & Rollinson, S. (2010). Food forensics: Methods for determining the authenticity of foodstuffs. *Trends in Food Science & Technology, 21*(12), 582–590.

Promeyrat, A., Sayd, T., Laville, E., Chambon, C., Lebret, B., & Gatellier, P. (2011). Early post-mortem sarcoplasmic proteome of porcine muscle related to protein oxidation. *Food Chemistry, 127*(3), 1097–1104.

Qiu, N., Liu, W., Ma, M., Zhao, L., & Li, Y. (2013). Differences between fertilized and unfertilized chicken egg white proteins revealed by 2-dimensional gel electrophoresis-based proteomic analysis. *Poultry Science, 92*(3), 782–786.

Rehbein, H. (1995). Food control by applied biochemistry of marine organisms: Comparison of proteins and metabolites from fish and invertebrate muscle. *Helgoländer Meeresun, 49*, 747–757.

Rhodes, C. N., Lofthouse, J. H., Hird, S., Rose, P., Reece, P., Christy, J., et al. (2010). The use of stable carbon isotopes to authenticate claims that poultry have been corn-fed. *Food Chemistry, 118*(4), 927–932.

Rodriguez, N., Ortiz, M. C., Sarabia, L., & Gredilla, E. (2010). Analysis of protein chromatographic profiles joint to partial least squares to detect adulterations in milk mixtures and cheeses. *Talanta, 81*, 255–264.

Rodríguez-Vázquez, R., Mato, A., López-Pedrouso, M., Franco, D., Sentandreu, M., & Zapata, C. (2020). Measuring quantitative proteomic distance between Spanish beef breeds. *Food Chemistry, 315*, Article 126293.

Roncada, P., Gaviraghi, A., Liberatori, S., Canas, B., Bini, L., & Greppi, G. F. (2002). Identification of caseins in goat milk. *Proteomics, 2*(6), 723–726.

Rossmann, A., Haberhauer, G., Holzl, S., Horn, P., Pichlmayer, F., & Voerkelius, S. (2000). The potential of multielement stable isotope analysis for regional origin assignment of butter. *European Food Research and Technology, 211*, 32–40.

Rout, P. K., & Verma, M. (2021). Post translational modifications of milk proteins in geographically diverse goat breeds. *Science Reports, 11*(1), 5619.

Sajid, M., Kawde, A. N., & Daud, M. (2015). Designs, formats and applications of lateral flow assay: A literature review. *Journal of Saudi Chemical Society, 19*(6), 689–705.

Sassi, M., Arena, S., & Scaloni, A. (2015). MALDI-TOF-MS platform for integrated proteomic and peptidomic profiling of milk samples allows rapid detection of food adulterations. *Journal of Agricultural and Food Chemistry, 63*(27), 6157–6171.

Savage, W. J., Richardson, R. I., Jolley, P. D., Hargin, K. D., & Steward, C. A. (1995). Investigation of methods to detect mechanically recovered meat in meat products—II: Gel electrophoresis. *Meat Science, 40*, 303–317.

Scarpeid, H. J., Kraal, K., & Hildrum, K. I. (1998). Identification of animal species in ground meat mixtures by multivariate analysis of isoelectric focusing protein profiles. *Electrophoresis, 19*, 3103–3109.

Schmidt, O., Quilter, J. M., Bahar, B., Moloney, A. P., Scrimgeour, C. M., Begley, I. S., et al. (2005). Inferring the origin and dietary history of beef from C, N and S stable isotope ratio analysis. *Food Chemistry, 91*(3), 545–549.

Sentandreu, M. A., Fraser, P. D., Halket, J., Patel, R., & Bramley, P. M. (2010). A proteomic based approach for detection of chicken in meat mixes. *Journal of Proteome Research, 9*, 3374–3383.

Sharma, R., Verma, A., Shinde, N., Mann, B., Gandhi, K., Wichers, J. H., et al. (2021). Adulteration of cow's milk with buffalo's milk detected by an on-site carbon nanoparticles-based lateral flow immunoassay. *Food Chemistry, 351*, Article 129311.

Sharma, R., Verma, A., Shinde, N., Mann, B., Gandhi, K., Wichers, J. H., et al. (2021). Adulteration of cow's milk with buffalo's milk detected by an on-site carbon nanoparticles-based lateral flow immunoassay. *Food Chemistry, 351*, Article 129311.

Silva, R. A., Bezerra, V. S., Pimentel Mdo, C., Porto, A. L., Cavalcanti, M. T., & Filho, J. L. (2016). Proteomic and peptidomic profiling of Brazilian artisanal 'Coalho' cheese. *Journal of the Science of Food and Agriculture, 96*(13), 4337–4344.

Skog, K. I., Johansson, M. A., & Jagerstad, M. I. (1998). Carcinogenic heterocyclic amines in model systems and cooked foods: A review on formation, occurrence and intake. *Food and Chemical Toxicology, 36*(9–10), 879–896.

Sørensen, E. S., Møller, L., Vinther, M., Petersen, T. E., & Rasmussen, L. K. (2003). The phosphorylation pattern of human αs1-casein is markedly different from the ruminant species. *European Journal of Biochemistry, 270*(17), 3651–3655.

Spink, J., & Moyer, D. C. (2011). Defining the public health threat of food fraud. *Journal of food science, 76*(9), R157–R162.

Stachniuk, A., Sumara, A., Montowska, M., & Fornal, E. (2021). Liquid chromatography-mass spectrometry bottom-up proteomic methods in animal species analysis of processed meat for food authentication and the detection of adulterations. *Mass Spectrometry Reviews, 40*(1), 3–30.

Thienes, C. P., Masiri, J., Benoit, L. A., Barrios-Lopez, B., Samuel, S. A., et al. (2019). Quantitative detection of beef contamination in cooked meat products by ELISA. *Journal of Aoac International, 102*, 898–902.

Toldra, F., Torrero, Y., & Flores, J. (1991). Simple test for differentiation between fresh pork and frozen/thawed pork. *Meat Science, 29*(2), 177–181.

Von Bargen, C., Brockmeyer, J., & Humpf, H. (2014). Meat Authentication: A New HPLC – MS/MS based method for the fast and sensitive detection of horse and pork in highly processed food. *Journal of Agricultural and Food Chemistry, 62*, 9428–9435.

Walker, G. S. (2017). Food authentication and traceability: An Asian and Australian perspective. *Food Control, 72*, 168–172.

Wang, J., Liang, Y., Omana, D.A., Kav, N.N., Wu, J. (2012). Proteomics analysis of egg white proteins from different egg varieties. *Journal Agriculture Food Chemistry, 60*(1), 272–82.

Wang, G., Zhou, G., Ren, H., Xu, R., Yang, Y., Guo, L., et al. (2018). Peptide biomarkers identified by LC–MS in processed meats of five animal species. *Journal of Food Composition and Analysis, 73*, 47–54.

Watzka, M., Buchgraber, K., & Wanek, W. (2006). Natural 15N abundance of plants and soil under different management practices in a montane grassland. *Soil Biology & Biochemistry, 38*, 1564–1576.

Yang, Y., Zheng, N., Yang, J., Bu, D., Wang, J., Ma, L., et al. (2014). Animal species milk identification by comparison of two-dimensional gel map profile and mass spectrometry approach. *International Dairy Journal, 35*, 15–20.

Yayla, M. E. A., Karakaya, F., Akcael, E., & Yucel, F. (2016). Development of a lateral flow immunoassay for the detection of pork components in raw meat. *Journal of Biotechnology, 231*, S50.

Zhang, L., Boeren, S., Hageman, J. A., van Hooijdonk, T., Vervoort, J., & Hettinga, K. (2015). Bovine milk proteome in the first 9 days: Protein interactions in maturation of the immune and digestive system of the newborn. *Plos One, 10*, Article e0116710.

Zhao, Y., Zhang, B., Chen, G., Chen, A., Yang, S., & Ye, Z. (2013). Tracing the geographic origin of beef in China on the basis of the combination of stable isotopes and multielement analysis. *Journal of Agricultural and Food Chemistry, 61*(29), 7055–7060.

Zuo, X., & Speicher, D. W. (2000). A method for global analysis of complex proteomes using sample prefractionation by solution isoelectrofocusing prior to two-dimensional electrophoresis. *Analytical Biochemistry, 284*, 266–278.

CHAPTER

11

Application of proteomics to the identification of foodborne pathogens

Ana G. Abril[a,c], *Tomás G. Villa*[a], *Pilar Calo-Mata*[b], *Jorge Barros-Velázquez*[b], *Mónica Carrera*[c]

[a]Department of Microbiology and Parasitology, Faculty of Pharmacy, University of Santiago de Compostela, Santiago de Compostela, Spain
[b]Department of Analytical Chemistry, Nutrition and Food Science, School of Veterinary Sciences, University of Santiago de Compostela, Lugo, Spain
[c]Department of Food Technology, Spanish National Research Council, Marine Research Institute, Vigo, Spain

11.1 Introduction

The spoiling of foods leads to substantial economic losses in the food industry; in this context, ensuring food safety and the quality control of foodstuffs is a principal need and a major issue facing the food control industry. To minimize microbiological hazards and preserve other food characteristics, such as organoleptic characteristics, nutritional value, and beneficial health properties that consumers demand, bacteria must be detected and identified in a rapid and efficient way. This will avoid the spread of foodborne outbreaks and risk management system capabilities. Pathogen identification at critical hazard points in the food chain provides information on transmission routes. On the other hand, it is necessary to better understand the physiology and ecology of pathogens to prevent spoilage and establish new fast and accurate methods for their identification (André et al., 2017).

Gram-positive bacteria, mainly *Firmicutes* species, are the principal spoilage bacteria in foodstuffs of animal origin. In particular, although lactic acid bacteria are beneficial in general, some species, such as *Bacillus*

Food Proteomics
DOI: https://doi.org/10.1016/B978-0-323-90889-4.00013-0

cereus and particularly *Clostridium botulinum*, are known as producers of possibly the most potent neurotoxin in nature and for being versatile spore formers with pathogenic characteristics. Furthermore, the species *Clostridium perfringens* is also remarkable as a toxin producer (André et al., 2017; Wells-Bennik et al., 2016). The production of toxins in foodstuffs and their consequences after ingestion result in symptoms ranging from mild gastrointestinal effects to the paralysis of muscles controlling breath, causing death, as occurs in botulism. Gram-negative bacteria cause approximately 69% of cases of bacterial food-borne disease. In addition, *Escherichia coli, Staphylococcus aureus, Salmonella* spp., *Campylobacter* spp., and *Listeria monocytogenes* are perhaps the main zoonosis-causing bacteria in foodstuffs and are often associated with sickness and disease (Abebe, Gugsa, & Ahmed, 2020). Other genera that are implicated in pathogeny and found in dairy products include *Streptococcus* species such as *Streptococcus uberis, Streptococcus agalactiae, Streptococcus dysgalactiae* and *Streptococcus bovis*, and *Staphylococcus aureus* has been identified as a mastitis producer in dairy products (Abril et al., 2020a; Carrera et al., 2017; Dumke, Hinse, Vollmer, Schulz, & Knabbe, 2015; Lundberg, Nyman, Unnerstad, & Waller, 2014; Quigley et al., 2013; Richards et al., 2014).

Food products sometimes constitute complex environments where bacteria suffer stress; however, bacteria have the ability to adapt via structural and physiological modifications. Genetic regulation and mutations are positively selected in a bacterial population for their benefits. Mechanisms that permit the growth of bacteria in such stressful environments are directly related to their virulence, which is also involved in immune escape and the infection of mammary cells (Alvarez-Ordóñez, Broussolle, Colin, Nguyen-The, & Prieto, 2015; Brüssow, Canchaya, Hardt, & Bru, 2004; Feiner et al., 2015).

Traditionally, food spoilage pathogen identification has been performed by culture-based techniques and the analysis of morphological, physiological, and biochemical pathogen characteristics. These methods lead to delays of several days or weeks before obtaining results, and dead cells or toxic substances cannot be detected. On the other hand, molecular techniques that are faster and more sensitive have been included, such as antibody-based assays, and DNA-based methods (Gopal et al., 2015). Nowadays, "foodomics" including genomics, transcriptomics, proteomics, and metabolomics, employed together with bioinformatics tools, can provide important information of the biomolecules produced by a food system or organism.

Proteomic tools achieve not only protein identification and quantification but also reveal protein expression, function, interaction, and structure information (Böhme, Fernández-No, Calo-Mata, & Barros-Velázquez, 2017). These techniques, including two-dimensional sodium dodecyl sulfate polyacrylamide gel electrophoresis (2-DE), have been employed since the 1970s and are still in use even though higher-sensitivity techniques such as mass spectrometry (MS) are available: all of these methods can be used together for microorganism identification and characterization. Gel-based analysis is carried out by comparing a migration pattern with reference gel patterns for microorganism identification. Although 2-DE is time-consuming and totally inadequate for rapid bacterial identification, this technique allows the analysis of protein expression, quantities, and post-translational modifications (PTMs) (Salazar, Wang, & Zhang, 2017). Actually, 2-DE is mainly used for the analysis of protein mixtures, to compare total protein abundance or to study the response of a given pathogenic strain to particular stressing conditions, including the physicochemical factors associated with bacterial growth, as well as the hassle of isolating the proteins of interest and carrying out the 2-DE and MS.

MS-based proteomics is a fast and accurate technique that enables the analysis of a variety of samples, such as bacteria, whole cells, cell lysates, or isolated proteins. Soft ionization techniques, such as matrix-assisted laser desorption/ionization (MALDI) and electrospray ionization (ESI), permit the analysis of large intact biomolecules. The MALDI ionization method is commonly coupled to time-of-flight mass spectrometry (MALDI-TOF MS), and ESI is usually preceded by liquid chromatography and coupled to mass spectrometry (LC-ESI-MS) for bacterial identification. The mass patterns resulting from MALDI-TOF MS resulting mass patterns are compared to databases to identify organism-specific fingerprints or proteins as biomarkers. After protein enzymatically digestion (bottom-up approach), selected proteins can be identified by the spectral profiles of peptides with the help of bioinformatics tools. MS in tandem (MS/MS) provides the fragmentation of selected peptides allowing the amino acid sequence characterization. Moreover, proteins can be analyzed by the spectral profiles derived from fragmentation of intact proteins (top–down approach) (Böhme et al., 2017).

The present chapter focuses on the applied proteomics in the identification of foodborne bacterial pathogens. The first part of the chapter explains general strategies and procedures to obtain proteome or protein of interest and the second part mainly describes MS techniques for peptide biomarker identification and target proteomics. Special attention is given to techniques such as MALDI-TOF and LC-MS/MS and their application in foodborne pathogen identification.

11.2 Proteomics strategies: discovery and targeted proteomics

Fig. 11.1 summarizes the common proteomic workflows based on two sequential proteomics approaches, the discovery/exploratory proteomics, and targeted proteomics, which are presented in detail.

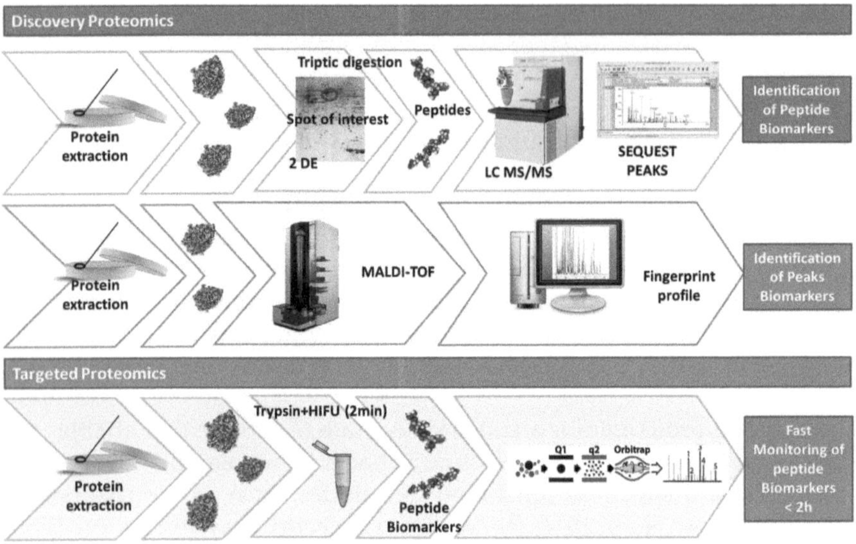

FIG. 11.1 Workflow of proteomics: discovery and targeted proteomics.

Discovery/exploratory proteomics is used to analyze a particular proteome to identify potential protein/peptide biomarkers (Khalilpour, Kilic, Khalilpour, Álvarez, & Yazdi, 2017). Commonly utilizing a bottom–up proteomics strategy, the proteins of interest are separated to reduce sample complexity and then enzymatically digested, commonly using trypsin, and the resulting peptides are analyzed by MS (Pandey et al., 2000). Bottom–up approaches can be divided into two different groups depending on whether the protein separation step involves a gel-based or gel-free approach. The 2-DE gel-based approach can be the method of choice for the separation of proteins from complex samples, where proteins are isolated and stained in-gel in different spots based on their pI and Mr properties (Rabilloud & Lelong, 2011). Then, individual spots of interest can be excised from the gel and digested into peptides with enzymes, and the resulting peptides are then analyzed by MS for protein identification. This gel-based method is the most relevant option for incompletely sequenced organisms, such as some fish species, in which the identification of proteins is based on the comparison of peptides from the protein spots of interest with orthologous proteins from related species or by *de novo* MS sequencing (Carrera, Canas, Pineiro, Vazquez, & Gallardo, 2007). Additionally, progress in the availability of specific sensitive detection methods makes 2-DE a good strategy for the determination of PTMs, such as glycosylation (Geng et al., 2015), phosphorylation (Mayer, Albrecht, & Schaller, 2015), and protein carbonylation (Pazos, da Rocha, Roepstorff, &

Rogowska-Wrzesinska, 2011). Several 2-DE image analysis software programs have been developed, such as PDQuest, Melanie, and Progenesis (Martinotti et al., 2016). In gel-free approaches, also known as shotgun proteomics, a complex mixture of proteins is directly digested with an enzyme, and the mixture of peptides obtained is separated by reversed-phase liquid chromatography (RP-LC), either alone or combined with a multidimensional LC technique such as strong cation exchange chromatography (SCX) (Zhang, Fonslow, Shan, Baek, & Yates III, 2013). The eluted peptides are analyzed and fragmented by tandem mass spectrometry (MS/MS) (Wolters, Washburn, & Yates, 2001; Carrera et al., 2017). Using protein database searching algorithms such as SEQUEST (Eng, McCormack, & Yates, 1994), Mascot (Perkins, Pappin, Creasy, & Cottrell, 1999), OMSSA (Geer et al., 2004), or X! Tandem (Craig et al., 2004), fragmentation spectra are assigned to putative peptide sequences, and the assignments are then validated with programs such as PeptideProphet (Keller, Nesvizhskii, Kolker, & Aebersold, 2002) or Percolator (Käll, Canterbury, Weston, Noble, & MacCoss, 2007). If the protein/peptides are not present in the database, peptides must be *de novo* sequenced (Shevchenko, Wilm, & Mann, 1997). Interpretation of MS/MS spectra is performed either manually or by computer-assisted programs, such as Byonic (Bern, Kil, & Becker, 2012), PEAKS (Ma et al., 2003), DeNovoX (Scigelova, Maroto, Dufresne, & Vázquez, 2007), and UVnovo (Robotham et al., 2016). Protein quantification is commonly required in certain discovery/exploratory proteomics investigations. The principal quantitative proteomics methods are isotope tagging by chemical reaction (e.g., isobaric tags for relative and absolute quantitation (iTRAQ), tandem mass tags (TMTs) or difference gel electrophoresis (DIGE)) (Mateos et al., 2015; Robotti et al., 2018; Stryiński et al., 2019); stable isotope incorporation via enzyme reaction (e.g., ^{18}O) (López-Ferrer, Ramos-Fernández, Martínez-Bartolomé, García-Ruiz, & Vázquez, 2006); metabolic stable isotope labeling (e.g., stable isotope labeling by/with amino acids in cell culture, SILAC) (Ong et al., 2002); and label-free quantification (e.g., measuring the intensity of peptides at the MS level) (Mueller et al., 2007). An alternative approach, top–down proteomics (Fornelli et al., 2018), characterizes the fragments produced by the dissociation of the intact proteins directly inside the mass spectrometer, avoiding the step of protein digestion. This intact protein analysis technique is currently available due to new dissociation methods and the high mass accuracy obtained in new high-resolution MS (HRMS) instruments (Carrera et al., 2015; Carrera, Canas, Vázquez, & Gallardo, 2010). The main goal of discovery/exploratory proteomics is to compare the resulting peptides and proteins by alignment search tools, such as BLAST (https://blast.ncbi.nlm.nih.gov/), to select particular peptide biomarkers, which will then be used in the second workflow step, namely, targeted proteomics.

Targeted proteomics is mainly used to monitor the peptide biomarkers selected in the discovery/exploratory phase with high sensitivity, accuracy, and reproducibility (Borrás & Savidó, 2017). In this scanning mode, the MS analyzer is focused on analyzing only the peptide/s of interest by selected/multiple-reaction monitoring (SRM/MRM) (Aebersold, Bensimon, Collins, Ludwig, & Sabido, 2016). Monitoring appropriate *m/z* pairs of precursor and fragment ions, called transitions, represents a sensitive MS mode for detecting and identifying peptide biomarkers. These are sensitive and selective methods that present a good signal-to-noise (S/N) ratio, an increased dynamic range, and high reproducibility (Lange, Picotti, Domon, & Aebersold, 2008). SRM/MRM procedures are commonly performed on triple-quadrupole mass spectrometers, scanning one or several peptide biomarkers or proteotypic peptides representing the protein of interest. However, the optimization of a definite SRM/MRM assay is a time-consuming process. More importantly, using this scanning mode, complete MS/MS spectra are not registered. The MS/MS spectrum of a peptide is tremendously important to verify its amino acid sequence. New procedures such as SRM-triggered MS/MS in quadrupole-ion trap (Q-IT) mass spectrometers (Aebersold et al., 2016), selected MS/MS ion monitoring (Carrera et al., 2011; Jorge et al., 2007), or parallel reaction monitoring (PRM) on IT or high-resolution Q-Orbitrap (Q-Exactive) instruments (Carrera, Gallardo, Pascual, González, & Medina, 2016), are alternative scanning modes that allow for sensitive monitoring of specific molecules to obtain complete structural information. The high scanning speed achievable in the IT and Orbitrap instruments allows the acquisition of MS/MS spectra in a fraction of a second, thus recording the obtained information from the complete spectra. High-confidence MS/MS spectra are registered due to the option of averaging the signal during acquisition. Targeted data-independent analysis (DIA), implemented as the sequential windowed acquisition of all theoretical fragment ion spectra (SWATH-MS) (Gillet et al., 2012), is an advanced MS mode that can identify and quantify large sets of proteins without the prerequisite to specify a set of proteins prior to acquisition. Targeted proteomics strategies combined with stable isotope dilution, ^{13}C- or ^{15}N-labeled absolute quantification peptide standards (AQUA) or concatenamer of standard peptides (QCAT) (Beynon, Doherty, Pratt, & Gaskell, 2005) are introduced to the sample as internal standards for absolute quantification of the proteins. Regarding bioinformatics, several software programs such as Skyline (Bereman, MacLean, Tomazela, Liebler, & MacCoss, 2012) and SRMCollider (Röst, Malmström, & Aebersold, 2012) are available for the analysis of targeted proteomics assays.

The following sections will demonstrate the utility of these proteomics approaches, discovery/exploratory proteomics, and targeted proteomics to lead to the identification of foodborne pathogens.

11.3 Discovery proteomics for the identification of foodborne pathogens

For protein identification, different studies that have been performed for pathogen identification by MS are shown in Table 11.1.

TABLE 11.1 Overview of discovery proteomics MALDI-TOF MS fingerprinting studies and LC-ESI-MS/MS analysis carried out on food spoilage pathogens.

LC-ESI-MS/MS	References
Clostridium perfringens	Sengupta et al. (2010)
S. aureus	Calo-Mata et al. (2016a), Calo-Mata et al. (2016b), and Carrera et al. (2017)
Bacillus cereus group	Dworzanski et al. (2006) and Pfrunder et al. (2016)
Yersinia pestis	Chenau et al. (2014)
E. coli	Lim et al. (2007), Fagerquist et al. (2014), and Jabbour et al. (2010)
Vibrio parahaemolyticus O3:K6 strain	Williams et al. (2004)
Listeria monocytogenes	Trost et al. (2005)
Streptococcus spp. strains associated with bovine mastitis	Abril et al. (2020b)
Alicyclobacillus acidoterrestris	Siciliano et al. (2019)
MALDI-TOF MS	References
Acinetobacter baumanii	Sousa et al. (2014) and Alvarez-Buylla et al. (2012)
Aeromonas hydrophila	Donohue, Smallwood, Pfaller, Rodgers, and Shoemaker (2006) and Dieckmann et al. (2010)
Bacillus spp.	Branquinho et al. (2014) and Fernández-No et al. (2013)
Campylobacter spp.	Mandrell et al. (2005), Bessède et al. (2011), Zautner et al. (2013), and Kolínská, Drevínek, Jakubů, and Zemlicková (2008)
Clostridium spp.	Grosse-Herrenthey et al. (2008) and Reil et al. (2011)
Cronobacter sakazakii	Zhu et al. (2011) and Stephan et al. (2010)
Escherichia coli	Siegrist et al. (2007), Novais et al. (2014), Matsumura et al. (2014), Clark et al. (2013b), Christner et al. (2014), and Khot and Fisher (2013)
Legionella spp.	Gaia, Casati, and Tonolla (2011), Moliner et al. (2010), and Pennanec, Dufour, Haras, and Réhel (2010)
Listeria spp.	Barbuddhe et al. (2008) and Jadhav, Sevior, Bhave, and Palombo (2014)

(Continued)

TABLE 11.1 Cont'd

LC-ESI-MS/MS	References
Mycobacterium spp.	Amlerová et al. (2014) and Adams, Salee, Dionne, Carroll, and Parrish (2015)
Salmonella enterica	Kuhns et al. (2012), Dieckmann and Malorny (2011), Dieckmann, Helmuth, Erhard, and Malorny (2008), and Sparbier et al. (2012)
Shigella spp.	Khot and Fisher (2013)
Staphylococcus aureus	Szabados, Woloszyn, Richter, Kaase, and Gatermann (2010), Böhme et al. (2012), Du, Yang, Guo, Song, and Wang (2002), Jackson, Edwards-Jones, Sutton, and Fox (2005), Wolters et al. (2011), Carbonnelle et al. (2007), Dubois et al. (2010), and Josten et al. (2013) Vibrio spp. Hazen et al. (2009), Dieckmann et al. (2010)
Yersinia enterocolitica	Lasch et al. (2010), Ayyadurai, Flaudrops, Raoult, and Drancourt (2010), and Stephan et al. (2011)
Morganella spp.	Fernández-No et al. (2010)
Proteus spp.	Fernández-No et al. (2010)
Lactobacillus sp. 30A	Fernández-No et al. (2010)
Staphylococcus xylosus	Fernández-No et al. (2010)
Vibrio parahaemolyticus	Hazen et al. (2009)
Acinetobacter calcoaceticus	Sousa et al. (2014)
Acinetobacter baumannii	Alvarez-Buylla et al. (2012) and Sousa et al. (2014)
Enterobacter cloacae	Elbehiry et al. (2017)
Klebsiella pneumonia	Elbehiry et al. (2017)

11.3.1 MALDI-TOF MS for the identification of foodborne pathogens

MALDI-TOF MS has been established as a promising technique for the characterization of microorganisms, and over the last decades, it has been employed in several studies with this aim (Böhme et al., 2016; Cheng, Chui, Domish, Hernandez, & Wang, 2016; Singhal, Kumar, Kanaujia, & Virdi, 2015). Moreover, the Food and Drug Administration definitively settled in 2016 that MALDI-TOF MS is indeed a reliable technique for both identification and, more importantly, for the discrimination of various related pathogens (Cheng et al., 2016). Methodologies such as sample preparation and protein extraction for bacteria, yeasts, and fungi have been reviewed due to the need and the importance of following a standardized protocol for spectral reproducibility (Clark, Kaleta, Arora, & Wolk, 2013a; Welker and Moore, 2011). Spectrum comparison and identification of bacterial strains is more efficient with consistent sample preparation, but conserved biomarker protein identification is also possible under different protocols (Böhme et al., 2017).

Commercial databases include a large number of bacterial strains with MALDI-TOF MS fingerprints that correspond to food spoilage pathogens. Table 11.1 shows an overview of foodborne pathogens analyzed and discriminated by MALDI-TOF MS. Species of genera such as *Escherichia, Yersinia, Proteus, Morganella, Salmonella, Staphylococcus, Micrococcus, Lactococcus, Pseudomonas, Leuconostoc,* and *Listeria* that have been implicated in food spoilage and contamination have been analyzed to obtain their spectral profiles and specific peak mass lists for comparison, which are freely available on the internet (http://bioinformatica.isa.cnr.it/Descr_Bact_Dbase.htm) (Böhme et al., 2017; Mazzeo et al., 2006). It is important to include spectral information of different strains isolated from different food matrices into the spectral libraries as well as their phenotypic and proteotypic profiles to understand their responses to environmental changes and stress conditions in a concrete food matrix. SpectraBank (Böhme et al., 2013) is a public spectral library database that any researcher may access that is positively used in the identification of bacteria associated with sea foodstuffs (Böhme et al., 2011).

Other studied bacteria, such as histamine-producing bacterial species, have been discriminated by MALDI-TOF MS fingerprinting (Fernández-No et al., 2010). Many studies have allowed the clear identification of bacterial strains isolated from food products at the species level and their differentiation from others from the same genus that are less pathogenic. These studies include the foodborne pathogen *Cronobacter sakazakii* in infant formulas (Stephan, Ziegler, Pflüger, Vogel, & Lehner, 2010; Zhu, Ratering, Schnell, & Wacker, 2011), enterotoxin-producing *Staphylococcus aureus* strains isolated from dairy products (Böhme et al., 2012), and *Vibrio parahaemolyticus*, the main causative agent of outbreaks of seafood-borne gastroenteritis (Böhme et al., 2011; Hazen et al., 2009).

Moreover, several bacteria, such as *Campylobacter jejuni, Escherichia coli, Salmonella enterica, Staphylococcus aureus, Acinetobacter baumannii* (Alvarez-Buylla, Culebras, & Picazo, 2012; Sousa et al., 2014), *Enterobacter cloacae,* and *Klebsiella pneumonia*, have been isolated at the genus and species levels. Data were compared with BD Phoenix system analysis, and 97% of bacterial isolates and 94.44% of fungal strains were correctly identified (Elbehiry et al., 2017).

MALDI-TOF MS has been employed in other studies as a confirmation tool for the identification of foodborne pathogens such as *Campylobacter* species (Bessède, Solecki, Sifré, Labadi, & Mégraud, 2011; Zautner et al., 2013), *Escherichia coli* (Christner et al., 2014; Khot & Fisher, 2013;

Matsumura et al., 2014), *Salmonella entérica* (Dieckmann and Malorny, 2011; Sparbier, Weller, Boogen, & Kostrzewa, 2012), *Staphylococcus aureus* (Josten et al., 2013), and *Enterobacter cloacae* (Almuzara et al., 2015; Khot & Fisher, 2013). Many studies have focused on the identification of seafood-associated bacteria isolated from food of sea origin, such as *Acinetobacter species, Campylobacter* species, *Staphylococcus* species, *Pseudomonas* species, and *Vibrio* species, by MALDI-TOF MS (Böhme et al., 2010, 2011, 2013).

Spectral variation comparison between two MALDI-TOF instruments, MALDI Biotyper from Bruker Daltonics and Vitek-MS from bioMerieux, for the determination of Viridans Group Streptococci at the species level has revealed that both instruments provide efficient identification of this organism, even the Mitis group (Angeletti et al. (2015a).

Several species of fungi have been identified properly by MALDI-TOF-MS from clinical samples and various food products (Lima & Santos, 2017). Species of yeast (Elbehiry et al. 2017) and a few genera of filamentous fungi, such as *Aspergillus* species (Alanio et al., 2011; DeCarolis et al., 2012), *Penicillium* species (Del Chierico et al., 2012) and *Fusarium* species (DeCarolis et al., 2012; Del Chierico et al., 2012), were studied using MALDI-TOF MS fingerprinting.

In addition, Tonacini and collaborators positively used the MALDI-TOF MS technique to detect Staphylococcal Enterotoxin Type B (SEB) in the culture broth of *S. aureus* isolated from a food sample. This toxin was biochemically characterized upon comparison with a purified SEB sample, thus confirming both the amino acid composition and the molecular mass (28.2 kDa). The toxin, which may be dangerous for human health, is not affected by pasteurization or other cooking manipulations (Tonacini et al., 2019). An accurate and reliable model has been developed by Bittar and colleagues based on MALDI-TOF MS and subsequent computer analysis to detect and characterize Panton–Valentine leukocidin cytotoxin in certain strains of *S. aureus* (Bittar, Ouchenane, Smati, Raoult, & Rolain, 2009). Additionally, the same technique has been used to detect the δ-toxin of this microorganism (Gagnaire et al., 2012).

Additionally, the analysis of biomolecular interactions in foodstuffs may be performed by mass spectrometry (BIA-MS); the technology uses the effect known as "surface plasmon resonance" (SPR) to detect toxin-antibody complexes, which had been previously fixed on sensor chips, followed by the appropriate MALDI-TOF MS analysis. In this context, Nedelkov and his colleagues were able to identify the SEB toxin both in milk and mushrooms at low concentrations (ca. 1 ng/mL; Nedelkov, Rasooly, & Nelson, 2000).

However, apart from these previously explained foci of the identification of foodborne pathogens, MALDI-TOF MS fingerprinting studies have been focused on the identification of clinical bacterial pathogens causing human diseases. Moreover, this method has been successfully applied for routine clinical analysis to directly and quickly identify strains directly from the patient's blood (Böhme et al., 2017).

If we take into account the reduction of labor costs, these new techniques may be contributing to achieving this goal. The comparison of MALDI-TOF MS fingerprinting results with reference databases is needed for excellent identification. Furthermore, the importance of the fingerprints of the reference proteins as well as reference data and of course the algorithms used for proper identification must be noted (Elbehiry et al., 2017; Pavlovic, Huber, Konrad, & Busch, 2013). On the other hand, MALDI-TOF MS fingerprinting analysis does not permit the determination of characteristic biomarkers or the identification of mixed cultures of two distinct species. In addition, specific peak mass determination corresponding to determined proteins by MALDI-TOF MS analysis is of great interest for bacterial pathogen control. These specific peak masses may be targeted in further approaches, thus allowing the detection of the corresponding species in a complex food matrix (Böhme et al., 2011).

The use of known "top–down proteomics" based on the analysis of intact proteins has indeed resulted in the possibility of resolving related species because of the different patterns of fragmentation, despite the peak masses being almost the same (Demirev et al., 2005; Böhme et al., 2011).

11.3.2 LC-ESI-MS/MS for the identification of foodborne pathogens

LC-ESI-MS/MS is more specific and has a much higher potential to search for characteristic peptides at all taxonomic levels than MALDI TOF MS. Specific peptides of virulence factors have been evaluated to detect non-pathogenic and pathogenic strains of the same species. Relatedly, the study of characteristic peptides enables the discrimination of several pathogenic bacterial species, such as *Clostridium perfringens* (Sengupta et al., 2010), the *Bacillus cereus* group (Dworzanski et al., 2006), and *Yersinia pestis* strains, in foodstuffs after whole cells are subjected to immune capture (Chenau et al., 2014). Moreover, peptide biomarkers were determined for shiga toxin, and *E. coli* toxin-producing serotypes have been analyzed and compared to nonpathogenic strains (Lim et al., 2007; Fagerquist et al., 2014; Jabbour et al., 2010). Furthermore, a study determined the unique biomarker that allows the identification of the pandemic *Vibrio parahaemolyticus* O3:K6 strain (Williams et al., 2004). *Listeria monocytogenes* and its virulence factors have also been studied by LC-ESI-MS/MS, allowing the differentiation of pathogenic strains (Trost et al., 2005).

To identify specific peptides of *S. aureus*, two proteins of this species have been studied. The trypsinized protein data obtained from

LC-ESI-MS/MS were compared with protein databases of bacteria such as *Streptococcus* spp., *Enterococcus* spp., *Bacillus* spp., *Listeria* spp., and Sal*monella* spp., and *Enterobacteriaceae*. Cysteine synthase and cysteine proteases staphopain A and staphopain B as well as 1-pyrroline-5-carboxylate dehydrogenase were identified as possible proteins for *S. aureus* differentiation directly in food products without the requirement of the previous culture of the bacteria. Moreover, the cysteine protease peptides AQKPVDNITQIIGGTPVVK, TESIPTGNNVTQLK, MTTYNEVDNLTK, YTINVSSFLSK, and KTGSPDYLLHFLEQKV are specific for *S. aureus* (Calo-Mata et al., 2016a). As *S. aureus* is a ubiquitous Gram-positive pathogenic microorganism that can cause a wide spectrum of human diseases, it has been deeply studied to determine a rigorous diagnosis. Other publications describe how virulent and nonvirulent *S. aureus* strains can be distinguished based on characteristic proteins. Corresponding tryptic peptides were analyzed by LC-ESI-MS/MS, and the results were compared with amino acid sequences available in public databases. Two peptides were found to be potentially useful for the detection of *S. aureus* (Calo-Mata, Böhme, Carrera, Caamaño-Antelo, & Gallardo, 2016b). Moreover, Carrera et al. (2017) recently reported, for the first time, the identification and characterization of 20 different foodborne strains of *S. aureus* using an easy and rapid protein preparation procedure followed by a shotgun proteomics approach and functional bioinformatics methodologies. The proposed goal was the identification of a typical and unique peptide biomarker for *S. aureus* based on 21 potential peptides that belonged to eight different proteins. From data obtained for more detailed analysis from proteins digested by trypsin, specific diagnostic peptides for *Bacillus cereus* could indeed be detected by LC-MS/MS (Pfrunder et al., 2016).

Fourteen different *Streptococcus* spp. strains associated with bovine mastitis were proteomically characterized. Protein mixtures were digested with trypsin, and the resulting peptides were analyzed by LC–ESI–MS/MS. Peptides that belong to proteins related to antibiotic resistance and the production of antimicrobial and virulence factors were studied. Comparison of our data with those of NCBI facilitated the identification of 134 *Streptococcus*-specific peptides, and two of them (EATGNQNISPNLTISNAQLNLEDKNK and DLWC*NM*IIAAK; C* carbamidomethylation of Cys and M* methionine oxidation) were specific to *S. dysgalactiae* (Abril et al., 2020a). Moreover, immunoproteomics was performed to identify cell surface-associated proteins in *Alicyclobacillus acidoterrestris* to define potential antigens with marked usefulness for the detection of such a bacterial contaminant (Siciliano, Uzzau, & Mazzeo, 2019).

By means of LC-ESI-MS/MS, the presence of a variety of bacterial toxins was determined, such as Shiga toxin, botulism, and *S. aureus* enterotoxins, the two main toxins from *C. difficile*, and toxins from *B. anthracis* (Silva, 2018).

Phage typing is a classic method employed for bacterial detection because of its high specificity to their hosts. Identification of a phage is a direct method of host identification. Biosensors based on receptor-binding proteins, as well as the study of their nucleic acids, the use of antibodies, and phage-display peptides, have been used for pathogen detection (Chanishvili, 2015; Lavigne, Ceyssens, & Robben, 2009; Rees & Voorhees, 2005; Richter, Janczuk-richter, Niedzió, Paczesny, & Ho, 2018; Singh, Poshtiban, & Evoy, 2013). On the other hand, techniques based on LC-ESI-MS/MS instrumentation have been used for pathogen identification. Bacteriophage lambda identification has been employed usingLC-ESI-MS/MS (Serafim, Ring, Pantoja, Shah, & Shah, 2017). A study aimed to analyze the proteomics of specific peptides of mastitis-causing streptococcal species for the identification of both phage and bacterial strains by LC-ESI-MS/MS. The phage peptide M*ATNLGQAYVQIM*PSAK was characterized as specific for *Streptococcus agalactiae*. In addition, there have been studies concerning the phylogeny of bacteriophages and the relationships of the peptides they encode with their respective infected bacteria (Abril et al., 2020b).

Indeed, a study using top–down proteomics based on the profiles of intact proteins was performed to differentiate *S. enterica* serovars (McFarland et al., 2014).

11.4 Targeted proteomics for the identification of foodborne pathogens

Targeted proteomics has potential applicability for the detection and quantification of microorganisms, their toxins, and antibiotic resistance markers in contaminated food. The last few years have seen the development of new technologies for toxin detection in foodstuffs, which also include MS-based methods, so targeted assays have been performed to analyze toxins in different food matrices. These include *Campylobacter jejuni* cytolethal distending toxin (CTD), *Staphylococcus aureus* enterotoxins (SEA, SEB, and SED), Shiga toxins from *Shigella dysenteriae* and EHEC strains (STX1 and STX2) and *Clostridium perfringens* epsilon toxin (ETX) (Shevchuk, O., 2021).

The applied MRM method based on LC-ESI-MS/MS has been reported to be valid for the detection of staphylococcal SEA and SEB in chicken meat, shrimp, green beans, and fruit juices or milk, among others. The versatility of this method and the overall advantages over other traditional methods rely on the fact that it is able to detect smaller amounts of toxins and that it can be directly applied to foodstuffs without prior purification. The method, however, has a warning point that centers on the fact that a high protein load in the sample does generate a high number of peptides that tend to suppress the electrospray. Andjelkovic and colleagues developed a method based on online SPE coupled to MRM to directly identify SEB and SEA in different food matrices by monitoring different labeled proteotypic peptides in six blinded extracts. The detection limits of SEA and SEB were approximately 8 and 4 ng/g, respectively (Andjelkovic, Tsilia, Rajkovic, De Cremer, & Van Loco, 2016). Moreover, the analysis of whole cells may be performed with LC-ESI-MS/MS; in this context, Carrera and her colleagues were able to unambiguously identify both toxins (SEA and SEB) plus virulence factors (i.e., LukF-PV and mazF) in *S. aureus* strains isolated in foodstuffs (Carrera et al., 2017). Peptides originating in bacteriophages from mastitis-producing *S. aureus* were analyzed as well (Abril et al., 2020c), and LC-ESI-MS/MS has also been used to detect toxins, proteins from bacteriophages, virulence factors, and antimicrobials for the characterization of *Streptococcus* spp., associated with *S. aureus*, and cases of mastitis (Abril et al., 2020b).

Six proteinaceous toxins produced by *C. perfringens* were detected and quantified (Duracova, Klimentova, Fucikova, & Dresler, 2018). Duracova and collaborators tested the developed PRM method in different food matrices and found that this kind of toxin is tightly adsorbed onto the matrices (>90% in most peptides), especially when dealing with animal tissues (Duracova, Klimentova, Myslivcova Fucikova, Zidkova, & Sheshko, 2019).

Moreover, to achieve higher sensitivity (e.g., when treating samples with low amounts of toxins or any other virulence factor), enrichment or pre-fractionation is necessary. Contamination by spores forming *Bacillus anthracis* may occur by contaminated food ingestion or aerosol inhalation. *B. anthracis* is genetically quite similar to *B. cereus* and *B. thuringiensis.* Chenau and colleagues developed a technique to discriminate B. anthracis in complex environmental samples (milk and soil). This approach utilizes a combination of immunocapture of intact spores and protein extraction; the protein fraction must be subjected to proteolysis followed by targeted MRM of the small acid-soluble spore protein (SASP-B) isoform, which is specific to *B. anthracis*. The same group has also developed a method to identify proteoforms exclusive to *B. anthracis,* resulting in the identification of 11 candidate markers that are unique to B. anthracis (Saleh, Staes, Deborggraeve, & Gevaert, 2019).

Other target techniques based on immune methods have permitted the identification of microbial pathogens. Therefore, immune extraction combined with PRM has been developed to absolutely quantify the toxins from *Ricinus communis,* i.e., ricin, SEB, and *Clostridium perfringens* epsilon toxin (ETX) in milk and tap water. Using stable isotope-labeled toxin standards spiked before immunocapture has led to specific peptide identification and quantification (Dupré et al., 2015). In addition, a multiplex immuno LC–SRM method has also been developed for the quantification of ricin, ETX, SEA, SEB, and SED toxins, shiga toxins from *Shigella dysenteriae* and entero-hemorragic *Escherichia coli* strains (STX1 and STX2), and *Campylobacter jejuni* cytolethal distending toxin (CDT) in complex human biofluids and food matrices, always showing high sensitivity. The lower limits of quantification (LOQs) of the assay were determined to be close to 1 ng•mL(-1). In addition, the method has also been used to monitor toxin stability in milk (Saleh et al., 2019; Shevchuk, 2021).

11.5 Concluding remarks and future directions

The food industry is constantly searching for new fast, reliable, and accurate methods to detect pathogens that are usable both in the laboratory and in the evaluation of possible foodborne pathogens in the final and marketable product. Most recent approaches, including discovery/exploratory proteomics, are used to analyze a particular proteome to identify potential protein/peptide biomarkers. Targeted proteomics is being revealed as a powerful tool for the identification of foodborne pathogens.

MALDI-TOF MS has been established as a promising technique for the characterization of microorganisms. However, LC-ESI-MS/MS, in contrast to MALDI TOF MS, is more specific and has a higher potential in searching for characteristic peptides at all taxonomic levels. This instrumental methodology enables the study and characterization of specific peptides of virulence factors to detect nonpathogenic and pathogenic strains of the same species and the discrimination of several pathogenic bacterial species by the specificity of such peptides.

Target proteomics in this chapter has been focused entirely on the detection of FBPs and particularly their toxins. This kind of strategy can be easily extended to other topics of foodomics, among others, and to studies directed to the characterization of allergens, wherein absolute quantification is required, as even trace amounts do represent a potential hazard to humans.

Author contributions

A.G.A. wrote the manuscript; A.G.A., T.G.V., P.C.-M., J.B.-V., and M.C. conceptualized, revised, and corrected the paper. M.C. co-supervised the work. All authors listed have made substantial, direct, and intellectual contributions to the work and approved the work for publication.

Funding

This work has received financial support from the Xunta de Galicia and the European Union (European Social Fund - ESF), from the Spanish Ministry of Economy and Competitivity Project AGL 2.013-48.244-R, and from the European Regional Development Fund (ERDF) (2007–2013). This work was also supported by the GAIN-Xunta de Galicia Project (IN607D 2017/01) and the Spanish AEI/EU-FEDER PID2019-103845RB-C21 project.

Conflicts of interest

The authors declare no conflicts of interest.

References

Abebe, E., Gugsa, G., & Ahmed, M. (2020). Review on major food-borne zoonotic bacterial pathogens. *Journal of Tropical Medicine, 2020*, 1–19.

Abril, A. G., Carrera, M., Böhme, K., Barros-Velázquez, J., Cañas, B., Rama, J. L., et al. (2020b). Characterization of bacteriophage peptides of pathogenic Streptococcus by LC-ESI-MS/MS: Bacteriophage phylogenomics and their relationship to their host. *Frontiers in Microbiology, 11*, 1241.

Abril, A. G., Carrera, M., Böhme, K., Barros-Velázquez, J., Rama, J. L. R., Calo-Mata, P., et al. (2020a). Proteomic characterization of antibiotic resistance, and production of antimicrobial and virulence factors in Streptococcus species associated with bovine mastitis. Could enzybiotics represent novel therapeutic agents against these Pathogens? *Antibiotics, 9*(6), 302.

Abril, G. A., Villa, G. T., Barros-Velázquez, J., Cañas, B., Sánchez-Pérez, A., Calo-Mata, P., et al. (2020c). Staphylococcus aureus exotoxins and their detection in the dairy industry and mastitis. *Toxins, 12*(9), 537.

Adams, L. L., Salee, P., Dionne, K., Carroll, K., & Parrish, N. (2015). A novel protein extraction method for identification of mycobacteria using MALDI-TOF MS. *Journal of Microbiological Methods, 119*, 1–3.

Aebersold, R., Bensimon, A., Collins, B. C., Ludwig, C., & Sabido, E. (2016). Applications and developments in targeted proteomics: From SRM to DIA/SWATH. *Proteomics, 16*, 2065–2067.

Alanio, A., Beretti, J. L., Dauphin, B., Mellado, E., et al. (2011). Matrix-assisted laser desorption ionization time-of flight mass spectrometry for fast and accurate identification of clinically relevant Aspergillus species. *Clinical Microbiology and Infection, 17*, 750–755.

Almuzara, M., Barberis, C., Traglia, G., Famiglietti, A., Ramirez, M. S., & Vay, C. (2015). Evaluation of matrix-assisted laser desorption ionization-timeof-flight mass spectrometry for species identification of nonfermenting gram-negative bacilli. *Journal of Microbiological Methods, 112*, 24–27.

Alvarez-Buylla, A., Culebras, E., & Picazo, J. J. (2012). Identification of Acinetobacter species: Is Bruker biotyper MALDI-TOF mass spectrometry a good alternative to molecular techniques? *Infection, genetics and evolution, 12*, 345–349.

Alvarez-Ordóñez, A., Broussolle, V., Colin, P., Nguyen-The, C., & Prieto, M. (2015). The adaptive response of bacterial food-borne pathogens in the environment, host and food: Implications for food safety. *International Journal of Food Microbiology, 213*, 99–109.

Amlerová, J., Studentova, V., & Hrabák, J. (2014). Identification of Mycobacterium spp. isolates using matrix-assisted laser desorption/ionization-time-of-flight mass spectrometry (MALDI-TOF MS). *Epidemiologie, Mikrobiologie, Imunologie: Casopis Spolecnosti pro Epidemiologii a Mikrobiologii Ceske Lekarske Spolecnosti JE Purkyne, 63*(3), 196–199.

Andjelkovic, M., Tsilia, V., Rajkovic, A., De Cremer, K., & Van Loco, J. (2016). Application of LC-MS/MS MRM to determine staphylococcal enterotoxins (SEB and SEA) in milk. *Toxins, 8*(4), 118.

André, S., Vallaeys, T., & Planchon, S. (2017). Spore-forming bacteria responsible for food spoilage. *Research in Microbiology, 168*(4), 379–387.

Angeletti, S., Dicuonzo., Avola A., Crea, F., Dedej, E., Vailati, F., Farina, C., et al. (2015a). Viridans group streptococci clinical isolates: MALDI-TOF mass spectrometry versus gene sequence-based identification. *Plos One, 10*, Article e0120502.

Ayyadurai, S., Flaudrops, C., Raoult, D., & Drancourt, M. (2010). Rapid identification and typing of Yersinia pestis and other Yersinia species by matrix-assisted laser desorption/ionization time-of-flight (MALDI-TOF) mass spectrometry. *BMC Microbioly, 10*, 285.

Barbuddhe, S. B., Maier, T., Schwarz, G., Kostrzewa, M., Hof, H., Domann, E., et al. (2008). Rapid identification and typing of Listeria species by matrix-assisted laser desorption ionization-time of flight mass spectrometry. *Applied and Environmental Microbiology, 74*, 5402–5407.

Bereman, M. S., MacLean, B., Tomazela, D. M., Liebler, D. C., & MacCoss, M. J. (2012). The development of selected reaction monitoring methods for targeted proteomics via empirical refinement. *Proteomics, 12*(8), 1134–1141.

Bern, M., Kil, Y. J., & Becker, C. (2012). Byonic: Advanced peptide and protein identification software. *Current protocols in bioinformatics, 40*(1), 13–20.

Bessède, E., Solecki, O., Sifré, E., Labadi, L., & Mégraud, F. (2011). Identification of Campylobacter species and related organisms by matrix assisted laser desorption ionization-time of flight (MALDI-TOF) mass spectrometry. *Clinical Microbiology and Infection, 17*, 1735–1739.

Beynon, R. J., Doherty, M. K., Pratt, J. M., & Gaskell, S. J. (2005). Multiplexed absolute quantification in proteomics using artificial QCAT proteins of concatenated signature peptides. *Nature methods, 2*(8), 587–589.

Bittar, F., Ouchenane, Z., Smati, F., Raoult, D., & Rolain, J. M. (2009). MALDI-TOF-MS for rapid detection of staphylococcal Panton–Valentine leukocidin. *International Journal of Antimicrobial Agents, 34*(5), 467–470.

Böhme, K., Morandi, S., Cremonesi, P., Fernández-No, I. C., Barros-Velázquez, J., & Castiglioni, B. (2012). Characterization of S taphylococcus aureus strains isolated from I talian dairy products by MALDI-TOF mass fingerprinting. *Electrophoresis, 33*(15), 2355–2364.

Böhme, K., Antelo, C. S., Fernández-No, I. C., Quintela-Baluja, M., Barros-Velázquez, J., Cañas, B., et al. (2016). Detection of foodborne pathogens using MALDI-TOF mass spectrometry. *Antimicrobial Food Packaging, 15*, 203–214.

Böhme, K., Fernández-No, I. C., Barros-Velázquez, J., Gallardo, J. M., Calo-Mata, P., & Caña, s B. (2010). Species differentiation of seafood spoilage and pathogenic gram-negative bacteria by MALDI-TOF mass fingerprinting. *Journal of Proteome Research, 9*, 3169–3183.

Böhme, K., Fernández-No, I. C., Barros-Velázquez, J., Gallardo, J. M., Cañas, B., & Calo-Mata, P. (2011). Rapid species identification of seafood spoilage and pathogenic Gram-positive bacteria by MALDI-TOF mass fingerprinting. *Electrophoresis, 32*, 2951–2965.

Böhme, K., Fernández-No, I. C., Calo-Mata, P., & Barros-Velázquez, J. (2017). *Proteomics of food spoilage pathogens*. In *Proteomics in Food Science: From Farm to Fork* (pp. 417–431). Amsterdam, The Netherlands: Elsevier Inc.

Böhme, K., Fernández-No, I. C., Pazos, M., Gallardo, J. M., Barros-Velázquez, J., et al. (2013). Identification and classification of seafood-borne pathogenic and spoilage bacteria: 16S rRNA sequencing versus MALDI-TOF MS fingerprinting. *Electrophoresis, 34*, 877–887.

Borràs, E., & Sabidó, E. (2017). What is targeted proteomics? A concise revision of targeted acquisition and targeted data analysis in mass spectrometry. *Proteomics, 17*(17-18), Article 1700180.

Branquinho, R., Sousa, C., Lopes, J., Pintado, M. E., Peixe, L. V., & Osório, H. (2014). Differentiation of Bacillus pumilus and Bacillus safensis using MALDI-TOF-MS. *Plos One, 9*(10), Article e110127.

Brüssow, H., Canchaya, C., Hardt, W., & Bru, H. (2004). Phages and the evolution of bacterial pathogens: From genomic rearrangements to lysogenic conversion. *Microbiology and Molecular Biology Reviews, 68*(3), 560–602.

Calo-Mata, P., Böhme, K., Carrera, M., Caamaño-Antelo, S., Gallardo, J., Barros-Velázquez, J., et al. (2016b). *Bacterial identification by LC-ESI-IT-MS/MS*. In *Microbes in the spotlight: Recent progress in the understanding of beneficial and harmful microorganisms*. (pp. 160–164): Universal Publishers.

Calo-Mata, P., Carrera, M., Böhme, K., Caamaño-Antelo, S., Gallardo, J. M., Barros-Velázquez, J., et al. (2016a). Novel peptide biomarker discovery for detection and identification of bacterial pathogens by LC-ESI-MS/MS. *Analitical Bioanalitical Techniques, 7*, 296.

Carbonnelle, E., Beretti, J. L., Cottyn, S., Quesne, G., Berche, P., Nassif, X., et al. (2007). Rapid identification of Staphylococci isolated in clinical microbiology laboratories by matrix-assisted laser desorption ionization-time of flight mass spectrometry. *Journal of clinical microbiology, 45*(7), 2156–2161.

Carrera, M., Böhme, K., Gallardo, J. M., Barros-Velázquez, J., Cañas, B., & Calo-Mata, P. (2017). Characterization of foodborne strains of Staphylococcus aureus by shotgun proteomics: Functional networks, virulence factors and species specific peptide biomarkers. *Frontiers in Microbiology, 8*, 1–15.

Carrera, M., Canas, B., López-Ferrer, D., Pineiro, C., Vázquez, J., & Gallardo, J. M. (2011). Fast monitoring of species-specific peptide biomarkers using high-intensity-focused-ultrasound-assisted tryptic digestion and selected MS/MS ion monitoring. *Analytical Chemistry, 83*(14), 5688–5695.

Carrera, M., Canas, B., Pineiro, C., Vazquez, J., & Gallardo, J. M. (2007). De novo mass spectrometry sequencing and characterization of species-specific peptides from nucleoside diphosphate kinase B for the classification of commercial fish species belonging to the family Merlucciidae. *Journal of Proteome Research, 6*(8), 3070–3080.

Carrera, M., Canas, B., Vázquez, J., & Gallardo, J. M. (2010). Extensive de novo sequencing of new parvalbumin isoforms using a novel combination of bottom-up proteomics, accurate molecular mass measurement by FTICR– MS, and selected MS/MS Ion monitoring. *Journal of proteome research, 9*(9), 4393–4406.

Carrera, M., Gallardo, J. M., Pascual, S., González, Á. F., & Medina, I. (2016). Protein biomarker discovery and fast monitoring for the identification and detection of Anisakids by parallel reaction monitoring (PRM) mass spectrometry. *Journal of proteomics, 142*, 130–137.

Carrera, M., Weisbrod, C., Lopez-Ferrer, D., Huguet, R., Gallardo, J. M., Schwartz, J., et al. (2015). Top-down, high-throughput of thermo-stable allergens using complementary MS/MS fragmentation strategies. *Thermo Fisher Scientific*, PN64488–EN 0615S.

Chanishvili, N. (2015). Bacteriophage-based products and techniques for identification of biological pathogens. In T. A. Camesano (Ed.), *Nanotechnology to Aid Chemical and Biological Defense* (pp. 17–33). Dordrecht: Springer.

Chenau, J., Fenaille, F., Simon, S., Filali, S., & Volland, H. (2014). Detection of Yersinia pestis in environmental and food samples by intact cell immunocapture and liquid chromatographytandem mass spectrometry. *Analytical Chemistry, 86*, 6144–6152.

Cheng, K., Chui, H., Domish, L., Hernandez, D., & Wang, G. (2016). Recent development of mass spectrometry and proteomics applications in identification and typing of bacteria. *Proteomics–Clinical Applications,, 10*(4), 346–357.

Christner, M., Trusch, M., Rohde, H., Kwiatkowski, M., Schlüter, H., Wolters, M., et al. (2014). Rapid MALDI-TOF mass spectrometry strain typing during a large outbreak of Shiga-toxigenic Escherichia coli. *Plos One, 9*, Article e101924.

Clark, A. E., Kaleta, E. J., Arora, A., & Wolk, D. M. (2013a). Matrix-assisted laser desorption ionization-time of flight mass spectrometry: A fundamental shift in the routine practice of clinical microbiology. *Clinical microbiology reviews, 26*(3), 547–603.

Clark, C. G., Kruczkiewicz, P., Guan, C., McCorrister, S. J., Chong, P., Wylie, J., et al. (2013b). Evaluation of MALDI-TOF mass spectroscopy methods for determination of Escherichia coli pathotypes. *Journal of Microbiological Methods, 94*, 180–191.

Craig, R., & Beavis, R. C. (2004). TANDEM: Matching proteins with tandem mass spectra. *Bioinformatics, 20*(9), 1466–1467.

DeCarolis, E., Posteraro, B., Lass-Florl, C., Vella, A., Florio, A. R., et al. (2012). Species identification of Aspergillus, Fusarium and Mucorales with direct surface analysis by matrix-assisted laser desorption ionization time-of flight mass spectrometry. *Clinical Microbiology and Infection, 18*, 475–484.

Del Chierico, F., Masotti, A., Onori, M., Fiscarelli, E., Mancinelli, L., et al. (2012). MALDI-TOF MS proteomic phenotyping of filamentous and other fungi from clinical origin. *Journal of Proteomics, 75*, 3314–3330.

Demirev, P. A., Feldman, A. B., Kowalski, P., & Lin, J. S. (2005). Top-down proteomics for rapid identification of intact microorganisms. *Analytical chemistry, 77*(22), 7455–7461.

Dieckmann, R., Helmuth, R., Erhard, M., & Malorny, B. (2008). Rapid classification and identification of Salmonellae at the species and subspecies levels by whole-cell matrix-assisted laser desorption ionization-time of flight mass spectrometry. *Applied and Environmental Microbiology, 74*, 7767–7778.

Dieckmann, R., & Malorny, B. (2011). Rapid screening of epidemiologically important Salmonella enterica subsp. enterica serovars by whole-cell matrixassisted laser desorption ionization-time of flight mass spectrometry. *Applied and Environmental Microbiology, 77*, 4136–4146.

Dieckmann, R., Strauch, E., & Alter, T. (2010). Rapid identification and characterization of Vibrio species using whole-cell MALDI-TOF mass spectrometry. *Journal of applied microbiology, 109*(1), 199–211.

Donohue, M. J., Smallwood, A. W., Pfaller, S., Rodgers, M., & Shoemaker, J. A. (2006). The development of a matrix-assisted laser desorption/ionization mass spectrometry-based method for the protein fingerprinting and identification of Aeromonas species using whole cells. *Journal of Microbiological Methods, 65*, 380–389.

Du, Z., Yang, R., Guo, Z., Song, Y., & Wang, J. (2002). Identification of Staphylococcus aureus and determination of its methicillin resistance by matrix-assisted laser desorption/ionization time-of-flight mass spectrometry. *Analytical Chemistry, 74*, 5487–5491.

Dubois, D., Leyssene, D., Chacornac, J. P., Kostrzewa, M., Schmit, P. O., Talon, R., et al. (2010). Identification of a variety of Staphylococcus species by matrix-assisted laser desorption ionization-time of flight mass spectrometry. *Journal of Clinical microbiology, 48*, 941–945.

Dumke, J., Hinse, D., Vollmer, T., Schulz, J., & Knabbe, C. (2015). Potential transmission pathways of streptococcus gallolyticus subsp. gallolyticus. *Plos One, 10*(5), Article e0126507.

Dupré, M., Gilquin, B., Fenaille, F., Feraudet-Tarisse, C., Dano, J., Ferro, M., et al. (2015). Multiplex quantification of protein toxins in human biofluids and food matrices using immunoextraction and high-resolution targeted mass spectrometry. *Analytical Chemistry, 87*, 8473.

Duracova, M., Klimentova, J., Fucikova, A., & Dresler, J. (2018). Proteomic methods of detection and quantification of protein toxins. *Toxins, 10*(3), 99. https://doi.org/10.3390/toxins10030099.

Duracova, M., Klimentova, J., Myslivcova Fucikova, A., Zidkova, L., Sheshko, V., et al. (2019). Targeted mass spectrometry analysis of clostridium perfringens Toxins. *Toxins, 11*(3), 177.

Dworzanski, J. P., J. P., Deshpande, S. V., Chen, R., Jabbour, R. E., et al. (2006). Mass spectrometry-based proteomics combined with bioinformatic tools for bacterial classification. *Journal of proteome research, 5*(1), 76–87.

Elbehiry, A., Marzouk, E., Hamada, M., Al-Dubaib, M., Alyamani, E., Moussa, I. M., et al. (2017). Application of MALDI-TOF MS fingerprinting as a quick tool for identification and clustering of foodborne pathogens isolated from food products. *New Microbiological, 40*(4), 269–278.

Eng, J. K., McCormack, A. L., & Yates, J. R. (1994). An approach to correlate tandem mass spectral data of peptides with amino acid sequences in a protein database. *Journal of the American Society for Mass Spectrometry, 5*(11), 976–989.

Fagerquist, C. K., Zaragoza, W. J., Sultan, O., Woo, N., Quiñones, B., et al. (2014). Top-down proteomic identification of Shiga toxin 2 subtypes from Shiga toxin-producing Escherichia coli by matrix-assisted laser desorption ionization-tandem time of flight mass spectrometry. *Applied and environmental microbiology, 80*, 2928–2940.

Feiner, R., Argov, T., Rabinovich, L., Sigal, N., Borovok, I., & Herskovits, A. A. (2015). A new perspective on lysogeny: Prophages as active regulatory switches of bacteria. *Nature Reviews Microbiology, 13*(10), 641–650.

Fernández-No, I. C., Böhme, K., Díaz-Bao, M., Cepeda, A., Barros-Velázquez, J., & Calo-Mata, P. (2013). Characterisation and profiling of Bacillus subtilis, Bacillus cereus and Bacillus licheniformis by MALDI-TOF mass fingerprinting. *Food microbiology, 33*(2), 235–242.

Fernández-No, I. C., Böhme, K., Gallardo, J. M., Barros-Velázquez, J., Cañas, B., & Calo-Mata, P. (2010). Differential characterization of biogenic amine-producing bacteria involved in food poisoning using MALDI-TOF mass fingerprinting. *Electrophoresis, 31*, 1116–1127.

Fornelli, L., Toby, T. K., Schachner, L. F., Doubleday, P. F., Srzentić, K., DeHart, C. J., et al. (2018). Top-down proteomics: Where we are, where we are going? *Journal of Proteomics, 175*, 3.

Gagnaire, J., Dauwalder, O., Boisset, S., Khau, D., Freydière, A. M., Ader, F., et al. (2012). Detection of Staphylococcus aureus delta-toxin production by whole-cell MALDI-TOF mass spectrometry. *Plos One, 7*(7), e40660.

Gaia, V., Casati, S., & Tonolla, M. (2011). Rapid identification of Legionella spp. by MALDI-TOF MS based protein mass fingerprinting. *Systematic and Applied Microbiology, 34*(1), 40–44.

Geer, L. Y., Markey, S. P., Kowalak, J. A., Wagner, L., Xu, M., Maynard, D. M., et al. (2004). Open mass spectrometry search algorithm. *Journal of Proteome Research, 3*(5), 958–964.

Geng, F., Huang, X., Majumder, K., Zhu, Z., Cai, Z., & Ma, M. (2015). Mass spectrometry and two-dimensional electrophoresis to characterize the glycosylation of hen egg white ovomacroglobulin. *Journal of Agricultural and Food Chemistry, 63*(37), 8209–8215.

Gillet, L. C., Navarro, P., Tate, S., Röst, H., Selevsek, N., Reiter, L., et al. (2012). Targeted data extraction of the MS/MS spectra generated by data-independent acquisition: A new concept for consistent and accurate proteome analysis. *Molecular & Cellular Proteomics, 11*(6), O111–016717.

Gopal, N., Hill, C., Ross, P. R., Beresford, T. P., Fenelon, M. A., & Cotter, P. D. (2015). The prevalence and control of Bacillus and related spore-forming bacteria in the dairy industry. *Frontiers in Microbiology, 6*, 1418.

Grosse-Herrenthey, A., Maier, T., Gessler, F., Schaumann, R., Böhnel, H., Kostrzewa, M., et al. (2008). Challenging the problem of clostridial identification with matrix-assisted laser desorption and ionizationtime-of-flight mass spectrometry (MALDI-TOF MS). *Anaerobe, 14*, 242–249.

Hazen, T. H., Martinez, R. J., Chen, Y., Lafon, P. C., Garrett, N. M., Parsons, M. B., et al. (2009). Rapid identification of Vibrio parahaemolyticus by whole-cell matrix-assisted laser desorption ionization-time of flight mass spectrometry. *Applied and Environmental Microbiology, 75*, 6745–6756.

Jabbour, R. E., Deshpande, S. V., Wade, M. M., Stanford, M. F., Wick, C. H., et al. (2010). Double-blind characterization of non-genome-sequenced bacteria by mass spectrometrybased proteomics. *Appl. Environ. Microbiol, 76*, 3637–3644.

Jackson, K. A., Edwards-Jones, V., Sutton, C. W., & Fox, A. J. (2005). Optimisation of intact cell MALDI method for fingerprinting of methicillin-resistant Staphylococcus aureus. *Journal of Microbiological Methods, 62*, 273–284.

Jadhav, S., Sevior, D., Bhave, M., & Palombo, E. A. (2014). Detection of Listeria monocytogenes from selective enrichment broth using MALDI-TOF mass spectrometry. *Journal of Proteomics, 97*, 100–106.

Jorge, I., Casas, E. M., Villar, M., Ortega-Pérez, I., López-Ferrer, D., Martínez-Ruiz, A., et al. (2007). High-sensitivity analysis of specific peptides in complex samples by selected MS/MS ion monitoring and linear ion trap mass spectrometry: Application to biological studies. *Journal of mass spectrometry, 42*(11), 1391–1403.

Josten, M., Reif, M., Szekat, C., Al-Sabti, N., Roemer, T., Sparbier, K., et al. (2013). Analysis of the matrixassisted laser desorption ionization-time of flight mass spectrum of Staphylococcus aureus identifies mutations that allow differentiation of the main clonal lineages. *Journal of Clinical Microbiology, 51*, 1809–1817.

Käll, L., Canterbury, J. D., Weston, J., Noble, W. S., & MacCoss, M. J. (2007). Semi-supervised learning for peptide identification from shotgun proteomics datasets. *Nature Methods, 4*(11), 923–925.

Keller, A., Nesvizhskii, A. I., Kolker, E., & Aebersold, R. (2002). Empirical statistical model to estimate the accuracy of peptide identifications made by MS/MS and database search. *Analytical Chemistry, 74*(20), 5383–5392.

Khalilpour, A., Kilic, T., Khalilpour, S., Álvarez, M. M., & Yazdi, I. K. (2017). Proteomic-based biomarker discovery for development of next generation diagnostics. *Applied Microbiology and Biotechnology, 101*(2), 475–491.

Khot, P. D., & Fisher, M. A. (2013). Novel approach for differentiating Shigella species and Escherichia coli by matrix-assisted laser desorption ionizationtime of flight mass spectrometry. *Journal of Clinical Microbiology, 51*, 3711–3716.

Kolínská, R., Drevínek, M., Jakubů, V., & Zemlicková, H. (2008). Species identification of Campylobacter jejuni ssp. jejuni and Campylobacter coli by matrix-assisted laser desorption/ionization time-of-flight mass spectrometry and PCR. *Folia Microbiologica, 53*(5), 403–409.

Kuhns, M., Zautner, A. E., Rabsch, W., Zimmermann, O., Weig, M., Bader, O., et al. (2012). Rapid discrimination of Salmonella enterica serovar Typhi from other serovars by MALDI-TOF mass spectrometry. *Plos One, 7*, e40004.

Lange, V., Picotti, P., Domon, B., & Aebersold, R. (2008). Selected reaction monitoring for quantitative proteomics: A tutorial. *Molecular Systems Biology, 4*(1), 222.

Lasch, P., Drevinek, M., Nattermann, H., Grunow, R., Stalmmler, M., Dieckmann, R., et al. (2010). Characterization of Yersinia using MALDI-TOF mass spectrometry and chemometrics. *Analytical Chemistry, 82*(20), 8464–8475.

Lavigne, R., Ceyssens, P., & Robben, J. (2009). Phage proteomics: Applications of mass spectrometry. In M. R. Clokie, & A. M. Kropinski. (Eds.), *Bacteriophages* (pp. 239–251). Totowa, NJ: Humana Press.

Lim, J. Y., Sheng, H., Seo, K. S., Park, Y. H., & Hovde, C. J. (2007). Characterization of an Escherichia coli O157:H7 plasmid O157 deletion mutant and its survival and persistence in cattle. *Applied and environmental microbiology, 73*, 2037–2047.

Lima, N., & Santos, C. (2017). MALDI-TOF MS for identification of food spoilage filamentous fungi. *Current Opinion in Food Science, 13*, 26–30.

López-Ferrer, D., Ramos-Fernández, A., Martínez-Bartolomé, S., García-Ruiz, P., & Vázquez, J. (2006). Quantitative proteomics using 16O/18O labeling and linear ion trap mass spectrometry. *Proteomics, 6*(S1), S4–S11.

Lundberg, Å., Nyman, A., Unnerstad, H. E., & Waller, K. P. (2014). Prevalence of bacterial genotypes and outcome of bovine clinical mastitis due to Streptococcus dysgalactiae and Streptococcus uberis. *Acta Veterinaria Scandinavica, 56*(1), 80.

Ma, B., Zhang, K., Hendrie, C., Liang, C., Li, M., Doherty-Kirby, A., et al. (2003). PEAKS: Powerful software for peptide de novo sequencing by tandem mass spectrometry. *Rapid Communications in Mass Spectrometry, 17*(20), 2337–2342.

Mandrell, R. E., Harden, L. A., Bates, A., Miller, W. G., Haddon, W. F., & Fagerquist, C. K. (2005). Speciation of Campylobacter coli, C. jejuni, C. helveticus, C. lari, C. sputorum, and C. upsaliensis by matrix-assisted laser desorption ionization-time of flight mass spectrometry. *Applied and Environmental Microbiology, 71*, 6292–6307.

Martinotti, S., & Ranzato, E. (2016). *2-DE Gel Analysis: The spot detection*. In *2-D PAGE Map Analysis* (pp. 155–164). New York, NY: Humana Press.

Mateos, J., Landeira-Abia, A., Fafián-Labora, J. A., Fernández-Pernas, P., Lesende-Rodríguez, I., Fernández-Puente, P., et al. (2015). iTRAQ-based analysis of progerin expression reveals mitochondrial dysfunction, reactive oxygen species accumulation and altered proteostasis. *Stem Cell Research & Therapy, 6*(1), 1–17.

Matsumura, Y., Yamamoto, M., Nagao, M., Tanaka, M., Machida, K., Ito, Y., et al. (2014). Detection of extended-spectrum-β-lactamase-producing Escherichia coli ST131 and ST405 clonal groups by matrix-assisted laser desorption ionization-time of flight mass spectrometry. *Journal of Clinical Microbiology, 52*, 1034–1040.

Mayer, K., Albrecht, S., & Schaller, A. (2015). *Targeted analysis of protein phosphorylation by 2D electrophoresis*. In *Plant Phosphoproteomics* (pp. 167–176). New York, NY: Humana Press.

Mazzeo, M. F., Sorrentino, A., Gaita, M., Cacace, G., Di Stasio, M., Facchiano, A., et al. (2006). Matrix-assisted laser desorption ionization-time of flight mass spectrometry for the discrimination of food-borne microorganisms. *Applied and Environmental Microbiology, 72*(2), 1180–1189.

McFarland, M. A., Andrzejewski, D., Musser, S. M., & Callahan, J. H. (2014). Platform for identification of Salmonella serovar differentiating bacterial proteins by top-down mass spectrometry: S. Typhimurium vs S. Heidelberg. *Analytical Chemistry, 86*, 6879–6886.

Moliner, C., Ginevra, C., Jarraud, S., Flaudrops, C., Bedotto, M., Couderc, C., et al. (2010). Rapid identification of Legionella species by mass spectrometry. *Journal of Medical Microbiology, 59*(3), 273–284.

Mueller, L. N., Rinner, O., Schmidt, A., Letarte, S., Bodenmiller, B., Brusniak, M. Y., et al. (2007). SuperHirn–a novel tool for high resolution LC-MS-based peptide/protein profiling. *Proteomics, 7*(19), 3470–3480.

Nedelkov, D., Rasooly, A., & Nelson, R. W. (2000). Multitoxin biosensor–mass spectrometry analysis: A new approach for rapid, real-time, sensitive analysis of staphylococcal toxins in food. *International Journal of Food Microbiology, 60*(1), 1–13.

Novais, Â., Sousa, C., de Dios Caballero, J., Fernandez-Olmos, A., Lopes, J., Ramos, H., et al. (2014). MALDI-TOF mass spectrometry as a tool for the discrimination of high-risk Escherichia coli clones from phylogenetic groups B2 (ST131) and D (ST69, ST405, ST393). *European Journal of Clinical Microbiology & Infectious Diseases, 33*(8), 1391–1399.

Ong, S. E., Blagoev, B., Kratchmarova, I., Kristensen, D., Steen, H., Pandey, A., et al. (2002). Stable isotope labeling by amino acids in cell culture, SILAC, as a simple and accurate approach to expression proteomics. *Molecular & cellular proteomics, 1*(5), 376–386.

Ortea, I., Canas, B., & Gallardo, J. M. (2009). Mass spectrometry characterization of species-specific peptides from arginine kinase for the identification of commercially relevant shrimp species. *Journal of Proteome Research, 8*(11), 5356–5362.

Pandey, A., & Mann, M. (2000). Proteomics to study genes and genomes. *Nature, 405*(6788), 837–846.

Pavlovic, M., Huber, I., Konrad, R., & Busch, U. (2013). Application of MALDI-TOF MS for the identification of food borne bacteria. *The Open Microbiology Journal, 7*, 135.

Pazos, M., da Rocha, A. P., Roepstorff, P., & Rogowska-Wrzesinska, A. (2011). Fish proteins as targets of ferrous-catalyzed oxidation: Identification of protein carbonyls by fluorescent labeling on two-dimensional gels and MALDI-TOF/TOF mass spectrometry. *Journal of Agricultural and Food Chemistry, 59*(14), 7962–7977.

Pennanec, X., Dufour, A., Haras, D., & Réhel, K. (2010). A quick and easy method to identify bacteria by matrix-assisted laser desorption/ionisation time-of-flight mass spectrometry. *Rapid Communications in Mass Spectrometry, 24*(3), 384–392.

Perkins, D. N., Pappin, D. J., Creasy, D. M., & Cottrell, J. S. (1999). Probability-based protein identification by searching sequence databases using mass spectrometry data. *Electrophoresis: An International Journal, 20*(18), 3551–3567.

Pfrunder, S., Grossmann, J., Hunziker, P., Brunisholz, R., Gekenidis, M. T., & Drissner, D. (2016). Bacillus cereus group-type strain-specific diagnostic peptides. *Journal of Proteome Research, 15*, 3098–3107. doi:10.1021/acs.jproteome.6b00216.

Quigley, L., O'sullivan, O., Stanton, C., Beresford, T. P., Ross, R. P., Fitzgerald, G. F., et al. (2013). The complex microbiota of raw milk. *FEMS Microbiology Reviews, 37*(5), 664–698.

Rabilloud, T., & Lelong, C. (2011). Two-dimensional gel electrophoresis in proteomics: A tutorial. *Journal of proteomics, 74*(10), 1829–1841.

Rees, J. C., & Voorhees, K. J. (2005). Simultaneous detection of two bacterial pathogens using bacteriophage amplification coupled with matrix-assisted laser desorption/ionization time-of-flight mass spectrometry. *Rapid Communications in Mass Spectrometry: An International Journal Devoted to the Rapid Dissemination of Up-to-the-Minute Research in Mass Spectrometry, 19*(19), 2757–2761.

Reil, M., Erhard, M., Kuijper, E. J., Kist, M., Zaiss, H., Witte, W., et al. (2011). Recognition of Clostridium difficile PCR-ribotypes 001, 027 and 126/078 using an extended MALDI-TOF MS system. *European Journal of Clinical Microbiology & Infectious Diseases, 30*, 1431–1436.

Richards, V. P., Palmer, S. R., Bitar, P. D. P., Qin, X., Weinstock, G. M., Highlander, S. K., et al. (2014). Phylogenomics and the dynamic genome evolution of the genus Streptococcus. *Genome Biology and Evolution, 6*(4), 741–753.

Richter, Ł., Janczuk-richter, M., Niedzió, J., Paczesny, J., & Ho, R. (2018). Recent advances in bacteriophage-based methods for bacteria detection. *Drug Discovery Today, 23*, 448–455.

Robotham, S. A., Horton, A. P., Cannon, J. R., Cotham, V. C., Marcotte, E. M., & Brodbelt, J. S. (2016). UVnovo: A de novo sequencing algorithm using single series of fragment ions via chromophore tagging and 351 nm ultraviolet photodissociation mass spectrometry. *Analytical Chemistry, 88*(7), 3990–3997.

Robotti, E., & Marengo, E. (2018). 2D-DIGE and fluorescence image analysis. *Methods in Molecular Biology, 1664*, 25–39.

Röst, H., Malmström, L., & Aebersold, R. (2012). A computational tool to detect and avoid redundancy in selected reaction monitoring. *Molecular & Cellular Proteomics, 11*(8), 540–549.

Salazar, J. K., Wang, Y., & Zhang, W. (2017). *Transcriptomics and proteomics of foodborne bacterial pathogens*. In *Applied Genomics of Foodborne Pathogens* (pp. 167–200). Cham: Springer.

Saleh, S., Staes, A., Deborggraeve, S., & Gevaert, K. (2019). Targeted proteomics for studying pathogenic bacteria. *Proteomics, 19*(16), Article 1800435.

Scigelova, M., Maroto, F., Dufresne, C., & Vázquez, J. (2007). High throughput de novo sequencing., http://www.thermo.com/ (Accessed 2 June 2021).

Sengupta, N., Alam, S. I., Kumar, B., Gautam, V., et al. (2010). Comparative proteomic analysis of extracellular proteins of Clostridium perfringens type A and type C strains. *Infection and immunity, 78*(9), 3957–3968.

Serafim, V., Ring, C. J., Pantoja, L., Shah, H., & Shah, A. J. (2017). Rapid identification of E. coli bacteriophages using mass spectrometry. *Journal of Proteomics and Enzymology, 6*(1), 1–5.

Shevchenko, A., Wilm, M., & Mann, M. (1997). Peptide sequencing by mass spectrometry for homology searches and cloning of genes. *Journal of Protein Chemistry, 16*(5), 481–490.

Shevchuk, O. (2021). Targeted proteomics for rapid and sensitive detection of foodborne pathogens. In A. Cifuentes (Ed.), *Comprehensive Foodomics*, (pp. 123–136): Elsevier, ISBN 9780128163962.

Siciliano, R. A., Uzzau, S., & Mazzeo, M. F. (2019). Proteomics for studying foodborne microorganisms and their impact on food quality and human health. *Frontiers in Nutrition, 6*, 104.

Siegrist, T. J., Anderson, P. D., Huen, W. H., Kleinheinz, G. T., McDermott, C. M., & Sandrin, T. R. (2007). Discrimination and characterization of environmental strains of Escherichia coli by matrix-assisted laser desorption/ionization time-of-flight mass spectrometry (MALDI-TOF-MS). *Journal of Microbiological Methods, 68*, 554–562.

Silva, C. J. (2018). Food Forensics: Using Mass Spectrometry To detect foodborne protein contaminants, as exemplified by shiga toxin variants and prion strains. *Journal of agricultural and food chemistry, 66*(32), 8435–8450.

Singh, A., Poshtiban, S., & Evoy, S. (2013). Recent advances in bacteriophage based biosensors for food-borne pathogen detection. *Sensors, 13*, 1763–1786.

Singhal, N., Kumar, M., Kanaujia, P. K., & Virdi, J. S. (2015). MALDI-TOF mass spectrometry: An emerging technology for microbial identification and diagnosis. *Frontiers in microbiology, 6*, 791.

Sousa, C., Botelho, J., Silva, L., Grosso, F., Nemec, A., Lopes, J., et al. (2014). MALDI-TOF MS and chemometric based identification of the Acinetobacter calcoaceticus-Acinetobacter baumannii complex species. *International Journal of Medical Microbiology, 304*(5-6), 669–677.

Sparbier, K., Weller, U., Boogen, C., & Kostrzewa, M. (2012). Rapid detection of Salmonella sp. by means of a combination of selective enrichment broth and MALDI-TOF MS. *European Journal of Clinical Microbiology & Infectious Diseases, 31*(5), 767–773.

Stephan, R., Cernela, N., Ziegler, D., Pflüger, V., Tonolla, M., et al. (2011). Rapid species specific identification and subtyping of Yersinia enterocolitica by MALDI-TOF mass spectrometry. *Journal of microbiological methods, 87*(2), 150–153.

Stephan, R., Ziegler, D., Pflüger, V., Vogel, G., & Lehner, A. (2010). Rapid genus-and species-specific identification of Cronobacter spp. by matrix-assisted laser desorption ionization-time of flight mass spectrometry. *Journal of Clinical Microbiology, 48*(8), 2846.

Stryiński, R., Mateos, J., Pascual, S., González, Á. F., Gallardo, J. M., Łopieńska-Biernat, E., et al. (2019). Proteome profiling of L3 and L4 Anisakis simplex development stages by TMT-based quantitative proteomics. *Journal of Proteomics, 201*, 1–11.

Szabados, F., Woloszyn, J., Richter, C., Kaase, M., & Gatermann, S. (2010). Identification of molecularly defined Staphylococcus aureus strains using matrix-assisted laser desorption/ionization time of flight mass spectrometry and the Biotyper 2.0 database. *Journal of Medical Microbiology, 59*(7), 787–790.

Tonacini, J., Stephan, D., Vogel, G., Avondet, M. A., Kalman, F., Crovadore, J., et al. (2019). Intact Staphylococcus enterotoxin SEB from culture supernatant detected by MALDI-TOF mass spectrometry. *Toxins, 11*(2), 101.

Trost, M., Wehmhöner, D., Kärst, U., Dieterich, G., Wehland, J., & Jänsch, L. (2005). Comparative proteome analysis of secretory proteins from pathogenic and nonpathogenic Listeria species. *Proteomics, 5*, 1544–1557.

Welker, M., & Moore, E. R. B. (2011). Applications of whole-cell matrix-assisted laser-desorption/ionization time-offlight mass spectrometry in systematic microbiology. *Systematic and Applied Microbiology, 34*, 2. –1.

Wells-Bennik, M. H., Eijlander, R. T., Den Besten, H. M., Berendsen, E. M., Warda, A. K., Krawczyk, A. O., et al. (2016). Bacterial spores in food: Survival, emergence, and outgrowth. *Annual Review of Food Science and Technology, 7*, 457–482.

Williams, T. L., Musser, S. M., Nordstrom, J. L., DePaola, A., & Monday, S. R. (2004). Identification of a protein biomarker unique to the pandemic O3:K6 clone of Vibrio parahaemolyticus. *Journal of Clinical Microbiology, 42*, 1657–1665.

Wolters, D. A., Washburn, M. P., & Yates, J. R. (2001). An automated multidimensional protein identification technology for shotgun proteomics. *Analytical Chemistry, 73*(23), 5683–5690.

Wolters, M., Rohde, H., Maier, T., Belmar-Campos, C., Franke, G., et al. (2011). MALDI-TOF MS fingerprinting allows for discrimination of major methicillin-resistant Staphylococcus aureus lineages. *International Journal of Medical Microbiology, 301*(1), 64–68.

Zautner, A. E., Masanta, W. O., Tareen, A. M., Weig, M., Lugert, R., Groß, U., et al. (2013). Discrimination of multilocus sequence typing-based Campylobacter jejuni subgroups by MALDI-TOF mass spectrometry. *BMC Microbiology, 13*(1), 1–8.

Zhang, Y., Fonslow, B. R., Shan, B., Baek, M. C., & Yates III, J. R. (2013). Protein analysis by shotgun/bottom-up proteomics. *Chemical Reviews, 113*(4), 2343–2394.

Zhu, S., Ratering, S., Schnell, S., & Wacker, R. (2011). Matrix-assisted laser desorption and ionization–time-of-flight mass spectrometry, 16S rRNA gene sequencing, and API 32E for identification of Cronobacter spp.: A comparative study. *Journal of Food Protection, 74*(12), 2182–2187.

CHAPTER

12

Peptidomic approach for analysis of bioactive peptides

Sol Zamuz[a], *Daniel Franco Ruiz*[a], *Mirian Pateiro*[a], *Ruben Dominguez*[a], *Paulo E.S. Munekata*[a], *Noemí Echegaray*[a], *María López Pedrouso*[b], *Jose M. Lorenzo*[a,c]

[a]Centro Tecnológico de la Carne de Galicia, Ourense, Spain
[b]Department of Zoology, Genetics and Physical Anthropology, University of Santiago de Compostela, Santiago de Compostela, Spain
[c]University of Vigo, Spain

12.1 Peptidomic: definition

The term peptidomics was first introduced in 2000 to describe the systematic, comprehensive, qualitative, and quantitative analysis of all peptides and small proteins (up to ~20 kDa) of a given biological system at a defined point of time. It represents peptide proteomics (Schulz-Knappe, Schrader, & Zucht, 2005) and serves as a complementary strategy to close the gap between proteomics and metabolomics. The peptidomic approach has been applied to several areas of research, including the identification and quantification of bioactive peptides derived from food proteins. Mass spectrometry (MS) is one of the most widely used analytical techniques in peptidomic, since it allows the detection and quantification of peptides in a complex matrix. To improve the range and sensitivity of this technique the complexity of the food samples must be reduced to the maximum. In this way, peptidomic sample analysis includes several steps which could be grouped into three key stages st (Fig. 12.1): sample preparation and protein isolation, peptide purification, and finally, detection and data

Food Proteomics
DOI: https://doi.org/10.1016/B978-0-323-90889-4.00011-7

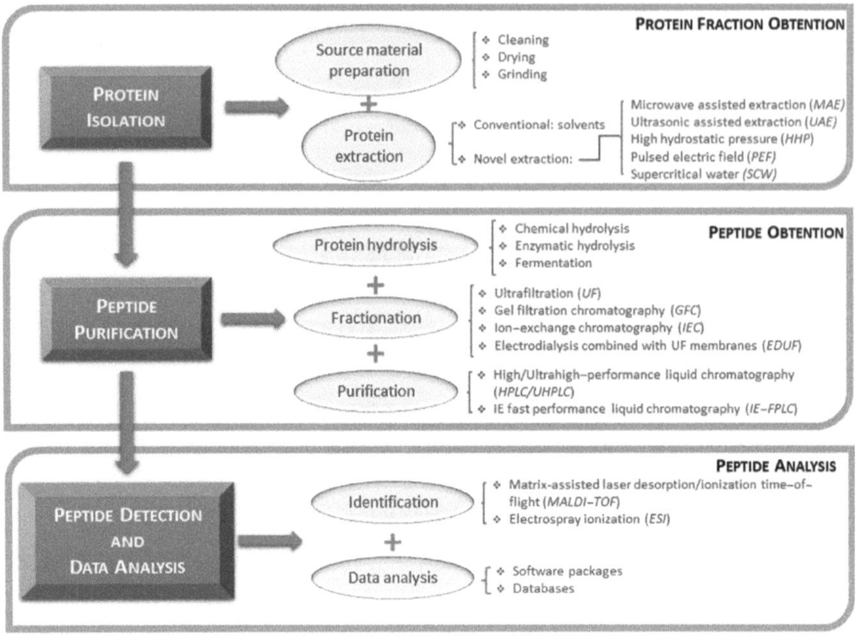

FIG. 12.1 Stages of peptidomic sample analysis.

analysis. Throughout this chapter, a description of the different analytical methods that have been developed to carry out each of these steps will be provided.

12.2 Sample preparation and protein isolation

Bioactive peptides can be obtained from different plants, animal, and marine protein sources (Tadesse & Emire, 2020). The choice of processing method to protein isolation depends on the type of matrix, the interactions among components, as well as nature and protein structure. However, in all cases, the samples have to be processed and the components of the raw material other than protein must be separated using different procedures, including cleaning, drying, and grinding of the matrix. Conventional methods of protein extraction are based on the use of organic and inorganic solvents to remove lipidic components, carbohydrates, and other soluble constituents. Protein obtention from animal and fish products involves cleaning the samples thoroughly with water to remove foreign materials and to separate the meat/flesh from the skin and bones. Afterwards, samples are ground and mixed with distilled water and different solvents (usually hexane), and finally, the extract is centrifugated to separate the lipids. In the case of plants and algae, the proteins are usually part of complex macromolecules comprised by carbohydrates, lipids, and fibers from the cell wall, among other components. The first step is to remove fat compounds with organic and inorganic solvents and the extract is cleaned, dried, and milled. Then the defatted extract is mixed with alkaline deionized water with (pH = 10) and centrifugated. The supernatant contains the protein that is concentrated using isoelectric precipitation by adding acid or base solution. Finally, protein extract is usually dried using a freeze dryer. In addition to the use of solvents, other processing technologies (Table 12.1) have been applied for protein extraction in the last decades (Tadesse & Emire, 2020). These novel technologies improved the analysis time, the yield and the peptides bioactivities, reducing the process costs.

TABLE 12.1 Sample processing techniques for the obtention of bioactive peptides in different plant, animal, and marine resources.

Protein source	Extraction method	References
Lotus seed	MAE	Gohi, Du, Zeng, Cao, and Zou (2019)
Rainbow trout	MAE	Ketnawa & Liceaga (2016)
		Nguyen, Jones, Kim, Martin-Gonzalez, and Liceaga (2017)
Sweet potato	MAE	Zhang et al. (2019b)
	HHP	Zhang & Mu, (2017)
Egg	UAE	Stefanović et al. (2014)
	PEF	Liu et al. (2019)
Milk and dairy derivates	MAE, UAE	Uluko et al. (2015)
	UAE	Madadlou, Sheehan, Emam-Djomeh, and Mousavi (2011)
	HHP	Bamdad, Shin, Shu, Nimalaratne, & Sunwoo (2017)
Tilapia	UAE	Kangsanant, Thongraung, Jansakul, Murkovic, and Seechamnanturakit (2015)
Wheat	UAE	Zhou, Ma, Yu, et al. (2013a)
Corn	UAE	Zhou, Ma, Ding, et al. (2013b)
Garlic	UAE	Huang, Ma, Peng, Wang, and Yang (2015)
Flaxseed	HHP	Franck et al. (2019)
Peanut	HHP	Dong et al. (2019)
Pinto bean	HHP	Garcia-Mora et al. (2016)

(Continued)

TABLE 12.1 Cont'd

Protein source	Extraction method	References
Macroalgae *Ulva*	PEF	Polikovsky et al. (2016)
Pine nut	PEF	Zhang et al. (2019b)
Codfish	SCW	Melgosa et al. (2021)
Red algae	SCW	Park, Jeong, and Chun (2019)
Tuna	SCW	Ahmed & Chun (2018)
Porcine blond	SCW	Álvarez, Tiwari, Rendueles, and Díaz (2016)

HPP, high hydrostatic pressure; MAE, microwave assisted extraction; PEF, pulsed electric field; SCW, subcritical water; UAE, ultrasonic assisted extraction.

Microwave-assisted extraction (MAE) involves the use of electromagnetic radiation in a frequency range from 300 MHz to 300 GHz to heat the solvents in contact with samples. The heating of the reaction system is faster than the traditional liquid extraction because the heating is due to the movement and collision of charged ions and through inter-and intramolecular friction, causing the breakdown of cell walls and membranes (Wang, Yu, Xing, & Li, 2017a). This fact also can change the protein conformation, weakening the bonds to the subsequent hydrolysis process. MAE spend less energy and solvent and it is more environmentally friendly than conventional extraction with solvents.

Ultrasonic assisted extraction (UAE) modifies the physicochemical properties of the samples due to the mechanical and thermal effects of the acoustic cavitation of the ultrasound. Two types of UAE are used in the food industry (Ozuna, Paniagua-Martínez, Castaño-Tostado, Ozimek, & Amaya-Llano, 2015): high intensity ultrasonic (16–100 kHz, power 10–1000 W/cm^2) and low intensity ultrasonic (100 kHz–1MHz, power $<1\ W/cm^2$). UAE modifies protein conformation by affecting hydrogen bonds and hydrophobic interactions, disrupting the structure of proteins, improving protein hydrolysis.

High hydrostatic pressure extraction (HHP) is a non-thermal technology in which isostatic pressure (100–1000 MPa) is applied for a few minutes (depending on the nature of samples) and transferred to liquid or partially liquid-based products. HPP process involves structural changes, as protein denaturation by inhibiting noncovalent interactions to a degree depending on the structural characteristics of proteins and the pressurization parameters. HHP improves peptide release and generation of bioactive peptides due to greater exposition of cleave sites made more accessible at subsequent treatments (Franck et al., 2019).

Pulsed electric field (PEF) is a non-thermal technology that involves applying electric fields strength of 10–50 kV/cm in the form of multiple short pulses (1–5ms) at frequencies of 0.2–0.4 MHz. The membrane permeabilization is achieved through the formation of aqueous pores (electroporation) when the induced transmembrane potential exceeds a critical value of 1volt and cells are disintegrated. In addition, this technique induces changes in the structure of proteins, and hydrolysis sites will be exposed (Chian et al., 2019; Tadesse & Emire, 2020). PEF process is chemical-free and energy-efficient.

Subcritical water (SWC) is a green technology for protein extraction that uses pressurized hot water at 100–374°C (temperatures between its atmospheric boiling point and critical point) and pressure above saturation value but less than the critical value (<22.1 MPa). At these subcritical points, the dielectric constant of water decreases to values similar to methanol, ethanol, and acetone at room temperature, with increasing temperature due to hydrogen bond dissociation. In these conditions, surface tension and viscosity of water decreases, allowing it to act as an effective solvent to dissolve organic substances that are insoluble at the atmospheric boiling point of water. Moreover, the water ionic product increases with temperature (three orders of magnitude higher than that at ambient conditions) leading to the increase of hydronium and hydroxide ions concentrations and allowing water to act as an acid or base catalyst (Melgosa et al., 2021).

12.3 Peptide purification

Once the protein concentrate has been obtained, the next step is to obtain the bioactive peptides and their subsequent separation. This stage includes three major processes to be described below: protein hydrolysis, peptide fractionation, and peptide purification.

12.3.1 Hydrolysis of proteins

Protein hydrolysates can be obtained by three different methods: chemical hydrolysis, enzymatic hydrolysis, and fermentation and each of them has its advantages and disadvantages.

Chemical hydrolysis is carried out by cleaving peptide bonds with acids or alkalis, being former ones preferred. Sulfuric acid (H_2SO_4) is the most commonly used acid in hydrolysis acid, butnitric acid (HNO_3), hydrochloric acid (HCl), phosphoric acid (H_3PO_4) and others as malic and oxalic acids are also used (Wang et al., 2017a). Although chemical hydrolysis is simple, quick, and can easily hydrolyze peptide bonds with high yield, usually requires high temperature, and samples are treated in extreme pH range, which could modify or destroyamino acids. Moreover, this procedure is insecure and environmentally unfriendly. Considering all disadvantages of chemical hydrolysis, the other hydrolysis methods have gained significant consideration to produce protein hydrolysates.

Enzymatic hydrolysis uses proteolytic enzyme preparations to cleave specific peptide bonds according to the desired peptides. The process is performed in a reactor containing deionized water, concentrated protein hydrolysate, and proteolytic enzymes. Buffers are used to maintain the optimum pH range during hydrolysis and temperature also has to be controlled. The reaction is maintained until a constant degree of hydrolysis is obtained and at the end of the process, the mixture is heated at a temperature beyond the value of enzyme inactivation and it is centrifugated. The supernatant contains peptides, which arestored below −20°C or dried by a freezer dryer (Tadesse & Emire, 2020). The different hydrolysate types obtained of enzymatic hydrolysis depending on used enzyme types because they have different peptide bond cleavage patterns, as well as, the hydrolysis conditions (pH, temperature, and time). Single, double, or multiple endogenous or exogenous enzymes can be used to release peptides of interest. The exogenous enzymes are preferred because they have a shorter hydrolysis time and the process is more controlled and reproducible. Food grade commercial enzymes widely used to generate peptides are obtained from microorganisms (Alcalase, Protamex, Neutrase, Subtilisin, Flavourzyme), animal origin (Pepsin, Trypsin, Chymotrypsin), and plant sources (Papain, Bromelain).

Fermentation is a cost-effective and eco-friendly method to produce bioactive peptides which involves inoculation of microbial strains into a reactor containing an adequate source of protein. The proteins are broken by the activity of microorganisms, caused by peptidases secreted by them during their growth. The production of protein hydrolysates is influenced by microbial growth parameters (inoculum conditions) and peptide content. Proteases secreted by microorganisms present several advantages over other proteases: low production cost, less time required for enzyme isolation, possibilities for mass production, and amenability of the organisms to genetic manipulations. Natural or added microorganisms with high proteolytic activity and specificity are required for cleaving the peptide bonds and producing a significant amount of bioactive peptides. Several starter cultures are employed for microbial fermentation of food proteins. For instance, proteases to produce bioactive peptides from different protein sources (marine, animal, or plant) are frequently obtained from lactic acid bacteria (*Lactobacillus* sp., *Pediococcus* sp., *Enterococcus* sp., *Streotococcus* sp), *Bacillus* sp., *Bifidobacterum* sp. and some fungi strains (Godinho et al., 2016).

12.3.2 Peptide fractionation

The protein hydrolysate obtained from hydrolysis results in a wide mixture of peptides with different physicochemical properties (molecular weight, net charge, amino acid composition) and different bioactive properties. Due to this fact, the next step in the analysis and the determination of bioactive peptides is the separation (or fractionation) of them, based on their physicochemical properties. This separation must be carried out by mild conditions as low temperature and pH value close to neutrality to maintain their structural and physicochemical characteristics. Moreover, a mixture of peptides is obtained together with other compounds (non-hydrolysate protein, lipids, and other aggregates), that they must be removed before the fractionation. Conventional technologies as filtration, centrifugation, precipitation, and use of adsorbents (diatoms earth and activated carbon) are used for this purpose. Afterward, depending on the peptide properties, they can be separated using techniques based on sieving (membrane filtration, gel filtration chromatography (GFC)), charge (pH-induced precipitation, electrodialysis, ion-exchange chromatography), detection of differences in hydrophobic interactions (solvent precipitation, reversed-phase chromatography) and/or affinity and special attribute molecular methods (affinity chromatography, immunoaffinity). The main technologies used in peptide fractionation are set out below (Marson, Belleville, Lacour, & Hubinger, 2020).

Membrane filtration is based on the separation of solutes dissolved in a fluid by the difference in molecular weights and particle size (Fig. 12.2). It can maintain protein stability and a high resolution is possible at low temperatures; in addition, it does not require solvents. This technique can be used at different levels depending on the size of molecules that can retain: microfiltration separate molecules with size between 0.05 and 10 mm and usually it is used for previous clarification, to concentrate or as pretreatment for other techniques; ultrafiltration (UF) separate macromolecules with molecular weight between 1 and 200 kDa; and nanofiltration (NF) separate particles with pore diameter less than 1 nm. UF is the main process used in the separation of macropeptides and nonhydrolyzed protein because UF molecular weight cut-offs (MWCOs) fit the size range of proteins and their fractions: membranes with MWCO t 1-10 kDa are suitable for the separation of bioactive peptides and membranes with low MWCO (<1 kDa) are used to concentrate peptides (Marson et al., 2020). The main problems with UF are poor selectivity and fouling. The poor selectivity of membrane involves that UF is restricted to desalination. Several strategies are being used for desalinization, as NF (allows permeating only monovalent mineral ions and water), use of active carbon or chromatographic separation. Concerning fouling, it is indispensable for fouling control strategies to maintain acceptable flux and solubility (Wang, Yu, Xing, & Li, 2017b).

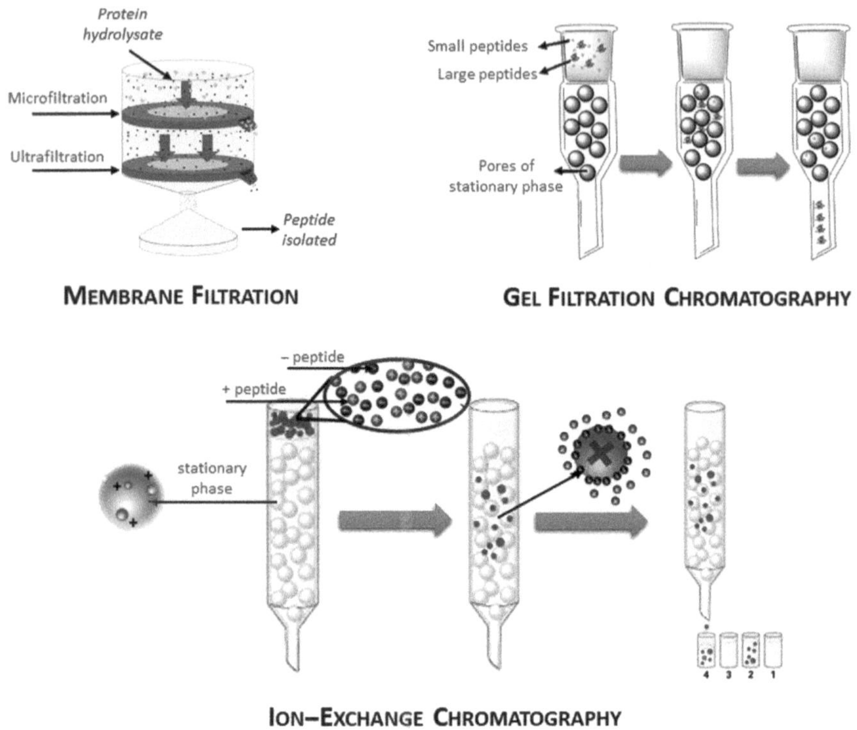

FIG. 12.2 Techniques to separation of peptides.

Another technique that separates molecules based on the size difference is GFC or size exclusion chromatography (Fig. 12.2). The column is packed with a solid-stationary phase that contains pores of 100–250 μm which permit to pass through them specified sized peptides dissolved in the mobile phase. Thus, smaller peptides enter the pores of the gel meanwhile larger peptides and other molecules are not retained. The first molecules have a longer retention time than the second ones (Büyükköroğlu, Dora, Özdemir, & Hizel, 2018). The elution conditions can be varied to suit the sample and the requirements for further steps. The buffer composition does not affect the GFC resolution, and this technique has high selectivity. GFC covers a molecular weight range from 100 to 8×10^7 Da separating, from peptides to very large proteins depending on pore size and properties of the stationary phase. The most used ones are dextran, dextran/bis-acrylamide, agarose, agarose/dextran, polyacrylamide, methacrylate, acrylamide, acrylamide/agarose, and cellulose among others.

As mentioned above, the separation of peptides will also depend on their charge and the possible interactions that may occur with the matrix. Some charged-based methods could be used to separate peptides and among these, the *ion-exchange chromatography* (IEC), which is an adsorption technique that retains the molecules on the column based on ionic interactions (Fig. 12.2). Peptides have functional groups and residues that can have positive or negative charges and IEC uses a stationary phase bearing either positively or negatively charged chemical groups to bind with an opposite charge. Thus, the stationary phase that containing positively (anion-exchange) or negatively (cation-exchange) charged groups attract negatively and positively charged peptides, respectively. Initially, samples are load in a low ionic strength mobile phase, and then bound peptides are eluted by displacement by increasing the ionic strength of the mobile phase. This process is usually performed via a gradient of increasing salt concentration but changes in pH and organic solvents (such as acetonitrile or methanol) are also used to modify the retentivity of different peptides (Velickovic, Ognjenovic, & Mihajlovic, 2012). Different IEC columns can be used depending on the target, sample, and necessary resolution. For example, DEAE Sephadex (weak anion exchanger) works pH ranges between 2 and 9; QAE Sephadex (strong anion exchanger) works pH ranges between 2 and 12; CM Sephadex (weak cation exchanger) works pH ranges between 6 and 10; and SP Sephadex (strong cation exchanger) works pH ranges between 4 and 13 (Wang et al., 2017b).

Moreover, *charge-based membrane techniques* can be used to separate peptides such as electrophoresis and electrodialysis. These techniques are based on the separation of electrically charged peptides by applying an electric field and depend on the conductivity value. In electrophoresis, when a potential difference is applied through a gel, the resulting differential potential permits that charged peptides will separate owing to their different electrophoretic mobility (Judd, 1994). In electrodialysis (ED), the separation is based on the migration of charged peptides through a membrane (ion-exchange membrane) after application of electric field (Dlask & Václavíková, 2017). Moreover, several techniques based on simultaneous seize and charge separation have been used also for the peptide separation. Among them, *electrodialysis combined with UF membranes* (EDUF) is one of the most promising, been considered for large-scale peptide production in substitution to several chromatographic operations. This technique is very selective, capable of separating targeted peptides from complex mixtures. EDUF consist of UF and ion-exchange membranes stacked together in a conventional electrodialysis cell so that UF allows the concentration and purification of peptides according to their sizes and ED allows a selective separation according to their electrical charges. In EDUF, no pressure is applied in the electrodialysis cell hence the electric potential difference is the only driving force of this technology. This process is regulated by the factors for ED, pH changes, and UF membrane properties (Dlask & Václavíková, 2017).

Finally, another promising technique for the separation of bioactive peptides is *membrane chromatography,* a separation technology based on the integration of membrane filtration and liquid chromatography into a single operation, although it is still not used as a conventional technique. The membranes are composed of specific functional groups onto their surface and peptides are adsorbed to these ligands during the flow. As the peptides are very close to the active binding site, the mass transfer resistance is reduced, and the binding capacity does not depend on flow rate. This is the main advantage of this technique: the improvement of the mass transfer due to the membranes with through pores minimizes the diffusion limitations associated with the use of beads (Boi, 2019).

12.3.3 Peptide purification

High-performance liquid chromatography (HPLC) has become a widely used, well-established tool for the analysis and purification of several biomolecules (Fig. 12.3). In HPLC, the mobile phase (liquid) passes through the column containing stationary phase at high pressure (up to 40 MPa). The mobile phase consists of a mixture of an organic solvent (to desorb peptides from the adsorbent surface), an ion-pairing reagent (to set the eluent pH and to interact with peptide and to enhance the separation) and another solvent of greater polarity (water, commonly).

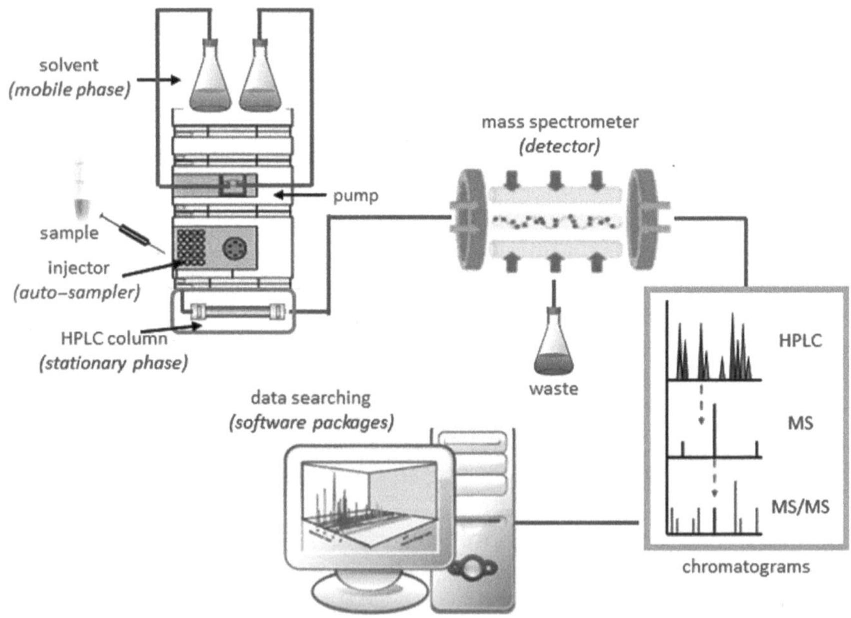

FIG. 12.3 HPLC-MS/MS technique to the identification of peptides.

A gradient elution (polarity of mobile phase increases or decrease at a controlled rate) is used to facilitate peak sharpness and resolution. The elution of peptides from the stationary phase is done when the solvent reaches the precise concentration necessary to cause desorption. Based on the nature of the stationary phase, HPLC is classified into normal-phase chromatography (polar stationary phase; NP) and reverse-phase chromatography (nonpolar stationary phase; RP). *Reverse-phase high-performance liquid chromatography* (RP-HPLC) is the most widely used chromatographic technique for the analysis and purification of bioactive peptides due to its high resolution, high sensitivity, ease of use, and short time of work. It can separate polypeptides of nearly identical sequences even those which differ by a single amino acid residue. This technique also can be used to fractionate peptides based on their hydrophobic characteristics. The separation is based on the selective interactions of peptides with the nonpolar stationary phase and a polar mobile phase (Conlon, 2007).

The usually employed organic solvents to peptide purification by RP-HPLC are acetonitrile, isopropanol, and ethanol. Acetonitrile (ACN) is the most used because of its volatility, low viscosity, and small ultraviolet adsorption at low wavelengths. Isopropanol has a high viscosity, and it is recommended to use a mixture of acetonitrile and isopropanol, and ethanol is often used for process scale purifications. A typical elution gradient has a slope of 0.5% to 1% per minute increase in organic solvent concentration, and extremely shallow gradients (0.05%–0.1% per minute) can be used to maximize resolution. t is recommended to avoid extremes percentages in organic solvent composition (no less than 3%–5% at the beginning and no more than 95% at the end). Trifluoroacetic acid (TFA) is the most common ion-pairing reagent because, as well as ACN, is volatile and has negligible ultraviolet adsorption at low wavelengths. TFA is normally used at concentrations between 0.05 to 0.1% (w/v) and this concentration can affect selectivity and resolution of chromatographic separation which provides another factor for optimizing peptide resolution. Other reagents used as ion-pairing are heptafluorobutyric acid (HFBA) to separate basic protein and triethylamine phosphate for preparative separations. Formic acid in concentrations of 10%–60% has proven to be more effective in liquid chromatography with mass detector than TFA because the latter one reduces the ion signal in the electrospray interface (Conlon, 2007).

The stationary phase of the columns consists of microspherical adsorbent particles, commonly of silica with hydrocarbon groups bonded at the surface to make them hydrophobic. Other synthetic polymers such as polystyrene-divinylbenzene are also used as stationary phases when it is necessary to separate at high temperature or in an extreme range of pH values. Indeed, they also allow using very acid or basic solutions to clean columns and to remove other proteins because they are not degraded at extreme pH values. The length of the hydrocarbon groups may also influence the effectiveness of separation. The usual lengths of these alkyl chains are C_4, C_8, and C_{18}. The C_{18} columns are generally used for peptides less than 5000 Da and C_4 columns for peptides larger than this value or small peptides with specific hydrophobicity. Phenyl columns can offer unique selectivity for some peptides and they exhibit shorter retention time and a different selectivity than C_{18} columns. On the other hand, column dimensions also affect the quality of the separation. Most bioactive peptide separations are carried out on columns with pore size about 300 Å and peptides with <~2000 MW may be separated on columns with 100 Å pore size. Small peptides are efficiently separated on columns of 15–25 cm length and 4.6 mm of internal diameter. The use of small diameter columns (0.075–2.1 mm) has increased because offers a reduction in the amount of solvent needed and increased sensitivity in terms of the detection limit. Very small diameter columns are used when coupling HPLC with MS. Another important chromatographic parameter to be considered in the process is the temperature, with affects peptide selectivity due to its effect on solvent viscosity, column backpressure, and retention time (D'Hondt et al., 2013).

The latest advances in instrumentation are focused on pushing the limits of peptide separation efficiency. Nowadays, ultra-high performance liquid chromatography (UHPLC) systems allow for backpressures of up to 120 MPa. The main difference is the size of used particles filled into the column. Indeed, particles with a size of 3 μm up to 5 μm are usual for HPLC, meanwhile particle sizes ≤ of 2 μm are used for UHPLC. The column dimensions range from 75 mm to 30 mm internal diameter and from 150 mm to 250 mm column length. UHPLC shows greater efficiency than HPLC, increasing resolution and the speed of separation, reducing the analysis time. Moreover, the sensitivity is improved because sharper peaks are obtained due to an enhancement in the signal-to-noise ratio. UHPLC also generates lower matrix effects than HPLC, which is attributed to the enhanced resolution between the analyte and matrix interferences (Howard, Kay, Pleasance, & Creaser, 2012).

Another chromatographic technique used for peptide purification is the *fast performance liquid chromatography on the ion-exchange columns* (IE-FPLC). FPLC is a form of liquid chromatography of medium pressure (typically 5 bar) with a high flow rate (1–5 mL/min) The stationary phase is a resin composed of small-diameter beads (generally cross-linked agarose) that are packed in glass or plastic columns and have high loading capacity and the mobile phase is an aqueous buffer solution. This technique allows a wide range of mobile phases and different stationary phases being the ion exchange the most common FPLC strategy on peptide separation (Pontis, 2017).

12.4 Peptide detection and data analysis

Following the consecutive analysis process, the next step is the detection, identification, and characterization of bioactive peptides. The standard method for their detection and identification is the liquid chromatography (HPLC or UHPLC) combined with a mass spectrometer (MS) or tandem mass spectrometer (MS/MS) detector. Finally, data are interpreted with bioinformatic tools using software packages and databases.

12.4.1 Detection and identification of peptides

The MS or MS/MS spectrometry are the standard technique to detect and identify bioactive peptides by their mass-to-charge ratio obtaining a mass spectrum. The peptides have to be ionized and then, they become positively charged without fragmentation or they break up into fragments with a positive charge. These ions are separated by accelerating produced by an electric or magnetic field and, they are detected by a mechanism capable of detecting charged particles (electron multiplier). The combination of HPLC (and/or UHPLC) with MS spectrometry increases the sensitivity concerning other detectors and reduces the need for chromatographic resolution because of the higher resolution capacity of the MS detectors. The trend toward faster analysis (few minutes) has led to the use the short columns (5–10 cm length) with 3 mm particle size and quick gradients.

The basic components of an MS spectrometer are ion source, mass analyzer, and detector. Two main ionization methods are used: matrix-assisted laser desorption/ionization (MALDI) and electrospray ionization (ESI). Other techniques as HPLC with evaporative light scattering detection (ELSD) and rapid resolution liquid chromatography-tandem mass spectrometry (RRLC-MS) are new possibilities in this context (Wang et al., 2017b).

Matrix-assisted laser desorption/ionization (MALDI) is a soft ionization (induces little or no fragmentation) that involves a laser striking a matrix to make the analyte molecules into the gas phase (Fig. 12.4). The peptides are embedded in an energy absorbent solution (matrix) and deposited on a solid surface (target) made of a conducting metal. Then, a very brief laser pulse irradiates the matrix that is rapidly heated and becomes vibrationally excited. The matrix molecules energetically ablated from the surface of the sample, absorbing the laser energy and volatilizing peptides into gaseous phase, that they are subsequently ionized as single protonated ions. The molar matrix-to-analyte ratio is normally adjusted into the range from 1000:1 to 100000:1 and the choice of the adequate matrix is key to the success in MALDI. In general, highly polar analytes work better combined with highly polar matrices and nonpolar analytes with

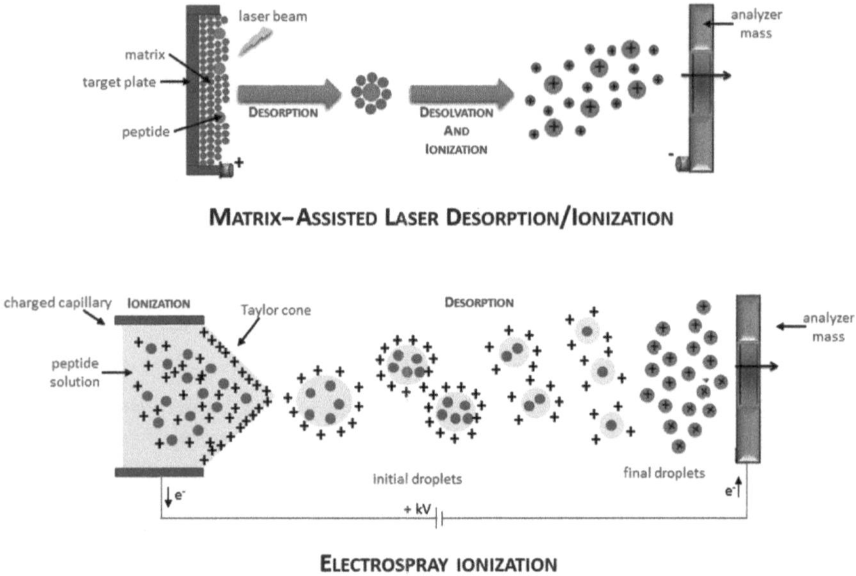

FIG. 12.4 Main ionization methods in mass spectrometry.

nonpolar matrices. The most commonly used matrixes for protein and peptides analysis are α-cyano-4-hydroxycinnamic acid, 2,5-dihydroxybenzoic acid, 3,5-dimethoxy-4-hydroxycinnamic acid, and 2,6-dihydroxyacetophenone (Duncan, Roder, & Hunsucker, 2008).

Electrospray ionization (ESI) is also a soft ionization but different MALDI, as produces multiple ions using an electrospray to create an aerosol by application of a high voltage to a liquid. This is the most common method used in bioactive peptides analysis (Fig. 12.4). The dilute peptide sample is injected into a metal capillary (typically 0.2 mm inner diameter, 60 mm in length, and heated to 100–300°C), where high voltage is applied and the sample solution is distorted into a Taylor cone that emits an aerosol of highly charged electrospray droplets. This spraying process is usually assisted by a coaxial inert sheath gas (dry N_2, Ar, or He) flow around the capillary which improves nebulization and allows solvent evaporation. As the solvent evaporates, the analyte molecules approach, repel, and finally, when the repulsion of the charges overcomes the surface tension, the drops burst (Coulomb Explosion). This process is repeated until the analyte is solvent-free remaining only the ions in their movement toward the mass analyzer (Banerjee & Mazumdar, 2012). ESI methodology can be used in either positive (ES+) and negative (ES−) ion mode: analytes with at least one positive elementary charge are detected in ES+, and analytes with at least one negative elementary charge can be detected in ES−. That way, singly or multiply protonated $[M + zH]^{z+}$ ions are generated in ES+

while singly deprotonated $[M - H]^-$ ions are dominant in ES−. ES+ is the most commonly mode used (Steckel & Schlosser, 2019). Among the sprayer modifications improving ESI, the most popular is the nanospray ionization which operates at low flow rates (<10 nL/min) requiring very low sample concentration (nmol/mL) and low volume. The charged droplets have a volume about 100–1000 times smaller than droplets produced by conventional ESI thus, a lower number of evaporation cycles is required (Banerjee & Mazumdar, 2012).

After the movement of the ions toward the mass analyzer, the ionized masses are separated based on their mass-to-charge ratio (m/z) using an electrical or magnetic field. These detectors work under low pressures (10^{-4} to 10^{-7} Pa) to permit the ions to pass through them efficiently and without interruption. There are many types of mass analyzers being the most commons: time-of-flight (TOF) mass analyzers, quadrupole (Q) mass analyzers, ion trap mass analyzers, and orbitrap mass analyzers. They can be used independently or in combination (multi-analyzer system (MS/MS)) in tandem or hybrid form. These analyzers can be classified as "in-time" or "in-space" depending on their operational parameters. "In-time" analyzers collections consecutively within a period of milliseconds to seconds while "in-space" analyzers collect and detect all ions simultaneously in non-scanning mode (Noor, Ahn, Baker, Ranganathan, & Mohamedali, 2021).

The TOF is one of the most common types of mass analyzers. Its basic principle relies on the fact that protonated ions are accelerated at a fixed potential and they can be separated according to their m/z ratio based on the length of time it takes them to travel through a flight tube of known length to reach the detector. The ions of different m/z are dispersed during their flight along a field-free drift path of known length. This m/z ratio may be measured by determining the time required for each ion to travel the length of the flight tube (TOF) and a characteristic mass spectrum called peptide mass fingerprint (PMF) is generated. Theoretically, all the ions are given the same initial kinetic energy, but the pulse is not felt by all ions to the same intensity, and therefore not all similar m/z values ions reach their ideal velocities. For this reason, a series of ring electrodes with high voltage (reflectron) is often placed at the end of the drift zone. These electrodes can repulse the ions back along the flight tube at a slightly displaced angle. Ions of different kinetic energy penetrate the reflectron to different depths before they get expelled into the opposite direction. Faster ions carrying more kinetic energy will travel a longer path than slower ones. In that way, the detector receives ions of the same mass at the same time. TOF is almost always used with a pulsed ion source and this is the reason that it is frequently coupled to MALDI because laser beam sources are frequently operated in pulsed mode (Boesl, 2017; Duncan et al., 2008). The MALDI-TOF mass spectrometric method is the backbone analysis for the generation of peptide profiles and generally is used for large sequences because peptides with molecular mass less than 500 Da fall within a low mass range where matrix interference is overwhelming (Agyei, Tsopmp, & Udenigwe, 2018).

Quadrupole (Q) mass analyzers consist of two pairs of parallel metal rods equidistant from each other and biased at equal and opposite potentials which acting as electrodes. These twin potentials contain a fixed direct current and alternating radiofrequency component (being 180° out of phase). The ions are accelerated through the quadrupole, hence any ion entering the quadrupole will have its trajectory deflected by the potential proportionally to its m/z value. At a fixed radio frequency value, only one specific m/z value will resonate with the field and be able to navigate to the end of the quadrupole and be detected. Ions with other m/z values will collide with the quadrupole, lose their charge, and not be detected. Quadrupoles can operate in two ways: SIM mode where the conditions are fixed for only one ion to pass with a certain m/z; and full SCAN mode where a determined range of masses are monitored (Mellon, 2003).

Ion trap mass analyzers employ similar principles as the quadrupole, and it consists of a ring-shaped electrode of a specific voltage sandwiched between two end-cap electrodes. The ions are introduced into the trap in a pulse mode and they enter the area between the electrodes through one of the end caps and an electric or magnetic field in the cavity causes the ions of certain m/z values to orbit in the space. To detect a specific ion, the radio frequency voltage is gradually increased. As the radio frequency voltage increases, oscillations of ions of a specific m/z ratio become unstable causing them to collide with the wall, and these ions are discharged through a hole in the end-cap electrode eliminating the possibility of traveling to and being detected by the detector (Mellon, 2003).

Orbitrap mass analyzers are a modification of an ion trap mass analyzer consisting of an outer barrel-like electrode and a coaxial inner spindle-like electrode that traps ions in an orbital motion around the spindle. Ions are injected or pulsed tangentially through a slot machined in one of the outer electrodes between the central and outer electrodes. When voltage is applied between both electrodes, the resulting electric field is linear along the axis, while the radial component of the field strongly attracts ions to the central electrode. Below, the electric field is increased by ramping the voltage on the inner electrode and the ions get squeezed toward the inner electrode until they reach the desired orbit inside the trap (moment in which the electric field is held constant and the field becomes static). Their frequency of rotation is directly related to their m/z value. The axial electric field caused by the special conical shape of electrodes pushes ions toward the widest part of the trap and the image current from the trapped ions is detected and converted to a mass spectrum using the Fourier transform of the frequency signal (Thomas, 2019).

Other mass analyzers used for the identification of peptides as *Magnetic sector* relies on the fact that magnetic fields can disperse ions in trajectories according to their m/z ratios in a manner that is analogous to how a glass prism disperses light into its various wavelengths. However, these analyzer shaves have been replaced, used only for small organic compounds as dioxins (Noor et al., 2021).

In a general sense, *tandem mass spectrometry* (MS/MS) refers to those multi-analyzer systems involving more of one mass spectrometer to increase specificity and resolution. Commonly, the ions are mass-filtered in the first analyzer and these may then undergo some form of reaction in subsequent reaction cells. The charged products emerging from this cell are then analyzed by the second mass analyzer. When the multi-analyzers are composed of the same type of analyzers they are called "tandem analyzers" and, when they are composed of different types, they are called hybrid instruments (Noor et al., 2021). The main tandem MS/MS includes *tripe quadrupole* (QqQ) and *time of flight/time of flight* (TOF/TOF) and the main hybrid MS/MS includes *quadrupole/time of flight* (Q/TOF). QqQ MS spectrometers are one of the most commonly used. Three sets of quadrupoles are used in sequence: the first set (Q1) scans across a range of m/z values or selectively filter ions of a selected m/z (precursor ion), the second set (Q2) is used exclusively as a collision cell to fragment the selected ions from Q1 and, the third set (Q3) scans the fragment ions (product ions) formed in Q2 to obtain a mass spectrum or to monitor a particular ion (Haag, 2016). TOF/TOF MS spectrometers combine two TOF analyzers in tandem and collision-induced dissociation (CID) is performed between the two TOF analyzers. Ions of different m/z are accelerated and separated in the first TOF based on their velocity. A timed-ion-selector allows passing ions of a particular m/z (precursor ion) based on their flight time. The selected ions are then decelerated before passing into a collision cell where they undergo CID. The resulting fragment ions (product ions) are re-accelerated into the second TOF and their TOF is measured to obtain a mass spectrum. TOF/TOF analyzer combined with MALDI ion sources (MALDI-TOF/TOF) is ideal for the analysis of peptides from tryptic digests (Haag, 2016). Q/TOF analyzer combines the high compound fragmentation efficiency of quadrupole and the rapid analysis speed and high mass resolution capability of TOF. This technology is closely similar to that of QqQ MS which Q3 has been replaced by a TOF (Allen & McWhinney, 2019).

Once the ions are separated, then they are passed to the detector, the final component of the mass spectrometer, which is responsible for detecting, amplifying, and determining the abundance of signals that arrive in the form of ion currents (Banerjee & Mazumdar, 2012). The first and the simplest detector for MS was the Faraday cup where ions are neutralized, and the resulting current is measured. This detector only detects and amplifies the ion current, hence it is used when ion flux is relatively large. However, the ion flux that traverses from the mass analyzer is quite small, thus, it has to be enlarged with an electron multiplier. The energetic ions strike a metal or semiconductor plate (conversion dynode) which is held at high voltage and emits secondary electrons. These electrons are accelerated and focused onto other dynodes. At each dynode, there is an increase in the number of electrons emitted. Finally, the output current is converted to a voltage signal which will be translated to an intensity value (the ordinate axis of the mass spectrum) using an analogue-to-digital converter. There are several types of electron multipliers being the multichannel plate detectors the most commonly used to identify peptides and they are used with Q and TOF analyzers. Besides, there are mass spectrometers where ions are detected by measurement of imagen current and converted to a mass spectrum using the Fourier transform of the frequency signal, as in the case of orbitrap.

12.4.2 Data analysis: bioinformatic or "in silico" tools

The results obtained of HPLC- or UHPLC-MS and/or HPLC- or UHPLC-MS/MS are interpreted with bioinformatic or in silico tools (software and hardware). Spectra generated from detectors depict the relationship between relative abundance or intensity of each ion of a peptide and its m/z value and digitalization of these analog data is possible to plot the electronic signals and to manipulate data. The PMFs generated by mass spectra are compared with online databases (NCBI, BIOPEP-UWM, BLAST, UniProt, etc.), which have storehouses of amino acid sequences that can be consulted for study and analysis of peptides resulting in their identification and quantification, and prediction of their biological activity. Some data analysis software tools are specialized for specific tasks while others provide multiple integrated tools in a single platform (trans-proteomic pipeline (TPP), OpenMS Proteomics Pipeline (TOPP), Central Proteomics Facilities Pipeline, SORCERER). Other pipelines combined these platforms with other individual tools (automated proteomics pipeline, Integrated Proteomics Pipeline (IP2), Proteomics Standard Initiative (PSI)) (Noor et al., 2021).

The data analysis is carried out using different bioinformatics tools (Table 12.2) and includes three steps (Noor et al., 2021): the conversion of raw MS data into formats that can be read by MS analysis software, spectral identification, and validation leading. The conversion of MS data into standard formats as mzXML, mzDATA, and mzML can be carried out by several converter tools such MSconvert, RawConverter, Mascot Distiller, ReAdW, MassWolf, and mzWiff, each compatible with one or more MS vendors. TPP and PSI work with MS data in their own file formats: pepXML (stores and analyzes peptides), prot XML (stores and analyzes

TABLE 12.2 Bioinformatic or in silico tools for analysis of mass spectrometry spectra.

Function		Software or informatic tools	Websites
MS data conversion		Mascot Distiller	http://www.matrixscience.com/distiller.html
		ReAdW	http://www.ionsource.com/functional_reviews/readw/t2x_update_readw.htm
		mzWiff	http://tools.proteomecenter.org/wiki/index.php?title=Software:mzWiff
		massWolf	http://tools.proteomecenter.org/wiki/index.php?title=Software:massWolf
		Msconvert	http://tools.proteomecenter.org/wiki/index.php?title=Msconvert_Capabilities
		RawConverter	http://fields.scripps.edu/rawconv/
Spectrum identification	Sequence database	Sequest (Thermo Fisher)	https://sourceforge.net/projects/sashimi/files/latest/download?source=files
		Mascot (Matrix Science)	http://www.matrixscience.com/distiller_download.html
		X! Tandem	https://www.thegpm.org/tandem/
		Morpheus	https://sourceforge.net/projects/morpheus-ms/
		ProbID	https://www.systemsbiology.org/research/
		MyriMatch	https://medschool.vanderbilt.edu/msrc/
		MassMatrix	http://www.massmatrix.net
		Comet	http://comet-ms.sourceforge.net/
		ProLuCID	https://github.com/bathyg/Prolucid_gui

(Continued)

TABLE 12.2 Cont'd

Function		Software or informatic tools	Websites
		MS-GF+	https://github.com/MSGFPlus/msgfplus
	Spectral matching	SpectraST (Spectra Search Tool)	https://sourceforge.net/projects/spectrastwindows/
		Bibliospec (MacCoss Lab)	https://skyline.ms/wiki/home/software/BiblioSpec/page.view?name=default
		X! Hunter (Global Proteome Machine)	http://www.thegpm.org/Hunter/index.html
		MyriMatch	https://medschool.vanderbilt.edu/msrc/
	De novo sequencing	PepNovo (CCMS)	http://proteomics.ucsd.edu/Software/PepNovo/
		PEAKS studio (Bioinformatics Solutions Inc.)	http://www.bioinfor.com/peaks-studio/
		Sequit (Proteome Factory)	https://www.proteome-factory.com/index.php/en/downloads
		Lutefisk	http://www.hairyfatguy.com/lutefisk/
	Hybrid method	GuntenTag (Yates Laboratory)	http://fields.scripps.edu/yates/wp/?page_id=17
		InsPecT	http://proteomics.ucsd.edu/software-tools/inspectms-alignment/
		SpecNets (Spectral Networks) (CCMS)	https://bix-lab.ucsd.edu/display/CCMStools/Spectral+Networks

Function	Software or informatic tools	Websites
Statistical analysis	PeptideProphet iProphet	https://moritz.systemsbiology.org/resources/software/
	PTMProphet Scaffold (Proteome Sofware)	http://www.proteomesoftware.com/products/scaffold/
	MAYU	http://proteomics.ethz.ch/muellelu/web/LukasReiter/Mayu/
	FDRAnalysis	https://omictools.com/fdranalysis-tool
	Probabilistic Consensus Scoring	https://www.openms.de/
	Semisupervised Expectation-Maximation (EM) Algorithm	http://tools.proteomecenter.org/software.php.
	Virtual Expert Mass Spectrometrist (VEMS)	http://www.yass.sdu.dk/

Data adapted from (Noor et al., 2021)

proteins) and analysisXML (stores all the identification and quantification of MS spectra). The next step is the MS spectra identification that can be carried out by different methods. Those commonly used are described below. In the sequence database search, the experimental fragment ion spectra are matched to the theoretical ion spectra of preferred peptides, and all possible peptide sequences with similar precursor m/z values are selected to investigate. Scores are calculated to examine the resemblance between both spectra (experimental and theoretical). Databases such as UniProt and NCBI use this procedure. Spectra correctly identify can be stored in spectral libraries which will be used for MS spectra identification step using the spectra matching method. This method eliminates the repeated rediscovery of proteins previously processed in MS/MS, enhancing the spectra identification. Some of the spectrum databases used are SRMAtlas, PeptideAtlas, and ProteomeXchange. When there is no or very limited information in databases, *de novo* sequencing is the method used: the sequence of the peptide is explicitly interpreted by manipulating peaks spacing from high-resolution fragment ion spectra and similar sequences are searched in a database using sequence alignment tools such as BLAST. This method is used to identify and characterize peptide modifications. Finally, a hybrid approach of conventional searching methods and *de novo* sequencing can be used. In this method, short sequence tags are generated using *de novo* sequencing and combined with the precursor m/z values of targeted peptides. This sequence is used for a sequence database search and only the theoretic spectrum of peptides with one or more sequence tags and similar m/z values will be considered. This method is more accurate and efficient than others because search space is limited. Once the peptides have been identified, a statistical validation is necessary to have robust confidence in the identifications. Different scores depict the measure of similarity between peptide sequences and spectra or between experimental and theoretical spectra. These scores are subjected to a statistical evaluation where probabilities to obtain the false discovery rate (FDR) is calculated. Two categories of FDR estimation are Target-Decoy searching and Empirical Bayes approaches. There are several statistical informatic tools and some are incorporated in searching tools such as Sequest, Mascot, X! Tandem, and MyriMatch.

12.5 In silico analysis of bioactive peptides

In addition to analytical techniques, there are also computational techniques to the analysis of bioactive peptides which allow to make an in-silico prediction of bioactive sequences encrypted in proteins and to know the structure-activity relationships of peptides previous to analytical research. This strategy overcomes the intensive cost and the time consumed by conventional methods. The databases used to identify peptides besides data about the amino acid sequences also contain information about the frequency of bioactive peptides occurrence in proteins and specific enzymes used to releasing peptides with biological activities so they can be used to predict the physicochemical and functional properties of peptides. These techniques also allow to analysis and characterize amino acid sequences of unknown proteins from new sources (Tadesse & Emire, 2020). Bioinformatic tools allow researchers to focus on a small number of peptides that are the most likely candidates of searched biological activities. The ability to predict physicochemical properties can help in the design of suitable purification steps for the analysis of the peptides of interest. At the same time, the ability to predict desirable and/or undesirable biological activities is important for the development of novel bioactive peptide sequences and the use of these peptides for functional foods or pharmaceutical applications.

The first step in the in silico analysis of peptides is the selection of proteins of known primary sequence and the acquisition of their amino acid sequences in databases such UniProtKB (UnitProt Consortium), Protein (National Center for Biotechnology Information; NCBI), or Protein Data Bank (Research Collaboratory for Structural Bioinformatics; RCSB). Then, an in silico digestion is conducted using protein/peptide cutting functions of protein digestion databases (such BIOPEP Enzyme action and ExPASy PeptideCutter). These databases generate a map of protein sequence with a list of cleaved peptides or an indication of cleaves sites for further processing. In addition, different parameters or descriptors are calculated with different informatic tools to know the frequency and the probability of peptide release from protein. The main tools and parameters used are the frequency of occurrence of peptide fragments, the potential to have a particular biological activity, the frequency with which peptide sequence is released from the protein by the chosen enzyme, and the theoretical degree of hydrolysis. The next step is the physicochemical, biological, and sensory characterization of the generated peptides using chemometric and cheminformatic techniques carried out with computational and statistical tools. These tools collect and analyze biochemical data and design models for understanding and predicting the behaviors of biochemical systems. This is possible because the presence and sequence arrangements of certain amino acid residues indicating the properties and potential activities of the peptides. The quantitative structure–activity relationship is done by regression analysis from the scores of several descriptors using partial least squares regression, principal components, and artificial neural networks. Some parameters used are z score (combines data of hydrophilicity/hydrophobicity, molecular size/bulkiness, electronic properties/charge); molecular surface-weighted holistic invariant molecular indices (information about amino acid/peptide-ligand interactions); Markovian chemical in silico design (to model intramolecular electron delocalization a time-base vibrational decay); vector of hydrophobic, steric and electronic properties; divided physicochemical property scores descriptor (vector of hydrophobic, steric, electronic properties and hydrogen bonds). Molecular docking studies are also used to predict peptide bioactivity. Finally, there are also bioinformatic tools for the identification of possible post-translational modifications (oxidation, methylation, acetylation) of amino acids residues such iPTMnet, STRAP PTM and jEcho. Table 12.3 shows a list of databases and informatic tools for the in-silico analysis of peptides (Agyei et al., 2018).

TABLE 12.3 Databases and bioinformatic tools for the in-silico analysis of bioactive peptides.

Function	Databases and informatic tools	Websites
Protein sequences	RCSB Protein Data Bank	https://www.rcsb.org/pdb/home/home.do
	UniProtKB	http://www.uniprot.org/
	NCBI Protein	http://www.ncbi.nlm.nih.gov/protein
	BIOPEP	http://www.uwm.edu.pl/biochemia/
	PepBank	http://pepbank.mgh.harvard.edu/
	BioPD	http://biopd.bjmu.edu.cn/
	SwePep	http://www.swepep.org/
	EROP-Moscow	http://erop.inbi.ras.ru/
	MilkAMP	http://milkampdb.org/
	PeptideDB	http://www.peptides.be/
	AMPer	http://marray.cmdr.ubc.ca/cgi-bin/amp.pl
In silico digestion and proteolytic enzymes	BIOPEP	http://www.uwm.edu.pl/biochemia/index.php/en/biopep
	PetideCutter	http://web.expasy.org/peptide_cutter/
	POPS	http://pops.csse.monash.edu.au/pops-cgi/index.php
	Enzyme Predictor	http://bioware.ucd.ie/~enzpred/Enzpred.php

Function	Databases and informatic tools	Websites
Identification and characterization of peptides and tools for chemometrics	PubChem	https://pubchem.ncbi.nlm.nih.gov/
	ProtParam	https://web.expasy.org/protparam/
	FoodDB	http://foodb.ca/
	ChEBI	https://www.ebi.ac.uk/chebi/
	AAindex	http://www.genome.jp/aaindex/
	HMDB	http://www.hmdb.ca/
	METLIN	https://metlin.scripps.edu/
	Peptigram	http://bioware.ucd.ie/peptigram/
Molecular docking of peptides	DOCK Blaster	http://blaster.docking.org/
	1-CLICK DOCKING	https://mcule.com/apps/1-click-docking/
	BSP-SLIM	https://zhanglab.ccmb.med.umich.edu/BSP-SLIM/
	SwissDock	http://www.swissdock.ch/
	FlexPepDock	http://flexpepdock.furmanlab.cs.huji.ac.il/
Prediction of bioactivity	PeptideRanker	http://bioware.ucd.ie/~compass/biowareweb/
	BIOPEP	http://www.uwm.edu.pl/biochemia/index.php/en/biopep
	AntiBP2	http://www.imtech.res.in/raghava/antibp2/
	PeptideLocator	http://bioware.ucd.ie/
Prediction of toxicity and/or allergenicity	ToxinPred	http://www.imtech.res.in/raghava/toxinpred/
	AlgPred	http://www.imtech.res.in/raghava/algpred/
	Allerdictor	http://allerdictor.vbi.vt.edu/
	EPIMHC	http://bio.dfci.harvard.edu/epimhc/
	SORTALLER	http://sortaller.gzhmu.edu.cn/
	ProPepper	https://propepper.net/

(Continued)

TABLE 12.3 Cont'd

Function	Databases and informatic tools	Websites
Prediction of tastant potential	BIOPEP	http://www.uwm.edu.pl/biochemia/index.php/en/biopep
	BitterDB	http://bitterdb.agri.huji.ac.il/dbbitter.php
	BitterPredit	http://bitterdb.agri.huji.ac.il/dbbitter.php#BitterPredict
	iPTMnet	https://research.bioinformatics.udel.edu/iptmnet/
Identification possible post-translational modofications	STRAP PTM	https://www.bumc.bu.edu/cardiovascularproteomics/cpctools/strap-ptm/
	jEcho	http://www.healthinformaticslab.org/supp/resources.php
Data adapted from (Agyei, D.; Tsopmp, A.; Udenigwe, 2018)		

12.6 Key findings

There is not any specific technique for the analysis of bioactive peptides and an appropriate method shall be chosen for either matrix and depending on the characteristics and peptides properties to be analyzed, as well as, the purpose of the analysis. The best strategy for the analysis of bioactive peptides is the combination of the use of bioinformatic tools and those analytical techniques which decrease the cost and the time consumed of conventional methods.

Acknowledgments

Sol Zamuz acknowledges the financial support PTA program (PTA2017-14156-I). Authors are members of the Healthy Meat network, funded by CYTED (ref. 119RT0568). Thanks to GAIN (Axencia Galega de Innovación) for supporting this project (grant number IN607A2019/01).

References

Agyei, D., Tsopmp, A., & Udenigwe, C. C. (2018). Bioinformatics and peptidomics approaches to the discovery and analysis of food-derived bioactive peptides. *Analytical and Bioanalytical Chemistry, 410*(15), 3463–3472. https://doi.org/10.1007/S00216-018-0974-1.

Ahmed, R., & Chun, B. S. (2018). Subcritical water hydrolysis for the production of bioactive peptides from tuna skin collagen. *Journal of Supercritical Fluids, 141*, 88–96. https://doi.org/10.1016/J.SUPFLU.2018.03.006.

Allen, D. R., & McWhinney, B. C. (2019). Quadrupole time-of-flight mass spectrometry: A paradigm shift in toxicology screening applications. *The Clinical Biochemist Reviews, 40*(3), 135. https://doi.org/10.33176/AACB-19-00023.

Álvarez, C., Tiwari, B. K., Rendueles, M., & Díaz, M. (2016). Use of response surface methodology to describe the effect of time and temperature on the production of decoloured, antioxidant and functional peptides from porcine haemoglobin by sub-critical water hydrolysis. *LWT, 73*, 280–289. https://doi.org/10.1016/J.LWT.2016.06.024.

Bamdad, F., Shin, S. H., Shu, J-W., Nimalaratne, c., & Sunwoo, H. (2017). Anti-inflammatory and antioxidant properties of casein hydrolysate produced using high hydrostatic pressure combined with proteolytic enzymes. *Molecules (Basel, Switzerland), 22*(4). doi:10.3390/MOLECULES22040609.

Banerjee, S., & Mazumdar, S. (2012). Electrospray ionization mass spectrometry: A technique to access the information beyond the molecular weight of the analyte. *International Journal of Analytical Chemistry, 2012*, 1–40. https://doi.org/10.1155/2012/282574.

Boesl, U. (2017). Time-of-flight mass spectrometry: Introduction to the basics. *Mass Spectrometry Reviews, 36*(1), 86–109. https://doi.org/10.1002/MAS.21520.

Boi, C. (2019). Membrane chromatography for biomolecule purification. *Current trends and future developments on (bio-) membranes: Membrane processes in the pharmaceutical and biotechnological field*, 151–166. https://doi.org/10.1016/B978-0-12-813606-5.00006-3.

Büyükköroğlu, G., Dora, D. D., Özdemir, F., & Hizel, C. (2018). Techniques for protein analysis. *Omics Technologies and Bio-Engineering: Towards Improving Quality of Life, 1*, 317–351. https://doi.org/10.1016/B978-0-12-804659-3.00015-4.

Chian, F. M., Kaur, L., Oey, I., Astruc, T., Hodgkinson, S., & Boland, M. (2019). Effect of pulsed electric fields (PEF) on the ultrastructure and in vitro protein digestibility of bovine longissimus thoracis. *LWT, 103*, 253–259. https://doi.org/10.1016/J.LWT.2019.01.005.

Conlon, J. M. (2007). Purification of naturally occurring peptides by reversed-phase HPLC. *Nature Protocols, 2*(1), 191–197. https://doi.org/10.1038/nprot.2006.437.

D'Hondt, M., Gevaert, B., Stalmans, S., Van Dorpe, S., Wynendaele, E., Peremans, K., et al. (2013). Reversed-phase fused-core HPLC modeling of peptides. *Journal of Pharmaceutical Analysis, 3*(2), 93–101. https://doi.org/10.1016/J.JPHA.2012.11.002.

Dlask, O., & Václavíková, N. (2017). Electrodialysis with ultrafiltration membranes for peptide separation. *Chemical Papers, 72*(2), 261–271. https://doi.org/10.1007/S11696-017-0293-6.

Dong, X., Li, J., Jiang, G., Li, H., Zhao, M., & Jiang, Y. (2019). Effects of combined high pressure and enzymatic treatments on physicochemical and antioxidant properties of peanut proteins. *Food Science & Nutrition, 7*(4), 1417–1425. https://doi.org/10.1002/FSN3.976.

Duncan, M. W., Roder, H., & Hunsucker, S. W. (2008). Quantitative matrix-assisted laser desorption/ionization mass spectrometry. *Briefings in Functional Genomics, 7*(5), 355–370. https://doi.org/10.1093/BFGP/ELN041.

Franck, M., Perreault, V., Suwal, S., Marciniak, A., Bazinet, L., & Doyen, A. (2019). High hydrostatic pressure-assisted enzymatic hydrolysis improved protein digestion of flaxseed protein isolate and generation of peptides with antioxidant activity. *Food Research International, 115*, 467–473. https://doi.org/10.1016/J.FOODRES.2018.10.034.

Garcia-Mora, P., Peñas, E., Frias, J., Zieliński, H., Wiczkowski, W., Zielińska, D., et al. (2016). High-pressure-assisted enzymatic release of peptides and phenolics increases angiotensin converting enzyme I inhibitory and antioxidant activities of pinto bean hydrolysates. *Journal of Agricultural and Food Chemistry, 64*(8), 1730–1740. https://doi.org/10.1021/ACS.JAFC.5B06080.

Godinho, I., Pires, C., Pedro, S., Teixeira, B., Mendes, R., Nunes, M. L., et al. (2016). Antioxidant properties of fish protein hydrolysates prepared from cod protein hydrolysate by Bacillus sp. *Applied Biochemistry and Biotechnology, 178*(6), 1095–1112. https://doi.org/10.1007/S12010-015-1931-5.

Gohi, B. F. C. A., Du, J., Zeng, H.-Y., Cao, X., & Zou, K.min. (2019). Microwave pretreatment and enzymolysis optimization of the lotus seed protein. *Bioengineering, 6*(2). doi:10.3390/BIOENGINEERING6020028.

Haag, A. M. (2016). Mass analyzers and mass spectrometers. *Advances in Experimental Medicine and Biology, 919*, 157–169. https://doi.org/10.1007/978-3-319-41448-5_7.

Howard, J. W., Kay, R. G., Pleasance, s., & Creaser, C. S. (2012). UHPLC for the separation of proteins and peptides. *Bioanalysis, 4*(24), 2971–2988. https://doi.org/10.4155/BIO.12.283.

Huang, L., Ma, H., Peng, L., Wang, Z., & Yang, Q. (2015). Enzymolysis kinetics of garlic powder with single frequency countercurrent ultrasound pretreatment. *Food and Bioproducts Processing, 95*, 292–297. https://doi.org/10.1016/J.FBP.2014.10.015.

Judd, R. C. (1994). Electrophoresis of peptides. *Methods in Molecular Biology (Clifton, N.J.), 32*, 49–57. https://doi.org/10.1385/0-89603-268-X:49.

Kangsanant, S., Thongraung, C., Jansakul, C., Murkovic, M., & Seechamnanturakit, V. (2015). Purification and characterisation of antioxidant and nitric oxide inhibitory peptides from Tilapia (Oreochromis niloticus) protein hydrolysate. *International Journal of Food Science & Technology, 50*(3), 660–665. https://doi.org/10.1111/IJFS.12680.

Ketnawa, S., & Liceaga, A. M. (2016). Effect of microwave treatments on antioxidant activity and antigenicity of fish frame protein Hydrolysates. *Food and Bioprocess Technology, 10*(3), 582–591. https://doi.org/10.1007/S11947-016-1841-8.

Liu, Y. F., Oey, I., Bremer, P., Silcock, P., Carne, A., & McConnell, M. (2019). Pulsed electric fields treatment at different pH enhances the antioxidant and anti-inflammatory activity of ovomucin-depleted egg white. *Food Chemistry, 276*, 164–173. https://doi.org/10.1016/J.FOODCHEM.2018.10.009.

Madadlou, A., Sheehan, D., Emam-Djomeh, Z., & Mousavi, M. E. (2011). Ultrasound-assisted generation of ACE-inhibitory peptides from casein hydrolyzed with nanoencapsulated protease. *Journal of the Science of Food and Agriculture, 91*(11), 2112–2116. https://doi.org/10.1002/JSFA.4438.

Marson, G. V., Belleville, M.-P., Lacour, S., & Hubinger, M. D. (2020). Membrane fractionation of protein hydrolysates from by-products: Recovery of valuable compounds from spent yeasts. *Membranes 2021, 11*(1), 23. https://doi.org/10.3390/MEMBRANES11010023.

Melgosa, R., Marques, M., Paiva, A., Bernardo, A., Fernández, N., Sá-Nogueira, I., et al. (2021). Subcritical water extraction and hydrolysis of cod (Gadus morhua) frames to produce bioactive protein extracts. *Foods, 10*(6), 1222. https://doi.org/10.3390/FOODS10061222.

Mellon, F. A. (2003). Mass spectrometry | Principles and instrumentation. *Encyclopedia of Food Sciences and Nutrition*, 3739–3749. doi:10.1016/B0-12-227055-X/00746-X.

Nguyen, E., Jones, O., Kim, Y. H. B., Martin-Gonzalez, F. S., & Liceaga, A. M. (2017). Impact of microwave-assisted enzymatic hydrolysis on functional and antioxidant properties of rainbow trout Oncorhynchus mykiss by-products. *Fisheries Science, 83*(2), 317–331. https://doi.org/10.1007/S12562-017-1067-3.

Noor, Z., Ahn, S. B., Baker, M. S., Ranganathan, S., & Mohamedali, A. (2021). Mass spectrometry–based protein identification in proteomics—a review. *Briefings in Bioinformatics, 22*(2), 1620–1638. https://doi.org/10.1093/BIB/BBZ163.

Ozuna, C., Paniagua-Martínez, I., Castaño-Tostado, E., Ozimek, L., & Amaya-Llano, S. L. (2015). Innovative applications of high-intensity ultrasound in the development of functional food ingredients: Production of protein hydrolysates and bioactive peptides. *Food Research International, 77*, 685–696. https://doi.org/10.1016/J.FOODRES.2015.10.015.

Park, J. S., Jeong, Y. R., & Chun, B. S. (2019). Physiological activities and bioactive compound from laver (Pyropia yezoensis) hydrolysates by using subcritical water hydrolysis. *The Journal of Supercritical Fluids, 148*, 130–136. https://doi.org/10.1016/J.SUPFLU.2019.03.004.

Polikovsky, M., Fernand, F., Sack, M., Frey, W., Müller, G., & Golberg, A. (2016). Towards marine biorefineries: Selective proteins extractions from marine macroalgae Ulva with pulsed electric fields. *Innovative Food Science & Emerging Technologies, 37*, 194–200. https://doi.org/10.1016/J.IFSET.2016.03.013.

Pontis, H. G. (2017). Protein and Carbohydrate Separation and Purification. *Methods for Analysis of Carbohydrate Metabolism in Photosynthetic Organisms*, 45–63. https://doi.org/10.1016/B978-0-12-803396-8.00003-X.

Schulz-Knappe, P., Schrader, M., & Zucht, H.-D. (2005). The peptidomics concept. *Combinatorial Chemistry & High Throughput Screening, 8*(8), 697–704. https://doi.org/10.2174/138620705774962418.

Steckel, A., & Schlosser, G. (2019). An organic chemist's guide to electrospray mass spectrometric structure elucidation. *Molecules (Basel, Switzerland), 24*(3). doi:10.3390/MOLECULES24030611.

Stefanović, A. B., Jovanović, J. R., Grbavčić, S.Ž., Šekuljica, N.Ž., Manojlović, V. B., Bugarski, B. M., et al. (2014). Impact of ultrasound on egg white proteins as a pretreatment for functional hydrolysates production. *European Food Research and Technology, 239*(6), 979–993. https://doi.org/10.1007/S00217-014-2295-8.

Tadesse, S. A., & Emire, S. A. (2020). *Production and processing of antioxidant bioactive peptides: A driving force for the functional food market*, Heliyon 6 (p. e04765): Elsevier Ltd. https://doi.org/10.1016/j.heliyon.2020.e04765.

Thomas, S. N. (2019). Mass spectrometry. *Contemporary practice in clinical chemistry*, 171–185. https://doi.org/10.1016/B978-0-12-815499-1.00010-7.

Uluko, H., Zhang, S., Liu, L., Tsakama, M., Lu, J., & Lv, J. (2015). Effects of thermal, microwave, and ultrasound pretreatments on antioxidative capacity of enzymatic milk protein concentrate hydrolysates. *Journal of Functional Foods, 18*, 1138–1146. https://doi.org/10.1016/J.JFF.2014.11.024.

Velickovic, T. C., Ognjenovic, J., & Mihajlovic, L. (2012). Separation of Amino Acids, Peptides, and Proteins by Ion Exchange Chromatography. *Ion Exchange Technology II: Applications, 9789400740266*, 1–34. https://doi.org/10.1007/978-94-007-4026-6_1.

Wang, X., Yu, H., Xing, R., & Li, P. (2017a). Characterization, preparation, and purification of marine bioactive peptides. doi:10.1155/2017/9746720

Wang, X., Yu, H., Xing, R., & Li, P. (2017). *Characterization, preparation, and purification of marine bioactive peptides*BioMed Research International*2017*: Hindawi Limited. https://doi.org/10.1155/2017/9746720.

Zhang, M., Huang, T.-S., & Mu, T.-H. (2019b). Production and in vitro gastrointestinal digestion of antioxidant peptides from enzymatic hydrolysates of sweet potato protein affected by pretreatment. *Plant Foods for Human Nutrition, 74*(2), 225–231. https://doi.org/10.1007/S11130-019-00724-Y.

Zhang, M., & Mu, T. H. (2017). Identification and characterization of antioxidant peptides from sweet potato protein hydrolysates by Alcalase under high hydrostatic pressure. *Innovative Food Science & Emerging Technologies, 43*, 92–101. https://doi.org/10.1016/J.IFSET.2017.08.001.

Zhang, S., Zhang, M., Xing, J., & Lin, S. (2019a). A possible mechanism for enhancing the antioxidant activity by pulsed electric field on pine nut peptide Glutamine-Tryptophan-Phenylalanine-Histidine. *Journal of Food Biochemistry, 43*(3). https://doi.org/10.1111/JFBC.12714.

Zhou, C., Ma, H., Ding, Q., Lin, L., Yu, X., Luo, L., et al. (2013a). Ultrasonic pretreatment of corn gluten meal proteins and neutrase: Effect on protein conformation and preparation of ACE (angiotensin converting enzyme) inhibitory peptides. *Food and Bioproducts Processing, 91*(4), 665–671. https://doi.org/10.1016/J.FBP.2013.06.003.
Zhou, C., Ma, H., Yu, X., Liu, B., Yagoub, A. E. G. A., & Pan, Z. (2013b). Pretreatment of defatted wheat germ proteins (by-products of flour mill industry) using ultrasonic horn and bath reactors: Effect on structure and preparation of ACE-inhibitory peptides. *Ultrasonics Sonochemistry, 20*(6), 1390–1400. https://doi.org/10.1016/J.ULTSONCH.2013.04.005.

Index

Page numbers followed by "*f*" and "*t*" indicate, figures and tables respectively.

Q

R

S

FSC
www.fsc.org
MIX
Papier | Fördert gute Waldnutzung
FSC® C083411

Kolibri 360 GmbH
Ferdinand-Jühlke-Straße 7
99095 Erfurt, Deutschland
produktsicherheit@kolibri360.de